Microorganisms for Sustainability

Volume 61

Microorganisms have been in existence since the origin of life on earth and can survive the most extreme habitats or conditions on earth. Microorganisms are involved in regulating biogeochemical cycles, maintaining plant and animal health, and sustaining the global food chain. Moreover, they play crucial roles in addressing the challenges of climate change and achieving the targets of Sustainable Development Goals (SDGs).

This multidisciplinary book series captures the role of microbes towards building a sustainable world, while encompassing cutting-edge technologies and current needs across various fields such as agriculture sustainability, bioremediation, restoration of degraded habitats and wastelands, and food security. Additionally, this series explores microbial applications in industries, and building their utilization in clean and green energy solutions. Furthermore, themes like microbial secondary metabolites, extremophilic microbes and modern omics, including next generation sequencing and metagenomics, are also covered in this series.

With contributions from researchers across the globe, this series addresses the important call of 'One Planet-One Health-One Future'. It comprises a collection of diverse volumes that provides insights for scientists, young researchers, educators and decision-makers in the government, private sector, and non-governmental organizations, empowering their efforts to achieve the global goals.

The series invites, evaluates, and accepts book proposals to ensure a diverse, inclusive, and evolving program. The final decision regarding acceptance rests with the series editor.

Peer-review

This book series follows a stratified review process. Proposals for individual volumes are reviewed by the series editor and then the editorial board members. On a case-to-case basis, external reviewers are also invited for further evaluation of the book proposal. Review of the chapters is the responsibility of the volume editor(s). A manuscript submission platform has been recently made available to the authors, volume editors and the series editor.

Venugopal Gopikrishnan
Radhakrishnan Manikkam
Kasem Soytong • Wafaa Mohamed Haggag
Editors

Harnessing Microbial Resources for Sustainable Agriculture

Springer

Editors
Venugopal Gopikrishnan
Centre for Drug Discovery
and Development
Sathyabama Institute of Science
and Technology
Chennai, Tamil Nadu, India

Radhakrishnan Manikkam
Centre for Drug Discovery
and Development
Sathyabama Institute of Science
and Technology
Chennai, Tamil Nadu, India

Kasem Soytong
Research Institute of Modern Organic
Agriculture
King Mongkut's Institute of Technology
Bangkok, Thailand

Wafaa Mohamed Haggag
Plant Pathology Department
National Research Centre
Dokki, Egypt

ISSN 2512-1901 ISSN 2512-1898 (electronic)
Microorganisms for Sustainability
ISBN 978-981-95-4816-3 ISBN 978-981-95-4817-0 (eBook)
https://doi.org/10.1007/978-981-95-4817-0

This Springer imprint is published by the registered company Springer Nature Singapore Pte Ltd.
The registered company address is: 152 Beach Road, #21-01/04 Gateway East, Singapore 189721, Singapore

Preface

In an era where the challenges of global food security, climate change, and environmental sustainability are increasingly pressing, the role of innovative agricultural practices has never been more critical. The overall contents in our new book titled *Harnessing Microbial Resources for Sustainable Agriculture*, aimed to bridge the gap between science and practice, providing a comprehensive exploration of microbial resources, can make to sustainable agriculture development.

The purpose of this work is to illuminate the intricate relationships between microorganisms and agricultural productivity, emphasizing their potential to increase soil health, improve crop resilience, and reduce dependency on chemical fertilizers and pesticides. By integrating foundational microbiological principles with practical applications, this work serves as a resource for undergraduate students, general readers, and professionals alike, all of whom are engaged in the pursuit of sustainable agricultural practices.

The significance of this work lies in its timely relevance to diverse audience. For students, it offers a basic understanding of microbial ecology and its applications in agriculture, fostering a new generation of conscious agricultural scientists and practitioners. For general readers, it highlights complex scientific concepts and underscores the importance of microbes in everyday agricultural practices. For professionals, it provides an evidence-based framework that can inform decision-making and policy development in sustainable agriculture.

The content of this book is organized around key themes that reflect the multifaceted role of microorganisms in agriculture. Each chapter combines technical rigor with accessible language, ensuring that readers from various backgrounds can engage with the content meaningfully. In our approach, we emphasize the importance of interdisciplinary collaboration, drawing insights from microbiology, agronomy, ecology, and environmental science. This holistic perspective not only enriches the reader's understanding but also underscores the interconnectedness of agricultural practices and ecological health.

As we embark on this journey to explore the potential of microbial resources, we invite readers to engage with the material critically and creatively. Our hope is that this textbook will inspire innovative thinking and practical solutions in the quest for sustainable agricultural systems that are resilient, productive, and environmentally sound. We extend our sincere gratitude to all the chapter contributors, researchers, educators, and practitioners whose work has informed this text. Their dedication to advancing knowledge in this vital field has been an invaluable resource. We also thank our readers for their commitment to fostering sustainable practices that will benefit both current and future generations.

Sincerely,

Chennai, Tamil Nadu, India — Venugopal Gopikrishnan
Chennai, Tamil Nadu, India — Radhakrishnan Manikkam
Bangkok, Thailand — Kasem Soytong
Dokki, Egypt — Wafaa Mohamed Haggag

Contents

1 **Innovation in Microbial Biotechnology for Organic Agriculture**.... 1
Kasem Soytong and Jiaojiao Song

2 ***Streptomyces* sp. Mitigates Biotic and Abiotic Stress Responses and Stimulates Plant Development** 21
Wafaa Mohamed Haggag, Venugopal Gopikrishnan, and Kasem Soytong

3 **Eco-friendly Fertilizers for Sustainable Agriculture: A Path Toward Greener Farming** 35
Wafaa M. Abd El-Rahim

4 **Challenges and Opportunities for the Use of Microorganisms in Sustainable Agriculture in the Future** 57
Wafaa M. Abd El-Rahim and Hassan Moawad

5 **Earthworm and Earthworm Cast-Associated Microbes for Sustainable Agriculture** 71
Justin Sheeba, Murali Monisha, Girupha Nishanth, Mekala Panneerselvam Karthik Prakash, Manikkam Radhakrishnan, Venugopal Gopikrishnan, and Ramesh Kumar Ashwin

6 **Bioprospects of Marine Microbes for Plant Growth Promotion and Disease Control** 89
Sandhya Rani Edadhala and Narasimha Golla

7 **Harnessing Plant-Associated Microorganisms for Increasing Agricultural Productivity and Biocontrol Potential in Harsh Environments**... 129
Fatma Aly Ahmed and Doaa S. Zaid

8 ***Chaetomium*: Goldmine for Sustainable Agriculture** 143
Kasem Soytong and Jiaojiao Song

9 Microbial Bioagents as Tools for Eliminating Toxigenic Fungi 173
Noran A. Al-Ansary, Wafaa Mohamed Haggag, and Medhat K. Ali

10 Remediating Heavy Metals from the Environment via Secondary Interactions Between Plants and Microorganisms 197
Samah Mostafa El-Sayed, Amr Said Mohamed, and Azza Abdel-Hamid Mohamed Mazher

11 Microbial Nanotechnology in Agricultural Applications 219
Jiaojiao Song and Kasem Soytong

12 *Bacillus* in Agriculture: Past, Present, and Future 251
Manigundan Kaari, Kishore Kumar Annamalai, Mekala Panneerselvam Karthik Prakash, Venugopal Gopikrishnan, and Radhakrishnan Manikkam

13 Microbial Biosurfactants for Agriculture Applications............ 271
Kishore Kumar Annamalai, Eshwarnath Vallipuram Sudhakar, Manigundan Kaari, Venugopal Gopikrishnan, Narasimha Golla, and Radhakrishnan Manikkam

14 Endophytes as Biological Control Agents of Plant Diseases 295
Ibrahim Elsaid Ibrahim Elshahawy and Wafaa Mohamed Haggag

15 The Role of Microorganisms in Activating Plant Immunity Against Viral Pathogens 335
Wafaa Mohamed Haggag, Ibrahim Elsaid Ibrahim Elshahawy, and Mohamed G. Farhat

16 Effects of Microbial Pesticides on Insect Control 355
Magda Mahmoud Sabbour

17 Value-Adding Agricultural Products by Microbial Intervention 375
Kamali Joshna Thayalan, Ponmurugan Karuppiah, Uma Eswaranpillai, Shyamkumar Rajaram, Manivannan Govindsamy, and Ganeshkumar Anbazhagan

18 Actinobacteria as Potential Biopesticides...................... 411
Sangeetha Murthy and Balagurunathan Ramasamy

19 The Potential of Microbes as Biofertilizers and Biopesticides for Sustainable Agriculture 429
Yenny Sariasih, Zainal Muktamar, and Nanik Setyowati

20 Portraying Plant Drought Responses and Its Mechanisms of Tolerance by Rhizobacteria................................ 451
Sakthi Uma Devi Eswaran and Lalitha Sundaram

Contributors

Wafaa M. Abd El-Rahim Department of Agricultural Microbiology, Agricultural and Biological Research Institute, National Research Centre (NRC), Giza, Egypt

Fatma Aly Ahmed Department of Medicinal and Aromatic Plants, Desert Research Center, Cairo, Egypt

Noran A. Al-Ansary Plant Pathology Department, National Research Centre, Giza, Egypt

Medhat K. Ali Plant Pathology Department, Faculty of Agriculture, Ain Shams University, Cairo, Egypt

Ganeshkumar Anbazhagan Department of Microbiology, Saveetha Medical College and Hospital, Saveetha Institute of Medical and Technical Sciences (SIMATS), Saveetha University, Thandalam, Chennai, Tamil Nadu, India

Kishore Kumar Annamalai Centre for Drug Discovery and Development, Sathyabama Institute of Science and Technology, Chennai, Tamil Nadu, India

Ramesh Kumar Ashwin Centre for Drug Discovery and Development, Sathyabama Institute of Science and Technology, Chennai, Tamil Nadu, India

Sandhya Rani Edadhala Department of Virology, Sri Venkateswara University, Tirupati, AP, India

Samah Mostafa El-Sayed Ornamental Plants and Woody Trees Department, Agricultural and Biological Research Institute, National Research Centre, Giza, Egypt

Ibrahim Elsaid Ibrahim Elshahawy Department of Plant Pathology, National Research Centre, Giza, Egypt

Sakthi Uma Devi Eswaran Research Scholar, Soil Biology and PGPR Lab, Department of Botany, Periyar University, Salem, Tamil Nadu, India

Uma Eswaranpillai Department of Botany, PSGR Krishnammal College for Women, Peelamedu, Coimbatore, Tamil Nadu, India

Mohamed G. Farhat Plant Pathology Department, National Research Centre, Giza, Egypt

Narasimha Golla Department of Virology, Sri Venkateswara University, Tirupati, AP, India

Venugopal Gopikrishnan Centre for Drug Discovery and Development, Sathyabama Institute of Science and Technology, Chennai, Tamil Nadu, India

Manivannan Govindsamy Department of Psychiatry (Research), Saveetha Medical College and Hospital, Saveetha Institute of Medical and Technical Sciences (SIMATS), Saveetha University, Thandalam, Chennai, Tamil Nadu, India

Wafaa Mohamed Haggag Plant Pathology Department, National Research Centre, Giza, Egypt

Manigundan Kaari Centre for Drug Discovery and Development, Sathyabama Institute of Science and Technology, Chennai, Tamil Nadu, India

Department of Applied Bioscience, Dong-A University, Busan, South Korea

Ponmurugan Karuppiah Department of Botany and Microbiology, College of Science, King Saud University, Riyadh, Saudi Arabia

Radhakrishnan Manikkam Centre for Drug Discovery and Development, Sathyabama Institute of Science and Technology, Chennai, Tamil Nadu, India

Azza Abdel-Hamid Mohamed Mazher Ornamental Plants and Woody Trees Department, Agricultural and Biological Research Institute, National Research Centre, Giza, Egypt

Hassan Moawad Department of Agricultural Microbiology, Agricultural and Biological Research Institute, National Research Centre (NRC), Giza, Egypt

Amr Said Mohamed Botanical Gardens Research Department, Horticulture Research Institute of Agricultural Research Center (ARC), Giza, Egypt

Murali Monisha Department of Biotechnology, Sathyabama Institute of Science and Technology, Chennai, Tamil Nadu, India

Zainal Muktamar Department of Soil Science, Faculty of Agriculture, University of Bengkulu, Bengkulu, Indonesia

Sangeetha Murthy Kriya Biosys Pvt. Ltd., Salem, Tamil Nadu, India

Girupha Nishanth Department of Biotechnology, Sathyabama Institute of Science and Technology, Chennai, Tamil Nadu, India

Mekala Panneerselvam Karthik Prakash Centre for Drug Discovery and Development, Sathyabama Institute of Science and Technology, Chennai, Tamil Nadu, India

Shyamkumar Rajaram Department of Biotechnology, Kamaraj College of Engineering and Technology, Virudhunagar, Tamil Nadu, India

Balagurunathan Ramasamy Vivekanandha Educational Institutions for Women, Tiruchengode, Tamil Nadu, India

Magda Mahmoud Sabbour Pests and Plant Protection Department, Agriculture and Biological Institute, National Research Centre, Giza, Egypt

Yenny Sariasih Department of Plant Protection, Faculty of Agriculture, University of Bengkulu, Bengkulu, Indonesia

Nanik Setyowati Department of Crop Production, Faculty of Agriculture, University of Bengkulu, Bengkulu, Indonesia

Justin Sheeba Centre for Drug Discovery and Development, Sathyabama Institute of Science and Technology, Chennai, Tamil Nadu, India

Jiaojiao Song Research Institute of Modern Organic Agriculture (RIMOA), King Mongkut's Institute of Technology Ladkrabang (KMITL), Bangkok, Thailand

King Mongkut's Institute of Technology Ladkrabang (KMITL), Bangkok, Thailand

Association of Agricultural Technology in Southeast Asia (AATSEA), Bangkok, Thailand

Kasem Soytong Research Institute of Modern Organic Agriculture (RIMOA), Bangkok, Thailand

King Mongkut's Institute of Technology Ladkrabang (KMITL), Bangkok, Thailand

Association of Agricultural Technology in Southeast Asia (AATSEA), Bangkok, Thailand

Eshwarnath Vallipuram Sudhakar Centre for Drug Discovery and Development, Sathyabama Institute of Science and Technology, Chennai, Tamil Nadu, India

Lalitha Sundaram Assistant Professor, Soil Biology and PGPR Lab, Department of Botany, Periyar University, Salem, Tamil Nadu, India

Kamali Joshna Thayalan Department of Botany, PSGR Krishnammal College for Women, Peelamedu, Coimbatore, Tamil Nadu, India

Doaa S. Zaid Department of Medicinal and Aromatic Plants, Desert Research Center, Cairo, Egypt

Chapter 1
Innovation in Microbial Biotechnology for Organic Agriculture

Kasem Soytong and Jiaojiao Song

Abstract Organic agriculture has expanded in many countries to encourage the consumption of safe food for human health and to rejuvenate agroecosystems in the surrounding environment. Agricultural inputs for organic agriculture must be based on scientific investigations, as bioproducts can be substituted for agrochemicals, leading to modern organic agriculture. Microbial biotechnology in agriculture has attracted increasing attention as an important strategy for developmental sustainability. The innovation of this research is the identification of new effective microbes for microbial fermentation, plant growth promotion, disease control, insect protection, and weed control, including integrated management, which can contribute to modern organic agriculture toward sustainable development goals (SDGs). Agricultural inputs for organic agriculture are key for crop and animal productivity and can be used as substitutes for agrochemicals, resulting in equal yields in terms of quality and quantity. Modern organic agriculture has been proposed as an improvement over traditional organic agriculture. Organic certification is required for organic products and foods to ensure that they are free from toxic agrochemicals and harmful human pathogens such as *Salmonella* spp. and *Escherichia coli*, and that they contain low levels of nitrates and heavy metals. Based on scientific findings, the use of agricultural inputs such as microbial decomposers, microbial fertilizers, microbial fungicides, microbial insecticides, and microbes for heavy metal remediation is proposed to support successful organic production of vegetables, fruits, and other crops in compliance with organic certification standards.

K. Soytong (✉)
Research Institute of Modern Organic Agriculture (RIMOA), King Mongkut's Institute of Technology Ladkrabang (KMITL), Bangkok, Thailand

Association of Agricultural Technology in Southeast Asia (AATSEA), Bangkok, Thailand

J. Song
Research Institute of Modern Organic Agriculture (RIMOA), Bangkok, Thailand

Association of Agricultural Technology in Southeast Asia (AATSEA), Bangkok, Thailand

King Mongkut's Institute of Technology Ladkrabang (KMITL), Prince of Chumphon Campus, Chumphon, Thailand

V. Gopikrishnan et al. (eds.), *Harnessing Microbial Resources for Sustainable Agriculture*, Microorganisms for Sustainability 61,
https://doi.org/10.1007/978-981-95-4817-0_1

Keywords Organic agriculture · Microbes · PGP · Disease and insect control

1.1 Introduction

Crop productivity is influenced by chemical fertilizers and pesticides, which can negatively affect soil health because of nutrient imbalances in cultivated soils. Chemical pesticides, including fungicides, insecticides, nematicides, and herbicides, have been used for years. Despite their advantages in agriculture, they are hazardous to humans, living organisms, and the environment. Chemical pesticides contaminate the atmosphere, soil, water, and food, leading to threats ranging from acute to chronic toxicity (Ahmad et al. 2024).

Organic agriculture or organic farming involves a sustainable agricultural system that considers the ecological environment, natural biological control of insect pests and diseases, and the use of biological fertilizers made from plant and animal debris. Modern organic agriculture has been developed to respond to environmental concerns and revitalize agroecosystems to produce food without toxic agrochemicals (chemical pesticides, synthetic fertilizers). Modern organic farming is successful on the basis of scientific research on agricultural inputs, which can be applied instead of agrochemical inputs and can compare productivity with that of conventional chemical agriculture. The agricultural inputs are mostly biological or microbial fertilizers, biological or microbial pesticides, etc. Innovative agriculture has recently focused on microbial biotechnology, which includes animal and crop production on organic farms. Beneficial microorganisms have been found to be powerful in revolutionizing agriculture to increase agricultural efficiency.

1.1.1 Microbial Fungicides

The microbial fungicides used as substitutes for chemical fungicides to control plant disease include *Bacillus amyloliquefaciens*, *Lactobacillus plantarum*, *Paecilomyces lilacinus*, *Trichoderma viride*, *Penicillium citrinum*, *Lactobacillus acidophilus*, *Bacillus subtilis*, *Bacillus pumilus*, *Clavispora lusitaniae*, *Bacillus velezensis*, *Pseudomonas fluorescens*, *Streptomyces tuirus*, *Trichoderma koningiopsis*, *Trichoderma asperellum*, *Trichoderma harzianum*, *Aureobasidium pullulans*, *Torulaspora indica*, *Metschnikowia pulcherrima*, *Aureobasidium pullulans* (Fenta and Mekonnen 2024), *Chaetomium globosum*, *Chaetomium cochlioides*, *Chaetomium elatum*, *Chaetomium luckownense*, *Chaetomium brasiliense*, *Chaetomium siamense*, and *Chaetomium bostrychoides* (Soytong et al, 2001). These microbial fungicides have been reported to control many phytopathogens associated with several diseases. In our experience in biocontrol research for plant disease control, most scientists have focused only on a single strain to develop a microbial fungicide, which has recently failed to control disease in the field because of abiotic and biotic factors. Other scientific investigations of scientists have not screened effective isolates of biocontrol agents well. The screening process for microbial fungicides should include the identification of species that are resistant to agrochemicals, as well as evaluations of their growth under different temperatures, salt concentrations, and pH conditions (both alkaline and acidic). In addition, the acute and dermal toxicity of the isolates should be assessed. This review focuses on the research and development of *Chaetomium* microbial fungicides and other related species.

Microbial fungicides are registered for plant disease control. Over the last decade, several microbial fungicides have been registered and patented as microbial fungicides. Several effective microbial fungicides have been formulated for commercial production. *Chaetomium*-microbial fungicides have been developed to support nonagrochemical cultivation and organic agriculture. The selected species of *Chaetomium* are scientifically safe for the environment, human beings, and ecology. *Chaetomium* microbial fungicides have been developed as broad-spectrum microbial fungicides to control several diseases, e.g., *Fusarium* wilt of tomato, *Phytophthora* rot of Durian, black pepper, Citrus, strawberry, etc. *Chaetomium* microbial fungicides were first patented for disease control and have been registered for commercialization at the Department of Agriculture in Laos, Thailand, Vietnam, and Cambodia. *Chaetomium* microbial fungicides are broad-spectrum microbial fungicides formulated from *Chaetomium globosum* and *Ch. cupreum*, which have been registered for commercialization in Thailand, Laos, Vietnam, and Cambodia for more than 10–15 years (Soytong et al. 2001).

Synergistic isolates of *Chaetomium brasiliense*, *Chaetomium cupreum*, *Chaetomium cochlioides*, *Ch. globosum*, *Chaetomium elatum*, *Chaetomium lucknowense*, *Chaetomium logirostre*, and *Chaetomium siamense* have been identified as broad-spectrum microbial fungicides or biological agents that are formulated as pellets, powders, and liquids to control plant diseases. A mixture of 22 strains of *Ch. cupreum* and *Ch. globosum* was patented as a broad-spectrum microbial fungicide under the international code 5 AO 1 N 25/12 (Patent No. 6266), issued in Thailand on February 22, 1994. These synergistic isolates have been shown to be resistant to agrochemical pesticides as follows: chemical insecticides such as chlorpyrifos 40, Cypermethyl 10, Cypermethyl 35, dichlorvos, endosulfan, DDVD 50, and parathion methyl 50; chemical fungicides such as copper oxychloride, carbendazim, dichloran 80, Mancozeb 80, metalaxyl 25, phosphoric acid, and sulfur 80; and chemical herbicides such as alachlor, glyphosate, and paraquat. Some strains of *Chaetomium* were found to be resistant to paraquat herbicides.

The microbial fungicide of *Chaetomium* is also compatible with chemical fertilizers such as 46-0-0, 16-20-0, 15-15-15+TE, 16-16-16+TE, 13-13-21, 9-24-24, 12-12-12+2MgO, 16-11-14+TE, 16-12-8, 16-16-8, 17-17-17+TE, 18-12-6, 18-46-0, 20-10-10+TE, 21-0-0, and 24-8-8+TE (Soytong et al. 2001). The microbial fungicides developed from *Ch. Globosum*, *Ch. cupreum*, and *Ch. lucknowense* have been confirmed to control wilt disease and dry root rot caused by *Fusarium oxysporum* in tea. These species can suppress the growth of colonies by 31–34% and decrease conidia formation by 67–75% after 30 days (Phong et al. 2016). This formulation has been shown to control the root rot of pomelo (*Phytophthora nicotianae*), which decreases the disease prevalence by 66–71% compared with that in nontreated controls (Hung et al. 2015). respectively (Charoenporn et al. 2010). The microbial fungicides which formulated from living spores Ch. globosum N0802, and Ch. lucknowense CLT showed the highest reduction of wilt incidence of tomato which were 44 and 36 %, respectively as compared to prochoraz that disease reduction only 21 % (Charoenporn et al, 2010).

1.2 Microbial Decomposers and Microbial Fertilizers

The support of the FAO for agricultural product demand would increase to 60% by 2030 (Mia et al. 2010). Food production must increase agricultural yield, and maintaining a safe environment is a major target in the twenty-first century (Berg, 2009). Chemical fertilizers are applied to promote the productivity of crops, but chemical fertilizers decrease soil fertility, which causes an imbalance of nutrients in the soil, resulting in pollution of the environment and living organisms (Sprent and Sprent 1990). However, biofertilizers are developed using potent microorganisms from rhizosphere soil as organic products that increase soil biodiversity and fertility in cultivated soil. It is ecologically friendly, nontoxic, and cost-effective and can be substituted for chemical fertilizers (Thomas and Singh 2019). The high-quality biofertilizers that use specific screening isolates of both fungi and bacteria differ in their bioactive abilities to produce enzymes, metabolites, and others to release essential plant nutrients, e.g., nitrogen, phosphorus, potassium, calcium, and magnesium, which are major elements, and copper, boron, zinc, chloride, and manganese, which are minor elements. Potent microorganisms are screened and used as biofertilizers, including fungi and bacteria, which can act as cyanobacteria and nitrogen-fixing bacteria; phytohormone-producing bacteria; and potassium- and phosphorus-solubilizing microbes in synergistic combinations, which regulate plant growth, such as amino acids and indole acetic acid (IAA), to increase crop productivity (Parikh and James 2012). Microbial biofertilizers have been shown to increase plant growth and yield by 10–40% (Kawalekar 2013).

1.2.1 Microbial Biofertilizers

Biofertilizers can be grouped into many types on the basis of their mode of action and their functions. The biofertilizers include phosphorus solubilizers, potassium solubilizers, nitrogen-fixing microbes, and plant growth-promoting rhizobacteria.

Nitrogen is a major element required for plant growth. Although the atmosphere contains about 80% nitrogen, plants cannot directly utilize this atmospheric form. Specific microorganisms can fix the nitrogen available to plants. They transform inert N_2 to organic forms (Reed et al. 2011). The exudates from the roots into the rhizosphere soil support nitrogen fixation by microbes, which enhances the ability of plants to substitute for chemical fertilizers. Many microorganisms, such as symbionts (*Rhizobium*), free-living bacteria (*Azotobacter*), and N_2-fixing bacteria associated with legumes, include *Rhizobium* and N_2-fixing bacteria associated with nonlegumes, e.g., *Rhodopseudomonas* spp., *Enterobacter* spp., *Bacillus* spp., *Rhodospirillum* spp., *Acetobacter* spp., and *Achromobacter* spp. (Meena et al. 2017).

1.2.1.1 Phosphate-Solubilizing Biofertilizers

The bacteria that solubilize phosphate can convert insoluble phosphate (HPO_4 and H_2PO_4) into the soluble form by chelation, ion exchange reactions, and organic acids. It can release organic acids that chelate cations to phosphate and change them to soluble forms (Sharma et al. 2013). Phosphorus-solubilizing microbes are found in *Bacillus* spp., *Enterobacter* spp., *Pseudomonas* spp., *Rhizobium* spp., *Aspergillus* spp., and *Penicillium* spp. (Anand et al. 2016). Compared with the untreated control, the phosphorus solubilizer *Aspergillus niger* as a biofertilizer significantly increased plant growth (Din et al. 2019).

1.2.1.2 Potassium-Solubilizing Biofertilizers

Potassium (K) is generally present in only 1–2% of the soil, which is available, and the mineral potassium form cannot be taken up by roots. Therefore, the continuous application of potassium to the soil is needed. Various bacterial species can produce organic acids, and chelation can change reactions from insoluble potassium to soluble potassium (Etesami et al. 2017). Potassium-solubilizing microorganisms, e.g., *Aspergillus niger*, *Arthrobacter*, *Bacillus* spp., *Cladosporium* spp., etc., have been identified (Sugumaran and Janarthanam 2007). *Bacillus pseudomycoides* is a potassium-solubilizing bacterium that increases K uptake during tea cultivation (Pramanik et al. 2019). *Aspergillus* spp. and *Penicillium* spp. effectively solubilize potassium (Xiafang and Weiyi 2002). *Thiobacillus thioparous* and *Thiobaciilus thioxidans* have been shown to oxidize sulfur to a usable form of sulfates in plants (Riaz et al. 2020).

The reduced form of sulfur compounds can pollute the environment but sulfur-oxidizing bacteria help protect the environment, which can eliminate sulfur pollution (Pokorna and Zabranska 2015). Zinc-solubilizing microbes, such as *Saccharomyces* spp., *Pseudomonas* spp., *Bacillus* spp., *Burkholderia* spp., and *Acinetobacter* spp., reportedly increase the amount of available Zn in the soil (Goteti et al. 2013).

Phosphorus- and potassium-solubilization methods have been developed for use as biofertilizers. The selected Actinobacteria strains—*Streptomyces chartreusis* AAS2, *Streptomyces flavogriseus* AAS7, *Streptomyces crystallinus* APSA1, and *Streptomyces* sp. APSA5—have demonstrated the ability to solubilize potassium, phosphorus and indole acetic acid (Jayanthi et al. 2023). These strains are being developed as phosphorus- and potassium-solubilizing decomposers.

The symbiotic association of mycorrhizae and roots is known as symbiosis, which increases the surface area of the root to absorb the available nutrients. Vesicular arbuscular mycorrhizae (VAMs) or endomycorrhizae are known as symbionts of fungi and plants (Douds et al. 2000). VAM fungi can absorb mobile nutrients, e.g., P, Ca, Mg, S, Cu, and Zn, in rhizosphere soil (Pathak et al. 2017). The famous species of VAM fungi are *Glomus Gigaspora*, *Acaulospora*, *Entrophospora*, *Sclerocystis*, and *Scutellospora* (Sullia et al. 1991).

Actinobacteria, as soil microorganisms associated with plants in rhizosphere soils, are known to release secondary metabolites, enzymes, growth-promoting substances, and siderophores to increase the growth of plants. Effective microbes can help improve soil by solubilizing nutrients and increasing nutrient uptake by plants (Khoshru et al. 2020). Plant growth-promoting rhizomicrobes (PGPRs), such as Actinobacteria, can be associated with either bacteria or fungi (Nouioui et al. 2018). Scientists have confirmed that PGPR exhibit excellent plant growth-promoting activity by releasing bioactive metabolites into rhizosphere soil for productivity (Wahyudi et al. 2019). PGPR play roles in phosphorus and potassium solubilization, indole acetic acid (IAA) production, nitrogen fixation, and siderophore production (Lee et al. 2012). Some PGPR have been reported to produce lignocellulases to decompose organic materials, resulting in nutrient recycling in soil (Das et al. 2007) and promoting the productivity of crops under abiotic stress (Cheng et al. 2018).

To improve the rhizosphere, plant growth-promoting microorganisms can reasonably revitalize soil fertility in agricultural production and facilitate the movement of available nutrients for plant growth development. Research has revealed that the potent fungi *Aspergillus oryzae* P13, *Aspergillus terreus* P1, *Aspergillus wentii* P9, *Emericella* sp. P19, *Chaetomium globosum* P20, *Chaetomium globosum* P26, *Chaetomium* sp. P11, *Eurotium* spp., *Trichoderma* sp. P17, and *Penicillium* spp. have the ability to produce amylase, cellulase, and ligninase. All the isolates produced at least one or two enzymes. Fungal species can produce amylases as follows: *Aspergillus oryzae* P13, *Aspergillus terreus* P1, *Aspergillus terreus* P6, *Aspergillus wentii* P9, *Aspergillus wentii* P22, *Chaetomium globosum* P20, *Chaetomium globosum* P26, *Emericella* sp. P19, *Eurotium* sp. P2, *Penicillium* sp. P12, and *Penicillium* sp. P15. The fungal isolates produced cellulase as follows: *Aspergillus oryzae* P13, *Aspergillus terreus* P1, *Chaetomium globosum* P20, *Emericella* sp. P19, *Eurotium* sp. P2, and *Penicillium* sp. P12. Ligninase is produced by *Aspergillus oryzae* P13, *Aspergillus terreus* P1, *Chaetomium globosum* P20, *Penicillium* sp. P3, *Trichoderma* sp. P23, *Emericella* sp. P19, and *Eurotium* sp. P5. These potential fungi can be developed as biosensors for fermentation processes to make compost. Fungal fertilizer is formulated from selected synergistic fungal species that produce amylase, cellulase, ligninase, and protease, e.g., *Achaetomium theilaviopsis*, *Arthrobotrys oligospora*, *Aspergillus sparsus*, *Emericella nidulans*, *Chaetomium lucknowense*, *Eurotium herbariorum*, and *Paecilomyces marquandii*. Mixing rice with sterilized bottom ash has been shown to increase plant height, number of tillers, grains per panicle, panicles per tiller, and panicle weight (Kowapradit et al. 2007).

The formulation of specific synergistic isolates of *Chaetomium* spp. has been developed as a biological fertilizer to promote growth in rice cultivation (Song and Soytong, 2017). Bacterial fertilizer developed from *Serratia marcescens* KMITL2020 is formulated as a growth stimulant, leading to increased growth parameters in rice production (Soytong et al. 2020).

Other fungal species involved in the development of biodecomposers include the following:

Aspergillus terreus releases cellulase and lignin to macerate cellulolytic and lignolytic materials during fermentation to produce compost (Jitkasemsuk et al. 2007). This isolate was also found to produce active metabolites against pathogens by Chaiyosanga et al. (2015). *A. terreus* has been reported to release the lumazine peptide penilumamide E, which is an active antimalarial (*P. falciparum*) with an IC_{50} of 2.83 μg/mL and is cytotoxic to human hepatoma and cholangiocarcinoma cell lines. *A. terreus* produces erreulactone A (Kim et al. 2003) and 4*R*-(hydroxymethyl)-4*R*-demethylterritin B (Peng, 1997), which suppress butyrolactone V and acetylcholinesterase activities and are antimalarials against *P. falciparum* K1 (Haritakun et al. 2010). *A. terreus* is recorded to produce terremides A and B, which actively suppress *Pseudomonas aeruginosa* and *Enterobacter aerogenes* (Wang et al. 2011); butyrolactone I, recorded as anti-H1N1; and asperterrestide A, recorded as inhibiting the influenza virus strains A/WSN/33 (H1N1) (Wang et al. 2011). This isolate has been shown to have synergistic effects with other enzymatic microorganisms to develop as a biological decomposer to degrade plant and animal debris during fermentation. *Neosartorya pseudofischeri* is further characterized by the active compounds chevalone F and 11-hydroxychevalone E, tryptoquivaline V, brasiliamide G, chevalones A–C, pyripyropene A, iso-chaetominine C, and lecanorin. In addition, 11-hydroxychevalone E actively inhibits *Escherichia coli* and *Salmonella enterica* (Paluka et al. 2020). *N. pseudofischeri* is an isolate that has been shown to produce cellulolytic enzymes and synergistic effects with other cellulolytic fungi to develop a mixture of microorganisms as a biological decomposer for compost production.

Eurotium chevalieri has been found to release the chevalones A–D, aszonapyrone B, eurochevalierine, sequiterpene, CJ-12662, aszonapyrone A, and ergosterol. These compounds are active against *P. falciparum* and *M. tuberculosis* (Kanolmedhakul et al. 2011). This isolate has also been shown to produce enzymes that can control *E. coli* and *Salmonella* sp. contamination in nature during the fermentation process and to synergize with other microorganisms to develop biological decomposers. Some species of *Chaetomium* can produce cellulolytic enzymes to degrade materials and wood. The specific isolates of *Chaetomium* spp. can produce natural active metabolites to act as antibiosis mechanisms against pathogenic humans and plants and promote ergosterol to increase soil fertility. It is reported to increase plant growth and yields, e.g., chili, citrus, corn, durian, kales, lettuce, potato, and tomato. *Chaetomium* can be developed as a microbial fertilizer that is registered in P. R. China (Song and Soytong, 2017).

The potential synergistic microorganisms of *Chaetomium* spp. and *Bacillus* spp. isolated and screened in China could promote the growth of plant crops produced from vegetables and tea. The use of microbial biofertilizers promotes the growth of citrus and peach, which are produced from *Bacillus* spp. and *Chaetomium* spp. (Song and Soytong, 2018). *Rhodospirillum centenum* is a bacterium that promotes the growth of rice. It is Gram-negative, mobile, and has a negative effect on gelatin activity. Many strains of *R. centenum* effectively produce enzymes, e.g., amylase, lipase, and protease. The rice seeds var. RD 41 and Pitsanulok 2 were inoculated with *R. centenum*, which resulted in good seed germination, and the shoot and root length and weights of the plants were determined. It has been developed as a microbial fertilizer to promote the growth of rice plants (Vareeket and Soytong, 2015).

1.2.2 Microbial Organic Fertilizer

Microbial organic fertilizers have developed from the synergistic combination of several isolates of enzymatic microbes that produce ligninase, cellulase, protease, amylase, and lipase to promote compost production during fermentation and degrade organic and inorganic nutrients to release available plant nutrients such as nitrogen, phosphorus, potassium, calcium, and magnesium. Copper, boron, molybdenum, iron, zinc, and manganese are necessary for plant growth. Microbial organic fertilizer containing *Ch. lucknowens*, *Aspergillus terreus*, *Arthrobotrys oligospora*, *Aspergillus oryzae*, *Pseudoeurotium zonatum*, *Pseudoeurotium ovale*, *E. nidulans*, *Emericella rogulosa*, *Mucor plumbeus*, *Penicillium variabile*, *T. harzianum*, and *T. hamatum* has been developed to be biodecomposers.

1.2.3 Liquid Organic Microbial Fertilizer

Liquid microbial fertilizer is mixed with synergistic strains of *Actinomyces*, *Bacillus subtilis*, *Saccharomyces cerevisiae*, *Rhodospirillum* spp., and *Rhodopseudomonas* spp., which are fermented in sterilized amino protein and fermented fish. N is high in nitrogen and amino acids and is rich in plant nutrients, which promote plant growth.

1.2.4 Microbial Active Substances (MAS)

MAS is a biological product formulated from metabolites of effective fungi, e.g., *Chaetomium* spp., *Trichoderma* spp., and *Emericella* spp. These microbial active substances are recommended for disease protection, such as root rot caused by *Phytophthora* spp., *Rhizoctonia* spp., *Sclerotium* spp., *Colletotrichum* spp., and *Fusarium oxysporum*. It is moderately controlled by bacterial wilt (*Pseudomonas solanacearum*), soft rot (*Erwinia carotovora*), and canker (*Xanthomonas campestris*).

A nanoelitor (Nano-EN) constructed from natural products of *E. nidulans* has been formulated as a natural product to induce immunity in rice plants. This isolate was proven to produce active compounds, e.g., emericellin, sterigmatocystin, and demethylsterigmatocystin. These metabolites exhibited antifungal activity against *M. oryzae* (rice blast). Nano-EN inhibited rice blast pathogens at ED_{50} values ranging from 4 to 9 µg/mL. Nano-EN functions as a plant vaccine by inducing the production of the sakuranetin phytoalexin in rice, thereby enhancing the plant's immunity to disease (Song et al. 2022).

1.2.5 Microbial Insecticides

Entomopathogenic fungi are found in agricultural habitats, dead insects, and conserved and disturbed rainforests. These include *Acremonium charticola*, *A. crissum*, *Beauveria bassiana*, *B. brongniartii*, *Cladosporium* spp., *Cordyceps* spp., *Hymenostilbe furcata*, *Hypocrella* spp., *Isaria cicadae*, *Isaria farinosus*, *Isaria fumosoroseus*, *Isaria ternuipes*, *Ophiocordyceps* spp., *Paecilomyces marquandii*, *Sporothrix indectorum*, *Stilbella buquetii*, *Torrubiella hemaipeterigena*, and *Verticillium* spp. These species are found in dead insects in Homoptera, Lepidoptera, and Hymenoptera (Aung et al. 2006a, b, 2008). Some of these species have been developed to be microbial insecticides, e.g., *Beauveria bassiana*, *Isaria farinosus*, *Paecilomyces marquandii*, and *Verticillium* spp.

Bacillus thuringiensis var. *kurstaki* (Bt) is a bacterium that is known as an entomopathogenic bacterium that controls insects such as Coleoptera, Diptera, Lepidoptera, Hymenoptera, Orthoptera, Isoptera, and some species of nematode. *B. thuringiensis* var. *kurstaki* was tested against the common cutworm (*Spodoptera litura*). The application of Bt at concentrations of 50 or 70 g/20 l of water resulted in the highest mortality of second- and third-stage larvae (Soytong and Nuanjamrat, 2001). It has been developed to be a microbial insecticide applied on organic farms.

1.2.6 Heavy Metal Remediation

Heavy metals contaminate soil, water, and agricultural products, which affects human health, and these metals are increased by the application of toxic agrochemicals and anthropogenic activities during modern industrialization. Air, water, and food are contaminated by toxic heavy metals, which pollute the environment (Mood et al. 2021). Heavy metals occur naturally and accumulate in the environment (Tchounwou et al. 2012). Food contamination with heavy metals affects human and animal health, living organisms, and biological diversity (Luo et al. 2011). It also contaminates food with heavy metals that are dangerous to humans, and a high dose of heavy metals can induce abdominal colic, kidney failure, and diarrhea (Tsai et al. 2017). Heavy metals at low doses are associated with neuropsychiatric disorders, fatigue, and intellectual function in children (Mazumdar et al. 2011). They react with biological systems by losing electrons and metal cations, forming the nucleophilic sites of vital molecules. Heavy metals reportedly affect various acute and chronic toxicological conditions, leading to birth defects, immune system dysfunction, kidney problems, nervous system problems, skin lesions, vascular damage, gastrointestinal diseases, dysfunction, and cancer (Gazwi et al. 2020). Heavy metals at high doses induce kidney failure, abdominal colic pain, and diarrhea (Tsai et al. 2017).

The amount of heavy metals that accumulate in soil has gradually increased, resulting in major environmental pollution. Heavy metal ions are normally nondegradable and remain in the soil until they reach high toxicity levels, which can decrease the growth of plants. Plants associated with plant growth-promoting rhizobacteria play an important role in increasing the number of growing plants and facilitating soil remediation to decrease heavy metal contamination via various mechanisms (Rajkumar et al. 2012). *B. subtilis* BS01 exhibited siderophores associated with remediation. *P. fluorescens* PF01 and *B. subtilis* BS01 strains exhibited phosphate-solubilizing activity. The phosphate solubilization activity of *P. fluorescens* was greater than that of *B. subtilis*. *P. fluorescens* produces siderophores and grows well in a 25% salt solution but exhibits poor growth at 10%. The concentrations of heavy metals found to accumulate in plants after treatment with *P. fluorescens* strains PF01, BS01, and PF01+BS01 were 1.5 µg/g, 1.6 µg/g, and 1.3 µg/g, respectively (Rajkumar et al. 2012). The bacterial isolates were inoculated into the soil, leading to a decrease in the level of heavy metals in the contaminated soil. *B. subtilis* BS01 and *P. fluorescens* PF01 have synergistic effects and have been developed as bioproducts for heavy metal remediation, which can be applied to contaminated wastewater and cultivated soils, especially for organic agriculture.

Gupa et al. (2022) reported that living organisms in the environment are affected by the toxicity of heavy metals in the soil, which can affect soil properties either indirectly or directly in agricultural ecology. PGPR are associated with biological remediation and have become a sustainable method for eradicating heavy metals that contaminate soil, water, and the environment via bioaccumulation, biotransformation, biosorption, chelation, precipitation, and siderophore production. Plant growth-promoting rhizobacteria are effective for the biological remediation of heavy metal contaminants in soil through the alteration of heavy metals. The tolerance of plants to heavy metal stress can be improved by the application of PGPR, which can increase the availability of nutrients in the soil by chelating ions and producing siderophores. Heavy metals are known to be nondegradable, and PGPR can help decrease heavy metals through bioremediation.

1.2.7 Organic Certification

Innovation in microbial biotechnology for organic agriculture is more successful for organic agricultural production and promotion. Organic certification must guarantee that consumers are getting safe food. It is a certification process for growers and producers of organic products and organic food (Labeling, article 30 of *Regulation (EU) 2018/848 of the European Parliament and of the Council of 30 May 2018 on organic production and labelling of organic products and repealing Council Regulation (EC) No 834/2007*). In general, agricultural inputs or organic agricultural production are key factors for successful organic agriculture. This review proposes modern organic agriculture, which involves the scientific investigation of agricultural inputs or organic agricultural production, including seed production, biofertilizers, biopesticides, and animal feed, to be used instead of agrochemicals. Organic certification is directly involved in business for organic agricultural products and food production, including growers, food processors, retailers, restaurants, and organic textiles (https://en.wikipedia.org/wiki/Organic_certification). The requirements for organic certification vary by country but generally involve adherence to production standards governing the growing, storage, processing, packaging, and shipping of agricultural products. These standards aim to avoid the use of synthetic chemical inputs—such as chemical fertilizers, pesticides, synthetic antibiotics, food additives, and irradiation (*"Organic Farming". United States Environmental Protection Agency. Archived from the original on June 24, 2003. Retrieved May 30, 2023*)—as well as genetically modified organisms (GMOs). Organic certification also typically requires farmland to remain free from prohibited chemical inputs for a designated period, often three or more years, and mandates strict separation of organic products from noncertified products. Compliance is periodically evaluated by accredited organic inspectors. The organic certification in each country is proven by the government for commercial purposes using the term organic, which is legally restricted. The average fee for organic certification is annually certified, and the certification fee ranges from approximately $400 to $2000/year in the United States and Canada, which depends on each agency and the size of organic farms. Financial assistance programs can help producers meet the requirements for organic certification (*"FAQ: Becoming a Certified Operation". United States Department of Agriculture: Agricultural Marketing Service. Archived from the original on 21 October 2014. Retrieved 4 November 2014*).

Organic inspectors are advised to maintain daily records of farming operations, including marketing data and all related activities. This information must be available for organic inspectors, who can perform specific tests for soil, water, and plants. The initial inspection of organic farms for certification requires verifying that the soil complies with basic regulations, including the absence of prohibited synthetic agrochemicals. Achieving this condition may take several years, depending on the farm's prior management practices. Consequently, conventional farms must undergo a transition period to meet organic quality standards, a process that typically takes 2–3 years (*"FAQ: Becoming a Certified Operation". United States Department of*

Agriculture: Agricultural Marketing Service. Archived from the original on 21 October 2014. Retrieved 4 November 2014, and "Public project results". CertCost. Archived from the original on 19 October 2015. Retrieved 30 October 2).

The Association of Agricultural Technology in Southeast Asia (AATSEA) is concerned with all regulations and requirements for organic certification with the same targets to promote non-toxic agricultural products and food to save human beings and living organisms for sustainable development goals. The AATSEA started to certify organic products in 2016, but it is more practical for organic inspection and certification on the basis of scientific knowledge. AATSEA concerns organic outputs that are delivered to consumers and supermarkets, e.g., vegetables, fruits, food processing products, animal products, etc., and that must not detect any toxic agrochemicals, human pathogens (*Salmonella* spp., *Escherichia coli*), or low nitrate or heavy metals. To clarify, the ATSEA has proposed a definition of nonagrochemical production (NAP) in the transition period to identify organic agriculture. The NAP explains that farmers must stop applying chemical fertilizers, chemical fungicides, chemical insecticides, chemical herbicides, and synthetic plant growth regulators for their organic production. The first target is to recover the imbalanced agroecosystem and surrounding environment and to improve biological diversity, increase the fertility of soil with increasing organic matter, increase natural plant nutrition, and increase the soil pH and electrical conductivity for plant production. NAP farmings normally still accumulate toxic chemicals and heavy metals in water, soil, and agricultural products. Growers are recommended to integrate conventional and cultural methods, including integrated crop management, and to increase biological activity, biodiversity, soil revitalization and remediation, and soil fertility through the use of microorganisms and other nonagrochemical methods. To meet organic standards in terms of both the quantity and quality of

Fig. 1.1 AATSEA requirement for organic certification

Fig. 1.2 AATSEA and Earthsafe powered by AATSEA organic certification

organic products, organic farmers are recommended to improve soil quality and apply biological and natural products, herb extracts, and other cultural methods as agricultural inputs. During the inspection process for organic certification, the NAP is converted to organic agriculture until outputs (products) are obtained that meet required standards, including the absence of toxic agrochemical residues, low nitrate levels, low heavy metal contents, and the absence of human pathogens (*Salmonella* spp. and *E. coli*) (Fig. 1.1).

Organic agriculture (OA) is an organic operation system that relies on a natural agroecosystem where negative environmental and social impacts, such as chemical synthetic fertilizers, chemical fungicides, chemical insecticides, chemical herbicides, synthetic plant growth regulators, veterinary drugs, and nongenetically modified seeds/organisms, must be questioned. Toxic agrochemicals are substituted by organic fertilizers, natural products, biological products, and natural substances for crop protection, including cultural practices, and integrated pests and crop management can be successfully promoted by organic agriculture.

Modern organic agriculture is promoted according to scientific evidence for all agricultural inputs in organic production, which enhances the maintenance of agroecosystems, biodiversity, soil fertility, and food safety for human health. Organic agricultural products must not contain toxic agrochemical residues, low nitrate contents, low heavy metal contents, or human pathogens (*Salmonella* sp. and *E. coli*) and are thus safe foods for confirmation by the ISO laboratory.

Agricultural inputs must be authorized by reliable institutes and certified as agricultural products for organic production. Fee for organic certification by AATSEA is waived for growers to promote them for sustainable living, except for chemical analysis and traveling expenses of organic inspectors. AATSEA organic inspectors must be trained to work as organic experts in the fields. These rules for organic certification, developed by the Association of Agricultural Technology in Southeast Asia (AATSEA), may be adopted or approved by any country that authorizes AATSEA to conduct organic certification (Fig. 1.2). Additionally, the Earthsafe Foundation, part of the Central Group and TOPS Supermarkets, has begun collaborating in Thailand's organic certification efforts. Under this system—referred to as Earthsafe powered by AATSEA—organic growers may have the opportunity to supply certified organic products directly to TOPS Supermarkets (Fig. 1.2).

1.3 Biological Techniques for Organic Crop Production

Biotechniques for organic agriculture have been developed with consideration of both abiotic and biotic factors. Abiotic factors include the surrounding environment of organic land, e.g., soil, water, wind, temperature, relative humidity, light intensity, wind breaks, and nutrient deficiency. Soil fertility is a critical factor, encompassing organic matter content, appropriate soil pH for nutrient uptake, electrical conductivity (EC), soil moisture and temperature, the oxygen–carbon dioxide balance, air and water drainage in the rhizosphere, and soil remediation to address heavy metal contamination, nitrate accumulation, and agrochemical residues. Thus, fertile soil must contain a full set of necessary nutrients for plant growth, e.g., nitrogen, phosphorus, potassium, calcium, sulfur, magnesium, copper, boron, molybdenum, iron, zinc, chloride, and manganese, in rhizosphere soil. It is necessary to revitalize, improve, and maintain rhizosphere soils under appropriate conditions for plant growth. High-quality microbial fertilizers are needed. The biotic factors influencing organic crop production include the living organisms present on farms, such as insect pests, plant diseases, weeds, economic and noneconomic crops, and plants used as windbreaks. Thus, a scientific basis for research on microbial fungicides, microbial insecticides, etc., must be proposed to growers. The organic inspector must provide recommendations for addressing all issues affecting successful organic

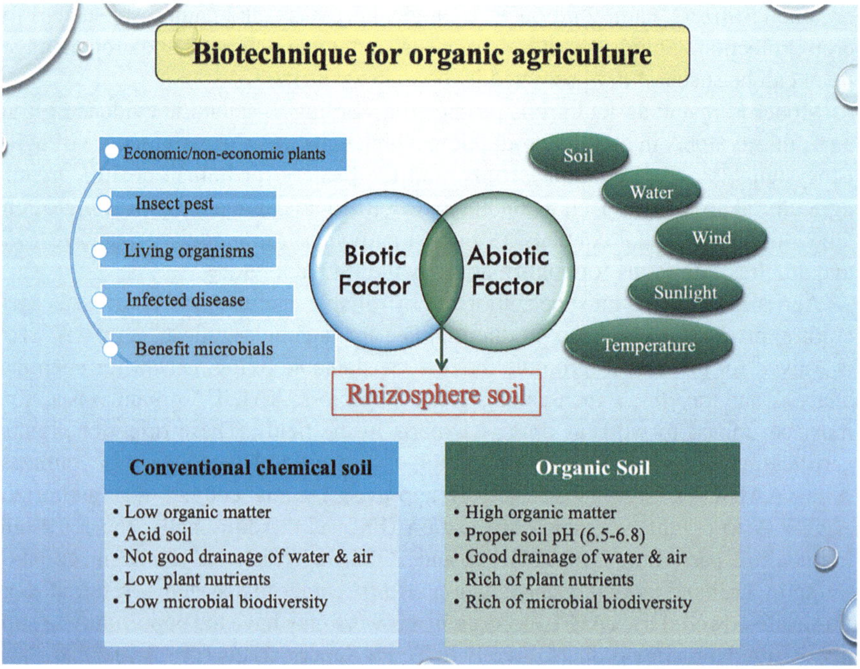

Fig. 1.3 Biotechnique for organic crop production 1

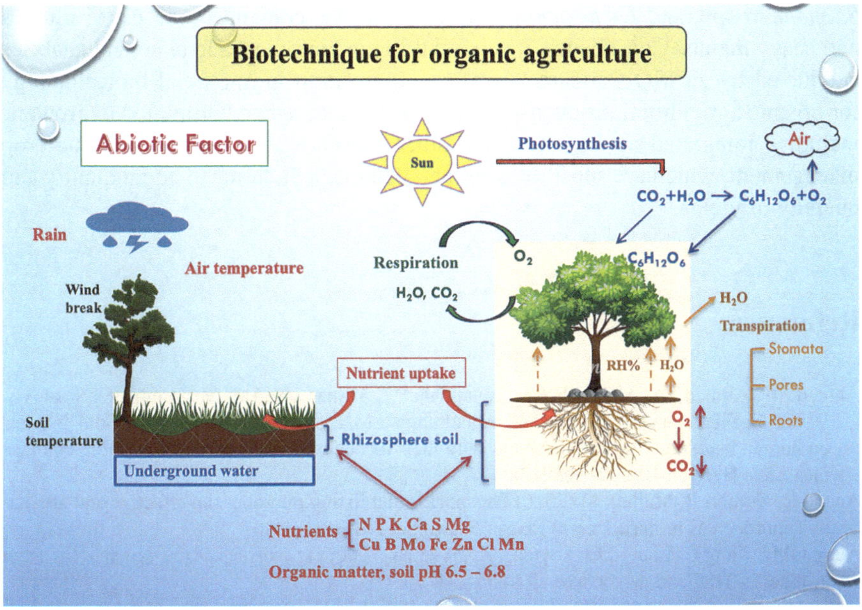

Fig. 1.4 Biotechnique for organic crop production 2

production (Figs. 1.3 and 1.4). This step is important for introducing microbial products (biodecomposers, biofertilizers, biofungicides, bioinsecticides, and bioremediation agents) to growers as substitutes for agrochemicals. It also includes guidance on integrated management practices, such as crop rotation, intercropping, mulching, harvesting, postharvest processing, and packing, to ensure the delivery of high-quality organic products to consumers and markets.

1.4 Conclusion Remarks

Modern organic agriculture must replace traditional organic agriculture, and more organic experts are needed to encourage the production of safe food for human health and to revitalize the environmental agroecosystem. More scientists are needed to perform more research on organic agriculture, which is based on scientific investigations to substitute toxic agrochemicals. Organic certification is important for ensuring that consumers have reliable organic products and delivery to markets. Organic certification is required for marketing to ensure that consumers are confident that it is free from toxic agrochemicals. Various certifying bodies have established regulations for organic certification, including the International Federation of Organic Agriculture Movements (IFOAM), the USDA Organic program, and others. The AATSEA organic regulation standard requires that organic products be free of detectable toxic agrochemicals and human pathogens, such as

Salmonella spp. and *Escherichia coli*, and that they contain low levels of nitrates and heavy metals. Compliance is verified through biological and chemical analyses conducted by an ISO-standard laboratory. Innovation in microbial biotechnology for organic agricultural production should be encouraged and promoted to growers, including integrated management and other alternative methods for organic crop management, which are substitutes for agrochemicals in terms of equal quality and quantity of yields.

References

Ahmad MF, Ahmad FA, Alsayegh AA, Zeyaullahc M, AlShahrani AM, Muzammil K, Saati AA, Wahab S, Elbendary EY, Kambal N, Abdelrahman MH, Hussain S (2024) Pesticides impacts on human health and the environment with their mechanisms of action and possible countermeasures. Heliyon 10(2024):e29128

Anand K, Kumari B, Mallick M (2016) Phosphate solubilizing microbes: an effective and alternative approach as biofertilizers. J Pharm Pharm Sci 2016(8):37–40

Aung OMJ, Kang C, Liang ZQ, Soytong K, Hyde KD (2006a) *Cordyceps nrciensis* sp. Nov from a spider in Thailand. Mycotaxon 97:235–240

Aung OM, Kang JC, Liang ZQ, Soytong K, Hyde KD (2006b) A new entomopathogenic species, *Hymenostilbe furcata*, parasitic on a hemiptera nymph in northern Thailand. Mycotaxon 97:235–240

Aung OM, Soytong K, Hyde KD (2008) Diversity of entomopathogenic fungi in rainforests of Chiang Mai Province, Thailand. Fungal Divers 30:15–22

Berg G (2009) Plant–microbe interactions promoting plant growth and health: perspectives for controlled use of microorganisms in agriculture. Appl Microbiol Biotechnol 84:11–18

Chaiyosang B, Kanokmedhakul K, Boonmak J, Youngme S, Kukongviriyapan V, Soytong K, Kanokmedhakul S (2015) A new lumazine peptide penilumamide E from the Aspergillus terreus. Natural Product Research 1–8

Charoenporn C, Kanokmedhakul S, Lin FC, Poeaim S, Soytong K (2010) Evaluation of bioagent formulations to control Fusarium wilt of tomato. Afr J Microbiol Res 9(36):5836–5844

Cheng YQ, Yang RJ, Lyu M, Wang SW, Liu XC, Wen Y, Song Y, Li JL, Chen Z (2018) IdeR, a DtxR family iron response regulator, controls iron homeostasis, morphological differentiation, secondary metabolism, and the oxidative stress response in *Streptomyces avermitilis*. Appl Environ Microbiol 84:e01503–e01518

Das M, Royer TV, Leff LG (2007) Diversity of fungi, bacteria, and actinomycetes on leaves decomposing in a stream. Appl Environ Microbiol 73:756–767

Din M, Nelofer R, Salman M, Khan FH, Khan A, Ahmad M, Jalil F, Din JU, Khan M (2019) Production of nitrogen fixing Azotobacter (SR-4) and phosphorus solubilizing *Aspergillus Niger* and their evaluation on *Lagenaria siceraria* and *Abelmoschus esculentus*. Biotechnol Rep 2019(22):e00323

Douds DD, Gadkar V, Adholeya A (2000) Mass production of VAM fungus biofertilizer. In: Mycorrhizal biology. Springer, Berlin/Heidelberg, pp 197–215

Etesami H, Emami S, Alikhani HA (2017) Potassium solubilizing bacteria (KSB): mechanisms, promotion of plant growth, and future prospects a review. J Soil Sci Plant Nutr 2017(17):897–911

Fenta L, Mekonnen H (2024) Microbial biofungicides as a substitute for chemical fungicides in the control of Phytopathogens: current perspectives and research directions. Scientifica (Cairo) 2024:5322696. https://doi.org/10.1155/2024/5322696

Gazwi HSS, Yassien EE, Hassan HM (2020) Mitigation of lead neurotoxicity by the ethanolic extract of Laurus leaf in rats. Ecotoxicol Environ Safe 192:110297. https://doi.org/10.1016/j.ecoenv.2020.110297

Goteti PK, Emmanuel LDA, Desai S, Shaik MHA (2013) Prospective zinc solubilizing bacteria for enhanced nutrient uptake and growth promotion in maize (*Zea mays* L.). Int J Microbiol 2013:1–7

Gupa N, Yadav KK, Kumar V, Prasad S, Pinto-Cabal MMS, Jeon BH, Kumar S, Abdellatti MH, Alsukaibia AKD (2022) Investigation of Heavy Metal Accumulation in Vegetables and Health Risk to Humans From Their Consumption. Front. Environ. Sci., Sec. Toxicology, Pollution and the Environment.Vol. 10 - 2022. https://doi.org/10.3389/fenvs.2022.791052

Haritakun R, Rachtawee P, Chanthaket R, Boonyuen N, Isaka M (2010) Butyrolactones from the fungus Aspergillus terreus BCC 4651. Chem Pharm Bull (Tokyo). 58(11):1545–8. https://doi.org/10.1248/cpb.58.1545.PMID:21048353

Hung PM, Wattanachai P, Kasem S, Poeaim S (2015) Biological control of *Phytophthora palmivora* causing root rot of pomelo using *Chaetomium* spp. Mycobiology 43(1):63–70

Jayanthi R, Kathireshan AK, Nepolean P, Gayathri G (2023) Characterization and evaluation of plant growth promoting potentials of Actinobacteria from tea ecosystem. Int J Agric Technol 19(5):2063–2078

Jitkasemsuk S, Lin F-C, Chaiwat T, Soytong K(2007) Enzymatic producing fungi for compost production. International conference on integration of science & technology for sustainable development, Bangkok, Thailand. 26–27 April 2007, 143

Kawalekar JS (2013) Role of biofertilizers and biopesticides for sustainable agriculture. J Bio Innov 2013(2):73–78

Kim WG, Cho KM, Lee CK, Yoo I D (2003) Terreulactones A, B, C, and D: Novel Acetylcholinesterase Inhibitors Produced by Aspergillus terreus II. Physicochemical Properties and Structure Determination. Journal of Antibiotics (56): 351–357

Khoshru B, Mitra D, Mahakur B, Sarikhani MR, Mondal R, Verma D, Pant K (2020) Role of soil rhizobacteria in utilization of an indispensable micronutrient—Zinc for plant growth promotion. J Crit Rev 7:4644–4654

Kowapradit S, Wattanachai Pongnak W, Soytong K (2007) Effect of bottom coal ash mixed with beneficial fungi on the growth of rice var Pathumthani. In: International conference on integration of science & technology for sustainable development, Bangkok, Thailand. 26–27 April 2007, pp 308–312

Lee J, Postmaster A, Soon HP, Keast D, Carson KC (2012) Siderophore production by actinomycetes isolates from two soil sites in Western Australia. Biometals 25:285–296

Luo C, Liu Chuanping, Wang Y, Liu X, Li F, Zhan G, Li X (2011) Heavy metal contamination in soils and vegetables near an e-waste processing site, south China. J. Hazadous Materials 186(1):481–490

Mazumdar M, Bellinger DC, Gregas M, Abanilla K, Bacic J, Needleman HL (2011) Low-level environmental lead exposure in childhood and adult intellectual function: a follow-up study. Environ Health 10(1):24. https://doi.org/10.1186/1476-069x-10-2

Meena VS, Mishra PK, Bisht JK, Pattanayak A (2017) Agriculturally important microbes for sustainable agriculture: volume 2: applications in crop production and protection. Springer, Berlin/Heidelberg

Mia MB, Shamsuddin Z (2010) Rhizobium as a crop enhancer and biofertilizer for increased cereal production. Afr J Biotechnol 2010(9):6001–6009

Mood MB, Naseri K, Zoya Tahergorabi Z, Khazdair MR, Sadeghi M (2021) Toxic mechanisms of five heavy metals: mercury, lead, chromium, cadmium, and arsenic. Front Pharmacol. Predictive Toxicology 12. https://doi.org/10.3389/fphar.2021.643972

Nouioui I, Carro L, García-López M, Meier-Kolthoff JP, Woyke T, Kyrpides NC, Pukall R, Klenk HP, Goodfellow M, Göker M (2018) Genome based taxonomic classification of the phylum Actinobacteria. Front Microbiol 9:2007

Paluka J, Kanokmedhakul K, Soytong M, Soytong K, Yahuafai J, Siripong P, Kanokmedhakul S (2020) Meroterpenoid pyrones, alkaloid and bicyclic brasiliamide from the fungus *Neosartorya hiratsukae*. Fitpterapia 142(104485):1–6
Parikh SJ, James BR (2012) Soil: the foundation of agriculture. Nat Educ Knowl 2012(3):2
Pathak D, Kumar M, Rani K (2017) Biofertilizer application in horticultural crops. In: Microorganisms for green revolution. Springer, Berlin/Heidelberg, pp 215–227
Peng FC (1997) Structure and anti-acetylcholinesterase activity of 4 alpha-(hydroxymethyl)-4 alpha-demethylterritrem B. J Nat Prod. 60(8):842–3. https://doi.org/10.1021/np9701236
Phong NH, Pongnak W, Soytong K (2016) Antifungal activities of *Chaetomium* spp. against Fusarium wilt of tea. Plant Protect Sci 52:10–17
Pokorna D, Zabranska J (2015) Sulfur-oxidizing bacteria in environmental technology. Biotechnol Adv 2015(33):1246–1259
Pramanik P, Goswami A, Ghosh S, Kalita C (2019) An indigenous strain of potassium-solubilizing bacteria Bacillus pseudomycoides enhanced potassium uptake in tea plants by increasing potassium availability in the mica waste-treated soil of North–east India. J Appl Microbiol 2019(126):215–222
Rajkumar M, Sandhya S, Prasad MNV (2012) Perspectives of plant-associated microbes in heavy metal phytoremediation. Biotehnology Advance Freitas 10(6):1562–1574
Reed SC, Cleveland CC, Townsend AR (2011) Functional ecology of free-living nitrogen fixation: a contemporary perspective. Annu Rev Ecol Evol Syst 2011(42):489–512
Riaz U, Mehdi SM, Iqbal S, Khalid HI, Qadir AA, Anum W, Ahmad M, Murtaza G (2020) Biofertilizers: eco-friendly approach for plant and soil environment. In: Bioremediation and biotechnology. Springer, Berlin/Heidelberg, pp 189–213
Sharma SB, Sayyed RZ, Trivedi MH, Gobi TA (2013) Phosphate solubilizing microbes: sustainable approach for managing phosphorus deficiency in agricultural soils. Springerplus 2013(2):587
Song J, Soytong K (2017) *Chaetomium* spp. as biological fertilizer for plant growth. Int J Agric Technol 13(6):941–951
Song JJ, Soytong K (2018) Research and development on bioproducts in China: a short communication. Int J Agric Technol 14(1):131–141
Song JJ, Soytong K, Kanokmedhakul S (2022) Control of rice blast disease caused by *Magnaporthe oryzae* by application of antifungal nanomaterials from *Emericella nidulans*. Plant Protect Sci 58:40–48
Soytong K, Nuanjamrat S (2001) Efficacyvtest of *Bacillus thuringiensis* var *kurstaki*, neem oil and rice bran oil to control commom cutwarm, *Spodoptera litura*. Special problem, King Mongkut's Institute of Technology Ladkrabang, Bangkok, 48 pp
Soytong K, Kanokmedhakul S, Kukongviriyapan V, Isobe M (2001) Application of *Chaetomium* species (Ketomium) as a new broad spectrum biological fungicide for plant disease control: a review article. Fugal Divers 7:1–15
Soytong K, Unthuraloet K, Danesh YR (2020) *Serratia marcescens* KMITL2020 as a plant growth stimulant and bacterial antagonist to control brown leaf spot ofrice caused by *Drechslera oryzae*. Int J Agric Technol 16(6):1515–1526
Sprent JI, Sprent P (1990) Nitrogen fixing organisms: pure and applied aspects. Nitrogen Fixing Org 1990(19):288
Sugumaran P, Janarthanam B (2007) Solubilization of potassium containing minerals by bacteria and their effect on plant growth. World J Agric Sci 2007(3):350–355
Sullia S (1991) Use of vesicular-arbuscular mycorrhiza (VAM) as biofertilizer for horticultural plants in developing countries. In Horticulture—New Technologies and Applications; Springer: Berlin/Heidelberg, Germany, 1991; pp. 49–53
Tchounwou PB, Yedjou CG, Patiola AK, Sutton D J (2012) Heavy Metals Toxicity and the Environment. EXS 101:133–164. https://doi.org/10.1007/978-3-7643-8340-4_6
Thomas L, Singh I (2019) Microbial biofertilizers: types and applications. In: Biofertilizers for sustainable agriculture and environment. Springer, Berlin/Heidelberg, pp 1–19

Tsai M-T, Huang S-Y, Cheng S-Y (2017) Lead poisoning can be easily misdiagnosed as acute porphyria and nonspecific abdominal painCase reports in emergency medicine 2017. Case Rep Emerg Med 2017(2):1–4. https://doi.org/10.3109/10408444.2013.768596

Vareeket R, Soytong K (2015) Evaluation of photosynthesizing bacteria for the growth of rice var. RD41. Int J Agric Technol (11)8:2257–2261

Wahyudi AT, Priyanto JA, Afrista R, Kurniati D, Astuti RI, Akhdiya A (2019) Plant growth promoting activity of actinomycetes isolated from soybean rhizosphere. Online J Biol Sci 19:1–8

Wang Y, Zheng J, Lui P, Wabg W, Zhu W (2011) Three New Compounds from Aspergillus terreus PT06-2 Grown in a High Salt Medium. Drugs 2011, 9(8), 1368–1378. https://doi.org/10.3390/md9081368

Xiafang S, Weiyi H (2002) Mechanism of potassium release from feldspar affected by the sprain Nbt of silicate bacterium. Acta Pedol Sin 2002(39):863–871

Chapter 2
Streptomyces sp. Mitigates Biotic and Abiotic Stress Responses and Stimulates Plant Development

Wafaa Mohamed Haggag, Venugopal Gopikrishnan, and Kasem Soytong

Abstract The increasing need of the population for food in limited agricultural land makes it necessary to increase crop production in an environmentally friendly manner. Approximately 20–40% of economic losses in agriculture are attributed to plant pathogens, which include biotic stressors such as fungal and bacterial infections, pests, weeds, and other agents that cause a variety of crop disorders. Abiotic stressors, such as high temperatures, drought, metal poisoning, and soil salinity, are severe constraints on crop productivity. Despite their high effectiveness and ease of use, agrochemicals pose a threat to the environment. A microbial consortium exists in the root zone, and this consortium of microorganisms can resist the impacts of biotic and environmental stresses on plants, resulting in sustainable agricultural productivity. To achieve sustainable crop productivity, rhizosphere and plant periphyton engineering, the best technology for enhancing crop production, requires the identification of a variety of microorganisms with different potentials. Beneficial microorganisms known as plant growth-promoting rhizobacteria reside close to plant roots and increase plant development. The gram-positive bacterium *Streptomyces* sp., often known as plant growth-promoting rhizobacteria, can increase plant development and resistance to adverse climatic conditions. *Streptomyces* sp. is renowned for its ability to produce a wide range of antimicrobials, as well as a large number of secondary and physiologically active metabolites. These metabolites are necessary for plants to withstand environmental stresses and can also function as biological pesticides by triggering plant defense mechanisms against pathogen invasion. Furthermore, they have a strong ability to increase plant development. To reduce the negative effects of biotic and environmental pressures,

W. M. Haggag (✉)
Plant Pathology Department, National Research Centre, Giza, Egypt

V. Gopikrishnan
Centre for Drug Discovery and Development, Sathyabama Institute of Science and Technology, Chennai, Tamil Nadu, India

K. Soytong
King Mongkut's Institute of Technology Ladkrabang (KMITL)- RIMOA, Bangkok, Thailand

V. Gopikrishnan et al. (eds.), *Harnessing Microbial Resources for Sustainable Agriculture*, Microorganisms for Sustainability 61,
https://doi.org/10.1007/978-981-95-4817-0_2

as well as climate change, this study proposes novel applications of beneficial *Streptomyces* and their active second-generation metabolites in agriculture.

Keywords *Streptomyces* sp. · Biotic · Abiotic stresses · Plant development

2.1 Introduction

Climate change has had a negative impact on global agriculture. Food security is at risk in developing countries due to climate change and various environmental and biotic stresses (Haggag 2024). Drought, salinity, heavy metals, and extreme temperatures are examples of abiotic stresses that are accelerated by global climate change, as are biotic stresses such as fungi, pests, bacteria, viruses, and weeds. The utilization of chemical pesticides and fertilizers in agriculture has harmed the environment (Ma et al. 2019). To combat the stresses caused by climate change, scientists are concentrating more on microorganisms that promote plant growth in an effort to create a viable, eco-friendly, and efficient substitute for conventional restoration and contemporary farming methods (Haggag and Ali 2019). Beneficial microorganisms known as "plant growth-promoting rhizobacteria" live close to plant roots and can affect plant growth without endangering the host plant. These bacteria infiltrate the plant's roots, stems, fruits, or flowers. According to Kim and Kwak (2023), they are crucial for regulating abiotic stress and fostering host growth.

The most common genus of actinobacteria and rhizobacteria that promote plant growth in the production of commercial antibiotics is *Streptomyces*. Numerous strains of *Streptomyces* generate enzymes, siderophores, volatile substances, antibiotics, plant hormones (such as IAA), and other active metabolites that can help plants grow and control bacteria, insects, weeds, and diseases. *Streptomyces* species can also protect soil against abiotic stresses such as salt, drought, and organic and inorganic pollutants and increase nutrient bioavailability (Haggag et al. 2017; Manullang and Chuang 2020; Mattei et al. 2022). Because *Streptomyces* produces secondary metabolites linked to biocontrol and plant growth stimulation, it has a wide range of potential agricultural uses.

Additionally, as the following articles demonstrate, these bacteria can help plants cope with abiotic stressors such as pollution, drought, and salinity. Owing to their ability to improve plant development and control disease, *Streptomyces* species are a good way to support sustainable agriculture by increasing crop yield while reducing the use of pesticides.

2.1.1 Streptomyces

A member of the Streptomycetaceae family and the order Actinomycetales (Kämpfer 2006), *Streptomyces* is a Gram-positive bacterium that is distinguished by the presence of guanine + cytosine (G + C) (Marios et al. 2023). There are more than 700 known species of *Streptomyces* bacteria (Euzéby 2008). Streptomycetes are distinguished by their intricate secondary metabolism. Between 5% and 23% (average: 12%) of the protein-coding genes associated with secondary metabolism are found in each *Streptomyces* species (Marios et al. 2023). More than two-thirds of all naturally occurring antibiotics, such as streptomycin, neomycin, cypemycin, bottromycin, grisemycin, and chloramphenicol, are produced by streptomycetes. Over the course of five continents, approximately 1000 *Streptomyces* species have been found in a range of terrestrial and aquatic habitats (Van der Meij et al. 2018). They are significant biological composters in soil.

2.1.2 *Bioprotection: Managing Biotic and Abiotic Stress*

Streptomyces has a number of known modes of action. Nonetheless, certain characteristics are species-specific (Alekhya et al. 2016). Some significant PGPR mechanisms are covered in the paragraphs that follow, with an emphasis on what is currently known about *Streptomyces*. These mechanisms fall into the following groups: bioprotection, biostimulation, and biofertilization (Jog et al. 2014) (Fig. 2.1).

2.1.2.1 *Streptomyces* Reduce Biotic Stress in Plants

Streptomyces is a genus with numerous industrial uses in agriculture that help with biological control and the stimulation *of plant growth.* Because of its crucial qualities, which include nutrient uptake, it is present in the root zone among microorganisms and has strong antimicrobial potential (Jog et al. 2014; Kumar et al. 2017). Additionally, *Streptomyces* strains can act as biological control agents in two important ways. They can either engage in parasitism or antibiosis, or they can compete with the pathogen for resources and space (Patel et al. 2018). They might also work at different points in the infection process to make the host plant they colonize resistant to disease (Katiyar et al. 2018). *Streptomyces* species can function as biocontrol agents to prevent a range of illnesses (Cao et al. 2016).

For example, rice is protected by *Streptomyces* from the fungus *Pyricularia oryzae* (Law et al. 2017), as well as from *Fusarium graminearum* in wheat (Palazzini et al. 2017), *Sclerotinia sclerotiorum* in lettuce (Chen et al. 2016), and *Pectobacterium carotovorum* in tomatoes (Dias et al. 2017). Its role in fostering more productive and sustainable agriculture is particularly noteworthy.

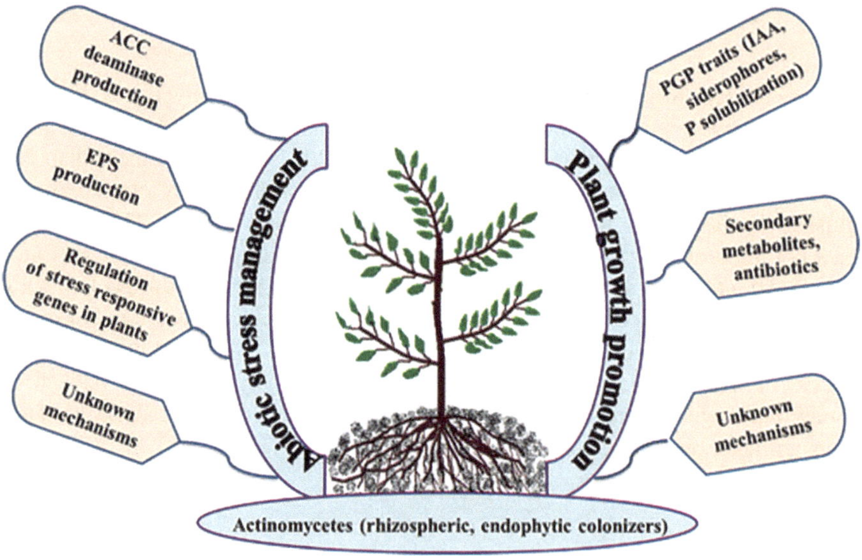

Fig. 2.1 A diagram showing how microorganisms that promote plant development work. The plant growth-promoting and abiotic stress-relieving characteristics of *Streptomyces* are conceptually presented. (Grover et al. 2016)

Antibiotics and Secondary Metabolites

The synthesis of many new and bioactive chemicals, including antibiotics and other active secondary metabolites that suppress infections, as well as significant and useful molecules for agriculture, depends on *Streptomyces* strains isolated from diverse environments (Bibb 2013; Pham et al. 2019).

Since *Streptomyces* species produce more than 70% of the natural compounds utilized in agriculture, their extensive secondary metabolism makes them highly desirable as microbial inoculants (Alam et al. 2022). *Streptomyces* species can exhibit beneficial functional traits in the field of plant health by generating specialized metabolites that increase the absorption of nutrients, promote plant growth, decrease biotic stress, increase resistance, and prevent pathogen or pest invasion and establishment through the use of pesticides (O'Sullivan et al. 2021; Park et al. 2023). Several products have been commercialized using *Streptomyces* microbial cells or their purified metabolites as biocontrol agents (Shi et al. 2020; Collinge et al. 2022).

Recent studies have shown that *Streptomyces* species can produce more than 150 thousand more specific compounds than are currently known (Donald et al. 2022; Lacey and Rutledge 2022). This metabolic aptitude is due to the exceptional diversity of biosynthesis-related gene clusters (BGCs), which are the highest of any known bacteria (Gavriilidou et al. 2022). A biosynthesis-related gene cluster is a group of genes that are physically adjacent to or contiguous to one another in the genome and that encode the enzymes required for the production of a certain specialized metabolite. Furthermore, genes linked to transport, self-resistance, or regulation of relevant metabolites may be included in the cluster (Kautsar et al. 2020).

Several studies have concentrated on identifying distinct microbial strains and specific compounds that can be utilized to fight plant diseases by separating novel *Streptomyces* species (spp.) and other MBCAs from plant compartments (Collinge et al. 2022; Wang et al. 2022). Endophytic *Streptomyces* isolates extracted from the roots of healthy wheat plants in western Australia, for example, reduced plant fungal infections and increased the expression of genes linked to plant defense (Belt et al. 2021). Ren et al. (2020) and Yang et al. (2020) reported that *Streptomyces* isolates possess the genes necessary for the synthesis of new natural and biological compounds as well as secondary metabolic products.

Suárez-Moreno et al. (2019) reported that *Streptomyces* A20 produces the antibiotics streptomycin D, E, and F, which protect rice from bacterial infections. After being separated from *Streptomyces* sp. NEAU-HV9, actinomycin D was shown to have strong antagonistic activity against *Ralstonia solanacearum* in tomato plants (Ling et al. 2020). The use of degradative enzymes, plant hormones, antibiotics, and antioxidants such as iron, zinc, phosphorus, potassium, hydrogen cyanide, guaiacol peroxidase, ascorbate peroxidase, glutathione reductase, and superoxide dismutase are examples of additional metabolites.

Fruit postharvest disease can also be effectively managed with the help of *Streptomyces* nonvolatile specialized metabolites. For example, the fungal pathogen *Botrytis cinerea* can prevent fruit rot in strawberries, tomatoes, and grapes by extracting the macrolides lucensomycin and 32,33-didehydroroflamycoin, also known as reveromycin polyketides, from *Streptomyces* spp. fermentation process broths (Kim et al. 2020; Nguyen et al. 2021). Additionally, in papaya, avocado, and mango fruits, the antimicrobial polymer epsilon-poly-L-lysine, which was isolated from several *Streptomyces* species, is widely used in the pharmaceutical and agricultural industries and is effective against postharvest diseases such as anthracnose (*Colletotrichum gloeosporioides*), blue mold (*Penicillium expansum*), and gray mold (*B. cinerea*) (Bai et al. 2022).

Certain plants and microorganisms that normally flourish in environments with minimal iron stress produce siderophores, which are tiny molecules that bind iron and move it across cell membranes into the plasma membrane for utilization in biological activities.

By efficiently utilizing limited amounts of iron, some bacteria can produce siderophores and survive in environments with limited iron (Pellegrini & Fernandez 2018; Wang et al. 2020). By repairing, providing, and transferring nutrients to plants, the detrimental effects of plant pathogens on plant health are reduced by preventing their growth and denying them critical elements for growth and energy production.

Numerous investigations have demonstrated that a microbe's resistance to infections is correlated with its capacity to generate and bind iron. Siderophores have been employed as secondary metabolites in industrial applications because of their involvement in a number of biological pathways (Khasheii et al. 2021).

A significant source of novel siderophore compounds with various chemical compositions and biological actions against harmful bacteria and fungi is the Gram-positive bacterium *Streptomyces* (Terra et al. 2020). A metabolomic study of extracts from liquid cultures of two *Streptomyces* isolates from the rhizosphere of medicinal plants revealed the formation of the catecholamine siderophore peptides streptobactin and tripinartin (Alinejad et al. 2024).

Additionally, they may colonize the root zones, endophytes, and lamellae of plants and improve soil fertility by fixing, supplying, and transporting nutrients to plants. To improve mineral uptake and change it from an unavailable form to an available form, these bacteria are used for soil remediation.

Certain bacteria need resources to thrive and manufacture enzymes, which are then used to make biologically active substances and proteins that have antibacterial properties (Pandey et al. 2005; Rico and Preston 2008). In plant tissues, L-glutamic acid may restore the core bacterial strain population and maintain microbial diversity, as claimed by Kim et al. (2021). Phenols, alkanes, hydrocarbons, aldehydes, ketones, esters, heterocyclic compounds, and aromatic derivatives are only a few diverse and complex volatile chemicals produced by bacteria (Tilocca et al. 2020).

Streptomyces species generate a variety of volatile chemicals that either directly or indirectly promote plant development. Plant growth is aided by volatile chemicals, which also stimulate plant-induced resistance and **block** and **stop** the growth of bacterial, viral, and fungal pathogens (Liu and Brettell 2019). Compounds that are easily transformed in terms of temperature, pH, moisture content, and soil texture are examples of abiotic soil properties that affect the production and operation of these compounds (Han et al. 2018).

The *Streptomyces* genus enhances the tolerance of plants to environmental stressors and reduces their harmful effects, the most important of which are bacterial, viral, fungal, insect, and weed diseases, by encouraging systemic-induced resistance (Tonelli et al. 2020). This resistance is caused by the generation of signals specific to systemic induced resistance, such as ethylene or jasmonic acid. In the sugarcane rhizosphere, Wang et al. (2021) discovered several strains of *Streptomyces*, including *Streptomyces griseorubiginosus* BTU6, which increased the expression of defense enzymes in sugarcane, initiating the resistance defense system.

Many *Streptomyces* strains have also been demonstrated to produce siderophores (Vijayabharathi et al. 2015), volatile organic compounds (VOCs) (Jones and Elliot 2017), or various antibiotics (Han et al. 2021; Terra et al. 2021) that increase plant disease resistance. Many *Streptomyces* strains, such as *S. yatensis*, *S. pactum*, *S. rochei*, *S. rubrogriseus*, *S. lincolnensis*, *S. hygroscopicus*, *S. antibioticus* strain M7, *S. albogriseolus* ND41, and *S. fimicarius* D153, can control nematodes by generating the antibiotics erinomycin, hygromycin, carboxamycin, vervinolin, and lincomycin (Mani et al. 2024). They also function as inducers of plant defense against nematodes. They have also demonstrated the process of promoting plant development and intrinsic potential.

2.1.2.2 *Streptomyces* Alleviate Abiotic Stress in Plants

The impact of climate change has caused significant concern about sustainable water supplies in several locations. Drought, the most damaging environmental stressor, directly affects plant seeds throughout every stage of their life, from propagation to terminal fruits (Zhang et al. 2022). Dry weather causes decreased photosynthesis, altered endogenous hormone levels, and increased plant osmotic regulators (Kuromori et al. 2022). Additionally, drought stress induces the production of reactive oxygen species such as superoxide radicals, hydrogen peroxides, and hydroxyl radicals, resulting in oxidative stress (Kerchev and Van Breusegem 2022).

The use of microorganisms that stimulate the growth of plants improves plant health and tolerance to environmental challenges such as drought, salinity, acidity, and heavy metal contamination, making them more sustainable than inorganic fertilizers in arid and salty environments. Numerous studies have shown that beneficial soil bacteria increase the development of plants by increasing mineral uptake, including nitrogen and phosphate fixation (Henneron et al. 2020).

Microorganisms play a well-established and well-known role in disease prevention and nutrient management and stimulate plant development. Nonetheless, the application of PGP bacteria for reducing plant stress caused by heat, osmotic stress, salt, drought, nutrient deficiencies, metal toxicity, and waterlogging is becoming increasingly important (Yao et al. 2010). Additionally, beneficial root microorganisms have been shown to respond to stress by altering plant metabolism and biosynthesis pathways. *Streptomyces* are important and widespread bacteria present in soil, plant roots, and aquatic environments. They are the principal source of antibiotics and physiologically active secondary metabolites (Quinn et al. 2020), which stimulate plant development and allow plants to adapt to and resist a wide range of environmental challenges. According to Gopalakrishnan et al. (2015), microbes can help plants cope with abiotic stressors by improving their metabolic and genetic capacities. Therefore, the application, screening, and selection of microorganisms that can withstand stress may be good ways to help crop plants in stress-prone locations overcome their productivity restrictions.

Rhizobacteria, which promote plant growth, are common microbial communities that surround the root system internally or externally. They can aggressively colonize plant roots, which can increase plant growth and development via direct or indirect methods. *Streptomyces* is a species of actinobacteria that can colonize root tissues and the rhizosphere with PGP activity (Thilagam and Hemalatha 2019).

Streptomyces have been shown to directly or indirectly promote plant development and stress tolerance by producing auxin (indole-3 acetic acid, IAA) and siderophores that cause plants to develop systemic resistance and control the root microbiome by producing signaling molecules or antibacterial compounds (Singh and Gaur 2017).

Streptomyces sp. isolated from the wheat rhizosphere, for example, have demonstrated that rhizobacteria promote plant development activities such as phosphate solubilization; IAA (indole-3-acetic acid) synthesis; siderophore, phytase, and chitinase activities; and sugar utilization in the rhizosphere (Jog et al. 2014). The ability of *Streptomyces lasalocidi* JCM 3373T (one of 36 *Streptomyces* strains) to alleviate salt stress in soybeans via the secretion of indole-3-carboxaldehyde (ICA1d) was tested (Lu et al. 2024). This was accomplished by regulating the expression of the stress-responsive genes GmPHD2 and GmWRKY54, as well as GmPIN2a and GmYUCCA6.

According to AbdElgawad et al. (2020), *Streptomyces* spp. isolated from the rhizosphere have been demonstrated to increase soil microbial populations, soil nutrients, and organic matter content in five legumes: soybean, bean, chickpea, lentil, and pea. Increased photosynthesis led to higher yields of legumes when *Streptomyces* sp. was added to the soil. *Streptomyces* sp. also increased the amount of nitrogen in the soil, legume tissues, and seeds, which activated essential nitrogen-metabolizing enzymes such as glutamine synthetase, nitrate reductase, and glutamate synthetase. Crop loss due to soil salinization is another increasing threat to agriculture worldwide; this problem is more pronounced on agricultural land in arid and coastal locations (Corwin 2021). Rapid salinization of agricultural land is largely caused by saline irrigation, limited precipitation, and excessive evapotranspiration. The evolution of crop farming has faced difficulties because of the abiotic stressors of salinity and drought. According to Nivedita et al. (2024), the beneficial interactions between many *Streptomyces* strains within consortia have the potential to improve rice plant resilience to salinity. To improve the ability of rice (*Oryza sativa* var. Swarna) to grow more easily and resist stress from drought, *Streptomyces* consortia enhance the process of photosynthetic oxidative function and proline concentration. According to Zayed et al. (2022), each strain may provide a unique set of stress-reducing traits, such as osmolyte production, ion homeostasis management, or antioxidative defense modification. The spiroketal polyketide pteridic acid H (1) and its isomer F (2), which *Streptomyces* manufactures to improve root growth under abiotic stress, effectively reduce salinity and drought stress.

The ability of the cell-free extract of Actinobacteria (Ac), which was isolated in a semiarid setting, to restore the decline in maize growth and enhance drought-induced physiological stress tolerance has been demonstrated (Mona et al. 2020). The production of second-generation metabolites and the antibiotic activity and antioxidant properties of three Ac isolates were assessed. Flavonoids, antioxidants, and antimicrobials were most abundant in isolate Ac3, which also possessed the capacity to create phytohormones and siderophores.

The development and growth of plants, along with additional physical and chemical processes, are significantly impacted by the pH of the soil (Msimbira and Smith 2020). Plants at high pH levels experience nutritional imbalance and inadequacy due to several stress factors associated with hydrogen ion toxicity (Msimbira and Smith 2020). The ability of *Streptomyces* strains CAI-127, KAI-32, and KAI-90 to promote the growth and development of rice and sorghum has been reported to occur at pH values between 5 and 13 (Gopalakrishnan et al. 2013).

The physiological and biochemical responses of plant cells are significantly restricted by cold stress (Hassan et al. 2021). An investigation into the effects of cold stress on the internal bacterial germs of two mulberry cultivars was also conducted: X792 is resistant to low temperatures, whereas DS is sensitive to them (Chen et al. 2022).

Proteobacteria and Actinobacteria are the most prevalent organisms under metal stress, but the uncontaminated site has a greater diversity of bacteria, according to an evaluation of the impact of stress from combined heavy metals, including cadmium, lead, and zinc, on bacterial diversity in rice fields (He et al. 2017).

Crucially, *Streptomyces* manufactures commercial products such as Actinovate® and Mycostop®, which have been used as biological fungicides and bactericides for an extended period of time. Recently, it has also been demonstrated that these chemicals may help plants adapt to drought and salinity (Fitzpatrick et al. 2018).

2.1.3 Recommendations and Prospects for the Future

Many environmental and biotic stresses can hinder a plant's ability to grow, develop, and produce. Many beneficial microorganisms coexist with plants and produce numerous metabolites that promote plant development, resilience to abiotic and biotic stressors, and adaptation to environmental circumstances. PGP bacteria are critical for increasing plant growth and development under stressful conditions, hence contributing to the development of sustainable agricultural systems. *Streptomyces* is one of the most essential bacteria because it helps plants develop defenses against biotic and different environmental influences. *Streptomyces* is a potential organism for the manufacture of novel natural chemicals, including antibiotics and bioactive substances utilized in agriculture. *Streptomyces* secretes many enzymes, antibiotics, and siderophores to reduce and suppress the growth of diseased plants and maintain plant health. Furthermore, under a range of stressors, it synthesizes and generates plant hormones to commence nutrient cycling. They can also have a positive synergistic connection with plants, leading to good health and plant nutrition. In soil, *Streptomyces* are a component of a microbial interaction network that helps plants cope with various stresses. Research on *Streptomyces* clearly suggests that organisms may interact with plants through a variety of means. The primary foundation of this promise is the application of environmentally friendly microorganisms that enhance plant development and manage pests. The application of biofertilizers may be a viable strategy to enhance the microbial state of the soil by promoting the natural soil microbiota, which affects the availability of nutrients and the breakdown of organic matter. A possible remedy for a more sustainable agricultural future is the use of biopesticides, which are collections of helpful microorganisms found in plants in the right formulations. However, many unknown and unexplored natural products produced by *Streptomyces* exist. As a result of the rapid growth in antimicrobial and insecticide resistance, the exploration and manufacture of new compounds and agricultural chemicals, particularly those based on the genome, must be pursued as one of the future goals for developing *Streptomyces* production. It is proposed that additional studies be performed to develop the use of such microorganisms in cleaner and more sustainable agriculture.

These investigations have aided in the creation of new *Streptomyces* that may result in elevated quantities of bioactive metabolites and reduce stress in the host plant. We conclude by demonstrating the ability of *Streptomyces* to release plant growth regulators and secondary metabolites that help plants overcome stresses from both abiotic and biotic sources. This is a potential alternative technique for plant development and crop production in the face of the biotic and environmental stresses caused by climate change.

References

AbdElgawad H, Abuelsoud W, Madany MMY, Selim S, Zinta G, Mousa ASM, Hozzein WN (2020) By increasing nitrogen availability and metabolism, Actinomycetes enrich the soil rhizosphere and improve the seed quality and productivity of legumes. Biomolecules 10:1675

Alam K, Mazumder A, Sikdar S, Zhao Y-M, Hao J, Song C et al (2022) Streptomyces: the biofactory of secondary metabolites. Front Microbiol 13. https://doi.org/10.3389/fmicb.2022.968053

Alekhya G, Gopalakrishnan S (2016) Exploiting plant growth-promoting Amycolatopsis sp. in chickpea and sorghum for improving growth and yield. Journal of Food Legumes 29(3&4): 225–231

Alinejad F, Gholam Khodakaramian G, Chandramowli D, Devreese B (2024) The catechol-type siderophore producer *Streptomyces* spp. with quorum-quenching activity was isolated from the rhizosphere of wild medicinal plants. Microbe 2:100033

Bai J-L, Wang H-H, Zhang J-M, Wu Q-P, Mo S-P, He Y-L et al (2022) Postharvest quality maintenance of wax apple and guava fruits by use of a fermented broth of an ϵ-poly-l-lysine-producing streptomyces strain. PLoS One 17(3):e0265457. https://doi.org/10.1371/journal.pone.0265457

Belt K, Foley RC, O'Sullivan CA, Roper MM, Singh KB, Thatcher LF (2021) A plant stress-responsive bioreporter coupled with transcriptomic analysis allows rapid screening for biocontrol of necrotrophic fungal pathogens. Front Mol Biosci 8. https://doi.org/10.3389/fmolb.2021.708530

Bibb MJ (2013) Understanding and manipulating antibiotic production in actinomycetes. Biochem Soc Trans 41(6):1355–1364. https://doi.org/10.1042/BST20130214

Cao P, Liu C, Sun P, Fu X, Wang S, Wu F, Wang X (2016) An endophytic Streptomyces sp. strain DHV3-2 from diseased root as a potential biocontrol agent against Verticillium dahliae and growth elicitor in tomato (*Solanum lycopersicum*). Antonie Van Leeuwenhoek. https://doi.org/10.1007/s10482-016-0758-6

Chen X, Pizzatti C, Bonaldi M, Saracchi M, Erlacher A, Kunova A, Cortesi P (2016) Biological control of lettuce drop and host plant colonization by rhizospheric and endophytic Streptomycetes. Front Microbiol 7:714

Chen CJ, Guo G, Li M, Liang XY, Gu YY (2022) Diversity of endophytic bacteria of mulberry (Morus L.) under cold conditions. Front Microbiol 13:923162

Collinge DB, Jensen DF, Rabiey M, Sarrocco S, Shaw MW, Shaw RH (2022) Biological control of plant diseases—what has been achieved and what is the direction? Plant Pathol 71(5):1024–1047. https://doi.org/10.1111/ppa.13555

Corwin D (2021) Climate change impacts on soil salinity in agricultural areas. Eur J Soil Sci 72:842–862

Donald L, Pipite A, Subramani R, Owen J, Keyzers RA, Taufa T (2022) Streptomyces: still the biggest producer of new natural secondary metabolites, a current perspective. Microbiol Res 13(3):418–465. https://doi.org/10.3390/microbiolres13030031

Euzéby JP (2008). Genus Streptomyces. List of Prokaryotic names with Standing in Nomenclature. Retrieved 2008-09-28

Fitzpatrick C, Copeland J, Wang PW, Guttman DS, Kotanen PM, Johnson MTJ (2018) Assembly and ecological function of the root microbiome across angiosperm plant species. Proc Natl Acad Sci USA 115:E1157–E1165

Gavriilidou A, Kautsar SA, Zaburannyi N, Krug D, Müller R, Medema MH et al (2022) Compendium of specialized metabolite biosynthetic diversity encoded in bacterial genomes. Nat Microbiol 7(5):726–735. https://doi.org/10.1038/s41564-022-01110-2

Gopalakrishnan S, Srinivas V, Sree Vidya M, Rathore A (2013) Plant growth promoting activities of Streptomyces spp. in sorghum and rice. Springer Plus 2:574

Gopalakrishnan S, Sathya A, Vijayabharathi R, Varshney RK, Laxmipathi Gowda CL, Krishnamurthy L (2015) Plant growth promoting rhizo-bia: challenges and opportunities. 3 Biotech 5:355–377

Grover M, Bodhankar S, Maheswari M, Srinivasarao C (2016) Actinomycetes as mitigators of climate change and abiotic stress. In: Plant growth promoting Actinobacteria. Springer, Singapore, pp 203–212

Haggag WM (2024) Plant biotic and abiotic stresses combinations and their management: review article. Int J Agric Technol 20(1):105–122

Haggag WM, Ali RR (2019) Microorganisms for wheat improvement under biotic stress and dry climate. Agric Eng Int CIGR J 21(2):118–126

Haggag WM, Tawfik MM, Abouziena HF, Abd El Wahed MSA, Ali RR (2017) Enhancing wheat production under arid climate stresses using bioelicitors. Pringer: Springer Nature; Gesunde Pflanzen 69(3):149–158

Han D, Yan D, Wang Q, Fang W, Wang X, Li J, Wang D, Li Y, Ouyang C, Cao A (2018) Effects of soil type, temperature, moisture, application dose, fertilizer, and organic amendments on chemical properties and biodegradation of dimethyl disulfide in soil. Land Degrad Dev 29:4282–4290

Hassan MA, Xiang C, Farooq M, Muhammad N, Yan Z, Hui X, Yuanyuan K, Bruno AK, Lele Z, Jincai L (2021) Cold stress in wheat: plant acclimation responses and management strategies. Front Plant Sci 12:676884

He, H.; Li,W.; Yu, R.; Ye, Z. Illumina-based analysis of bulk and rhizosphere soil bacterial communities in paddy fields under mixed heavy metal contamination. Pedosphere 2017, 27, 569–578

Henneron L, Kardol P, Wardle D, Cros C, Fontaine S (2020) Rhizosphere control of soil nitrogen cycling: a key component of plant economic strategies. New Phytol 228:1269–1282

Jog R, Pandya M, Nareshkumar G, Rajkumar S (2014) Mechanism of phosphate solubilization and antifungal activity of Streptomyces spp. isolated from wheat roots and rhizosphere and their application in improving plant growth. Microbiology 160:778–788

Jones SE, Elliot MA (2017) *Streptomyces* exploration: competition, volatile communication and new bacterial behaviors. Trends Microbiol 25:522–531

Kämpfer P (2006) The family Streptomycetaceae, part I: taxonomy. In: Dworkin M, Falkow S, Rosenberg E, Schleifer KH, Stackebrandt E (eds) Prokaryotes, pp 538–604. https://doi.org/10.1007/0-387-30743-5_22. isbn:978-0-387-25493-7

Kautsar SA, Blin K, Shaw S, Weber T, Medema MH (2020) BiG-FAM: the biosynthesis-related gene cluster families database. Nucleic Acids Res 49(D1):D490–D497. https://doi.org/10.1093/nar/gkaa812

Kerchev P, Van Breusegem F (2022) Improving oxidative stress resilience in plants. Plant J 109:359–372

Khasheii B, Mahmood P, Mohammadzadeh A (2021) Siderophores: importance in bacterial pathogenesis and applications in medicine and industry. Microbiol Res 250:126790. https://doi.org/10.1016/j.micres.2021.126790

Kim D, Kwak Y (2023) The endophytic Streptomyces population induced by L-glutamic acid enhances plant resilience to abiotic stresses in tomato. Front Microbiol 14. https://doi.org/10.3389/fmicb.2023.1180538

Kim JD, Kang JE, Kim BS (2020) Postharvest disease control efficacy of the polyene macrolide lucensomycin produced by Streptomyces plumbeus strain CA5 against gray mold on grapes. Postharvest Biol Technol 162:111115. https://doi.org/10.1016/j.postharvbio.2019.111115

Kumar V, Kumar M, Sharma S, Prasad R (2017) Probiotics and plant health. Springer, Singapore

Kuromori T, Fujita M, Takahashi F, Yamaguchi-Shinozaki K, Shinozaki K (2022) Intertissue and interorgan signaling in drought stress response and phenotyping of drought tolerance. Plant J 109:342–358

Lacey HJ, Rutledge PJ (2022) Recently, discovered secondary metabolites from Streptomyces species. Molecules 27(3):887. https://doi.org/10.3390/molecules27030887

Law JWF, Ser HL, Khan TM, Chuah LH, Pusparajah P, Chan KG et al (2017) The potential of Streptomyces as biocontrol agents against the rice blast fungus, Magnaporthe oryzae (Pyricularia oryzae). Front Microbiol 8:3

Ling L, Han X, Li X, Zhang X, Wang H, Zhang L, Cao P, Wu Y, Wang X, Zhao J (2020) A *Streptomyces* sp. NEAU-HV9: isolation, identification, and potential as a biocontrol agent against *Ralstonia solanacearum* of tomato plants. Microorganisms 8:351

Liu H, Brettell LE (2019) Plant defense by VOC-induced microbial priming. Trends Plant Sci 24:187–189

Lu L, Liu N, Fan Z, Liu M, Zhang X, Tian J, Yanjun Y, Lin H, Huang Y, Kong Z (2024) A novel PGPR strain, *Streptomyces lasalocidi* JCM 3373^{T}, alleviates salt stress and shapes root architecture in soybean by secreting indole-3-carboxaldehyde. Plant Cell Environ. https://doi.org/10.1111/pce.14847

Ma Y, Vosatka M, Freitas H (2019) Editorial: beneficial microbes alleviate climatic stresses in plants. Frontiers Plant Science 16:595. https://doi.org/10.3389/fpls.2019.00595

Mani J, Kandasamy D, Vendan RT, Sankarasubramanian H, Mannu J, Nagachandrabose S (2024) Unlocking the potential of *Streptomyces* species as promising biological control agents against phytonematodes. Physiol Mol Plant Pathol 134:102465. https://doi.org/10.1016/j.pmpp.2024.102465

Manullang W, Chuang H (2020) *Streptomyces* sp. mitigates the abiotic stress response and promotes plant growth. J Plant Prot Res 60(3):263–274. https://doi.org/10.24425/jppr.2020.133955

Marios N.; Hesketh, Andrew; Frangou, Nikoletta; Mossialos, Dimitris; Van de Peer, Yves; Oliver, Stephen G.; Amoutzias, Grigorios D. (2023)."A Panoramic view of the genomic landscape of the genus Streptomyces. *Microbial* Genomics 9 (6). doi:https://doi.org/10.1099/mgen.0.001028

Mattei V, Motta A, Saracchi M, Kunova A, Cortesi P, Pizzatti C et al (2022) Wheat seed coating with *Streptomyces* sp. strain DEF39 spores protects against Fusarium head blight. Microorganisms 10:1536. https://doi.org/10.3390/microorganisms10081536

Mona W, Hassan YM, Mohamed Mahmoud SM, Nashwa H, Al-Maghrabi OA, Samy S, Saleh AM, Hamada AE (2020) A bioactive fraction from Streptomyces sp. enhances maize tolerance against drought stress. J Microbiol Biotechnol 30(8):1156–1168. https://doi.org/10.4014/jmb.2003.03034

Msimbira LA, Smith DL (2020) The roles of plant growth promoting microbes in enhancing plant tolerance to acidity and alkalinity stresses. Front Sustain Food Syst 4:106

Nguyen HTT, Park AR, Hwang IM, Kim J-C (2021) Identification and delineation of the mechanism of action of the antifungal agent reveromycin and its new derivative isolated from Streptomyces sp. JCK-6141. Postharvest Biol Technol 182:111700. https://doi.org/10.1016/j.postharvbio.2021.111700

Nivedita S, Behera SS, Behera PK, Parwez Z, Giri S, Behera HT, Ray L (2024) Salt-resistant Streptomyces consortia promote growth of rice (*Oryza sativa* var. Swarna) alleviating salinity and drought stress tolerance by enhancing photosynthesis, antioxidant function, and proline content. Microbe 4:100124

O'Sullivan CA, Belt K, Thatcher LF (2021) Tackling control of a cosmopolitan phytopathogen: sclerotinia. Front Plant Sci 12. https://doi.org/10.3389/fpls.2021.707509

Park E-J, Jang H-J, Park J-Y, Yang YK, Kim MJ, Park CS et al (2023) Efficacy evaluation of streptomyces nigrescens KA-1 against the root-knot nematode meloidogyne incognita. Biol Control 179:105150. https://doi.org/10.1016/j.biocontrol.2023.105150

Pellegrini P, Fernández RJ (2018) Crop intensification, land use, and on-farm energy use efficiency during the worldwide spread of the green revolution. Proc Natl Acad Sci 115(10):2335–2340

Pham JV, Yilma MA, Feliz A, Majid MT, Maffetone N, Walker JR, Kim E, Cho HJ, Reynolds JM, Song MC (2019) A review of the microbial production of bioactive natural products and biologics. Front Microbiol 10:1404

Quinn G, Banat A, Abdelhameed A, Banat I (2020) Streptomyces from traditional medicine: sources of new innovations in antibiotic discovery. J Med Microbiol 69:1040–1048

Ren H, Shi C, Zhao H (2020) Computational tools for discovering and engineering natural product biosynthetic pathways. Science 23:100795

Shi L, Wu Z, Zhang Y, Zhang Z, Fang W, Wang Y et al (2020) Herbicidal secondary metabolites from actinomycetes: structure diversity, modes of action, and their roles in the development of herbicides. J Agric Food Chem 68(1):17–32. https://doi.org/10.1021/acs.jafc.9b06126

Singh S, Gaur R (2017) Endophytic Streptomyces spp. underscore induction of defense regulatory genes and confers resistance against Sclerotium rolfsii in chickpea. Biol Control 104:44–56

Suárez-Moreno ZR, Vinchira-Villarraga DM, Vergara-Morales DI, Castellanos L, Ramos FA, Guarnaccia C, Degrassi G, Venturi V, Moreno-Sarmiento N (2019) Plant-growth promotion and biocontrol properties of three Streptomyces spp. isolates to control bacterial Rice pathogens. Front Microbiol 10:290. https://doi.org/10.3389/fmicb.2019.00290

Terra L, Ratcliffe N, Castro H, Vicente A, Dyson P (2021) Biotechnological potential of *Streptomyces* siderophores as new antibiotics. Curr Med Chem 28(7):1407–1421

Thilagam R, Hemalatha N (2019) Plant growth promotion and chili anthracnose disease suppression ability of rhizosphere soil actinobacteria. J Appl Microbiol 126:1835–1849

Tilocca B, Cao A, Migheli Q (2020) Scent of a killer: microbial volatilome and its role in the biological control of plant pathogens. Front Microbiol 11:41

Tonelli ML, Figueredo MS, Rodriguez J, Fabra A, Ibanez F (2020) Induced systemic resistance-*like* responses elicited by rhizobia. Plant Soil 448:1–14

Van der Meij A, Willemse J, Schneijderberg MA, Geurts R, Raaijmakers JM, van Wezel GP (2018) Interand intracellular colonization of Arabidopsis roots by endophytic actinobacteria and the impact of plant hormones on their antimicrobial activity. Antonie Van Leeuwenhoek 111(5):679–690

Vijayabharathi R, Sathya A, Gopalakrishnan S (2015) Plant growth promoting microbes from herbal vermicompost. In: Egamberdieva D, Shrivastava S, Varma A (eds) Plant-growth-promoting rhizobacteria (PGPR) and medicinal plants. Springer, Cham

Wang, Z.; Yu, Z.X, Solanki M, Yang LT, Xing YX, Dong DF, Li YR (2020). Diversity of sugarcane root-associated endophytic *Bacillus* and their activities in enhancing plant growth. J Appl Microbiol 128:814–827

Wang Z, Solanki MK, Yu Z-X, Anas M, Dong D-F, Xing Y-X, Malviya MK, Pang F, Li Y-R (2021) Genome characteristics reveal the biocontrol potential of actinobacteria isolated from sugar cane rhizosphere. Front Microbiol 12:797889

Wang B, Li L, Lin Y, Shen D, Shao X, Zhong C et al (2022) Targeted isolation of biocontrol agents from plants through phytopathogen coculture and pathogen enrichment. Phytopathol Res 4(1):19. https://doi.org/10.1186/s42483-022-00124-2

Yang Z, He J, Wei X, Ju J, Ma J (2020) Exploration and genome mining of natural products from marine *Streptomyces*. Appl Microbiol Biot 104:67–76

Yang Z, Qiao Y, Konakalla NC, Strøbech E, Harris P, Peschel G, Agler-Rosenbaum M, Weber T, Andreasson E, Ding L (2023) *Streptomyces* alleviates abiotic stress in plants by producing pteridic acids. Nat Commun 14:7398

Yao L, Wu Z, Zheng Y, Kaleem I, Li C (2010) Growth promotion and protection against salt stress by *Pseudomonas putida* Rs-198 on cotton. EurJ Soil Biol 46:49–54

Zayed MT, Tiwari P, Ansari WA, Kumar SC, Kumar M, Chakdar H, Srivastava AK, Singh UB, Saxena AK (2022) Biopriming with a consortium of *Streptomyces araujoniae* strains modulates defense response in chickpea against Fusarium wilt. Front Microbiol 13. Article 998546

Zhang H, Zhu J, Gong Z, Zhu J (2022) Abiotic stress responses in plants. Nat Rev Genet 23:104–119

Chapter 3
Eco-friendly Fertilizers for Sustainable Agriculture: A Path Toward Greener Farming

Wafaa M. Abd El-Rahim

Abstract The agricultural sector plays a critical role in feeding the global population, but its reliance on synthetic fertilizers has contributed to environmental degradation, soil depletion, and greenhouse gas emissions. As the demand for sustainable practices grows, eco-friendly fertilizers are emerging as vital tools for promoting agricultural productivity while preserving the environment. In the quest for sustainable agriculture, eco-friendly fertilizers have emerged as crucial alternatives to synthetic fertilizers. Understanding the biological processes involved in soil health and plant nutrition can help us understand the role of these organic fertilizers in promoting a greener farming system. This review explores the potential of eco-friendly fertilizers to increase soil health, reduce environmental impact, and contribute to long-term agricultural sustainability.

Keywords Chemical fertilizers · Pollution · Eco friendly fertilizers · Soil health · Green farming

3.1 The Need for Eco-friendly Fertilizers

Conventional fertilizers, which are rich in synthetic nitrogen, phosphorus, and potassium (NPK), have driven agricultural yields for decades. However, their overuse has led to soil degradation, water pollution from runoff, and a significant carbon footprint. According to the Food and Agriculture Organization (FAO), nearly 30% of agricultural land is classified as degraded. To address this, farmers and researchers are increasingly turning to sustainable alternatives that improve soil structure, increase biodiversity, and reduce dependency on chemical inputs (Nooeaid 2020; Singh 2017; Vasa 2021).

W. M. Abd El-Rahim (✉)
Department of Agricultural Microbiology, Agricultural and Biological Research Institute, National Research Centre (NRC), Giza, Egypt

V. Gopikrishnan et al. (eds.), *Harnessing Microbial Resources for Sustainable Agriculture*, Microorganisms for Sustainability 61,
https://doi.org/10.1007/978-981-95-4817-0_3

3.2 Biological Concept: Soil Microbial Communities and Nutrient Cycling

3.2.1 Soil Microbial Communities

Soils are home to a complex community of microorganisms, including bacteria, fungi, and archaea, which play pivotal roles in nutrient cycling. These organisms help decompose organic matter, fix atmospheric nitrogen, and enhance the soil structure. The health and diversity of these microbial communities are critical for efficient nutrient uptake by plants (Nooeaid 2020; Singh 2017).

3.2.2 Nutrient Cycling

Nutrient cycling refers to the movement and exchange of organic and inorganic matter back into the production of living matter. The key nutrients for plants—nitrogen (N), phosphorus (P), and potassium (K)—are cycled through various biogeochemical processes in the soil. The following describes how eco-friendly fertilizers fit into this system:

3.2.2.1 Organic Matter Decomposition

Organic fertilizers, such as compost or manure, provide a source of organic matter that enhances microbial activity. As these materials decompose, they release nutrients in a form that plants can readily absorb (Manimozhi 2012; Bhardwaj 2014).

3.2.2.2 Nitrogen Fixation

Certain eco-friendly fertilizers, such as those containing legumes (like clover), can increase nitrogen availability through symbiotic relationships with nitrogen-fixing bacteria (e.g., *Rhizobium*). These bacteria convert atmospheric nitrogen into ammonia, a form that plants can utilize.

3.2.2.3 Mycorrhizal Associations

Many eco-friendly fertilizers encourage the growth of mycorrhizal fungi, which form symbiotic relationships with plant roots. These fungi extend the reach of the root system, improving the ability of plants to access nutrients, particularly phosphorus and water (Basu 2021; Abioye 2023).

3.3 Types of Eco-friendly Fertilizers

3.3.1 Compost

Composting is far from a "new" process. Its applications range from the "managed decomposition" of biological matter and putrescible wastes to the stabilization of sewage sludge for land application and product manufacturing. Currently, there is an additional and very important perspective and application for composting—environmental stewardship. In addition to waste handling attributes and the generation of valuable products, composting plays an important role in augmenting soil properties (Rastogi et al. 2023). Moreover, compost is decomposed organic material that enriches soil with nutrients and improves its structure and moisture retention. When vermicompost is produced through the decomposition of organic waste by earthworms, it is rich in essential nutrients and growth hormones. It improves soil aeration, enhances nutrient absorption, and supports sustainable waste management by recycling agricultural residues (Bargaz 2018; Nosheen 2021).

3.3.1.1 Definition and Importance of Composting

Composting today is increasingly seen as a primary component of integrated waste management strategies. The goals of composting drive site-specific programs and define how composting fits as part of a greater goal. Composting is both an end process and a means to achieve a greater goal. As an end product, compost provides an alternative to landfilling or burning for organic residues. For many, compost is the ultimate end product. However, composting offers many other benefits. Composting is a way of treating organic residues and creating a saleable product, such as animal bedding. Composting is a way to reduce residual mass and volume that requires handling and regulations (Manea et al. 2024; Waqas et al. 2023).

3.3.1.2 The Microbiology of Composting

The backbone of composting is the work of microorganisms. Microbial activity causes the energy stored in organic matter to be released, and the organic materials are broken down during the process. The primary agents involved in compost breakdown are bacteria and fungi, which are aided by other organisms, such as actinomycetes, protozoa, nematodes, and earthworms. Bacteria specialize in breaking down simple sugars, oils, starches, and proteins released by fungi and actinomycetes. These organisms can break down complex molecules such as cellulose, lignin, complex fats, and complex nitrogen compounds well (Qiu et al. 2021; Wang et al. 2023). The process by which microorganisms break down organic matter is enhanced by several conditions. High levels of oxygen, a moisture content of 60%, and temperatures between 130 °F and 150 °F can greatly favor the breakdown process.

Exceptionally high moisture content or low oxygen levels result in a process called anaerobic respiration, where the special bacteria in the compost begin to respire in a way that does not need oxygen. A properly maintained compost will develop a complex food chain, and the community of organisms will become increasingly varied over time. Different organisms specialize during each phase of the composting process. If bacteria cannot survive in high-temperature ranges in compost, they die and are replaced by thermophilic bacteria that are best suited to living in the environment in which they are created. As the compost cools, the types of bacteria needed for lower temperatures flourish, leading to the development of other organisms, such as fungi, molds, insects, and earthworms. A variety of microbial activities create compost with more plant-available nutrients (Ho et al. 2022; Li et al. 2021).

3.3.1.3 Environmental and Agricultural Benefits of Composting

The environmental benefits of composting are numerous and diverse. This process of recycling food and other organic waste avoids the addition of solid waste to already burdened landfills. Landfills are the third largest source of human-related methane release; methane is 21 times more effective than carbon dioxide at trapping heat in the atmosphere. Composting also reduces the need for chemical fertilizers, which are made with fossil fuels and can contribute to ecosystem and human health problems through runoff. In both fields and gardens, compost has been found to improve soil quality by increasing structure, fertility, water-holding capacity, and biological activity. These benefits can lead to increased plant and crop productivity (Bremaghani 2024; Nordahl et al. 2023).

A variety of studies reflect the potential of compost for increasing crop yields. A review of field trials revealed that fields fertilized with compost produced 50–100% higher yields than those fertilized with synthetic chemicals. Another study identified a soil compaction threshold above which crop yields would decrease. Through experiments with compost, researchers have reported that soil compaction decreases below the threshold, opening possibilities for future research. In addition, various studies in the developing world have investigated the relationship between compost use and crop yield. It has been reported that compost use is associated with an increase in agricultural productivity of an average of 13%. Crucially, increased yields are greatest when the soil is poor and infertile; global agriculture mostly occurs in poorly fertile soils. This research provides support for compost's particular promise for small-scale farms that have limited funds for inputs (Kheiralipour et al. 2024; Wang et al. 2022a, b).

3.3.1.4 Composting Techniques and Best Practices

There are multiple ways to manage compost, but almost all of them depend on allowing microorganisms to break down organic matter, with or without the addition of worms. In the United States, households and community gardens often use two types of composting techniques: hot composting is highly managed and involves turning the pile every few days to move decomposing matter from the center to near the edges of the pile. Frequent turning or use of tumblers adds oxygen and speeds up microbial action. Compost is typically produced in 3–6 months in a backyard setting. Cold composting is less common and can take anywhere from 6 months to 2 years to decompose fully. Turning the pile is a low priority and is often forgotten, so slow decomposition occurs. Vermicomposting involves decomposition with the use of worms that eat kitchen scraps. This process is often performed indoors via commercial systems, making it useful in urban environments where outdoor space may be limited (Xie et al. 2023; Amuah et al. 2022). Regardless of technique, some basic factors must be considered when making compost. To support healthy microbial life, compost requires air, water, fuel, and greens. Proper ratios of browns and greens, adequate moisture, and appropriate aeration are key elements to consider when attempting to maximize beneficial microbial proliferation and more efficiently break down food waste. A compost pile that reeks or attracts pests is one sign that something has gone wrong, but with care and practice, anyone can learn to make low-odor compost effectively. One simple way to gauge compost activity is to measure the pile temperature. Many species of bacteria involved in the decomposition process produce heat as a byproduct, and a pile that is "cooking" usually produces good, low-odor compost (Dsouza et al. 2021; Manea et al. 2024).

3.3.2 Green Manure

Cover crops are grown and then plowed back into the soil to add nutrients and organic matter. Green manure involves growing specific crops, such as clover or legumes, and plowing them back into the soil to increase fertility. This practice improves nitrogen content, suppresses weeds, and prevents soil erosion, contributing to long-term soil health (Iqbal 2020).

3.3.2.1 Types of Green Manure Crops

Green manure crops are sown and plowed into the soil to provide a range of benefits, including soil fertility and structural improvement, disease control, and organic matter addition. Green manure crops can be categorized by their growth habit: legumes, which are nitrogen-fixing, and nonlegumes, which vary in their nitrogen contribution. Legumes are usually the green manure of choice for an organic system owner because they supply nitrogen and can also bring many other benefits.

Legumes can be further categorized by their growth habits before they are plowed into the ground, such as self-regenerators, good bulk builders, and good ground coverers (Iderawumi and Kamal 2022; Larkin 2022). Some common legumes used for green manure include alfalfa, vetch, and clover. These plants provide many benefits, such as a high nitrogen supply and good-quality organic matter. Some advantages include providing good food for insects and bees, fixing nitrogen, being deep-rooted, being good weed suppressants, controlling nematodes, and being rich in minerals. Hard-seeded green manure and cover crops have seeds that can lie in the soil for long periods of time after they are shed before they sprout. These are usually perennial plants that are held in the soil weed bank until they find the right time to sprout. They can be very useful in a farming system, as even after incorporating the ground cover crop with a tractor, they will continue germinating and multiplying as they are constantly being brought back to the surface whenever they start to put their seeds up into the profile. They are often used as cover crops or green manure to stop the re-establishment of different weed species later in the year, which always seems to follow a particular crop due to the cropping cycle (Pál and Zsombik 2023; Gao et al. 2024).

3.3.2.2 Benefits and Impact of Green Manure

Green manure is a crop that is grown to be plowed in or left on the surface of the soil, primarily for soil improvement and soil fertility management. More specifically, when a green manure crop is plowed into the soil, it enriches the soil by adding organic matter and nutrients. Although green manure has been grown on a modest scale in most parts of the world for centuries, it is only in the past decade that the ability of green manure to support complex biological processes in soil has become of interest to scientists and funding agencies in many countries. This chapter examines the impact that green manure can have on the soil biological processes that underpin the provision of ecosystem services for sustainable agriculture. In particular, this chapter focuses on the contribution of green manure to the development of agroecological systems that make more efficient use of natural resources and reduce the need for high inputs of inorganic fertilizer and pesticides (Lyu et al. 2024; Lei et al. 2022). The primary benefit of green manure is to improve the fertility and productive capacity of the soil in which it is grown. In this respect, the benefits can be direct or indirect. The direct benefits of green manures are mainly through the capture of nutrients for subsequent release, which leads to enhanced soil fertility. The harvesting of nutrients from green manure is achieved either through nutrient capture and subsequent release at the time of crop degradation and mineralization or through nutrient displacement to plant roots. Fallowing or incorporating green manure is a practice that can address soil degradation and erosion, reduce plant nutrition, sequester carbon, reduce soil moisture, and improve general soil quality (Lei et al. 2022; Iderawumi and Kamal 2022).

3.3.2.3 Practical Applications and Techniques of Green Manure

Several practical applications exist for the use of green manure, directly translating the crop management techniques that are already in place. For example, with respect to planting and managing, most green manure crops discussed in this chapter follow the same crop establishment techniques and are usually planted by broadcasting or drilling seeds via the same method as the subsequent cash crop or via the use of a specialized cover crop seeder. Specialized and successful termination methods such as mowing, crushing, or roller crimpers are all feasible options that directly depend on the type of green manure plant grown. Whether a green manure crop is best used as an intercrop, relay winter cover crop, or field cover crop will ultimately be decided on the basis of the specific needs of the farming system (Gatsios et al. 2021; Wang et al. 2022a, b). To ensure the efficacy of green manure benefits, following the recommended planting and incorporation techniques is vital. In doing so, it may even be feasible to utilize current farming equipment by using green manure cover crop attachment, further simplifying the transition process. The proper timing of the incorporation of a green manure crop is also important since establishment, moderate growth, and active vegetative growth periods are more beneficial in terms of soil benefits. If these periods are not suitable for practical reasons, the crops may need to be terminated early to prevent seed set or excessive root growth. To monitor the expected soil and further improve the biology of a green manure plant, biological tests and analyses can be performed. Furthermore, plant population assessment can provide a quick indication of how well a crop has grown. Regular soil testing and monitoring is an excellent way to determine soil improvements over time. Farmers are responsible for continuously performing and interpreting soil analysis, especially through open-minded consultations, to assess safety and feasibility while enhancing their livelihoods (Kaneko et al. 2023; Iderawumi and Kamal 2022).

3.3.3 Bone Meal

A natural source of phosphorus derived from animal bones that promotes root development and flowering.

3.3.3.1 The Role and Applications of Bone Meal in Agriculture

Bone meal is the product obtained through the grinding or crushing of animal bones to fine particles and is one of the earliest organic fertilizers used in agriculture. Historically, bone meal has been used in agriculture, providing animal protein and mineral sources, refining plant nutrients, and improving soil characteristics. Although the main suppliers of bone meal are slaughterhouses associated with the meat industry, an increasing number of dedicated bone meal producers grind only bones, both raw and pretreated; thus, bone meal is also known as a meat industry byproduct. Bone meal, as an organic amendment, has the potential to enhance the physical and chemical properties of soils to benefit the plants growing in them. With today's concerns about synthetic chemical fertilizer inputs, both environmentally friendly and in relation to food quality and safety, interest in the composition of organic amendments, such as bone meal, has increased. In concrete terms, this material can constitute a legal and feasible alternative for the fertilization of peatland, replacing up to 20% of total mineral nitrogen. The present review highlights the role and applications of bone meal in agriculture, advancing the evidence base provided by the four reviews reportedly held to date. The topics covered include an introduction to bone meal, applications for plant nutrition, the contribution of bone meal to the biogeochemical cycles of the soil, and the ecological and socioeconomic potential of bone meal amid the current food supply chain (Stępień et al. 2021; Piash et al. 2023).

3.3.3.2 Nutritional Composition of Bone Meal

Bone meal is a mixture of minerals derived from animal bones and consists mainly of calcium phosphate. When used as fertilizer in agriculture, it contains both macronutrients such as phosphorus and calcium and numerous trace minerals. Although these minerals are the main components of bone meal around the world, there is variation in the content of minerals in different countries according to the bone type and processing source. Bone meal contains 1–4% phosphorus but also contains 20–27% calcium, 0.5–3% nitrogen, 0.1–0.25% potassium, and 6–20% proteins, with the remaining mass being aggregates of organic and inorganic residuals. Compared with popular animal byproducts used in agriculture as fertilizers, such as manure, poultry litter, or biosolids, bone meal has been shown to have desirable nutrient contents at high levels in some cases. These nutrients are critical for the development of roots, stems, flowers, and fruits in plants. The bioavailability of bone meal is important to ensure the timely release of nutrients and the least potential environmental impact. Phosphorus is present in bone meal forms, which include orthophosphate, condensed phosphates, or organic phosphate. Although the percentage of water-soluble phosphorus added to the soil from bone meal is very low, 5–20% is still considered immediately available, as are the condensed phosphates, or in the form of organically complexed phosphate that is released during weathering processes. Calcium is responsible for pH neutralization associated with organic fertilizer materials. Bone meal is alkaline, which makes it ideal for correcting acidic soils with low pH. Calcium from the bone meal reacts to neutralize the acidic effect of both organic and synthetic nitrogenous fertilizers added to the soil. Nutrient balance is also important for optimizing soil microbial activity and soil physical characteristics. Organic fertilizers are often the ideal fertilizer balance, optimizing the soil microbiota and not favorably selecting groups of soil organisms. Phosphorus in the soil is removed primarily by abiotic soil adsorption and fixation via precipitation with calcium, aluminum, and iron compounds, respectively. Leaching of phosphorus may occur in calcareous soils, but the overall phosphorus leaching potential is much lower than that of nitrogen (Piash et al. 2023; Załuszniewska and Nogalska 2022).

3.3.3.3 Benefits and Uses in Agriculture

Bone meal is often used as a soil supplement and crop fertilizer to provide plants with necessary nutrients and encourage healthy growth. Milled animal bones are reportedly the raw material for the preparation of this product, which is rich in calcium and phosphorus, which are among the main minerals present in bone minerals. The benefits of bone meal for promoting the healthy growth of plants are primarily through its positive influence on the development of the root system and the overall vigor and health of plants, which translates into an increase in crop yield and improvement in quality. Bone meal is often applied in ornamental plant cultivation. It is used in the flower industry for the production of bulbs or perennial plants, as well as in the cultivation of fruit-bearing plants such as raspberries, where it is recommended, especially in orchards of high ecological value, for use in chemical fertilizers, or in organic farming, where the lack of mineral phosphorus biofertilizers that certified objects can be used is an indicator to include rock and bone fertilizers in organic farming. Assumptions concerning the organic production of fruit in orchards, defined as dairy, were adopted on this basis. Bone and rock meal is valuable for the practical application of the methodology. Physical, chemical, and microbiological analyses revealed a positive impact of both tested bone meals on soil quality. Understanding the beneficial effects of bone meal on soil structure and microbiological activity would be especially beneficial for modern agriculture, which is an essential issue due to the growing demand for the production of renewable raw materials and our understanding of the need for sustainability. This, in turn, would encourage economic advantages through the reduced expense of the costs related to the contribution of valuable minerals. Cheaper forms of phosphorus fertilizers are particularly important because of fluctuations in the global fertilizer market in recent years. On the basis of this study, we believe that bone meal is a valuable product with advantages that are easily comparable to those of commonly used phosphorus fertilizers (Borges et al. 2023; Nogalska and Załuszniewska 2021; Simons et al. 2023).

3.3.3.4 Application Methods and Dosage

Bone meal can be applied to the soil surface; this technique is called broadcasting or top dressing. It is also possible to mix bone meal into the planting holes prior to planting. Coating seeds with bone meal is another possibility. Soluble bone meal preparations can be applied as a liquid to leaves or soil, possibly via irrigation. Generally, bone meal application before planting increases the availability of phosphorus over the growth period. Soil properties, however, play a significant role in this regard. If soil has a high fixation potential, most of the phosphorus is bound by the soil as insoluble iron, aluminum, or calcium phosphates and is fixed within a few days. In this case, availability over a long period would only be guaranteed if the bone meal was applied during the growing period. However, in soils with high residual phosphorus solubility, phosphates from bone meal are easily dissolved and available for plant uptake. The required dosage of bone meal can be determined on the basis of soil tests and the requirements of the crops. Crops such as wheat, potatoes, and sugar beets, for example, need relatively large amounts of phosphorus. If previous applications to soil are known, years of cropping without addition to soil tests or field trials can be used as additional indicators. However, field trials should be carried out frequently to account for changing soil properties and field management practices. Broadcasting is the most commonly used application technique. The suggested quantities of bone meal are usually applicable to this method. However, future research might also consider adjusting the quantities in different types of soils or changes in field management. Further research to assess whether amounts or methods of bone meal application are adjustable, we can examine the following points: recommended quantities for broadcast application might not be applicable if bone meal is incorporated because of better utilization effects. A practical approach for accurate doses of bone meal according to crop and band characteristics should be developed. The soil's available phosphorus level must be considered and incorporated into the final decision for application dosage. The application rates should be adjusted according to the changes in the soil phosphorus levels. The integrated application or coapplication of bone meal with other soil amendments and/or organic matter in agriculture should be evaluated. The use of bone meal in combination with other phosphorus fertilizers needs to be evaluated (Tiwari et al. 2023; Enguwa et al. 2024; Kumawat et al. 2022).

3.3.4 *Fish Emulsion*

Fish emulsion is a liquid fertilizer made from fish by-products and is rich in nitrogen and trace minerals.

3.3.4.1 The Science and Application of Fish Emulsion in Agriculture

Fish emulsion is an organic fertilizer derived from whole fish or fish by-products, liquefied by the addition of water and stabilized by an acid to a low pH, with the resulting emulsion containing significant degraded protein and oil fractions. Fish has been used since antiquity as essential nourishment, and fish by-products from the fishing industry have contributed to fertilizer and animal food for centuries. However, the present renewed interest in fish emulsion can be partly attributed to an increasing demand for organic and natural fertilizers in modern agriculture, as well as for economic and eco-friendly use of natural resources. Furthermore, the depletion of land-based fish resources has led to the overexploitation of marine resources to produce fishmeal, thus increasing its price further. As a result, there is a global necessity to exploit fish and fish waste productively before the industry becomes more dependent on land-based fisheries due to environmental measures controlling open-water capture (Jaies et al. 2024; Zhang et al. 2023). Fish emulsion contains the major elements of N, P, and K, as well as other essential elements to a lesser degree. The minor element content is due to the unique composition of the fish and its processing conditions. Fish emulsion also contains organic matter and carbon sources, which increase and enhance beneficial microbial activity, thereby improving overall soil health and plant growth. Through fermentation and hydrolysis processes, the high-energy macromolecules (proteins, carbohydrates, and lipids) present in fish are enzymatically hydrolyzed, broken up, and released in a form that can be readily taken up and utilized by plants and beneficial microorganisms. Macromolecular aggregates (proteins, peptides, and polypeptides) dissolve in water, forming a solution of amino acid chains of peptides and free amino acids. In addition to its natural occurrence in fish, other environmental contributors to the presence of beneficial forms of nitrogen in fish emulsion include ammonia and a lesser amount of urea (Maulu et al. 2021; Chen et al. 2023).

3.3.4.2 Nutritional Benefits and Mechanisms of Fish Emulsion

Fish emulsion is a liquid organic fertilizer manufactured through enzymatic hydrolysis of fish waste or entire fish. The substance holds high levels of nitrogen and phosphorus, potassium, and essential trace minerals, and it is often alkaline. Nutritional elements such as nitrogen, phosphorus, and potassium increase crop yield and promote more rapid vegetative growth and, thus, are classified as primary nutrients, whereas trace elements are assumed to empower the crops to overcome plant diseases and viruses and fulfill their normal crop growth patterns. The N, P, and K absorbed by plants are mostly utilized for the development of growth and photosynthesis. Plants need several other trace nutrients besides major nutrients that are absorbed directly by plants, and these are useful for increasing the growth of plants. Microbial activity improves soil fertility, especially by increasing the organic matter content of mineral soils. Farmers also reported crop stress reduction by the use of fish emulsion in stressed conditions since it increases the availability of essential nutrients to crops and improves soil nutrient quality when compared to the control or conventional inputs. Secondarily, emulsion suspensions of fish have been shown to be useful for plant growth and stress management arising from biotic and abiotic factors (Muscolo et al. 2022; Ciriminna et al. 2022; Zhang et al. 2023). Upon application, fish emulsion can provide a plethora of benefits for crop management and soil health. First, it is a rich source of nitrogen, phosphorus, and potassium; hence, it promotes plant vegetative growth. The presence of other essential elements and growth regulators may improve crop development when combined with fish emulsion. Second, it induces microbial activity in the rhizosphere as well, enhancing soil fertility. Several disease incidence cancellations after the application of fish emulsion can be attributed to the induced systemic resistance of the plant to diseases. Third, with its better-quality nutrients and enzymes in fish protein, it may increase root respiration and produce a healthy and robust rooting system of plants. This could impact other factors like plant growth hormones and plant metabolites for plant growth. Moreover, the emulsions can enhance plant growth and development actively, as they are rich in NPK, micronutrients, and growth hormones. Additionally, the application of fish emulsion leads to a higher initial yield of vegetables, peanuts, pecans, and other crops. When applied before transplanting in the field to any crop, it ensures plant hardiness and earlier setting of fruit. Due to the increased market demand for toxin-free and safe food, extensive studies are required to explore the potential of using fish emulsions as an effective fertilizer in organic farming. Fish emulsions showed controlling advantages and are a cheaper option that is used for additional and effective nutrient management. In fact, it is the Earth's most comprehensive fertilization strategy. With the use of small amounts of the product, it is possible to cover extensive areas. Because of the presence of these nutrients in fish proteins, microorganisms consume the fish remains and liberate the nutrients when added to the soil; hence, they are readily available to hungry plants that are lacking nutrients and are not able to manufacture chlorophyll. In comparison to chemical fertilizers, the development of plants in various stages can be enhanced by fish fertilizer (Panda and Das 2024; Panday et al. 2024; Matthews et al. 2024). This is produced by the enzymatic hydrolysis of fish waste, which produces high-quality nutrients and bulk fertilizers. Regular applications on a weekly to biweekly schedule will ensure the desired nutrients are available. If you are applying it to the root zone, use fish emulsion and kelp.

3.3.4.3 Application Techniques and Best Practices

Fish-based fertilizer products are highly adaptable and can be applied using a wide variety of crop management methods and growing systems. The main application methods include supplementing fruit and vegetable crops through drip/micro irrigation systems for horticultural production, soil application on pasture, hayfields, and turf, and foliar application on flowering potted and bedding plants (Araujo et al. 2022). Certain application regimes for fish-based fertilizers in agriculture have consistently produced superior results. Some key best practices include diluting the concentrate at ratios ranging from 1:5 to 1:20 and stirring thoroughly to ensure optimal nutrient delivery to crops. Because some crops prefer a slightly acidic soil environment, occasional application of fish-based fertilizer diluted to a ratio of 1:30 or lower may help maintain favorable pH conditions for soil-based, pH-sensitive crops. Owing to its high quality, some users have also employed fish-based fertilizer in hydroponic gardening systems and as a water-conditioning agent at full strength, demonstrating its versatility and excellent production standards (Baruch et al. 2024).

Technology Transfer to Fruit and Vegetable Market Gardens: In the heart of the valley, two case study farmers have adopted specific techniques that have improved their incomes. In the first case study, within a vegetable market garden setting, the farmer implemented a basic starter dilution of 1:20 and applied 10 liters of the end solution fish unique as a tonic to supplement the general nutrient needs of the cabbage and beetroot seedlings applied directly over the seedlings at the time of handling. In the second case study, foliar applications combining seaweed and fish unique were introduced in a strawberry production system. The best results were achieved using a dilution ratio of 1:10, applied every three weeks as a nutrient tonic (Hansen et al. 2022; Kabui et al. 2024). Application of fish unique product for commercial agriculture will change from year to year; the above suggestions outlining frequency of applications are based on using a crop life cycle on average from plant transplant or inversion to maturity or commencement of fruiting. Repeat application is determined and decided on by the grower and should be regulated based on the specific health of the soil for the added benefit of the crop. Ideal application of the product stems from its use in an integrated production system, which takes into account concurrent application of new technologies and the enhancement of nitrogen use efficiency. Management of wilderness and pasture production additionally requires consideration of the overfeeding-the-bush concept, aligning with the principles of sustainable agriculture (Krishnani et al. 2022; Zajicek et al. 2023). While fish unique may be used as a starter nutrient, the field transfer of national fertilizer programs does not yet include the option to incorporate fish nutrients. Best practice indicates dilution ensures better coverage over foliage, and in the case of application on foliage, the use of pressurized equipment suggests the nozzle pressure be kept below 20 psi to be consistent with best practice and not to over-penetrate surfaces of items sprayed. For full canopy, this application method also means that using 70 L/ha of product at a 1:10 ratio means that 700 L of fish unique is not put out at a time. Also, spraying towards the cooler part of the day is best (Warrender et al. 2024; Tang et al. 2023).

3.3.5 Seaweed Extracts

Seaweed extracts provide a range of micronutrients and growth hormones, enhancing plant growth and resistance to stress. Seaweed-based fertilizers are gaining popularity for their ability to stimulate plant growth, improve resistance to drought, and enhance nutrient uptake. Rich in trace minerals, seaweed fertilizers support soil rejuvenation and crop vigor (Verma 2022; Hakim 2021).

Seaweeds create a natural resource with a broad range of uses and applications across multiple sectors, including food, animal feed, crop enhancement, bioactive compounds, and pharmaceuticals. In their organic form, seaweeds are a commonly eaten food and are regularly used in kelp farming. Traditionally, people in Scotland and Ireland have used seaweeds to produce seafood for hundreds of years. In parts of the Pacific Islands, especially Hawaii, traditionalists have been using seaweeds to create tasty supplements. Seaweeds developed more and more in the early 1990s as an attraction toward various functional foods. Many commercial, small, and expanding companies have also recently incorporated several mixtures of seaweeds.

(El-Beltagi et al. 2022; Brakel et al. 2021). With increasing focus on sustainable resources, seaweeds are environmentally friendly. In comparison, seaweeds are vast, easy to grow, and have high yields—thus contributing to the extraction of profitable by-products related to the process. Current trends in the seaweed extraction industry often focus on seaweed carbohydrates, notably from macro- and micronutrients such as alginates, cellulose, and carrageenans. Over the last number of decades, substantial growth has had a significant global economic impact. Recently, an increased emphasis has been placed on seaweed phenolics and polysaccharides. Estimates show a prospective seaweed extract market of about 478 million dollars. Present seaweed extractions have been demonstrated to have the properties of anti-oxidation, anti-cancer, anti-inflammatories, antivirals, etc. There are, in general, traditional ancient uses for seaweeds in the cosmetics industry. The promising applications and potential advantages accrued from seaweed extracts for industrial needs and other applications will also be explored in significant detail. The study introduces several areas of seaweed use, both historically and in recent years (World Health Organization 2022; Deepika et al. 2022).

3.3.5.1 Seaweed Extracts

These extracts provide a range of micronutrients and growth hormones, increasing plant growth and resistance to stress. Seaweed-based fertilizers are gaining popularity for their ability to stimulate plant growth, improve resistance to drought, and increase nutrient uptake. Seaweed fertilizers rich in trace minerals support soil rejuvenation and crop vigor (Verma 2022; Hakim 2021).

Seaweeds are natural resources with a broad range of uses and applications across multiple sectors, including food, animal feed, crop enhancement, bioactive compounds, and pharmaceuticals. In their organic form, seaweeds are a commonly eaten food and are regularly used in kelp farming. Traditionally, people in Scotland and Ireland have used seaweed to produce seafood for hundreds of years. In parts of the Pacific Islands, especially Hawaii, traditionalists have used seaweeds to create tasty supplements. Seaweeds developed increasingly more in the early 1990s as an attraction for various functional foods. Many commercial, small, and expanding companies have also recently incorporated several mixtures of seaweed (El-Beltagi et al. 2022; Brakel et al. 2021). With increasing focus on sustainable resources, seaweeds are becoming more environmentally friendly. In comparison, seaweeds are vast, easy to grow, and have high yields—thus contributing to the extraction of profitable byproducts related to the process. Current trends in the seaweed extraction industry often focus on seaweed carbohydrates, notably macro- and micronutrients such as alginates, cellulose, and carrageenans. In recent decades, substantial growth has had a significant global economic impact. Recently, increased emphasis has been placed on seaweed phenolics and polysaccharides. The prospective seaweed extract market is estimated to cost approximately 478 million dollars. Seaweed extracts have been demonstrated to have antioxidative, anticancer, anti-inflammatory, and antiviral properties, among other properties. In general, traditional ancient uses of seaweed in the cosmetics industry have been identified. The promising applications and potential advantages of seaweed extracts for industrial needs and other applications will also be explored in significant detail. This study introduces several areas of seaweed use, both historically and in recent years (World Health Organization 2022; Deepika et al. 2022).

3.3.5.2 Chemical Composition and Nutritional Value

Seaweed extracts are known to consist of diverse beneficial components, such as polysaccharides, including agar, alginates, carrageenans, cellulose, laminarin, mannitol, and many small fractions of saccharides, polypeptides, and proteins with distinctive amino acid contents and antioxidant properties; dietary fibers; vitamins; minerals; and soluble and insoluble brown pigments. Several of the components of brown seaweed are hydrocolloid fractions (Qiu et al. 2022). The specific polysaccharides present in the hydrocolloid fraction of some brown seaweeds are uniquely gelled with cations to form alginates, carrageenans, and agar and are nonpectin in nature. Alginates are linear copolymers of 1,4-linked β-D-mannuronic acid and α-L-guluronic acid residues. Carrageenans consist of D-galactose repeating monomers connected by α-1,3- and α-1,4-glycosidic linkages and include three main classes on the basis of the number of sulfate ester groups per repeating disaccharide unit and the type of glycosidic linkage. Agar polysaccharides are composed of alternating 3-linked β-D-galactopyranose and 4-linked 3,6-anhydro-α-L-galactopyranose. Scientific research has indicated that these compounds, which are also referred to as marine hydrocolloids, can affect health through various mechanisms in the digestive system, particularly the large intestine. There are considerable variations in polysaccharide concentrations within and between seaweed species. These differences in polysaccharide content and occurrence can guide the selection of seaweed species that are likely to have the greatest potential when seaweed-derived components are used for functional food and pharma applications (Akter et al. 2024; Moawad et al. 2022).

3.3.6 Organic Fertilizers

Derived from natural materials such as compost, manure, and bone meal, organic fertilizers release nutrients slowly, improving soil fertility over time. They enhance microbial activity and increase organic matter, fostering better water retention and reducing soil erosion (El-Ramady 2018; Sathya 2017; Itelima 2018).

3.3.6.1 Biofertilizers

Biofertilizers include living microorganisms, such as nitrogen-fixing bacteria (e.g., *Rhizobium*) and mycorrhizal fungi, which increase nutrient availability to plants. By fostering symbiotic relationships with crops, biofertilizers reduce the need for synthetic inputs, promoting healthier growth and resilience against pests (Verma 2022; Hakim 2021).

3.3.6.2 Visual Examples

- Composting process: A diagram illustrating the stages of composting, showing the breakdown of organic material by microorganisms and the final product rich in nutrients.
- Nitrogen cycle: A visual representation of the nitrogen cycle highlighting the roles of nitrogen-fixing bacteria, the conversion of ammonia to nitrates, and uptake by plants.
- Mycorrhizal network: An illustration showing the relationship between plant roots and mycorrhizal fungi, emphasizing how fungi extend the root system and improve nutrient absorption (Verma 2022; Hakim 2021; El-Ramady 2018; Mazid 2014).

Ways to Study or Understand the Topic in Depth

1. Field Experiments: Experiments comparing the effects of eco-friendly fertilizers and synthetic fertilizers on crop yield, soil health, and microbial diversity were conducted.
2. Microbial Analysis: Soil samples can be used to analyze microbial communities through techniques such as DNA sequencing, which can provide insights into the health and diversity of soil organisms.
3. Literature Review: This review explores scientific articles and publications on sustainable agriculture, focusing on the benefits and limitations of various eco-friendly fertilizers.
4. Workshops and Courses: Participating in workshops or online courses focused on organic farming practices, soil health, and agroecology.
5. Community Gardening: Get involved in community gardening projects that emphasize the use of organic practices, allowing hands-on experience with eco-friendly fertilizers.

By understanding the biological principles behind eco-friendly fertilizers and their role in sustainable agriculture, we can promote practices that enhance soil health, support biodiversity, and contribute to food security in an environmentally responsible manner (Mazid 2014; Pindi 2012; Stojanović 2018).

3.4 Benefits of Eco-friendly Fertilizers

3.4.1 *Environmental Sustainability*

Eco-friendly fertilizers reduce chemical runoff, minimizing the contamination of water bodies and mitigating eutrophication. They also decrease greenhouse gas emissions associated with synthetic fertilizer production and application, contributing to climate change mitigation (Pindi 2012; Stojanović 2018).

3.4.2 *Soil Health and Biodiversity*

By enhancing soil microbial communities and organic matter, eco-friendly fertilizers improve soil structure and nutrient cycling. This promotes healthier crops and reduces the risk of soil degradation.

3.4.3 *Cost-Effectiveness and Economic Benefits*

Although the initial cost of eco-friendly fertilizers may be relatively high, their long-term benefits, such as improved soil fertility and reduced input needs, lead to relatively high economic returns. Additionally, organic farming practices can attract premium prices in the market, increasing farmer income (Mahmud 2021, Akhtar 2022).

3.4.4 *Resilience to Climate Change*

Eco-friendly fertilizers help crops withstand environmental stress, such as drought or extreme temperatures. By enhancing soil moisture retention and nutrient availability, they increase the resilience of agricultural systems to climate variability.

3.5 Challenges and Limitations

Despite their benefits, the widespread adoption of eco-friendly fertilizers faces several challenges. Limited availability, high production costs, and a lack of awareness among farmers hinder large-scale implementation. Additionally, the effectiveness of eco-friendly fertilizers can vary with soil type, crop variety, and climatic conditions. Addressing these barriers requires collaborative efforts from governments, researchers, and agricultural organizations to provide training, subsidies, and policy support (Sandeep 2015; Hajiboland 2017).

3.6 Future Directions and Innovations

Emerging technologies are enhancing the effectiveness of eco-friendly fertilizers. Innovations such as the use of nanofertilizers, microbial consortia, and bioengineered crops could revolutionize sustainable agriculture. Governments and agricultural bodies are also investing in research and development to promote eco-friendly farming practices (Abbamondi 2016; Viruel 2014).

3.7 Conclusion

Eco-friendly fertilizers represent a promising path toward achieving sustainable agriculture. By prioritizing soil health, reducing environmental impact, and fostering resilience, these fertilizers align with global efforts to ensure food security while preserving the planet. To fully harness their potential, stakeholders must address existing challenges and promote widespread adoption through policy, education, and innovation. In doing so, the agricultural sector can pave the way for a greener, more sustainable future.

References

Abbamondi GR (2016) Plant growth-promoting effects of rhizospheric and endophytic bacteria associated with different tomato cultivars and new tomato hybrids. Chem Biol Technol Agric 3(1). https://doi.org/10.1186/s40538-015-0051-3

Abioye OM (2023) The role of biofertilizers in sustainable agriculture: an eco-friendly alternative to conventional chemical fertilizers. Appl Sci Eng Prog. https://doi.org/10.14416/j.asep.2023.07.001

Akhtar N (2022) Improvement of plant responses by Nanobiofertilizer: a step toward sustainable agriculture. Nano 12(6):965–965. https://doi.org/10.3390/nano12060965

Akter A, Sobuj MKA, Islam MS, Chakroborty K, Tasnim N, Ayon MH et al (2024) Seaweed polysaccharides: sources, structure and biomedical applications with special emphasis on antiviral potentials. Future Foods 10:100440

Amuah EEY, Fei-Baffoe B, Sackey LNA, Douti NB, Kazapoe RW (2022) A review of the principles of composting: understanding the processes, methods, merits, and demerits. Org Agric 12(4):547–562

Araujo GS, Silva JWAD, Cotas J, Pereira L (2022) Fish farming techniques: current situation and trends. J Mar Sci Eng 10(11):1598

Bargaz A (2018) Soil microbial resources for improving fertilizers efficiency in an integrated plant nutrient management system. Front Microbiol 9. https://doi.org/10.3389/fmicb.2018.01606

Baruch EM, Yarnell SM, Grantham TE, Ayers JR, Rypel AL, Lusardi RA (2024) Mimicking functional elements of the natural flow regime promotes native fish recovery in a regulated river. Ecol Appl 34(6):e3013

Basu A (2021) Plant growth promoting Rhizobacteria (PGPR) as green bioinoculants: recent developments, constraints, and prospects. Sustainability 13(3):1140–1140. https://doi.org/10.3390/su13031140

Bhardwaj D (2014) Biofertilizers function as key player in sustainable agriculture by improving soil fertility, plant tolerance and crop productivity. Microb Cell Factories 13(1). https://doi.org/10.1186/1475-2859-13-66

Borges WL, Garcia JP, Teixeira PC, Polidoro JC (2023) Cowpea, corn, and soybean yield using fertilizer based on the bone meal in the eastern Amazon. Rev Bras Ciênc Agrár 18(1):e2577–e2577

Brakel J, Sibonga RC, Dumilag RV, Montalescot V, Campbell I, Cottier-Cook EJ et al (2021) Exploring, harnessing and conserving marine genetic resources toward a sustainable seaweed aquaculture. Plants People Planet 3(4):337–349

Bremaghani A (2024) Utilization of organic waste in compost fertilizer production: implications for sustainable agriculture and nutrient management. Law Econ 86:86–98

Chen X, Shang S, Yan F, Jiang H, Zhao G, Tian S et al (2023) Antioxidant activities of essential oils and their major components in scavenging free radicals, inhibiting lipid oxidation and reducing cellular oxidative stress. Molecules 28(11):4559

Ciriminna R, Scurria A, Pizzone DM, Calabrò PS, Muscolo A, Mauriello F, Pagliaro M (2022) Economic and technical feasibility of Anchois Fert organic fertilizer production. Curr Res Green Sustain Chem 5:100315

Deepika C, Wolf J, Moheimani N, Hankamer B, von Herzen B, Rao AR (2022) Utilization of seaweeds in the Australian market–commercialization strategies: current trends and future prospects. In: Sustainable global resources of seaweeds volume 1: bioresources, cultivation, trade and multifarious applications. Springer International Publishing, Cham, pp 265–294

Dsouza A, Price GW, Dixon M, Graham T (2021) A conceptual framework for incorporation of composting in closed-loop urban controlled environment agriculture. Sustainability 13:2471

El-Beltagi HS, Mohamed AA, Mohamed HI, Ramadan KM, Barqawi AA, Mansour AT (2022) Phytochemical and potential properties of seaweeds and their recent applications: a review. Mar Drugs 20(6):342

El-Ramady H (2018) Nanofertilizers vs. biofertilizers: new insights. Env Biodivers Soil Secur 2(1):40–50. https://doi.org/10.21608/jenvbs.2018.3880.1029

Enguwa KBP, Horn LN, Awala SK (2024) Comparative effect of different irrigation levels and soil amendments on cabbage productivity in semi-arid Central Namibia. Irrig Drain 73:538–556

Gao X, He Y, Chen Y, Wang M (2024) Leguminous green manure amendments improve maize yield by increasing N and P fertilizer use efficiency in yellow soil of the Yunnan-Guizhou plateau. Front Sustain Food Syst 8:1369571

Gatsios A, Ntatsi G, Celi L, Said-Pullicino D, Tampakaki A, Savvas D (2021) Legume-based mobile green manure can increase soil nitrogen availability and yield of organic greenhouse tomatoes. Plants 10(11):2419

Hajiboland R (2017) Environmental and nutritional requirements for tea cultivation. Folia Hortic 29(2):199–220. https://doi.org/10.1515/fhort-2017-0019

Hakim S (2021) Rhizosphere engineering with plant growth-promoting microorganisms for agriculture and ecological sustainability. Front Sustain Food Syst 5. https://doi.org/10.3389/fsufs.2021.617157

Hansen LS, Sorgho R, Mank I, Nayna Schwerdtle P, Agure E, Bärnighausen T, Danquah I (2022) Home gardening in sub-Saharan Africa: a scoping review on practices and nutrition outcomes in rural Burkina Faso and Kenya. Food Energy Secur 11(3):e388

Ho TTK, Le TH, Tran CS, Nguyen PT, Thai VN, Bui XT (2022) Compost to improve sustainable soil cultivation and crop productivity. Case Stud Chemi Environ Eng 6:100211

Iderawumi AM, Kamal TO (2022) Green manure for agricultural sustainability and improvement of soil fertility. Farming Management 7:1–8

Iqbal R (2020) Potential agricultural and environmental benefits of mulches—a review. Bulletin of the National Research Centre/Bulletin of the National Research Center 44(1). https://doi.org/10.1186/s42269-020-00290-3
Itelima JU (2018) A review: Biofertilizer—A key player in enhancing soil fertility and crop productivity 2(1). https://www.pulsus.com/scholarly-articles/a-review-biofertilizer%2D%2Da-key-player-in-enhancing-soil-fertility-and-cropproductivity.pdf
Jaies I, Qayoom I, Saba F, Khan S (2024) Fish wastes as source of fertilizers and manures. In: Fish waste to valuable products. Springer Nature Singapore, Singapore, pp 329–338
Kabui KK, Rawson A, Athmaselvi KA (2024) Selected fermented foods of Manipur, India: traditional preparation methods, nutritional profile, and health benefits. Food Chem Adv:100864
Kaneko M, Kato N, Hattori I, Matsuoka M, Vendramini JM (2023) Seeding and harvesting times and climate conditions are important for improving nitrogen and Fiber contents of green manure Sunn hemp. Sustainability 15(9):7103
Kheiralipour K, Brandão M, Holka M, Choryński A (2024) A review of environmental impacts of wheat production in different agrotechnical systems. Resources 13:93
Krishnani KK, Boddu VM, Chadha NK, Chakraborty P, Kumar J, Krishna G, Pathak H (2022) Metallic and nonmetallic nanoparticles from plant, animal, and fisheries wastes: potential and valorization for application in agriculture. Environ Sci Pollut Res 29(54):81130–81165
Kumawat L, Yadav M, Yadav SL (2022) Chapter-7 modern concepts of fertilizers, application and their management technologies. Advanced innovative Technologies in Agricultural Engineering for sustainable. Agriculture:121
Larkin RP (2022) Long-term effects of compost amendments and brassica green manures in potato cropping systems on soil and crop health and productivity. Agronomy 12:2804
Lei B, Wang J, Yao H (2022) Ecological and environmental benefits of planting green manure in paddy fields. Agriculture 12:223
Li L, Xu L, Hu Z, Yu JC (2021) Enhanced mass transfer of oxygen through a gas–liquid–solid interface for photocatalytic hydrogen peroxide production. Adv Funct Mater
Lyu H, Li Y, Wang Y, Wang P, Shang Y, Yang X et al (2024) Drive soil nitrogen transformation and improve crop nitrogen absorption and utilization-a review of green manure applications. Front Plant Sci 14:1305600
Mahmud AA (2021) Biofertilizers: a nexus between soil fertility and crop productivity under abiotic stress. Curr Res Environ Sustain 3:100063–100063. https://doi.org/10.1016/j.crsust.2021.100063
Manea EE, Bumbac C, Dinu LR, Bumbac M, Nicolescu CM (2024) Composting as a sustainable solution for organic solid waste management: current practices and potential improvements. Sustainability 16(15):6329
Manimozhi K (2012) Eco-friendly fertilizers for sustainable agriculture. Int J Sci Res 2(11):255–257. https://doi.org/10.15373/22778179/nov2013/81
Matthews S, Siddiqui Y, Ali A (2024) Unleashing the power of biostimulants for enhanced crop growth, productivity, and quality: a comprehensive review. J Plant Nutr 48:703–725
Maulu S, Nawanzi K, Abdel-Tawwab M, Khalil HS (2021) Fish nutritional value as an approach to children's nutrition. Front Nutr 8:780844
Mazid M (2014) Future of biofertilizers in Indian agriculture: an overview. Int J Agric Food Res 3(3). https://doi.org/10.24102/ijafr.v3i3.132
Moawad MN, El-Sayed AA, Abd El Latif HH, El-Naggar NA, El-Din NGS, Tadros HR (2022) Chemical characterization and biochemical activity of polysaccharides isolated from Egyptian Ulva fasciata Delile. Oceanologia 64(1):117–130
Muscolo A, Mauriello F, Marra F, Calabrò PS, Russo M, Ciriminna R, Pagliaro M (2022) AnchoisFert: a new organic fertilizer from fish processing waste for sustainable agriculture. Global Chall 6(5):2100141
Nogalska A, Załuszniewska A (2021) The effect of meat and bone meal (MBM) on crop yields, nitrogen content and uptake, and soil mineral nitrogen balance. Agronomy:11–2307

Nooeaid P (2020) Eco-friendly polyvinyl alcohol/Polylactic acid Core/Shell structured Fibers as controlled-release fertilizers for sustainable agriculture. J Polym Environ 29(2):552–564. https://doi.org/10.1007/s10924-020-01902-9

Nordahl SL, Preble CV, Kirchstetter TW, Scown CD (2023) Greenhouse gas and air pollutant emissions from composting. Environ Sci Technol 57(6):2235–2247

Nosheen S (2021) Microbes as biofertilizers, a potential approach for sustainable crop production. Sustainability 13(4):1868–1868. https://doi.org/10.3390/su13041868

Pál V, Zsombik L (2023) Effect of common vetch (*Vicia sativa* L.) green manure on the yield of corn in crop rotation system. Agronomy 14(19)

Panda SK, Das S (2024) Potential of plant growth-promoting microbes for improving plant and soil health for biotic and abiotic stress management in mangrove vegetation. Rev Environ Sci Biotechnol 23(3):801–837

Panday D, Bhusal N, Das S, Ghalehgolabbehbahani A (2024) Rooted in nature: the rise, challenges, and potential of organic farming and fertilizers in agroecosystems. Sustainability 16:1530

Piash MI, Uemura K, Itoh T, Iwabuchi K (2023) Meat and bone meal biochar can effectively reduce chemical fertilizer requirements for crop production and impart competitive advantages to soil. J Environ Manag 336:117612

Pindi PK (2012) Liquid microbial consortium- a potential tool for sustainable soil health. J Fertilizers Pesticides 3(4). https://doi.org/10.4172/2155-6202.1000124

Qiu Z, Li M, Song L, Wang C, Yang S, Yan Z, Wang Y (2021) Study on nitrogen-retaining microbial agent to reduce nitrogen loss during chicken manure composting and nitrogen transformation mechanism. J Clean Prod 285:124813

Qiu SM, Aweya JJ, Liu X, Liu Y, Tang S, Zhang W, Cheong KL (2022) Bioactive polysaccharides from red seaweed as potent food supplements: a systematic review of their extraction, purification, and biological activities. Carbohydr Polym 275:118696

Rastogi M, Verma S, Kumar S, Bharti S (2023) Soil health and sustainability in the age of organic amendments: a review. Int J Environ Clim Change 13:2088–2102

Sathya A (2017) Plant growth-promoting actinobacteria: a new strategy for enhancing sustainable production and protection of grain legumes. 3 Biotech 7(2). https://doi.org/10.1007/s13205-017-0736-3

Simons AM, Ahmed M, Blalock G, Nesin B (2023) Indigenous bone fertilizer for growth and food security: a local solution to a global challenge. Food Policy 114:102396

Singh RP (2017) Eco-friendly nitrogen fertilizers for sustainable agriculture. Springer eBooks:227–246. https://doi.org/10.1007/978-981-10-3638-5_11

Stępień A, Wojtkowiak K, Kolankowska E (2021) Use of meat industry waste in the form of meat-and-bone meal in fertilizing maize (*Zea mays* L.) for grain. Sustainability 13:2857

Stojanović SS (2018) Microbial fertilizers: a comprehensive review of current findings and future perspectives. Span J Agric Res 16(1):e09R01–e09R01. https://doi.org/10.5424/sjar/2018161-12117

Tang J, Tong X, Chen Y, Wu Y, Zheng Z, Kayitmazer AB et al (2023) Deposition and water repelling of temperature-responsive nanopesticides on leaves. Nat Commun 14(1):6401

Tiwari H, Pandey V, Babu R (2023) Chapter-22 basic principles of crop production. Kaptan Baboo Syed Tazeen Zaidi Dr. Manoj Kumar Mishra Prabhas Kumar Shukla, 298

Vasa TN (2021) Recovery of struvite from wastewaters as an eco-friendly fertilizer: review of the art and perspective for a sustainable agriculture practice in India. Sustain Energy Technol Assess 48:101573–101573. https://doi.org/10.1016/j.seta.2021.101573

Verma KK (2022) Recent trends in nano-fertilizers for sustainable agriculture under climate change for global food security. Nano 12(1):173–173. https://doi.org/10.3390/nano12010173

Viruel E (2014) Inoculation of maize with phosphate solubilizing bacteria: effect on plant growth and yield. J Soil Sci Plant Nutr 14:819–831. https://doi.org/10.4067/s0718-95162014005000065

Wang P, Han S, Lin Y (2023) Role of microbes and microbial dynamics during composting. In: Current developments in biotechnology and bioengineering. Elsevier, pp 169–220

Waqas M, Hashim S, Humphries UW, Ahmad S, Noor R, Shoaib M et al (2023) Composting processes for agricultural waste management: a comprehensive review. Processes 11(3):731

Warrender D, Connell C, Jones E, Monteux S, Colwell L, Laker C, Cromar-Hayes M (2024) Mental health deserves better: resisting the dilution of specialist pre-registration mental health nurse education in the United Kingdom. Int J Ment Health Nurs 33(1):202–212

World Health Organization (2022) Report of the expert meeting on food safety for seaweed–Current status and future perspectives. Rome, 28–29 October 2021

Xie S, Tran HT, Pu M, Zhang T (2023) Transformation characteristics of organic matter and phosphorus in composting processes of agricultural organic waste: research trends, vol 6. Mater Sci Energy Technol, pp 331–342

Zajicek P, Corbin J, Belle S, Rheault R (2023) Refuting marine aquaculture myths, unfounded criticisms, and assumptions. Rev Fish Sci Aquac 31(1):1–28

Załuszniewska A, Nogalska A (2022) The effect of meat and bone meal (MBM) on phosphorus (P) content and uptake by crops, and soil available P balance in a six-year field experiment. Sustainability 14:2855

Zhang J, Akyol Ç, Meers E (2023) Nutrient recovery and recycling from fishery waste and byproducts. J Environ Manag 348:119266

Chapter 4
Challenges and Opportunities for the Use of Microorganisms in Sustainable Agriculture in the Future

Wafaa M. Abd El-Rahim and Hassan Moawad

Abstract This article is a crucial starting point for comprehending the importance, difficulties, and future developments related to the application of microorganisms in sustainable agriculture. In line with the UN's Sustainable Development Goals, this highlights the vital role that microbes play in many facets of human life. It also emphasizes how important bacteria are to environmental bioremediation, renewable energy, and food production. Additionally, the introduction highlights how agricultural intensification threatens global food security by negatively affecting soil health and microbial diversity (Mukhopadhyay. Front Microb 14:1293302, 2023).

Keywords Environmental bioremediation · Renewable energy · Soil health

Additionally, this section outlines the main obstacles and difficulties encountered when using microbial tools in agriculture, focusing on the restrictions associated with transferring the effectiveness of microbial products from controlled to field settings. This approach addresses practical difficulties in the use of microbial inoculants, such as their short shelf life, incompatibility with agrochemicals, and storage and transportation concerns, as well as the uneven effectiveness of microbial products in various soils, crops, and environments (Batista and Singh 2021). This thoroughly reviews the opportunities and challenges of using microorganisms for sustainable agriculture and sets the stage for the following parts.

W. M. Abd El-Rahim (✉) · H. Moawad
Department of Agricultural Microbiology, Agricultural and Biological Research Institute, National Research Centre (NRC), Giza, Egypt

V. Gopikrishnan et al. (eds.), *Harnessing Microbial Resources for Sustainable Agriculture*, Microorganisms for Sustainability 61,
https://doi.org/10.1007/978-981-95-4817-0_4

4.1 Importance of Microorganisms in Sustainable Agriculture

Restoring soil health and increasing crop output moves away from concentrating on individual soil bacteria and toward a comprehensive microbiome approach supporting sustainable farming methods (Ray et al. 2020). However, because uncontrolled biotic and abiotic variables in agricultural settings can affect the performance of microbial inoculants, scaling up the efficacy of microbial products from controlled circumstances to field conditions presents issues. The broad use of microbe-based products in agriculture is further hampered by practical difficulties such as short shelf lives, incompatibilities with agrochemicals, and storage and transportation problems (Batista and Singh 2021).

4.1.1 Nitrogen Fixation

One necessary process that involves transforming atmospheric nitrogen into a form that plants can easily use is nitrogen fixation by microbes. This procedure contributes significantly to improving soil fertility and encouraging robust plant growth, which decreases reliance on artificial fertilizers and is consistent with sustainable agricultural principles (Soumare et al. 2020). More than 60% of the fixed nitrogen on Earth is derived from biological nitrogen fixation (BNF), which is a crucial component of sustainable agriculture and a possible way to reduce the use of nitrogen fertilizers and their detrimental effects on the environment (Aasfar et al. 2021).

Azotobacter species significantly contribute to BNF, producing compounds that promote plant growth and provide nonleguminous plants with substantial amounts of nitrogen. Studies on the genetic modification of *Azotobacter* to improve its capacity to fix nitrogen and colonize plants have yielded promising results, underscoring the possibility of further developments in sustainable agriculture. The creation of effective biofertilizers and their incorporation into sustainable farming methods depend on understanding the mechanisms and capabilities of nitrogen-fixing microorganisms such as *Azotobacter*.

4.1.2 Biocontrol of Plant Pathogens

Using microorganisms to control plant diseases is a viable path toward sustainable agriculture. The well-studied genus *Streptomyces* has demonstrated great promise as a biological control agent for plant diseases. However, environmental variables such as temperature, humidity, and soil pH affect its effectiveness and may not be very persistent in the soil, requiring repeated applications. For adoption to be successful, further issues such as mass production, regulatory approval, formulation, and integration with current farming techniques must be resolved (Khan et al. 2023a, b). Similarly, plant pests, such as bacterial and fungal diseases, insects, and nematodes that consume plants, are susceptible to biocontrol by root-associated bacteria. These bacteria can increase plant resistance to pests and diseases at a systemic level, emphasizing their potential for the biocontrol of various rhizosphere targets (Lee et al. 2022). These findings highlight the necessity of continued study and development to overcome these obstacles and fully utilize the promise of microorganism-based biocontrol for sustainable agriculture.

4.1.3 Improving Soil Health

A key component of sustainable agriculture is enhancing soil health; microorganisms are essential. Microorganisms support nutrient cycling by breaking down organic matter and improving the soil structure, maintaining soil fertility and total agricultural productivity (Ray et al. 2020). Plant growth, improved nutrient utilization, and pest and phytopathogen management are just a few advantages of the complex microbial population or microbiome in plants and soil. As a result, a more comprehensive approach to the microbiome is being adopted to improve agricultural productivity and restore soil health.

Additionally, by releasing a variety of root exudates and carbon-rich chemicals into the rhizosphere, agricultural activities such as crop diversification shape local microbial species and impact microbial diversity (Gupta et al. 2022). Since microorganisms can lessen the adverse effects of chemical fertilizer and pesticide applications, sustainable agriculture must comprehend how various agricultural techniques affect soil microbial ecosystems.

4.2 Current Applications of Microorganisms in Agriculture

Microorganisms are currently used in many agricultural applications, such as biostimulants, biological pest management, and biofertilizers. These microbial compounds are essential for controlling pests, increasing plant development, and improving agricultural production. However, there are difficulties in transferring the effectiveness of microbial products from controlled to field settings, which can result in uneven outcomes for various crops, soils, and habitats (Batista and Singh 2021). Several factors, including native microbial populations, harsh weather, poor soil properties, and ambient and soil contaminants, can impact the effectiveness of microbial inoculants. The broad use of microbial products in agriculture is further hampered by practical difficulties such as short shelf life, incompatibility with agrochemicals or farming equipment, a lack of suitable recommendations, and problems with storage and transportation (Gupta et al. 2022).

One potential strategy in contemporary agriculture is the use of soil microbial populations to produce healthy and sustainable crops. A recommended agricultural approach is to optimize the root-associated microbiome in crop production by protecting plants and supplying nutrients. Agronomic management and various ecological conditions, including salinity, drought, nutrient deficiencies, pollution, soil erosion, diseases, pests, and invasive plants, impact the interactions between microbial populations and crops. These findings open the door for more studies and advancements in this area by highlighting the opportunities and challenges related to the existing uses of microorganisms in agriculture.

4.2.1 Biofertilizers

By providing vital nutrients to plants and increasing soil fertility through the actions of beneficial microbes, biofertilizers are critical to sustainable agriculture. Because they can solve issues associated with traditional fertilizer use, these environmentally benign substitutes for chemical fertilizers are becoming increasingly critical. Research has emphasized the importance of creating customized biofertilizer products via tactics that improve particular microbial activities associated with soil nutrient dynamics. The effective use of biofertilizers also depends on establishing closeness with growers, guaranteeing regulatory clarity, and promoting cooperation between farmers, scientists, advisors, and legislators (Bargaz et al. 2018).

Additionally, the use of biofertilizers is beneficial for sustainable agriculture since it improves plant health metrics and increases crop yield, soil health, and protection from adverse situations (Chaudhary et al. 2022). These findings highlight how biofertilizers can help meet the demands of expanding people in the future while maintaining agricultural yields over the long term.

4.2.2 Biological Pest Control

As a crucial part of integrated pest management (IPM), biological pest control uses naturally occurring microorganisms to reduce weeds, diseases, and insect pests in agriculture. By reducing damage to nontarget creatures and ecosystems, this method provides a sustainable and eco-friendly substitute for chemical pesticides (Vimala Devi et al. 2012). Since 2006, microbial pesticides have been registered for commercial use in India, where agricultural loss from insect pests is estimated to be worth Rs. 200 billion annually. This is due to the desire to minimize reliance on pesticides. Products that use microbes, including *Bacillus thuringiensis*, *Beauveria bassiana*, *Verticillium lecanii*, *Metarhizium anisopliae*, and nuclear polyhedrosis virus, to control insect pests are included in the registration.

Agricultural practices, especially the use of organic amendments, significantly impact the diversity of soil microbial communities, which are essential for preserving crop yield and quality (Gupta et al. 2022). Biological control, bioremediation, and the use of beneficial soil microorganisms are emphasized as ways to improve food security through sustainable farming methods and reduce the negative impacts of agrochemicals in soil. By providing substitutes for harmful agrochemicals, these microbial compositions maximize agricultural yields while reducing damage to the environment.

4.2.3 Biostimulants

Microorganism-derived biostimulants, among others, are essential for sustainable agriculture because they increase crop performance, stress tolerance, and plant growth. These biostimulants are an essential component of integrated methods for sustainable agricultural management since they help sustainably improve agricultural production and plant resilience (Sun and Shahrajabian 2023). Biostimulants have been demonstrated to increase nutrient uptake, promote root development, stimulate phytohormone production, and reduce the need for inorganic mineral fertilizers. They are also environmentally benign and have no discernible adverse effects on fruit quality or yield (Shahrajabian et al. 2021). They also benefit sustainable farming practices since they increase drought tolerance, deter pests, speed up root development, increase fertilization, strengthen root protection, and detoxify heavy metals.

The application of biostimulants—especially those made from microorganisms—is becoming more widely acknowledged as a sustainable and prosperous method of crop supplementation that helps restore damaged ecosystems and mitigates the environmental issues caused by overfertilization. The need for long-term research and standardization procedures in this area is highlighted by the possibility that variations in the content of biostimulant products could lead to inconsistent observed results and make efforts to identify the mechanisms behind their effects more difficult.

4.3 Challenges in Harnessing Microorganisms for Sustainable Agriculture

Several issues hamper the broad use of microbiological tools in sustainable agriculture. Scaling up the effectiveness of microbial products from controlled to field environments is a significant challenge. The broader use of microbial inoculants by farmers is limited by inconsistent outcomes across different soils, crops, or habitats since uncontrolled biotic and abiotic factors can influence microbial inoculant success. In addition to impeding the broader adoption of microbe-based products, practical difficulties, such as short shelf life, incompatibility with agrochemicals, a lack of suitable guidelines, and storage and transportation problems, exist (Batista and Singh 2021). The functioning of agricultural systems and natural ecosystems can also be impacted by stress factors such as salinity, drought, nutrient deficiencies, pollution, soil erosion, disease, pests, and invasive species caused by intensive farming practices and agrochemicals. A viable strategy to address the difficulty of sustainable and healthy crop production is to take advantage of the agroecosystem services provided by soil microbial communities; nevertheless, this approach involves addressing the intricacies of several ecological aspects and agronomic management (Gupta et al. 2022).

4.3.1 Regulatory Hurdles

Approval procedures, standardization, and compliance requirements are some of the obstacles that may prevent the broader use of microbe-based solutions in agriculture (Batista and Singh 2021). While excessively lenient rules may result in unacceptable hazards to people or the environment, inconsistent regulatory demands can cause evaluation processes to become convoluted and time-consuming. Therefore, to facilitate microbial biodiversity in crop production while maintaining a sufficient degree of environmental and human safety, it is necessary to modify and harmonize the current legislation. Furthermore, the low uptake of microbial goods, especially in developing nations, has been attributed to insufficient regulatory frameworks and inefficient quality control procedures (Sundh et al. 2021).

4.3.2 Lack of Awareness and Education

According to Gupta et al. (2022), soil microbial populations and biological activities are significantly impacted by shifts in land use patterns and nonscientific agricultural methods. The complex interaction between plants and microbes is further highlighted by how crop diversity and the release of root exudates affect the structure of the microbial population. Similarly, Mukhopadhyay et al. (2023) emphasized the critical function of *Bacillus* species, which promotes plant growth (PGP), in the bioaugmentation of soil for sustainable farming. It is clear that improving educational initiatives and raising awareness of the potential of microorganisms in agriculture are essential to tackling the issues of inadequate knowledge and advancing sustainable farming methods.

4.3.3 *Technical Limitations*

Several technical issues that impede the practical use of microbial formulations in agriculture must be resolved for maximum effectiveness. The selectivity of biofertilizers is a significant drawback that could restrict their compatibility with specific chemical fertilizers or pesticides, impacting nutrient management and integrated pest management initiatives (Khan et al. 2023a, b). Furthermore, inadequate survival, poor carrier selection, or inefficient delivery techniques limit the efficacy of bacterial strains in field settings. Microbial strains that exhibit advantageous functions even in harsh environments and have a competitive advantage over local microflora must be carefully chosen. Furthermore, real-world obstacles, including short shelf-life, incompatibility with agrochemicals, and storage and transportation problems, make it more difficult for microbial inoculants to be widely used in agriculture (Batista and Singh 2021).

Scientists, agricultural practitioners, and policymakers must collaborate and conduct ongoing research and development to overcome these obstacles and promote sustainable agricultural practices. To guarantee the viability and usefulness of microbial formulations on a broader scale, it is imperative to investigate and capitalize on their potential advantages while aggressively resolving their disadvantages. The secret to promoting safe and sustainable practices is reorienting research toward creating formulations based on bioinoculants with wide temperature and elevation ranges.

4.4 Opportunities for Advancing Microbial Applications in Agriculture

Knowledge-sharing platforms, research and development funding, and biotechnological advancements are all strongly related to opportunities for expanding microbial uses in agriculture. There is room for improvement, even with the difficulties in transferring the effectiveness of microbial products from controlled to field settings. Funding for research and development is essential for fostering innovation, increasing the use of microorganisms, and maximizing their benefits for agricultural sustainability. Furthermore, cooperative platforms and knowledge exchange activities are crucial to address disparities in microbial product performance among various soils, crops, and settings and to share best practices (Batista and Singh 2021).

Additionally, taking advantage of the agroecosystem services provided by soil microbial populations offers a viable strategy for maintaining the biosphere while producing crops sustainably. More effective utilization of beneficial microbial services is possible through strategies to maximize the role of the root-associated microbiome in crop production, such as nutrient supply and plant defense. These tactics are essential for addressing long-term economic and environmental issues without sacrificing crop quality or yields. Furthermore, agronomic management and ecological factors impact the interactions between microbial communities and crops, underscoring the possibility for novel agricultural techniques to improve agricultural system functionality (Gupta et al. 2022).

4.4.1 Biotechnology Innovations

In the upcoming decades, microbial biotechnology will be crucial in the fight against climate change, according to Verstraete et al. (2021). On the basis of fundamental ideas in microbial ecology, researchers have focused on the possible important roles that microbial technology may play in natural and artificial environments. Furthermore, Batista and Singh (2021) highlighted the limitations and difficulties in the broad international use of microbial products in agriculture, including the inability to scale up their efficacy from controlled to field conditions; the uneven effectiveness of microbial products across various soils and crops; and real-world issues with shelf-life, agrochemical compatibility, and insufficient regulatory frameworks.

4.4.2 Research and Development Funding

Funding for research and development (R&D) is essential for expanding the use of microorganisms in agriculture. Beachy (2014) highlights the importance of continuously maintaining a research agenda, especially in nations where research and agriculture are distinct fields. This support has made intense scientific endeavors, product development, and the application of microbial solutions possible. Moreover, higher agricultural productivity has been associated with basic and translational research investments. However, over the past 20 years, financing for these initiatives has declined despite the growing demand for agricultural products, underscoring the need for more substantial global funding commitments.

Batista and Singh (2021) focused on the practical difficulties associated with the use of microbial inoculants, including their short shelf life, incompatibility with agrochemicals, storage and transportation concerns, and the importance of R&D financing (Batista and Singh 2021). They also noted that several uncontrollable biotic and abiotic factors may impact the variable effectiveness of microbial products in the field. These elements include native microbial populations, harsh weather, unfavorable soil properties, and soil and environmental contaminants. Furthermore, industrialized and developing nations face very different barriers to the broader use of microbe-based products in agriculture. These barriers include awareness, legal frameworks, and commercial and scientific backing, among others.

4.4.3 Knowledge Exchange Platforms

Platforms for exchanging knowledge are essential for addressing the potential problems associated with the use of microorganisms in sustainable agriculture. These platforms help spread best practices, technical innovations, and knowledge by facilitating cooperation, information exchange, and capacity building. These platforms facilitate the sharing of experiences and insights by bringing together practitioners, researchers, and specialists, which helps ensure that microorganisms are used effectively in agriculture (Batista and Singh 2021). A promising strategy in sustainable agriculture is to utilize the agroecosystem services provided by soil microbial populations. To maximize the role of the root-associated microbiome in crop production through nutrient supply and plant protection, strategies for the efficient exploitation of beneficial microbial services and low-input biotechnology are maintained (Gupta et al. 2022). By offering a forum for debating and implementing such tactics, these platforms aid in advancing and adopting sustainable farming methods.

4.5 Emerging Trends in Microbial Agriculture

New developments in microbial agriculture are shaping the direction of sustainable farming. Owing to the potential of microorganisms, synthetic biology techniques and precision microbial farming are transforming agricultural methods. These cutting-edge techniques and technologies present new opportunities to increase crop yields, increase the effectiveness of nutrient utilization, and manage pests and phytopathogens. The relevance of different microbial consortia in fostering plant growth and reestablishing soil health is highlighted by the growing trend away from single, specialized soil bacteria and toward a holistic approach to the microbiome (Ray et al. 2020).

However, despite their enormous potential, several obstacles prevent the widespread use of microbial products in agriculture. Among the main obstacles are the influence of uncontrollable biotic and abiotic elements, inconsistent outcomes across various soils and crops, and scaling up the effectiveness of microbial products from controlled to field environments. The application of microbial inoculants is further hindered by practical limitations, such as short shelf life, agrochemical incompatibility, and storage and transportation problems, especially in developing nations (Batista and Singh 2021). To fully utilize microorganisms in sustainable agriculture, several issues must be resolved.

4.5.1 Precision Microbial Farming

To maximize farming operations, precision microbial farming combines microbial technologies, data analytics, and modern agricultural methods. This method emphasizes focused and effective farming by using microorganisms to improve sustainable and resource-efficient agricultural systems (Bargaz et al. 2018). Understanding the ways in which microorganisms may affect nutrient dynamics, developing cooperative scientific and industrial relationships, and emphasizing being close to growers to gain their acceptance through in-field demonstrations and customized reports are all essential to the success of precision microbial farming. The success of the upcoming generation of agro-inputs based on microbial inoculants also depends on clear regulations and a cooperative attitude among farmers, scientists, advisors, and legislators.

Additionally, scaling up efficacy from controlled conditions to field conditions; inconsistent efficacy across soils, crops, or environments; and practical limitations, such as limited shelf life and incompatibility with agrochemicals, are all issues associated with the use of microbial tools in agriculture (Batista and Singh 2021). These limitations differ between established and developing nations. While developing regions face obstacles, including high costs, a lack of scientific support, and insufficient regulatory frameworks, developed regions face concerns connected to variable efficacy, affecting wider acceptance. For precision microbial farming to reach its full potential in sustainable agriculture, these issues must be resolved.

4.5.2 Synthetic Biology Approaches

Through the creation of genetic circuits, pathway engineering, and biobased product synthesis, synthetic biology has the potential to transform microbial agriculture completely. Rapid developments in genome design, synthesis methods, and molecular tools such as CRISPR/Cas9 have made applying synthetic biology in agriculture easier. This has made it possible to synthesize entire bacterial genomes and create intricate multigene constructions, which have revolutionary potential for creating novel microbial solutions to agricultural problems. Furthermore, synthetic biology makes it possible to restructure the genomes of plants and microorganisms that promote plant growth and improve communication between them, which eventually helps to create "microbe-friendly" crops and sustainable farming practices (Goold et al. 2018).

Additionally, the creation of bioformulations, such as nanobiopesticides and nanobiofertilizers, presents chances to increase the effectiveness, stability, and delivery of microbial tools in agriculture. The nanoencapsulation technique, which uses nanoparticles to transfer microorganisms that promote plant growth in a controlled way, has demonstrated promise in increasing plant biomass accumulation and preventing the growth of pathogens. However, prior to their widespread use, the toxicity of nanoparticles to humans and the environment must be thoroughly examined (Batista and Singh 2021).

4.6 Future Directions and Prospects

There are potential problems with the future use of microbial solutions in agriculture. To achieve the broad adoption and influence of microorganisms in determining the direction of agriculture, policy support, technological developments, and cooperative efforts are needed as the need for sustainable agricultural methods continues to grow. Effective management of microbiomes in soils and body systems is crucial, according to Verstraete et al. (2021), who also stress the need for improved nutrient and energy techniques and more clearly defined ecological principles. Furthermore, Batista and Singh (2021) emphasized the practical difficulties in using microbial inoculants, such as their short shelf life, incompatibility with agrochemicals, and storage- and transportation-related problems. To fully realize the promise of microbial technology for agricultural development and environmental sustainability, these issues must be resolved.

4.6.1 Integration of Microbial Solutions in Mainstream Agriculture

There are advantages and disadvantages of using microbial treatments in conventional agriculture. Scaling up efficacy from controlled to field circumstances is one of the main obstacles because the results could vary depending on the soil, crop, or environment (Batista and Singh 2021). Several factors, including uncontrolled biotic and abiotic impacts, native microbial communities, harsh weather, and poor soil properties, can impact the efficiency of microbial solutions in agricultural contexts. For integration to succeed, practical obstacles must also be resolved, including short shelf life, incompatibility with agrochemicals, a lack of suitable recommendations, and storage and transportation problems.

However, incorporating soil microbial populations into agriculture has encouraging prospects for environmentally friendly agricultural production. Without sacrificing yields or producing quality, taking advantage of the agroecosystem services provided by soil microbial communities can help find sustainable environmental and financial solutions (Gupta et al. 2022). Strategies for leveraging low-input biotechnology and beneficial microbial services are advised to maximize the contribution of root-associated microbiomes to crop productivity, particularly in light of global climate change. These findings highlight how mainstream microbial solutions in agriculture have the potential to revolutionize the field and improve agricultural system sustainability.

4.6.2 Policy Support for Microbial Agriculture

Policy support is essential for microbial solutions to be widely adopted and used responsibly in agriculture. Policy frameworks, incentives, and regulatory actions are crucial in establishing a supportive environment for incorporating microorganisms into agricultural operations, as noted by Batista and Singh (2021). However, scaling up the effectiveness of microbial products from controlled to field circumstances is difficult, as the outcomes vary depending on the soil, crop, and environment. Various factors, including native microbial communities, harsh weather, and poor soil properties, can influence microbial inoculant success in complex agricultural settings. The adoption of these materials is further hampered by practical issues such as their short shelf life and incompatibility with agrochemicals.

Furthermore, Sundh et al. (2021) stressed the importance of regulatory actions in guaranteeing microbiological product effectiveness, quality assurance, and intellectual property protection. These authors suggest that regulatory regimes strike a balance between diversification driven by variations in intended use and intended effects and harmonization on the basis of the shared characteristics and dangers of the groupings of microorganisms. The authors also support reforming and harmonizing laws to streamline assessments and product approvals and eventually increase the use of microbial biodiversity in agricultural production while preserving environmental and human safety.

4.7 Conclusion

In conclusion, microorganisms have advantages and disadvantages in sustainable agriculture. Changes in land use patterns, residue management, and farming techniques all impact the dynamics of the microbial population in soil, which substantially affects soil microbial diversity (Gupta et al. 2022). Additionally, scaling up efficacy from controlled conditions to field conditions, inconsistent results across different soils and crops, and practical issues such as limited shelf life and incompatibility with agrochemicals are all obstacles to the use of microbial tools in agriculture (Batista and Singh 2021).

Overcoming these obstacles will require teamwork, creativity, and legislative support to utilize the potential of microorganisms to fully promote sustainable farming methods. To maximize the advantages of microorganisms while reducing their drawbacks, it is also essential to comprehend and measure how agricultural activities affect soil microbial communities.

In summary, microbial agriculture has a bright future in sustainable farming. Because soil microbes can operate as sources or sinks of carbon, controlling the microbiomes in soils and body systems is essential for environmental sustainability (Verstraete et al. 2021). Key ecological principles must be established for effective microbiome management, allowing for the assessment and control of contemporary biotechnologies. Furthermore, methods such as converting wastewater into drinking water and producing microbial proteins use microorganism resource efficiency, underscoring their potential to help with resource management and climate change issues.

Additionally, developing agricultural policies and creating sustainable practices requires understanding how agricultural practices affect soil microbial communities and functional groups (Gupta et al. 2022). Crop diversification and the impact of plant root exudates on microbial community structure highlight the potential to increase land productivity and foster microbial diversity in agricultural systems. Overall, the information gained from these discoveries helps direct agriculture toward a more sustainable future by highlighting the vital role that microbial technologies play in both natural and artificial ecosystems.

References

Aasfar A, Bargaz A, Yaakoubi K, Hilali A, Bennis I, Zeroual Y, Meftah Kadmiri I (2021) Nitrogen fixing Azotobacter species as potential soil biological enhancers for crop nutrition and yield stability. Front Microbiol 12:628379

Bargaz A, Lyamlouli K, Chtouki M, Zeroual Y, Dhiba D (2018) Soil microbial resources for improving fertilizers efficiency in an integrated plant nutrient management system. Front Microbiol 9:1606

Batista DB, Singh BK (2021) Realities and hopes in the application of microbial tools in agriculture. Microb Biotechnol 14(4):1258–1268

Beachy NR (2014) Building political and financial support for science and technology for agriculture. Phil Trans R Soc B Biol Sci 369(1639):20120274

Chaudhary P, Singh S, Chaudhary A, Sharma A, Kumar G (2022) Overview of biofertilizers in crop production and stress management for sustainable agriculture. Front Plant Sci 13:930340

Goold HD, Wright P, Hailstones D (2018) Emerging opportunities for synthetic biology in agriculture. Genes 9(7):341

Gupta A, Singh UB, Sahu PK, Paul S, Kumar A, Malviya D, Singh S, Kuppusamy P, Singh P, Paul D, Rai JP, Singh HV, Manna MC, Crusberg TC, Kumar A, Saxena AK (2022) Linking soil microbial diversity to modern agriculture practices: a review. Int J Environ Res Public Health 19(5):3141

Khan S, Srivastava S, Karnwal A, Malik T (2023a) Streptomyces as a promising biological control agents for plant pathogens. Front Microbiol 14:1285543

Khan A, Singh AV, Gautam SS, Agarwal A, Punetha A, Upadhayay VK, Kukreti B, Bundela V, Jugran AK, Goel R (2023b) Microbial bioformulation: a microbial assisted biostimulating fertilization technique for sustainable agriculture. Front Plant Sci 14:1270039

Lee JH, Anderson AJ, Kim YC (2022) Root-associated bacteria are biocontrol agents for multiple plant pests. Microorganisms 10(5):1053

Mukhopadhyay M, Mukherjee A, Ganguli S, Chakraborti A, Roy S, Choudhury SS, Subramaniyan V, Kumarasamy V, Sayed AA, El-Demerdash FM, Almutairi MH, Şuţan A, Dhara B, Kumar Mitra A (2023) Marvels of bacilli in soil amendment for plant-growth promotion toward sustainable development having futuristic socioeconomic implications. Front Microbiol 14:1293302

Ray P, Lakshmanan V, Labbé JL, Craven KD (2020) Microbe to microbiome: a paradigm shift in the application of microorganisms for sustainable agriculture. Front Microbiol 11:622926

Shahrajabian MH, Chaski C, Polyzos N, Petropoulos A, S. (2021) Biostimulants application: a low input cropping management tool for sustainable farming of vegetables. Biomolecules 11(5):698

Soumare A, Diedhiou AG, Thuita M, Hafidi M, Ouhdouch Y, Gopalakrishnan S, Kouisni L (2020) Exploiting biological nitrogen fixation: a route toward a sustainable agriculture. Plants 9(8):1011

Sun W, Shahrajabian MH (2023) The application of arbuscular mycorrhizal fungi as microbial biostimulant, sustainable approaches in modern agriculture. Plants 12(17):3101

Sundh I, Del Giudice T, Cembalo L (2021) Reaping the benefits of microorganisms in cropping systems: is the regulatory policy adequate? Microorganisms 9(7):1437

Verstraete W, Yanuka-Golub K, Driesen N, De Vrieze J (2021) Engineering microbial technologies for environmental sustainability: choices to make. Microb Biotechnol 15(1):215–227

Vimala Devi PS, Ranga Rao GV, Gopalakrishnan S, Sivakumar G (2012) Environmental impact of microbial pesticides 426:261–272

Chapter 5
Earthworm and Earthworm Cast-Associated Microbes for Sustainable Agriculture

Justin Sheeba, Murali Monisha, Girupha Nishanth, Mekala Panneerselvam Karthik Prakash, Manikkam Radhakrishnan, Venugopal Gopikrishnan, and Ramesh Kumar Ashwin

Abstract Earthworms are important ecological engineers, and they enhance sustainable agriculture via soil–ecosystem interactions. Earthworms flourish in microbe-abundant habitats by utilizing innate defense systems and antibacterial peptides to thwart microbial threats. Processed organic matter, or vermicast (more popularly referred to as worm cast), is a highly nutritious soil amendment that can increase soil productivity and maintain the multitude of lichens and bacteria that are necessary for plant nutrition and soil fertility. The earthworm gut is a dynamic bioreactor housing a highly diverse microbial community that is engaged in organic matter decomposition and nutrient mineralization, as well as pollutant transformation processes. Soil microbial communities perform essential functions, such as carbon sequestration, nitrogen fixation, and denitrification, in regulating the gut microbiota. Moreover, earthworms enhance soil aggregation, which helps retain water, reduces erosion, and promotes nutrient stabilization, increasing ecosystem resilience over the long term. Earthworms and their microbiota findings highlight the potential for sustainable agriculture, which could replace chemical fertilizers and pesticides. In addition, earthworm casts and intestines have also been shown to harbor microbial diversity with functional groups capable of producing phytohormones, siderophores, and biopesticides to increase plant growth as well as pathogen suppression. The earthworm and its associated microbes combine both conventional and high-tech agricultural practices, thus providing a sustainable, efficient, and earth-friendly agricultural system. This chapter describes the noncapitalized niches

J. Sheeba · M. P. Karthik Prakash · M. Radhakrishnan · V. Gopikrishnan (✉) · R. K. Ashwin
Centre for Drug Discovery and Development, Sathyabama Institute of Science and Technology, Chennai, Tamil Nadu, India

M. Monisha · G. Nishanth
Department of Biotechnology, Sathyabama Institute of Science and Technology, Chennai, Tamil Nadu, India

V. Gopikrishnan et al. (eds.), *Harnessing Microbial Resources for Sustainable Agriculture*, Microorganisms for Sustainability 61,
https://doi.org/10.1007/978-981-95-4817-0_5

of earthworms and their idealism to create mutualistic relevance to sustainable agricultural enhancement practices.

Keywords Earthworms · Soil productivity · Sustainable agriculture

5.1 Introduction

Earthworms are soil macroinvertebrates in the Annelida phylum, class Chaetopoda, and order Oligochaeta, whose evolution began during the pre-Cambrian era. There are approximately 23 families, 700 genera, and more than 7000 earthworm species worldwide. Earthworms are crucial to many ecosystem processes and services, including decomposition, nutrient cycling, and climate regulation. They are also commonly used as bioindicators of soil biodiversity and health (Helen et al. 2021). Earthworms are divided into three critical functional groups or ecological groups that determine their impacts on soil ecosystems and their microbial communities: (i) epigeic earthworms, which live at the soil surface and are litter feeders; (ii) endogeic earthworms, which are dwelling worms that create horizontal tunnels within the soil while consuming mineral soil and decomposing organic matter, thus displaying geophagous characteristics; and (iii) anecic worms, which build permanent vertical burrows and transport litter into their burrows for microbial predecomposition, thus placing their casts at the burrow mouths (Regina et al. 2019).

Earthworms, by their feeding habits, can be divided into two types: geophagous and detritivorous. Earthworms feed on decomposed organic materials, such as plant litter, rotting vegetables, fruits, plant roots, and animal dung, on or near the soil surface. On the other hand, geophagous earthworms consume soil due to their horizontal tunnels in the deep Earth (Nitin et al. 2021). Widely viewed as farmer friends, earthworms are an enormous contributor to ecosystems. They are important for agroecosystem management. They contribute to soil nutrient dynamics by transporting nutrients from deep soil layers to the surface, where they are more easily absorbed by plants. They also help in the aeration of soil, which provides soil with appropriate conditions to penetrate roots, ultimately improving the fertility of the soil and leading to increased productivity of crops (Akhila and Entoori 2022). The presence of a minimum amount of crop residue, the density of the earthworm population, and the fertilization rate determine the effectiveness of earthworms in increasing crop yields (Sharma et al. 2017).

5.2 Earthworm

The earthworm is a common saprozoic organism that encounters an environment that is full of microorganisms, and some earthworms may have pathogenic potential to the earthworm itself. To survive under these conditions, they have evolved efficient defenses against invading microorganisms. Unlike vertebrates, for example, earthworms do not have true antibodies, and their innate immune systems are poor at protecting them from incoming foreign materials (Jay et al. 2005). Studies suggest that there should be a number of active antibacterial proteins and peptides within the earthworm, ideographically unlike immunoglobulin (Ig), which is involved in the defense mechanism against the actions of pathogens (Kauschke et al. 2007). The inner part, with resilience, is the core of the defense network of earthworms and can be considered concentric circles. Earthworms employ defense mechanisms to protect themselves from foreign microorganisms introduced as they move between soil layers. Most microorganisms originate from the soil—the earthworm's primary habitat—and the worm's defense system functions in an export-oriented manner to resist invading microbes. Only the antibiotic barrier and natural immunity play leading roles in the defense mechanism of earthworms. Like in vertebrates, immunity in invertebrates has been shown to operate via both humoral and cellular defenses. Research on insects has shown that the extraction of many ABPs from insects has been successful (Jay et al. 2005).

5.3 Earthworm Cast

Earthworm casts are bioorganic engineered structures that are built by earthworms through the decay of soil organic substances (Gopikrishnan et al. 2020). Vermicasts, or earthworm casts, are regarded as organic soil amendments that significantly influence soil physical, chemical, and biological factors, consequently rendering them living and fertile media (Omokaro et al. 2024). Earthworms excrete substances called casts, which contain high levels of important nutrients, including nitrogen, phosphorus, potassium, and magnesium, that increase the fertility of the soil.

5.3.1 Fermentation

These casts also contain microorganisms that increase in size as the organic matter in the earthworm digestive tract becomes degraded (Akhila and Entoori 2022). Microbes in earthworm casts have reduced microbial biomass carbon but increased metabolic activity on the basis of a relatively high metabolic quotient (qCO_2). These findings suggest that casts represent hotspots of microbial metabolism with important implications for soil nutrient cycling (Lejoly et al. 2024). They provide microorganisms with favorable environmental conditions, including reduced stress and increased nutrient availability, analogous to a hotspot of microbial activity (Aira et al. 2019). Understanding how earthworms improve the soil structure affects the porosity and water-holding capacity of soil aggregates.

Through nutrient-laden mucus deposition and the defecation of bacteria housed within casts, they promote nutrient cycling (Kokilamani et al. 2024). Because of the high nutrient content of castings (for example, in ammonium (NH4+), copper (Cu), iron (Fe), manganese (Mn), and phosphorus (P), as well as better soil moisture and pH), these castings are very fertile environments. Research indicates that chemical components found in earthworm casts lead to a fast-acting nutrient cycle, allowing for improved soil fertility (Santos et al. 2021). This is why casts contribute to the health of soil, especially in agricultural systems, by helping improve structure and water retention and fostering microbial diversity. Earthworm casts are also known as plant-genuines, as microorganisms inside earthworm casts produce phytohormones and siderophores and suppress soil-borne pathogens, which also minimizes the use of chemical fertilizers (Chaoui et al. 2003). There are initially high ammonium levels, but fast nitrification results in relatively stable levels of both nitrogen forms due to the protection of organic matter in dry casts. Over time, the initial physical protection of nutrients within casts is traded for a progressively disintegrating aggregate structure. The composition of earthworm casts varies and reflects the soil habitat, which is derived from different habitats with various soil types containing defined ecological niches. The activity of earthworms creates microhabitats that support microbial communities that are essential for organic matter breakdown, allowing for soil carbon preservation and reduced loss through mineralization (Wang et al. 2021). Cast effects on soil may play a role by controlling inputs of carbon (C) by increasing rates of decomposition and biological activity in the drilosphere (the soil environment surrounding the earthworm) and by protecting carbon by stabilizing C in soil aggregates through a myriad of pathways. This process increases nutrient cycling and serves to maintain soil formation processes, simultaneously improving the soil structure, which benefits long-term soil health and carbon sequestration (Santos et al. 2021). The microbial assemblage in earthworm casts is composed of several functional groups—nitrifiers, cellulolytic organisms, and nitrogen-fixing bacteria—thus adding to the biological activity of the soil. Soils need aggregates to stabilize soil carbon, preventing rapid decomposition and facilitating long-term carbon sequestration, a process in which earthworms play an important role (Gautam et al. 2024). This stabilization function of mycorrhizae reduces soil compaction and erosion, and a heterogeneous soil structure makes water and nutrient retention easier (Iordache 2023).

Such casts and soil cleanliness, as well as sustainable agriculture, provide many health benefits, but all these factors play multifaceted roles if they can only help improve the quality of life of the soil. They are important for promoting nutrient cycling and microbial diversity and act as stabilizing agents for soil aggregates. They influence root structure, water holding capacity, and pH buffering, all of which improve plant growth and fertility. Earthworm casts constitute an important nutrient source in the soil, and the resident microbial communities occupying these microbial refuges vary to increase their metabolic activity in terms of decomposition, nutrient mineralization, and pathogen suppression, resulting in favorable soil ecosystems (Aira and Domínguez 2022).

5.4 Earthworm Gut

The earthworm gut is the most heterogeneous and variable environment for soil microorganisms and functions as a "soil bioreactor" and microniche for advancing the soil microbiota. This will be key to helping us address these important questions of soil health and nutrient cycling, as earthworms (an important representative model) carry a unique gut environment and have a significant impact on microbial processes (Kliszcz et al. 2024). Microbiological weathering of soil has been shown to be facilitated by earthworm gut microbiomes, and it is becoming clearer that this process drives certain aspects of soil weathering. It is also involved in soil-forming processes (Hu et al. 2018). Earthworm gut microorganisms are linked to various ecological roles of earthworms. The soil microbial community structure and function may be determined in the intestine of earthworms that ingest and digest materials (Zhang et al. 2022).

Microorganisms that are gut associated with earthworms, as well as soil microorganisms, are important components of the soil ecosystem that play important roles in element cycling, organic matter decomposition, pollutant degradation, soil fertility enhancement, and crop yield enhancement (Ma et al. 2016). Earthworm host gut microbiome diversity is critical for nutrient metabolism, immunity, and physiology. Gut-associated bacteria are present in worm casts and play important roles in a number of soil ecosystem functions, such as nitrogen fixation and Verm remediation (Liu et al. 2019). The gut of an earthworm hosts millions of decomposer microbes that confer significant physiological and ecological benefits to the host during the processing of a cellulose-dominated diet. While the food source was shown to influence the earthworm gut microbiome, the core microbiome was relatively unchanged (Sapkota et al. 2020). Soil particles can affect the gut microbiota when the gut of earthworms digests them. Consequently, the gut microbial communities of earthworms are highly affected by climatic factors and the environmental characteristics (such as pH, SOC, TN, and TP) of the adjacent soil (Thakuria et al. 2010; Zhang et al. 2022). According to Cordero and Datta (2016), the interplay among the numerous microbes colonizing the earthworm gut plays a crucial role in shaping its community structure and diversity. The gut microbiota enhances the adaptation of earthworms to temperature and food resources via cellulose degradation. The gut of earthworms is an anaerobic medium that thrives in an aerated environment with consistently wet conditions and stable nutrient content. It operates as a biological Renart, where some swallowed microorganisms can be either lysed or removed (Mirleau et al. 2024). Denitrification in the earthworm gut contributes to N_2O emissions by earthworms; denitrifiers can be cultured in high numbers in the earthworm gut, and denitrification can occur in the gut casts of earthworms. Most denitrifiers have the ability to both produce and consume N_2O, and the net emission of N_2O during denitrification is regulated by multiple parameters, including pH, growth phase, and nitrate and electron donor concentrations (Horn et al. 2003).

5.5 Sustainable Agriculture

Food has always served as a bedrock of humanity—the agrarian revolution is credited with allowing food surpluses that were a fundamental requirement for societies to develop into cities and communities. Agriculture, especially for those living in remote India, is a significant aspect of livelihood in a country where a large segment of the population still relies on using the land for their sustenance. Owing to diversifying agricultural practices over the years, a variety of methods, each with unique feature differences and challenges, have emerged. Traditional farming, which was the first model for agriculture, was based on indigenous knowledge systems and cultural practices. It stresses harmony with nature and draws on ancient methods. However, it has relatively low efficiency and, as such, often fails to meet the needs of a rapidly growing population. In stark contrast, contemporaneous agriculture responded to the new world by adopting transformative technologies, including chemical fertilizers and large-scale mechanization. Some of these innovations significantly improved crop yields and food production but often at a high environmental cost, with soil degradation, water pollution, and great loss of biodiversity. In light of the disadvantages associated with both conventional and contemporary agriculture, sustainable farming has arisen as a viable solution. The goal of this method is to attain high productivity while still protecting the Earth. Farmers can conserve natural resources, promote biodiversity, and maintain ecological health via sustainable practices such as crop rotation, organic agriculture, and integrated pest management. Resilience is another key attribute of sustainable agriculture, which can shape farming systems so that they are flexible to STRESSORS (e.g., climate change) and can provide a continuous supply of food for current and future generations. Ultimately, it fundamentally paves the way for a fairer, more environmentally friendly agrarian system that works toward the prosperity of both humans and the planet. One of the major occupations in India is agriculture. Indigenous knowledge-based traditional farming practices rely heavily on natural resources, manual labor, and minimal technological intervention. This method is characterized by a system of local inputs, crude implements, organic manure, manual management, cattle husbandry for grazing, pesticide application in the field, and cultural practices. However, with changing agricultural needs and climate change, these centuries-old practices are under greater pressure (Sahoo et al. 2024). Weeds, diseases, pest infestations, and climate-related challenges are all increasing threats to many human environments, and they threaten to reduce crop and livestock productivity in these regions. These issues have only deteriorated with the limited technology adoption in traditional farming in the past four decades, causing periodic disruptions to agricultural yields. If this trend continues, even greater difficulties are predicted over the next 25 years. The agricultural sector is gradually declining in India, which seriously affects production capacity and impacts the functions of the ecosystem (Mohanraj et al. 2016).

Both rainfed and irrigated farming systems are now severely threatened, with yields expected to decline without creative resource conservation. The excessive use and depletion of ecological resources, particularly soil and water, have resulted in severe environmental degradation, thereby intensifying concerns regarding the long-term sustainability of agriculture. To address these challenges, integrating sustainable practices, advanced technologies, and prudent resource management is crucial. These practices are likely to minimize damage, maximize throughput, and ultimately safeguard the future of agriculture in India (Lamba et al. 2021). Sustainable and resilient farming systems are pivotal for our food and rural livelihoods, particularly in the face of increasing pressures from population growth and climate change. While these practices have been a tradition in India, they are facing many challenges, which are hindering their effectiveness. These range from low productivity to write-offs owing to access to finance, field losses, and environmental degradation to meeting demand by a growing young population rising to reach 40% of the total. On the other hand, technological innovations in agriculture are designed to solve these challenges. Technological agriculture increases efficiency and productivity improvements and reduces environmental impact by incorporating modern tools and techniques (Soma and Nuckchady 2021). In the world's rich economies, such innovations have made farming a profitable and sustainable business, even if farmers sometimes complain. In numerous developing countries, however, farming remains at the subsistence stage, with producers having trouble achieving profit stability and breaking free of resource limitations. This sharp contrast illustrates the vital role of innovation in shaping new landscapes for agriculture across the world (Yates 1964). Modern agriculture, which became dominant around the mid-1950s, flourished in countries such as India but largely within regions such as Punjab, Haryana, and Western Uttar Pradesh. Its arrival has given rise to world factors, such as elevated agricultural output, decreased work requirements, and proper land use. In addition, great progress in genetic modification has allowed for bushier crops with greater resistance to weeds and pests. However, the large-scale expansion of modern methods—monocropping supplemented by heavy application of herbicides, pesticides, and chemical fertilizers—has led to a rapid loss of soil fertility. Entrepreneurs can read more about these techniques and the short-term and potentially long-term side effects of soil health through data each day before marking the soil until October 2023. Pesticide use has been a major driver of the increased productivity of agriculture, leading to increased food security. However, there is growing concern regarding the unsustainable nature of their use because of their negative impacts on human and environmental health. The overuse or misuse of pesticides can lead to soil and water contamination, harm nontarget organisms, and create long-term health risks. Similarly, traditional chemical and synthetic fertilizers, which affect plant health to increase crop yields, have contributed to several environmental crises that impact soil degradation, water pollution, and biodiversity loss. These fertilizers also have harmful effects on crop quality and pose health risks to consumers. With increasing emphasis on sustainable agriculture, there is a dire need to find environmentally friendly substitutes that maintain productivity but create fewer adverse effects (Sarbani et al. 2022). Sustainable development is development that can be maintained. IOS discovered the idea of sustainable growth, which is defined as growth that produces higher consumption and higher standards of living without depleting them or compromising their availability for future generations. This is done while meeting our needs now and preserving natural resources to support the needs of future generations for prosperity and well-being (Sabina Begum 2015). This is why sustainable agriculture is a cornerstone for achieving global food security. With the global population expected to reach nearly 10 billion by 2050, demand for food is expected to increase. Therefore, because there is always dynamic coverage for the same successful outcome, sustainable farming practices are necessary to ensure a consistent and reliable food supply. It also speaks to important subjects such as monocultures vs. polycultures, hybrid crops, the criteria for choosing plants and livestock, integrated pest management, and farm preparedness for extreme weather, such as droughts and floods. It also examines diversification into farm tourism and adding value to a product before it is sold (John Mason 2003).

This approach integrates the best of the old and embraces new methods of sustainable modern agriculture while preserving ecological integrity in the process of stimulating the production system. Agriculture has several serious problems, including climate change, depletion of soil health, nutrient deficiency, excessive use of chemicals and pesticides, and heavy metal loads in soil. Moreover, global population growth continues to lead to increased demand for food, which places even more pressure on agricultural systems. It aims to heal the earth through time-honored methods and modern technology, rebuild the ecological balance and biodiversity, make sustainable farming truly sustainable, and guarantee stability between agriculture and natural ecosystems. In recent years, global urbanization, increasing populations, and greater impacts and effects of climate change have caused the demand worldwide for high-quality and sustainable agricultural food to accelerate significantly more than ever before. Owing to these pressures, food shortages and the human population expansion process have been emphasized as fundamental challenges for long-term sustainable development. These challenges necessitate novel approaches and joint initiatives toward food security and sustainability for generations to come (Nur Maisarah Mohamad Sarbani et al. 2022). While sustainable agriculture aims to minimize the environmental impact of food production, both traditionally and modernly, it can harm the environment and society and ensure long-term productivity. It emphasizes ecological balance, crop productivity, and natural resource conservation. Sustainable agriculture protects the environment, reduces pollution, and enhances biodiversity. It stresses the responsible use of natural resources, including water, pesticides, and raw materials. The continuous use of chemical inputs for food production has severely undermined the resistance of plants, destroying the bioenergy structure of the soil. This has resulted in decreased crop production, lower plant productivity, and reduced overall soil fertility and agricultural productivity, which have become long-term limitations to sustainable agricultural practices (G. Vasilevski 2003).

The agriculture industry has an established presence as a leading domain for data use throughout its lifecycle via precision agriculture, with more advanced data and technology use than ever before. Advances in the domain are driven by new developments in information and communication technologies, including the Internet of Things (IoT). This transformative change has disrupted conventional agriculture and offered new opportunities, challenging opportunities (Sharma et al. 2021). When applied properly, this technology can potentially reward farmers with improved productivity and profit to advance sustainable food security, particularly with Internet of Things (IoT) technology. However, barriers such as restricted access to the internet, expensive integration, and low awareness of its usage still prevent the shift to smart farming. Therefore, more determination should be made in this field (Debangshi et al. 2023). To promote sustainability in agriculture, several innovative strategies are used, such as climate-smart agriculture (CSA), organic farming, biodynamic agriculture, sustainable intensification, and regenerative agriculture. IFS, precision agriculture, INM, and IPM are some of the modern and advanced practices. Such approaches aim to make the best use of resources, minimize the impact on the environment, and increase productivity. The impacts of their adoption have revealed a promising ability to enhance soil health, control pests effectively, and maintain an ecological balance to guarantee sustainable food production systems for future generations and to combat the impacts of climate change (Muhie 2022).

5.6 Microbes in Agriculture for Sustainability

Soil is a rich and diverse environment for microorganisms, which play important roles in sustainable agriculture. These microorganisms help maintain the health of soils, decompose organic matter, cycle nutrients, and enhance the physical structure of soils to help plants withstand stress. A healthy and diverse microbial community sustains ecosystems because microorganisms interact with plants to ensure the bioavailability of nutrients, decrease disease susceptibility, and improve soil fertility. Thus, a balanced microbiome is crucial for the development of stable, productive, and environmentally sustainable agricultural systems. The soil microbiome, root-derived compounds, and agricultural crop productivity hubs are improved. By optimizing these traits, researchers and farmers can facilitate proper plant growth, improve yield, and promote ecosystem resilience. Its goal is to affect the soil structure, utilize beneficial microbes, and optimize farming techniques to develop a more integrated system that supports vigorous plants and efficiently uses resources (Anju Kumari et al. 2014). Soil microbes play a fundamental role in sustainable agriculture because they carry out many key processes that contribute to soil health and crop productivity. These microorganisms are important in nutrient cycling, where they break down organic matter to return crucial nutrients back into the soil for plant use. Nitrogen-fixing bacteria are particularly important, as they convert nitrogen from the atmosphere into a form that plants can readily absorb, thus reducing the need for chemical fertilizers. They also improve the soil structure by creating a glue-like substance that bonds soil particles together, which increases the WHC and aeration of the soil, both of which promote root growth. The role of microbes in the sustainability of agriculture and environmental development focuses on the successful utilization of several microbial applications for enhancing agricultural production along with environmental quality. Microbial inoculants and enzymes can have a significant effect on the restoration and conservation of degraded natural and agricultural ecosystems. Microbial inputs also promote greater microbial diversity, which improves soil quality and increases the yield of crops while sustaining agriculture. In addition, they play a critical role in rehabilitating degenerated ecosystems, thus being beneficial for the continued health of agricultural and natural ecosystems. Soil microbiomes also reduce pest and disease pressure through pathogen suppression; certain community group members either outcompete harmful pathogens or produce natural compounds that are growth inhibitors of pathogens, both of which reduce chemical pesticide use. It also aids in plant growth by producing hormones and enzymes that promote growth. Additionally, microbes reduce greenhouse emissions as they decompose organic matter in a way that minimizes the release of harmful gases and buries carbon in the ground. They also decompose organic waste, adding nutrients to the soil and reducing agricultural waste. Taken together, these microbial actions result in a more resilient, efficient, and sustainable agricultural system. Biopesticides based on microbes offer long-term protection to crops during their growing season, thus providing an alternative green method for pest management. These biopesticides are not as toxic as chemical pesticides and therefore do not affect plant growth while preventing harmful side effects on crops. They operate by infecting insect pests with disease, causing devastating mortality in these pest populations. This selective nature significantly reduces the adverse effects on friendly organisms and the ecosystem, providing a more balanced, natural biopesticide that reduces multiple applications of chemicals and traditional synthetic pesticides while promoting overall healthier and more sustainable agricultural practices (Thakur et al. 2020). Many of these microbes are linked to sustainable agriculture through soil health by improving soil chemistry and enabling nutrient release and absorption. Some of the most important microbial groups include bacteria, fungi, and actinobacteria. Microorganisms negatively impact the environment and create a favorable habitat for plants, reducing the use of synthetic fertilizers and turning them to the "green" industry. One of the main roles of the plant microbiome is the establishment of mutualistic relationships between plants and microorganisms. These relationships encompass nitrogen-fixing prokaryotes (e.g., the root nodule bacterium *Rhizobium*), arbuscular mycorrhizal fungi (AMF), and phosphorus-solubilizing bacteria, which are critical to the mechanisms by which plants obtain key nutrients. Although nitrogen is a vital element for plant growth, plants cannot directly utilize atmospheric nitrogen. Nitrogen-fixing bacteria—particularly *Rhizobium*—convert atmospheric nitrogen into bioavailable organic forms, thereby allowing plants to assimilate and use it for growth and development. On the other hand, arbuscular mycorrhizal fungi help with phosphorus uptake, and phosphate-solubilizing bacteria improve soil availability. These symbiotic relationships are necessary to provide a steady source of nutrients to plants, leading to healthier crops and sustainable farming processes.

Nitrogen (N) plays an important role in the growth, development, and well-being of plants. It constitutes a significant part of amino acids, proteins, and chlorophyll and is vital for plant metabolism and photosynthesis. Atmospheric nitrogen (N_2) makes up a large part of the atmosphere, but plants are unable to use atmospheric nitrogen directly. Rather, they depend on other forms of nitrogen, specifically nitrate (NO_3^-) and ammonium ions (NH_4^+), the types they take up through their roots. This process of transforming atmospheric nitrogen into these useful forms is referred to as nitrogen fixation. This happens naturally via biological and abiotic processes. Bacteria obtain nitrogen (N_2) through a process known as biological nitrogen fixation, which occurs in the root nodules of leguminous plants, where N_2 is converted into ammonia. Other bacteria in the soil will then convert this ammonia into nitrates, where the nitrates can be taken up by plants. This natural process is important for keeping the soil nitrogen-rich, which is important to ensure that plants have a ready supply of this important nutrient. However, these nitrogen-fixing bacteria, including *Burkholderia* spp., *Rhizobium* spp., *Pseudomonas* spp., *Bradyrhizobium* sp., *Bacillus* spp., *Frankia* spp., *Enterobacter* spp., *Azospirillum* spp., *Azotobacter* spp., and *Gluconacetobacter* spp., have been discovered in association with diverse host plants. These nitrogen-fixing endophytic microbes are important in the context of plant growth, crop yield, and soil health and are important for sustainable agricultural practices. *Azospirillum* spp. are free-living nitrogen-fixing bacteria that can promote the growth of nonleguminous crops (Rana et al. 2023). Phosphate-solubilizing bacteria (PSB) promote the efficient utilization of phosphate, facilitating more sustainable cropping systems and promoting the prudent use of mineral nutrients. In addition, there is a growing biotechnological effort to increase the ability of roots to acquire phosphate from preexisting soil phosphate pools and to increase the fraction of fertilizer phosphate that crops are able to absorb. This is shifting toward the cooptimization of PSB, fertilizer, and crop root processes to improve not only the phosphate efficiency but also the overall performance of the agricultural system [2]. In this holistic mindset, the goal is to optimize the overall fungal–plant–soilroot interaction and then optimize the transfer of nutrients from the soil to the plant and from the root to the soil, with a focus on what is available rather than what has been removed. These efforts are crucial at the same time for ensuring high agricultural productivity and environmental sustainability by reducing dependence on the overuse of mineral fertilizers and improving the use of natural soil resources (Aasfar et al. 2021). The existence of heavy metals and persistent organic pollutants are considerable environmental hazards that cause environmental pollution and damage. The infiltration of heavy metals through untreated industrial waste not only affects soil but also makes agricultural land impossible to sustain, severely hampering crop yield and quality production. The other group, which is mostly related to the Dawkins Ph group, is from the incineration and combustion of different products, such as fossil fuels, leading to the release of harmful pollutants as a result of both primary emissions and secondary pollutants that cause environmental degradation. This reflects the need for sustainable waste management practices and pollution control measures to address their negative impacts (Tufail et al. 2022). Some bacteria that are resistant to heavy metals can relieve plants affected by heavy metals. Some heavy metal-resistant bacteria that can be found in soil are *Pseudomonas* spp., *Bacillus* spp., and *Alcaligenes* spp. Arbuscular mycorrhizal fungi (AMF) are important bioinoculants in sustainable agriculture, as they help establish a beneficial symbiotic association with a variety of crop plants. AMF enhance soil health and quality through three main approaches: improving the soil structure; altering plant physiological processes; and facilitating biotic interactions. Taken together, these actions enhance plant growth, performance, and productivity. Glomalin production by AMF promotes the formation of soil aggregates, which improves the soil structure. They also increase nutrient absorption, thus increasing soil fertility and crop productivity. Moreover, the overall physiology of plants also changes to cope with biotic stressors (working against the environment), such as pathogens and weeds, and abiotic stresses (salt, drought, extreme temperatures, and heavy metal pollution). As biocontrol agents, AMF mitigate plant pathogens and promote synergistic interactions among other microorganisms inhabiting the rhizosphere and aboveground areas. Furthermore, AMF are important in land restoration and help seedlings establish better. Together, these functions significantly improve plant health, growth, and overall productivity (Kalamulla et al. 2022). Arbuscular mycorrhizal fungi (AMF) produce intricate networks with adjacent plants, allowing for the exchange and distribution of vital resources via linked root networks [3]. Some important species of AMF include *Glomus* spp., *Rhizophagus irregularis*, *Acaulospora* spp., *Entrophospora* spp., and *Scutellospora* spp. The soil hosts a rich diversity of naturally occurring microorganisms that are essential for maintaining plant health and productivity.

Among these are microbes uniquely equipped to aggressively target and eliminate plant pathogens present in the soil and on seeds, which often threaten crop health by reducing yields and quality. These beneficial microorganisms, collectively referred to as biocontrol agents, include a wide range of bacteria, fungi, and even viruses, each utilizing distinct mechanisms to combat harmful pathogens. For example, bacterial genera such as *Bacillus*, *Pseudomonas*, and *Streptomyces* are renowned for their ability to produce antibiotics, enzymes, and bioactive compounds that inhibit the growth of pathogens. Similarly, fungal biocontrol agents such as *Trichoderma* and *Gliocladium* employ strategies such as outcompeting pathogens for resources, directly parasitizing them, or producing metabolites that disrupt their growth and development. Even viruses, including bacteriophages and mycoviruses, play a role in controlling pathogens by infecting and neutralizing specific bacterial or fungal threats. In addition to protecting plants from soil- and seed-borne diseases, these biocontrol agents contribute to a more sustainable agricultural system by reducing reliance on synthetic pesticides. By enhancing soil biodiversity and resilience, they promote a balanced ecosystem that supports long-term crop productivity. The use of these natural microbial allies allows farmers to minimize environmental impacts while improving both the health and quality of their crops (Ram et al. 2018) (Fig. 5.1).

The plant-associated microbiome can also serve as a biocontrol agent, offering a natural and effective way to protect plants from harmful pathogens. This biocontrol is achieved through several mechanisms employed by the microbiota, such as the production of siderophores, volatile organic compounds, enzymes, and antibiotics. Siderophores, for example, bind to iron in the soil, making it less accessible to pathogens and thereby inhibiting their growth. Volatile compounds emitted by microbes can suppress the growth of pathogens or directly kill them, whereas

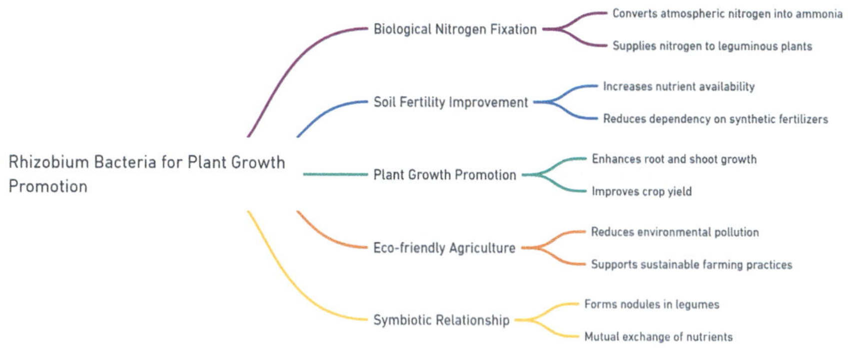

Fig. 5.1 Plant growth-promoting abilities of *Rhizobium*-associated bacteria

enzymes produced by certain microbes degrade pathogen cell walls, neutralizing their impact. Antibiotics synthesized by the microbiota target and eliminate harmful microbes in the environment of plants. Additionally, the microbiome contributes to plant health by modulating phytohormone levels, which regulate various growth and defense processes in plants. These combined actions of the microbiota not only increase plant resilience to biotic stress but also contribute to sustainable agricultural practices by reducing the need for chemical pesticides.

These microbes support plants together with disease-causing pathogens as robust biocontrol agents against the injury caused by pathogenic bacteria and fungal strains. These beneficial microorganisms help fight off plant diseases caused by bacteria and fungi, leading to healthier and more productive crops. As a result, diverse formulations of microbial inoculants have been designed and are commercially available. They include targeted strains of positive microbes that help with nutrient uptake, pathogen activity suppression, and stimulated phytostimulation. They are widely applicable to numerous agricultural and horticultural systems to increase yields, increase plant quality, and decrease reliance on chemical pesticides and fertilizers. The use of microbial inoculants in farming has potential for sustainable crop production, ecosystem health, and soil biofertility in the long run. Microbial inoculants protect plants via different mechanisms, resulting in plants being healthy and having high yields. Their main mode of action involves the release of hydrolytic enzymes, which are secreted in an extracellular manner. They dissolve the cell walls of harmful microbes, which inactivates them and prevents infections. These microbes also compete with pathogens for vital nutrients in the soil or plant environment, essentially starving the pathogens of what is necessary to live and grow. Moreover, secondary metabolites secreted by these microbial inoculants can be toxic and can even inhibit the growth of crop-pathogenic microbes at low concentrations. These secondary metabolites are then used to inhibit the growth, reproduction, and general function of pathogens, providing additional protection to the plant. These integrated approaches make microbial inoculants effective agents that promote plant wellness, minimizing the need for chemical interventions in agriculture and horticulture.

Owing to the wide range of compounds with diverse biological activities present in earthworms, castings, and the gut, biotechnological opportunities exist. To date, scientists have focused on discovering new microbes in unusual and underexplored ecosystems. Among these, actinobacteria are notable secondary metabolite producers with diverse bioactive compounds of industrial importance. Actinobacteria possess bioactive metabolites but are less investigated in earthworms, earthworm guts, and earthworm casts. The reserve that describes earthworms along with their native microbes can be exploited for endophytic actinobacteria bioprospecting (this review).

5.7 Application

As (facultative) members connecting the host earthworm gut with its adjacent terrestrial environment, the earthworm gut is nutrient rich, which is conducive to earthworm colonization. Everyone has long known that the earthworm is an ecological engineer, but only in the past decade have scientists begun to understand that the worm gut and its microbiota differ, depending on soil conditions, and that they are an integral part of the maintenance of earthworm metabolism and the transformation of nutrients (Zhou et al. 2019). This was largely made possible by rapid advances in methods for isolating bacteria and screening for organisms, which include both culturable and unculturable organisms through classical isolation approaches as well as next-generation sequencing technologies. For example, the content of aerobic and anaerobic bacteria in the gut of the worm is 12–20 and 10–4000 times greater than that in the soil (Horn et al. 2003). It is generally agreed that the majority of tonsillar bacteria of the worm are derived from the soil next to the location of the worm. However, owing to its unique anaerobic conditions, the earthworm intestine creates a niche environment supporting the growth of anaerobic and/or facultative anaerobic bacteria. This produces the vast differences between the bacterial populations in the worm gut and those in the soil. The concentrations of dominant nutrient components (i.e., carbon, nitrogen, phosphorus, and sulfur) in the worm gut were reported to be 2–5 times greater than those in the adjacent soil. Consequently, the nutrients and amount of bacteria present in the gut of the worm are sufficient for the transformation of pollutants obtained through the soil, such as heavy metals, organic pollutants (hydrocarbons, e.g., petroleum), microplastics, and antibiotics (Wang et al. 2019). We highlight the reciprocal effects of pollutants on the activity and diversity of bacteria present in the gut of worms (Chao et al. 2019). Consequently, here, we reviewed recent advances in understanding the intestinal bacteria of the worm and the ecological service of intestinal bacteria from the perspective of the nutrient cycle and chemical transformation.

5.8 Symbiosis

One of the main subjects of soil microbial ecology is the study of the symbiotic relationships between microorganisms and invertebrates. Symbiosis refers to the coexistence of two or more species, one of which has a higher degree of organization than the others and hosts the second one inside either their cells or special organs or inside their digestive tracts. The tremendous diversity of soil microbes and animals suggests diverse partner combinations, including gut symbioses. The symbionts of the digestive tract of termites, for example, are known, as are their putative functions. The intestinal microorganisms of other terrestrial invertebrates (diplopoda, wood lice, insect larvae or imagos, earthworms) have also been the subject of microbiological studies. Direct or indirect evidence for such groupings of microorganisms differing in composition from those found in soil and related substrates is obtained from isolation on selective media and from studies of molecular genetics (here, we use the term broadly to encompass studies on the basis of microbial phylogeny as well as in vitro genetic engineering). We did not confirm the functional relationships of these microbial communities with the symbionts of the host. Hence, exploring the matter and structure of invertebrate digestive tract microbial populations and their activity in their houses is an important factor in soil ecology. The functional role of microorganisms in earthworm intestines has been addressed in very few articles. Nitrous oxide (N_2O)-producing and methane-oxidizing bacteria have been shown to contribute to greenhouse gas (GHG) production. Therefore, although earthworms are key components of mesofauna and provide significant "ecosystem engineering" functions in soil, the taxonomic composition and functions of the microbial community remain unclear.

References

Aasfar A, Bargaz A, Yaakoubi K, Hilali A, Bennis I, Zeroual Y, Meftah Kadmiri I (2021) Nitrogen-fixing Azotobacter species as potential soil biological enhancers for crop nutrition and yield stability. Front Microbiol 12

Aira M, Domínguez J (2022) Earthworm activity and its effects on soil health: a review. Soil Biol Biochem 130:20–30

Aira M et al (2019) Microbial hotspots in earthworm casts: implications for soil nutrient dynamics. Appl Soil Ecol 139:56–64

Akhila K, Entoori KM (2022) Earthworm casts as organic amendments: impacts on soil fertility. J Soil Health 15(3):45–59

Begum S (2015) Impact of agricultural modernization on sustainable livelihood among the tribal. Int J Res Hum Art Literat 3(5):55–66

Chao H, Kong L, Zhang H, Sun M, Ye M, Huang D, Zhang Z, Sun D, Zhang S, Yuan Y, Liu M, Hu F, Jiang X (2019) The metoprolium guillelmi gut is a hospitable microenvironment for the potential transmission of antibiotic resistance genes. Sci Total Environ 669:353–361

Cordero RJ, Datta S (2016) Microbial community structure in earthworm guts and its ecological implications. Environ Microbiol Rep 8(4):495–503

Debangshi U, Sadhukhan A, Dutta D, Roy S (2023) Application of smart farming Technologies in Sustainable Agriculture Development: a comprehensive review on present status and future advancements. Int J Environ Clim Chang 13:3689–3704

Gautam S et al (2024) Earthworm activity and carbon stabilization in soils: a comprehensive review. Soil Sci Soc Am J 88(1):1–12

Gopikrishnan K et al (2020) Bioorganic structures formed by earthworms: impacts on soil properties. Soil Res 58(2):123–134

Helen A et al (2021) Earthworms as bioindicators of soil health: a global perspective. Ecol Indic 121:107–115

Horn M et al (2003) Denitrification processes in earthworm guts and their implications for nitrogen cycling. Soil Biol Biochem 35(6):847–853

Hu Y et al (2018) The role of earthworm gut microorganisms in soil weathering and genesis. Geoderma 314:35–42

Iordache M (2023) Chemical composition of earthworm casts as a tool in understanding the earthworm contribution to ecosystem sustainability—a review. Plant Soil Environ 69(6):247–268

Jay N et al (2005) Antibacterial proteins in earthworms: mechanisms of defense against microbial threats. J Invertebr Pathol 90(1):43–50

Kalamulla R, Sandaruwan D, Karunarathna SC, Stephenson SL, Tibpromma S, Elgorban AM, Al-Rejaie S, Yapa PN, Suwannarach N (2022) Assessment of community dynamics of Arbuscular Mycorrhizal fungi in the Rice (*Oryza sativa* L.) rhizosphere and potential application as biofertilizer. Sustainability (Switzerland) 14(24)

Kauschke E et al (2007) The immune system of earthworms: insights into antibacterial peptides. Comp Biochem Physiol B Biochem Mol Biol 147(3):369–375

Kliszcz J et al (2024) The gut microbiome of earthworms: a dynamic environment for microbial interactions. Appl Soil Ecol 168

Kokilamani AL et al (2024) The impact of earthworm casts on soil structure and nutrient cycling. J Agric Sci 162(2):145–156

Kumari A, Sharda A, Singh H (2014) Role of microorganisms in sustainable agriculture: a review. Int J Curr Microbiol App Sci 3(10):564–580

Lamba S, Rani S, Kumar N (2021) Application of innovative space technology approaches to the sustainability of agricultural systems in the developing world. Remote Sens Lett 12(3):315–324

Lejoly J et al (2024) Metabolic activity in earthworm casts: implications for microbial dynamics in soil ecosystems. Appl Soil Ecol 168

Liu L et al (2019) Contribution of gut bacteria to nitrogen cycling processes in earthworms: a review. Front Microbiol 10

Ma Y et al (2016) Earthworm gut microbiota and its role in nutrient metabolism and crop yield enhancement. Nat Commun 7(1)

Mason J (2003) Sustainable agriculture. Landlinks Press

Mirleau P et al (2024) The role of earthworm gut microbiota in nutrient metabolism and soil health: a review article. Microb Ecol 68(1):12–24

Mohanraj I, Ashokumar K, Naren J (2016) Field monitoring and automation using IoT in the agriculture domain. Procedia Comput Sci 93:931–939

Muhie SH (2022) Novel approaches and practices to sustainable agriculture. J Agric Food Res 10:100446

Nitin G et al (2021) Feeding habits of earthworms and their impact on soil fertility: a review study from India's agricultural fields. Soil Biol Biochem 152

Omokaro E et al (2024) The role of vermicasts in enhancing soil quality for sustainable agricultural practices. Environ Sci Pol 45(2):78–85

Ram RM, Ghosh A, Kumar S (2018) Biocontrol agents in agriculture: protecting plants and promoting sustainability. Indian J Agric Sci 88(5):686–695

Rana KL, Kour D, Kaur T, Negi R, Devi R, Yadav N, Rai PK, Singh S, Rai A, Yadav A, Sayyed R, Yadav AN (2023) Endophytic nitrogen-fixing bacteria: untapped treasure for agricultural sustainability. J Appl Biol Biotechnol 11:1–21.cha

Regina R et al (2019) Ecological roles of different functional groups of earthworms in agroecosystems: a review article. Agric Ecosyst Environ 275

Sahoo D, Mohanty P, Mishra S, Behera M (2024) Does climate-smart agriculture technology improve the subjective Well-being of farmers? Evidence from microlevel data in Odisha. India Farming Syst 3:100124

Santos D et al (2021) Chemical composition and nutrient cycling dynamics in earthworm casts: a review article on implications for agricultural sustainability practices. Soil Res 59(3)

Sapkota R et al (2020) Influence of diet on the gut microbiome composition of earthworms: implications for nutrient cycling in soils. Microb Ecol 79(3):689–700

Sarbani M, Maisarah N, Yahaya N (2022) Advanced development of biofertilizer formulations using microorganisms as inoculants for sustainable agriculture and the environment—a review. Malaysian J Sci H Tech 8:92–101

Sharma DK, Tomar S, Chakraborty D (2017) Role of earthworm in improving soil structure and functioning. Curr Sci:1064–1071

Sharma D, Nawab AZB, Alam M (2021) Integrating M-health with IoMT to counter COVID-19. In: Raza K (ed) Computational intelligence methods in COVID-19: surveillance, prevention, prediction and diagnosis, pp 373–396

Soma T, Nuckchady B (2021) Communicating the benefits and risks of digital agriculture technologies: perspectives on the future of digital agricultural education and training. Front Commun 6:762201

Thakur N, Kaur S, Tomar P, Thakur S, Yadav AN (2020) Microbial biopesticides: current status and advancement for sustainable agriculture and the environment. New and Future Developments in Microbial Biotechnology and Bioengineering: Trends of Microbial Biotechnology for Sustainable Agriculture and Biomedicine Systems: Diversity and Functional Perspectives:243–28221

Thakuria D et al (2010) Environmental factors affecting the gut microbiota of earthworms: a comprehensive analysis across different habitats. Environ Microbiol Rep 2(5)

Tufail MA, Iltaf J, Zaheer T, Tariq L, Amir MB, Fatima R, Asbat A, Kabeer T, Fahad M, Naeem H, Shoukat U, Noor H, Awais M, Umar W, Ayyub M (2022) Recent advances in bioremediation of heavy metals and persistent organic pollutants: a review. Sci Total Environ 850:157961

Wang J, Nie Y, Dai H, Wang M, Cheng L, Yang Z, Chen S, Zhao G, Wu L, Guang S, Xu A (2019) Parental exposure to TiO_2 NPs promotes the multigenerational reproductive toxicity of cd in *Caenorhabditis elegans* via bioaccumulation of cd in germ cells. Environ Sci Nano 6(1332)

Wang Y et al (2021) Earthworm activity enhances soil structure and nutrient cycling: a review on sustainable agriculture practices worldwide. Soil Sci Soc Am J 85(6)

Yates PL (1964) The need for agriculture is a disadvantage in a growing world. Int J Agrarian Aff 4(3):150–174

Zhang Y et al (2022) Interactions between earthworm gut microbiota and soil microbial communities: implications for ecosystem functioning under climate change scenarios. Soil Biol Biochem 174

Zhou G-W, Yang X-R, Sun A-Q, Li H, Lassen SB, Zheng B-X, Zhu Y-G (2019) Mobile incubator for iron (III) reduction in the gut of the soil-feeding earthworm Pheretima guillelmi and interaction with denitrification. Environ Sci Technol 53(8):4215–4223

Chapter 6
Bioprospects of Marine Microbes for Plant Growth Promotion and Disease Control

Sandhya Rani Edadhala and Narasimha Golla

Abstract Marine environments are rich in microbial diversity and present an untapped reservoir of bioactive compounds and metabolic pathways. This chapter explores the bioprospects of marine microbes and presents a promising frontier for enhancing plant growth and controlling diseases. Marine bacteria, fungi, and actinomycetes promote plant growth, including phosphate solubilization, nitrogen fixation, the synthesis of antimicrobial chemicals, and siderophore production. Plant diseases can be significantly controlled by various plant mechanisms, such as biofertilizers, plant growth-promoting *rhizobacteria* (PGPR), antagonism, induced systemic resistance (ISR), biocontrol products, biological soil amendments, biological seed treatments, synthesis of lytic enzymes, and integrated pest management (IPM). This chapter also explores the applications of marine microbes in agriculture as biofertilizers, biostimulants, biocontrol agents, bioremediation agents, and climate change adaptation agents. Additionally, it addresses future challenges and available prospects. This chapter highlights the untapped potential of marine bioprospects in the development of eco-friendly agricultural practices and the role of marine microbes in enhancing plant growth and providing sustainable disease management solutions, offering a viable alternative to chemical fertilizers and pesticides.

Keywords Bioprospecting · Marine microorganisms · Plant growth promotion · Integrated pest management

6.1 Introduction

The vast and diverse marine ecosystem covers approximately 70% of the planet's surface. It includes oceans, seas, coral reefs, estuaries, and coastal areas, each with its own unique characteristics and life forms. The major marine ecosystem

S. R. Edadhala · N. Golla (✉)
Department of Virology, Sri Venkateswara University, Tirupati, AP, India

V. Gopikrishnan et al. (eds.), *Harnessing Microbial Resources for Sustainable Agriculture*, Microorganisms for Sustainability 61,
https://doi.org/10.1007/978-981-95-4817-0_6

components include oceans such as the Pacific, Atlantic, Indian, Southern, and Arctic, seas, and coral reefs. Estuaries, where freshwater meets saltwater, and coastal areas such as beaches and mangroves also play important roles in marine ecosystems. The key features and functions of the marine ecosystem include biodiversity, primary production by phytoplankton, food webs, nutrient cycling, habitat provision, climate regulation, and human interactions. However, marine ecosystems face major threats such as pollution, climate change, overfishing, and habitat destruction. Conservation efforts such as marine protected areas, sustainable fishing practices, pollution control, climate action, and research and education are crucial for protecting and restoring marine ecosystems (Das et al. 2006). Bioprospecting, often known as biodiversity prospecting, is the process of finding and bringing to market novel natural goods, which means that scientists are searching for naturally occurring compounds that may benefit humans. These goods could originate from microbes, plants, animals, or other natural sources (Beattie et al. 2011). The hope is that by commercializing and producing items that people can utilize for their gain, these discoveries will contribute to society, as these materials are frequently utilized in medications, cosmetics, and other items. The process of finding and exploiting natural resources with potential advantages for people is known as bioprospecting. These resources are gathered and placed to benefit the community (Weiss and Eisner 1998).

Marine microbes such as *Pseudomonas fluorescens*, *Bacillus subtilis*, and *Azospirillum brasilense* promote plant growth through nutrient solubilization, phytohormone production, and biocontrol. They suppress plant diseases, produce phytohormones, increase the soil content, increase nutrient availability, and reduce oil pollution (Lugtenberg et al. 2013). Nutrients such as nitrogen, iron, and phosphorus are crucial for bacterial development, but their lack in agricultural soils can lead to poor development. Marine microbes, such as *cyanobacteria*, are essential for the fixation of nitrogen and the production of phytohormones such as auxin, cytokinin, and gibberellin (Mishra et al. 2021). These growth-promoting bacteria regulate plant growth processes, cell division, and differentiation. Understanding siderophore synthesis by bacteria can lead to sustainable agriculture and improved soil fertility.

Marine microbes can control plant diseases through biofertilizers, promoting soil nutrient availability and causing systemic resistance. Researchers are developing commercially viable marine microbe-based agricultural biocontrol products (Timofeeva et al. 2023). Biological soil amendments and seed treatments can increase soil health and reduce the incidence of soil-borne illnesses (Shu et al. 2022). Integrated disease management can improve efficiency and sustainability by integrating marine microbe-based techniques into pest management programs. Scientists are exploring novel microbial strains and application strategies for maximum effectiveness in agricultural settings. Marine microbes offer a sustainable alternative to traditional agriculture because of their unique biochemical properties (Dara 2019). Research is now focusing on their use as biofertilizers and biocontrol agents, enhancing crop productivity and reducing chemical inputs.

6.2 Marine Microbes for Plant Growth Promotion

Marine microbes have been recognized for their potential to promote plant growth through various mechanisms, such as nutrient solubilization, phytohormone production, and biocontrol. Some notable examples of marine microbes that can act as plant growth promoters are as follows:

6.2.1 Pseudomonas *spp.*

6.2.1.1 *Pseudomonas fluorescens*

The positive effects of *P. fluorescens* as a biocontrol agent in agriculture are becoming more widely recognized. Recent investigations have shown its efficiency in suppressing a range of plant diseases, such as bacteria, nematodes, and fungi. The ability of bacteria to generate secondary metabolites, including phytohormones and antibiotics, increases their effectiveness in biocontrol. These metabolites provide systemic resistance in plants and impede the growth of infections, strengthening the inherent defensive mechanisms of plants (Negi et al. 2022). According to Rai et al. in 2017, *P. fluorescens*, as a biocontrol agent, is a sustainable substitute for conventional pesticides, improving environmental health and decreasing the ecological influence of agricultural production. Optimizing *P. fluorescens* usage for biocontrol and understanding the underlying mechanisms that allow it to protect crops from diseases are the main goals of continuing research in the field (Yanes and Bajsa 2016).

6.2.1.2 *Pseudomonas aeruginosa*

Pseudomonas aeruginosa produces plant growth-promoting substances such as indole-3-acetic acid (IAA), which can increase nutrient uptake, and phosphorus solubilizers such as gluconic acid, which convert atmospheric nitrogen to ammonia (NH_3) and produce hydrocyanic acid (HCN), which can exert biological control by inducing systemic resistance. These bacteria are especially well suited for use as biological control agents and biofertilizers because they can obtain nutrients from a variety of root exudates (Lugtenberg et al. 1999) and are widely distributed in soils, especially in the rhizosphere, where they reproduce more quickly than other rhizosphere bacteria do, allowing them to maintain high population density and to obtain a wide range of metabolites that they can use to exert antagonistic or inhibitory activity against phytopathogens; these bacteria can have diazotrophic capacity, produce auxins and siderophores, and solubilize phosphorus (Haas and Keel 2003).

6.2.2 Bacillus *spp.*

6.2.2.1 *Bacillus subtilis*

Bacillus subtilis produces various phytohormones, enzymes, and antimicrobial compounds that aid in plant growth and protection against pathogens. Both soil fertility and ecosystem stability are at risk because of the increasing dependence of farmers on chemical fertilizers. Owing to their numerous advantages for improving soil conditions and crop productivity, plant growth-promoting *Rhizobacteria*, or plant growth promotors (PGPRs), are at the vanguard of sustainable agriculture. By producing lipopeptides, forming biofilms, and inducing systemic resistance, *Bacillus subtilis* is a prevalent plant growth promoter in soil that is essential for providing plants with biotic and abiotic stress endurance. *Bacillus* species remove metals from soil as part of bioremediation systems (Mahapatra et al. 2022).

6.2.2.2 *Bacillus licheniformis*

Bacillus licheniformis produces plant growth-promoting hormones and improves soil structure and nutrient availability. Lichens stimulate plant development while inhibiting the growth of bacterial and fungal diseases. Moreover, research suggests that lichens are abundant, naturally occurring sources of biocontrol agents and plant growth-promoting substances that are employed in agriculture (Sansinenea 2019).

6.2.3 Azospirillum *spp.*

6.2.3.1 *Azospirillum brasilense*

This nitrogen-fixing bacterium is often used to increase the nitrogen content in the soil, which is crucial for plant growth. *Azospirillum* employs several mechanisms to promote the growth of plants. Various *Azospirillum* species produce plant hormones that change the growth pattern of plant roots, and frequently damaged roots develop more branches and fine root hairs, which may improve the ability of plants to absorb nutrients and water (Fukami et al. 2018). In addition to these modifications, *Azospirillum* changes the forms of plant nutrients, such as phosphorus and nitrogen, to increase their availability to plants. However, the amount of nitrogen biologically fixed for beneficial crop plants is debatable (Bashan and De-Bashan 2010). Additionally, this microbe produces antioxidants that shield plant roots from environmental challenges such as flooding and drought. The reduction in plant disease caused by *Azospirillum* can also indirectly increase plant progress and stimulate the plant immune system to fight against pathogen infections; this is a technique called "induced systemic resistance" (Fukami et al. 2018).

6.2.4 Vibrio *spp.*

6.2.4.1 *Vibrio natriegens*

A recently discovered ocean bacterium called *V. natriegens* reaches a dense population of cells on glucose and has the reputation for being tolerant of high salt concentrations, which is a significant benefit when aiming for high cell density and product concentration and rapid growth and the ability to produce plant growth-promoting substances such as indole-3-acetic acid (IAA) (Midhun and Arun 2023).

6.2.5 Thalassospira *spp.*

6.2.5.1 *Thalassospira xiamenensis*

They are involved in nitrogen fixation and enhance nutrient uptake in plants, oxygen requirements, nitrate reduction, and denitrification abilities, as well as their ability to utilize six different carbon sources as plant growth promotors (Liu et al. 2007).

6.2.6 Rhodococcus *spp.*

6.2.6.1 *Rhodococcus fascians*

This fungus is known to produce cytokinins, which are important for cell division and growth in plants. *R. fascians* can promote plant growth by producing plant growth modulators, including auxin and cytokinin, as well as by breaking down ethylene via 1-amino-cyclopropane-1-carboxylic acid deaminase (Francis et al. 2016). *R. fascians* affects hormones other than cytokinin and auxin. The maintenance and multiplication of meristematic cells depend on the inhibition of gibberellic acid, which regulates cellular differentiation (Simón-Mateo et al. 2006).

6.2.7 Marinobacter *spp.*

6.2.7.1 *Marinobacter hydrocarbonoclasticus*

Marinobacter hydrocarbonoclasticus contributes to the cycling of nutrients and, through increasing nutrient availability, can promote plant development. A variety of *Marinobacter* strains are proficient in breaking down acyclic isoprenoid molecules and both polycyclic and aliphatic-aromatic hydrocarbons. They are commonly found to degrade hydrocarbon organisms in a wide range of contaminated marine settings, indicating their significant role in reducing oil pollution (Branchu et al. 2017).

6.2.8 Planococcus *spp.*

6.2.8.1 *Planococcus maritimus*

Planococcus produces various enzymes and metabolites that can improve plant growth and stress tolerance. Terpenoids are responsible for the aroma of eucalyptus; the tastes of ginger, cloves, and cinnamon; the yellow color of sunflowers; and the red color of tomatoes (Waghmode et al. 2020). The extensive and most diverse families of naturally occurring molecules are called terpenes or isoprenoids. They are primarily found in plants; however, larger classes, including sterols and squalene, are also found in animals and are responsible for the fragrance, taste, and pigments of plants (Alamgir and Alamgir 2018).

6.2.9 Halomonas *spp.*

Halomonas elongata produces exopolysaccharides that aid in soil aggregation and water retention, which are beneficial for plant growth. Because they can convert NO_3 to NO_2, *Halomonas elongata* can develop anaerobically when NO3 is present (Vreeland et al. 1980). To defend itself against environmental stresses, *Halomonas elongata* produces ectoine to improve production efficiency in bioreactors (Fatollahi et al. 2021).

6.2.9.1 *Cyanobacteria* spp.

Cyanobacteria belong to the genera *Anabaena* and *Nostoc* and can be used in agriculture for various applications. They can fix atmospheric nitrogen, produce growth-promoting substances, increase nutrient availability, and act as biostimulants. They also produce antimicrobial compounds that protect plants from pathogens. *Cyanobacteria* can be applied as biofertilizers, biostimulants, or biopesticides, and their effectiveness depends on environmental conditions such as temperature, light, and soil conditions. Examples of *cyanobacteria* in agriculture include *Anabaena*, *Nostoc*, *Spirulina*, and *Oscillatoria* (Kumar et al. 2019). The tabular column features marine microbes, their role in plant growth promotion, and their utility in disease control (Table 6.1).

6.3 Plant Growth Promotion

Nutrients such as fixed nitrogen, iron, and phosphorus increase bacterial production. A lack of these elements in agricultural soils can lead to poor plant development and increased numbers of plant crops. Farmers are turning to synthetic sources of phosphate and nitrogen, but these sources can be costly and harmful to human health and the environment. Exploring marine microbes for new agricultural products may offer more sustainable alternatives for plant growth.

6.3.1 Nitrogen Fixation

Nitrogen fixation is crucial for life, as it converts molecular nitrogen into essential compounds needed for various organic substances (Fig. 6.1). This process is vital for agriculture, fertilizer production, and the production of nitrogen-containing chemicals such as explosives and medicines. The chemical process known as "nitrogen fixation" transforms molecular nitrogen (N_2) into nitrogen compounds such as ammonia (NH_3) (Burris and Wilson 1945). The nitrogenase protein complex catalyzes the reduction of nitrogen gas to ammonia in *Cyanobacteria*, specifically in a specialized cell called the heterocyst. The production and activity of the nitrogenase complex are regulated by genetic and environmental factors such as oxygen levels and concentrations of ammonia and nitrogen species. The heterocyst cell is crucial for nitrogenase function because of its sensitivity to oxygen (Li et al. 2003). *Diazotrophs*, including Cyanobacteria, which can increase soil fertility, play a key role in biological nitrogen fixation (Berde et al. 2022). Various bacteria and archaea also contribute to nitrogen fixation in different environments. *Cyanobacteria*, such as *Trichodesmium*, are important nitrogen fixers in marine ecosystems and contribute to the carbon and nitrogen cycles. Overall, nitrogen fixation is essential for various biological processes and the overall health of ecosystems (Wong 2024). Key mechanisms and steps involved in nitrogen fixation: Symbiotic nitrogen fixation is a process in which nitrogen-fixing bacteria such as *Rhizobium* and *Bradyrhizobium* form a partnership with plants, especially legumes. This process involves bacteria entering the roots to form nodules where they convert atmospheric nitrogen into ammonia via the enzyme nitrogenase. Plants provide bacteria with carbohydrates, while the produced ammonia is used to enrich soil with nitrogen or support plant growth through organic compounds such as amino acids. Symbiotic nitrogen fixation is a crucial process in enriching soil with nitrogen and supporting plant growth. (Beringer and Johnston 1984). Nonsymbiotic nitrogen fixation by free-living soil bacteria such as *Azotobacter*, *Clostridium*, and *cyanobacteria* converts atmospheric nitrogen to ammonia via the nitrogenase enzyme complex. This process enhances

Table 6.1 The role of marine microbes in plant growth promotion

S. No	Marine microbes	Role in plant growth promotion	Role in disease control	Reference
1.	*Pseudomonas aeruginosa*	Fixes nitrogen and converts it into ammonia Solubilizes phosphates for plant growth Produces growth hormones for root elongation and nutrient uptake	Produces antibiotics like phenazine and pyocyanin Inhibits plant pathogen growth and produces secondary metabolites and enzymes Effective against root diseases caused by soil-borne fungi	Lugtenberg et al. (1999)
2.	*Bacillus subtilis*	Promoting root growth and cell elongation Enhances differentiation Solubilizes phosphate for faster, healthier plant development	Controlling plant pathogens and inhibiting root rot and leaf blight Enhancing immune system response	Mahapatra et al. (2022)
3.	*Streptomycin* spp.	Produces siderophores for iron-chelating compounds Role in plant growth hormones like auxins, cytokinins, and gibberellins, for root elongation and cell division	"Effective biocontrol agent" Produces antifungal compounds Inhibits fungal growth Effective against plant diseases	Watve et al. (2001)
4.	*Azospirillum* spp.	Nitrogen conversion enhances crop growth Converts atmospheric nitrogen into ammonia Produces phytohormones Stimulates root elongation and nutrient uptake	Antagonistic against soil-borne pathogens Produces antimicrobial compounds Induces systemic resistance	Fukami et al. (2018)
5.	*Rhizobium* spp.	"Symbiotic relationship with leguminous plants" Converts atmospheric nitrogen into ammonia Enriches soil with nitrogen Promotes sustainable farming	Inhibits pathogenic microorganisms in the root zone and prevents root rot and wilt	Mabrouk et al. (2018)

(continued)

Table 6.1 (continued)

S. No	Marine microbes	Role in plant growth promotion	Role in disease control	Reference
6.	*Trichoderma harzianum*	Phytohormone production Promotes plant growth Enhances nutrient uptake Stimulates root development and cell division	"Controlling plant pathogens through mycoparasitism" Produces antibiotics to prevent infections	Tyśkiewicz et al. (2022)
7.	*Psedomonas fluroscens*	Solubilization of phosphate in soil Crucial for plant growth and energy transfer Produces growth hormones like auxins, gibberellins, and cytokinins Stimulates root elongation and cell division	"Suppressing plant pathogens and reducing diseases" Triggers systemic resistance Reduces root rot and leaf blight	Negi et al. (2016)
8.	*Penicillium* spp.	Solubilizing phosphate enhances plant growth Enhances crop yields Releases growth hormones Promotes root elongation	Antifungal compounds: Preventing diseases Improve plant health Reduce chemical fungicide use Create a healthier root environment	Altaf et al. (2018)
9.	*Enterobacter* spp.	"Nitrogen conversion and soil fertility" Generates nitrogen from the atmosphere into ammonia Crucial for plant production and nitrogen fixation Solubilizes phosphate	Antibiotics in plant pathogen control Inhibit the growth of plant pathogens Control diseases like bacterial blight and fungal infections Act as biocontrol agents	El Enshasy et al. (2020)
10.	*Klebsiella*	"Gibberellins and auxins generate growth hormones" Enhance plant growth and yields Promote vitamins uptake and root formation	Plant health benefits Inhibits pathogen growth Prevents diseases like root rot and leaf spots Competes with pathogens for resources Promotes healthier plants	Eid et al. (2021)

(continued)

Table 6.1 (continued)

S. No	Marine microbes	Role in plant growth promotion	Role in disease control	Reference
11.	*Serratia marcescens*	Hormones in plant growth Produce growth-promoting hormones Stimulate root elongation and cell division Optimize nutrient uptake Enhance plant health and yield	Chitinase mechanism Produces enzymes for chitin breakdown Inhibits pathogenic fungi Reduces fungal diseases	Kshetri et al. (2019)
12.	*Bacillus pumilus*	Growth hormones in plant development Stimulate root growth and cell division Improve plant development Solubilize phosphate	Produces surfactin and iturin and disrupts fungi cell membranes Protects plants from diseases	Gutiérrez-Mañero et al. (2001)
13.	*Actinomyces* spp.	Enhances soil fertility Supports plant growth Contributes to healthier, vigorous plants Reduces synthetic fertilizer needs	Antibiotics that inhibit soil-borne pathogen growth protect plants from diseases like root rot	Avilala and Golla (2019)
14.	*Halomonas* spp.	"Salt-living microorganism enhances plant salt tolerance" Produces osmoprotectants and stress-related compounds Beneficial in saline soils	Inhibits plant pathogen growth and protects crops from diseases Acts as a biocontrol agent	Desale et al. (2014)
15.	*Vibrio* spp.	*Vibrio* spp.'s impact on plant nutrition Enhances nutrient uptake Improves root growth and function Promotes healthier plants and higher yields Produces growth hormones	*Vibrio* sp. bacteriocins: Plant disease control Inhibits pathogenic bacteria growth Disrupts cell membranes and metabolic processes Effective biological control agent	Midhun and Arun (2023)

(continued)

Table 6.1 (continued)

S. No	Marine microbes	Role in plant growth promotion	Role in disease control	Reference
16.	*Marinobacter* spp.	Plant growth enhancement Solubilizes phosphate Produces growth hormones Promotes root development Facilitates efficient nutrient absorption Enhances plant health and resilience	Antimicrobial compounds in plant health Protects plants from soil diseases Reduces nutrient competition Inhibits pathogenic microorganisms	Branchu et al. (2017)
17.	*Planococcus* spp.	Root system's role in plant health Promotes nutrient absorption and growth hormones Enhances plant productivity	Produces antibiotics to inhibit soil pathogenic microorganisms Controlling diseases like root rot and wilts	Waghmode et al. (2017)
18.	*Pseudoalteromonas* spp.	"Growth hormone production enhances plant growth" Enhances nutrient absorption Solubilizes insoluble phosphate compounds Increases phosphorus availability for plant growth	Antimicrobial peptides in crop protection Inhibits pathogen growth Protect crops from diseases Acts as a natural biocontrol agent	Parrilli et al. (2021)
19.	*Shewanella* spp.	Promoting sustainable agriculture Solubilizes phosphorus Enhances energy transfer, photosynthesis, and root development Fixes nitrogen Reduces chemical fertilizer use	Plant antimicrobial compounds and protects against root rot and fungal infections Competes with harmful microbes for nutrients and space Enhances soil-borne disease resistance	Timofeeva et al. (2022a, b)
20.	*Flavobacterium* spp.	Produces growth hormones like auxins and cytokinins Enhances plant vigor and root system strength Improves root structure and nutrient uptake Leads to better growth and higher yields	Suppresses harmful bacteria and fungi Protects plants from diseases Acts as a biocontrol agent	Bernardet and Bowman (2006)

(continued)

Table 6.1 (continued)

S. No	Marine microbes	Role in plant growth promotion	Role in disease control	Reference
21.	*Arthrobacter* spp.	Hormones for plant growth Generate cytokinins and auxins Promote root and shoot development Auxins aid nutrient absorption, branching, and root elongation Cytokines aid phosphate solubilization, leaf growth, and cell division	Produces antimicrobial compounds Inhibits harmful pathogen growth in soil Protects crops from soil-borne diseases Competes with pathogens for nutrients and space Enhances plant defense mechanisms	Jones and Keddie (2006)
22.	*Nocardiopsis* spp.	Produces siderophores, increasing iron availability in iron-deficient environments Enhances plant nutrient uptake, improving root and shoot development Efficient use of soil resources benefits plant growth	Produces antibiotics to inhibit pathogenic microbes Controls plant diseases like root rot and leaf blights Serves as a biocontrol agent, reducing disease incidence	Goodfellow and Williams (1983.)
23.	*Micrococcus* spp.	Enhancing plant growth Improves nutrient uptake Promotes root development Produces growth hormones	Protects plants from soil-borne pathogens Disrupts harmful microbes, reducing root rot and blight Serves as a natural biocontrol agent, maintaining soil microbial balance	Guo et al. (2020)
24.	*Rhodococcus* spp.	Produces growth hormones like auxins and phosphate solubilization Promotes plant development and nutrient absorption Aids in phosphate solubilization, especially in phosphorus-limited soils	Inhibits pathogenic bacteria and fungi in soil Protects plants from root rot and fungal infections Competes with harmful pathogens in the rhizosphere	Francis et al. (2006)

(continued)

Table 6.1 (continued)

S. No	Marine microbes	Role in plant growth promotion	Role in disease control	Reference
25.	*Mycobacterium* spp.	Role in nitrogen fixation Converts atmospheric nitrogen into ammonia Supports sustainable agricultural practices	Protects plants from soil-borne diseases Competes with pathogenic microorganisms for resources Reduces disease development Promotes overall plant health	Swarnalakshmi et al. (2016)
26.	*Saccharomyces cerevisiae*	Produces growth hormones and enhances nutrient uptake Promotes root and plant growth yields	Produces antimicrobial peptides that protect plants from pathogens Inhibiting growth and spreading diseases	Pereira et al. (2021)
27.	*Candida* spp.	Enhances plant nutrient absorption Improves enzyme secretion Develops root system Enhances growth and productivity in nutrient-deficient soils	Antimicrobial compounds in plants Inhibits pathogenic bacteria and fungi Protects plants from diseases Reduces reliance on chemical pesticides	Calderone and Fonzi (2001)
28.	*Debaryomyces hancenii*	Improving plant growth in salt-affected soils Produces osmotic regulators Produces compatible solutes Produces growth hormones	Antimicrobial compounds in plants Protects plants from pathogens Inhibits harmful growth Reduces disease incidence	Kurtzman and Fell (2006)
29.	*Aureobasidium pullulans*	Plant growth hormone overview Promotes plant health by solubilizing phosphate Enhances nutrient uptake Supports growth in limited phosphorus-rich soils	Antifungal compounds in plant health Inhibits pathogenic fungi growth Protects plants from diseases like powdery mildew and root rot Acts as a biocontrol agent in the root system	Zalar et al. (2008)

(continued)

Table 6.1 (continued)

S. No	Marine microbes	Role in plant growth promotion	Role in disease control	Reference
30.	*Rhodosporidium toruloides*	Plant production hormone Enhances plant development Adapts to environmental conditions Improves nutrient uptake Supports plant health in nutrient-poor soils	Biocontrol agent enhances plant health Competes with pathogenic microorganisms for nutrients Provides space for plant health	Wang et al. (2016)
31.	*Cladosporium* spp.	"Promoting plant growth and health" Promotes growth hormones Solubilizes phosphate Enhances nutrient uptake in phosphorus-limited soils	Biocontrol agent in plant health Competes with pathogenic fungi Promotes plant health and disease resistance	Bensch et al. (2015)
32.	*Alternaria* spp.	Growth hormones and plant health Regulates root elongation and cell division Enhances nutrient uptake Promotes healthier, productive plants	Antimicrobial compounds in plant disease control Inhibit harmful pathogens Control leaf spots and blights Act as a biocontrol agent, limiting pathogenic microorganisms' resource competition	Thomma (2003)
33.	*Mortierella* spp.	Solubilization of phosphate Produces growth hormones like auxins and gibberellins Regulates plant growth processes Aids in phosphate solubilization Enhances nutrient uptake Supports plant growth in phosphorus-deficient soils	Inhibit the growth of pathogenic fungi Protect plants from diseases like root rot and leaf blight Act as biocontrol agents, enhancing plant health and disease resistance	Wagner et al. (2013)

(continued)

Table 6.1 (continued)

S. No	Marine microbes	Role in plant growth promotion	Role in disease control	Reference
34.	*Aspergillus niger*	Regulating root and shoot development Enhances nutrient uptake Improves plant health and productivity	Inhibits pathogenic microorganisms Protects plants from diseases Acts as a biocontrol agent	Samson et al. (2014)
35.	*Fusarium* spp.	"Produces growth hormones in low-phosphorus soils" Accelerates plant growth Improves nutrient uptake	Prevents pathogenic fungus formation Shields plants against root rot Serves as a biocontrol agent Improves plant health and disease resilience	Sushma et al. (2013)
36.	*Chaetomium* spp.	Stimulate plant growth processes like root elongation and shoot development Improve nutrient uptake by encouraging root development and expanding absorption surface area	Inhibits the growth of bacteria and fungi Defends plants from diseases Acts as a biocontrol agent, increasing plant vitality and resilience to disease	Moya et al. (2016)
37.	*Botrytis cinerea*	Growth hormones in plants Controls plant growth processes Improves nutrient uptake Crucial for photosynthesis and energy transfer	Protects plants from diseases and acts as a biocontrol agent Competes with pathogenic fungi for resources	Orozco-Mosqueda et al. (2023)
38.	*Verticillium* spp.	Produces growth hormones like auxins and gibberellins Increases absorption surface area and encourages root growth	Protects against diseases and acts as a biocontrol agent Prevents bacteria and fungi growth	Li et al. (2018)
39.	*Phytophthora* spp.	Produces growth hormones and solubilizes phosphate, improves nutrient intake, and supports plant growth in low-phosphorus soils	Protects plants from pathogenic fungi Promotes plant health by competing with pathogenic microbes	Sharma et al. (2015)

(continued)

Table 6.1 (continued)

S. No	Marine microbes	Role in plant growth promotion	Role in disease control	Reference
40.	*Colletotrichum* spp.	Plant growth hormones Controls plant growth Increases absorption surface area Encourages root growth Enhances plant health and productivity	Antifungal compounds in plant health, inhibit fungal growth Protects plants from diseases Acts as a biocontrol agent Enhances plant health and disease resistance	Svoboda et al. (2021)
41.	*Phoma* spp.	Plant growth hormones Produce auxins and gibberellins Control plant health and nutrient intake Support shoot growth and root elongation	Prevents plant diseases Reduces illness frequency Enhances plant health and disease resistance	Avinash and Rai (2014)
42.	*Sclerotinia sclerotiorum*	Growth hormone development Regulates plant growth Improves nutrient uptake Enhances absorption surface area	Protects plants from pathogens Inhibits bacteria and fungi growth Acts as a biocontrol agent, reducing disease incidence	Chen et al. (2021)
43.	*Botryosphaeria* spp.	Growth hormones and phosphorus support plant growth Improves energy transfer Enhances root development Enhances photosynthesis Improves nutrient uptake	Antifungal compounds in plant health Protects plants from diseases and acts as a biocontrol agent Competes with pathogenic microorganisms	Khan et al. (2020)
44.	*Alternaria alternaria*	Plant growth regulation Produces growth hormones, auxins, and gibberellins Regulates plant health Enhances nutrient uptake through root development	Protects plants from diseases and acts as a biocontrol agent Reduces pathogen spread	Prakash et al. (2022.)

(continued)

Table 6.1 (continued)

S. No	Marine microbes	Role in plant growth promotion	Role in disease control	Reference
45.	*Phytophthora infestans*	Improves nutrient uptake Promotes plant growth Creates growth hormones like auxins and gibberellins	Protect plants from harmful fungi and microbes as biocontrol agents Enhance plant health and resource management	Kumbar et al. (2019)
46.	*Ulocladium* spp.	Produce plant growth hormones Release gibberellins and auxins Vital for plant development and nutrient intake Encourage root growth and increase surface area Improve plant health and productivity	Protect plants from pathogens and inhibit growth Act as biocontrol agents and reduce pathogen spread	Yadav and Saxena (2018)
47.	*Trichoderma viride*	Plant health improvement Enhances root biomass Improves nutrient uptake and absorption Synthesizes phytohormones for root development	Enhance plant health and protect against root rot diseases	Stewart and Hill (2014)
48.	*Aspergillus flavus*	Used as soil pretreatment Produces bioactive chemicals for plant growth Enhances soil fertility and health Enhances organic matter and beneficial microbes	Generates antimicrobial chemicals in plant health and serves as a biocontrol agent	Hung and Rutgers (2016)
49.	*Cyanobacteria*	Converts atmospheric nitrogen into ammonia Enhances soil fertility and growth nutrients Solubilizes phosphate for energy transfer, photosynthesis, and root development	Inhibit bacteria and fungi growth and act as biocontrol agents Reduce harmful pathogens Support healthier plant growth	Kumar et al. (2019)

(continued)

Table 6.1 (continued)

S. No	Marine microbes	Role in plant growth promotion	Role in disease control	Reference
50.	*Pantoea agglomerans*	Solubilize phosphate for photosynthesis, energy transfer, and root development Includes auxins, gibberellins, and cytokinins	Lower disease frequency Improve plant health Function as biocontrol agents Restrict pathogen spread	Premachandra et al. (2016)

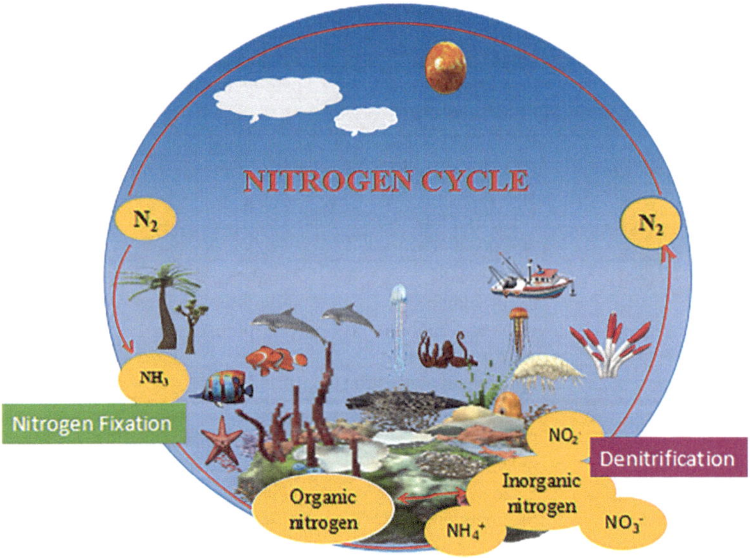

Fig. 6.1 Nitrogen cycle

nitrogen availability in the soil, which benefits plant growth and soil fertility via different farming methods. Fixing nitrogen in plants such as legumes and nonleguminous plants reduces the need for synthetic nitrogen fertilizers, promoting sustainable agricultural practices (Flora et al. 2021).

This process supports plant growth, enhances soil fertility, and plays a critical role in sustainable agriculture by reducing dependency on chemical fertilizers (Fig. 6.2). Understanding and leveraging nitrogen fixation can significantly improve crop productivity and environmental health.

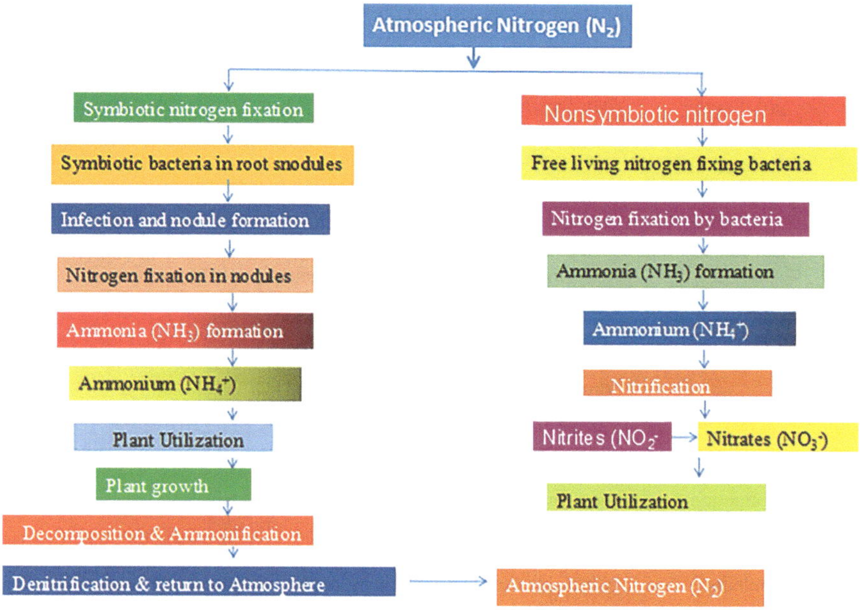

Fig. 6.2 Steps involved in nitrogen fixation

6.3.2 Phosphate Solubilization

Phosphorus, which exists in soil either in salts of minerals or organic molecules, is an essential ingredient for plant growth. As a viable substitute for chemical fertilizers, plant growth-promoting bacterial inoculants can be used as biofertilizers. During the synthesis of organic acids, bacteria can solubilize phosphate complexes in the soil, increasing the amount of phosphorus that is accessible for plant absorption. Endophytic bacteria can yield gluconic acid, solubilizing inorganic phosphate and enhancing the growth of plants (Billah et al. 2019). Marine bacteria such as *Pseudomonas* and *Bacillus* species can solubilize phosphates, increasing their availability to plants and promoting root development (Rawat et al. 2021).

Phosphorus deficiency is a major issue in agriculture worldwide, affecting more than 40% of arable land. To address this problem, fertilizers are commonly used, but interest in the use of biofertilizers as more sustainable and cost-effective solutions is increasing. Phosphate-solubilizing microorganisms (PSMs) are promising options, as they can convert insoluble phosphate into a form that plants can readily use. These organisms, including various bacteria and fungi, can live in the rhizosphere or within plant tissues. PSMs employ different mechanisms to solubilize and release phosphorus while also promoting plant growth through the production of hormones and biocontrol agents (Emami-Karvani and Chitsaz-Esfahani 2021) (Table 6.2).

The following are the primary processes that lead to phosphate solubilization:

Table 6.2 Phosphate-solubilization promoters and mechanisms

S. No	Phosphate-solubilization promotors	Mechanism	Reference
1.	Organic acid production	PSMs produce organic acids, including lactic acid, citric acid, oxalic acid, and gluconic acid. These acids help solubilize phosphate minerals by lowering the soil's pH	Rashid et al. (2004)
2.	Enzyme secretion	PSMs secrete enzymes that hydrolyze organic phosphorus molecules into inorganic phosphate, including phosphatases and phytases	Silva et al. (2023)
3.	Proton efflux	Protons (H^+), which are released into the soil by certain microbes, can directly solubilize phosphate minerals by raising the acidity of the soil. The synthesis of organic acid is often linked to this process	Kuhad et al. (2011)
4.	Siderophore production	Siderophores have sequestration of iron, which can decrease the development of insoluble iron-phosphate complexes and increase phosphate availability	Li and Liu (2021)
5.	Chelation	Chelation agents produced by microorganisms bind to metal ions, including calcium, iron, and aluminum	Ramke and Jeyaraman (2022)
6.	Carbon source utilization	Microorganisms generate acids and other substances that aid in the solubilization of phosphate by using carbon sources, such as sugars	Rawat et al. (2021)

Microorganism phosphate solubilization is a vital natural process that increases phosphorus availability and promotes plant growth. By comprehending and using these mechanisms, agricultural methods have become more efficient and sustainable, lowering the need for artificial fertilizers and enhancing soil health.

6.3.3 Production of Phytohormones

Major phytohormone groups, such as auxin, cytokinin, and gibberellin, are produced by marine bacteria such as *Agrobacterium*, *Pseudomonas*, *Azospirillum*, *Bacillus*, *Enterobacter*, *Pseudomonas*, *Proteus*, *Klebsiella*, and *Mycorrhizal*. These growth-promoting bacteria function as diazotrophs for nitrogen fixation and as producers of phytohormones by supporting the indole-3-acetic acid pathways that biosynthesize auxin and cytokinin. They are vital to agriculture and biotechnology since they are essential to plant growth (Oyedele et al. 2024).

6.3.3.1 Auxins

Some marine bacteria, such as *Azospirillum* spp., *Cyanobacteria*, *Microalgae*, and fungi, produce auxins, which are plant hormones that regulate several plant growth processes and engage with different plant signaling pathways. Auxins have a direct effect on plant root architecture, facilitating improved nutrient and water uptake by plants under a variety of stressful circumstances (Tariq and Ahmed 2022). Through biosynthetic pathways, auxins, significantly indole-3-acetic acid (IAA), are produced from tryptophan. Tryptophan aminotransferase converts tryptophan to indole-3-pyruvic acid (IPA), and the YUCCA enzyme subsequently converts IAA to IAA. This is the first step in the critical pathway. Auxins are involved in tropisms such as phototropism and gravitropism in plants, as well as promoting cell elongation, preserving apical dominance, and stimulating root initiation (Chandler 2015).

6.3.3.2 Cytokinin

Cytokinins are produced by certain marine microbes, such as *Rhodopseudomonas* spp., *Azospirillum* spp., *Anabaena* spp., *Nostoc* spp., and microalgae, and helps in the proliferation, growth, and diversity of plants. During the processes of the proliferation and expansion stages of leaf cell growth, CKs enhance cell growth and promote cell division. Cytokinin decreases the buildup of sugars during leaf senescence, increases the synthesis of chlorophyll, and increases the amount of time that leaves are photosynthetic (Salimi and Hamedi 2021).

Cytokinins are plant hormones derived from adenine through biosynthesis pathways involving enzymes such as isopentenyl transferase. They play crucial roles in cell division, shoot initiation, delay of leaf senescence, and nutrient mobilization in plants. Cytokinins promote growth and development by stimulating various processes in different plant tissues (Mok 2019).

6.3.3.3 Gibberellin

Marine microbes such as *Azospirillum* spp., *Acinetobacter* spp., *Anabaena* spp., *Nostoc* spp., *Aspergillus* spp., and *Penicillium* spp. can synthesize gibberellins (GAs), which are plant hormones that control a variety of functions related to plant growth, organ development, and reactions to the environment. These stages include the germination of seeds, stem elongation, leaf expansion, and the transition to flowering (Verma et al. 2019).

Gibberellins are produced through biosynthesis pathways involving enzymes such as ent-kaurene synthase and GA 20-oxidase, leading to active forms such as GA1 and GA3. In plants, they promote stem elongation by stimulating cell division and elongation, aid in seed germination by breaking dormancy and inducing enzyme synthesis, and regulate flowering and floral organ development (Hedden and Kamiya 1997).

6.3.3.4 Ethylene

Some marine bacteria, including some strains of *Vibrio* and *Pseudomonas*, create ethylene. The amount of ethylene in plants growing in coastal or marine environments is influenced by this microbial ethylene (Jha et al. 2012). Marine microbes, such as certain bacteria and fungi, play crucial roles in influencing ethylene production and regulation in plants. Ethylene is a plant hormone involved in various physiological processes, and marine microbes can impact ethylene mechanisms in plants through direct production of ethylene, modulation of plant ethylene production, interaction with plant roots, and induction of systemic resistance (Bacon and White 2016). Some marine bacteria produce ethylene themselves, influencing ethylene levels in plants growing in marine or coastal environments. Additionally, many marine bacteria produce 1-aminocyclopropane-1-carboxylic acid (ACC) deaminase, which breaks down the precursor of ethylene in plants, reducing ethylene production and alleviating stress responses in plants. Through their interactions with plant roots, marine bacteria can improve plant growth and stress tolerance by regulating the levels of hormones, particularly ethylene. Additionally, they have the ability to strengthen plant resilience to both biotic and abiotic stressors by fostering systemic resistance. Marine fungi such as endophytic fungi and bacteria such as *Pseudomonas* spp. are examples of microbes that affect ethylene processes in the marine environment. These bacteria can increase overall plant development, encourage the growth of roots and shoots, reduce stress-induced reactions, and increase crop output and quality via agricultural techniques, particularly in coastal and saline-prone areas (Hyde et al. 1998).

6.3.4 Siderophore Production

Siderophores are molecules produced by marine microbes and plants to chelate iron. Plant growth-promoting bacteria in the rhizosphere help plants absorb iron by synthesizing and releasing siderophores. This study explored different bacterial strains that enhance plant growth by producing siderophores; examined the chemical nature of siderophores and their biosynthesis and degradation; and reviewed methods for analyzing siderophore synthesis by bacteria. The benefits of these compounds to plants can support the development of microbiological preparations for sustainable agriculture and improved soil fertility. Siderophore-producing bacteria include *Azotobacter*, *Azospirillum*, *Bacillus*, *Methylobacterium*, *Nocardia*, *Pseudomonas*, *Rhodococcus*, *Serratia*, and *Streptomyces* (Timofeeva et al. 2022a, b). Siderophores are essential for plants to acquire iron, and they include various types, such as catecholates, hydroxamates, carboxylates, and mixed types. Plants can produce siderophores themselves or rely on symbiotic relationships with rhizosphere microbes to access these essential compounds. This microbial symbiosis is crucial for plants such as maize, barley, and wheat, which produce *phytosiderophores* to chelate iron from the soil. The mechanism of siderophore-mediated iron uptake involves secretion, complex formation, and uptake by specific transporters on the root cell membrane (Neilands 1995). Siderophore production is tightly regulated by the iron status of the plant, with transcriptional and hormonal mechanisms playing key roles. The benefits of siderophore production include increased iron uptake, increased stress tolerance, and improved soil health through increased soil fertility. Challenges remain in understanding the diversity of siderophore production across plant species and in harnessing siderophores for biotechnological applications in agriculture (Ahmed and Holmström 2014).

6.3.5 Stress Tolerance

The environmental stresses that affect plants include water scarcity, drought, extreme temperatures, salinity, and air pollution. "Plant stress physiology" is the term used to describe how plants function in unfavorable environments. Auxins, gibberellins, cytokinins, abscisic acid, ethylene, salicylic acid, jasmonates, brassinosteroids, and strigolactones are the main hormones that plants generate, and they play significant roles in directing plant mechanisms associated with infections and abiotic stresses. Some marine microbes help plants withstand abiotic stresses and heavy metal contamination by producing stress-alleviating compounds (Verma et al. 2016).

Plants have various mechanisms to cope with environmental stresses such as drought, salinity, and cold. Physiological mechanisms include stomatal regulation, osmotic adjustment, leaf morphology adjustments, and root system alterations. Biochemical mechanisms involve the production of antioxidants, heat shock proteins, and compatible solutes (Chaves et al. 2009). Molecular mechanisms include stress-responsive gene expression, signal transduction pathways, and microRNAs. Genetic mechanisms focus on stress-tolerant varieties and genetic modifications. Integrated stress tolerance involves crosstalk between pathways and epigenetic regulation. Examples of stress tolerance mechanisms include drought tolerance with deep root systems and osmotic adjustment, salinity tolerance with sodium ion sequestration and compatible solute accumulation, cold tolerance with antifreeze proteins, and altered membrane lipid composition (Sreenivasulu et al. 2007).

6.4 Disease Control Mechanisms

Plant diseases can be significantly controlled by various mechanisms in marine bacteria. They can be applied in the following ways:

6.4.1 *Biofertilizers*

Certain marine microorganisms, such as phosphate-solubilizing and nitrogen-fixing bacteria, increase the availability of nutrients in soil. They facilitate the transformation of atmospheric nitrogen into a form in which plants can use and solubilize phosphate to increase its availability to plant roots and increase plant yield and development without the use of chemical fertilizers, which reduces pollution in the environment (Timofeeva et al. 2023). Natural fertilizers called biofertilizers are made of living microorganisms that increase soil fertility and encourage plant development. A variety of methods are essential for controlling the spread of plant diseases. Biofertilizers include beneficial microorganisms that compete with diseases for nutrients and space, generate antimicrobial chemicals, cause systemic resistance in plants, manufacture lytic enzymes, produce siderophores from biofilms on plant roots, and induce global tolerance in plants. BF indirectly supports disease resistance by enhancing overall plant health and growth (Avis et al. 2008). Examples of biofertilizers include *Rhizobium* spp., *Trichoderma* spp., *Bacillus* spp., and *Azospirillum* spp., each offering unique disease control benefits. *Rhizobium* bacteria improve plant nutrition and resistance to diseases in legume crops, whereas *Trichoderma* fungi have biocontrol properties and induce systemic resistance in plants. *Bacillus* species produce antibiotics, form biofilms, and induce systemic resistance against various pathogens, and *Azospirillum* bacteria promote plant growth and health, reducing plant susceptibility to diseases (Harman et al. 2004).

6.4.2 *Plant Growth-Promoting* Rhizobacteria *(PGPR)*

Rhizobacteria are found in many aquatic environments. By generating phytohormones, improving nutrient uptake, and fostering systemic resistance against infections, plants populate the rhizosphere—the area of soil affected by root secretions—and encourage plant growth (Verma et al. 2019a, b). Plant growth-promoting rhizobacteria (PGPR) are beneficial bacteria that increase plant growth through direct and indirect mechanisms. Direct mechanisms include nutrient solubilization, nitrogen fixation, iron chelation, enzyme production, and organic compound production. These mechanisms promote root growth and overall plant development. Indirect mechanisms involve the biocontrol of pathogens through antibiotic production, lytic enzymes, competition for resources, and induced systemic resistance. PGPR can also help plants tolerate abiotic stresses such as drought and salinity (Lugtenberg and Kamilova 2009). Examples of PGPR include *Rhizobium* spp., *Pseudomonas fluorescens*, *Azospirillum brasilense*, and *Bacillus subtilis*, each with specific mechanisms and effects on plant growth and health. By understanding and utilizing these mechanisms, PGPR could be effectively integrated into sustainable agricultural practices to improve crop yields and resilience

(Bhattacharyya and Jha 2012). PGPR play crucial roles in enhancing plant production and health through various beneficial mechanisms.

6.4.2.1 Antagonism

Some marine microorganisms are hostile to plant diseases. They can produce antimicrobial chemicals, induce systemic resistance in plants, or compete with pathogenic fungi, bacteria, and nematodes to prevent their development and reproduction. Examples include several marine fungal and bacterial species that can inhibit the growth of common plant infections, such as *Phytophthora*, *Rhizoctonia*, and *Fusarium* (Hajek and Eilenberg 2018). According to Heisey et al., antagonism in plants involves interactions between plants and other organisms that are harmful to plants. This can occur through allelopathy, where plants release chemicals to inhibit the growth of neighboring plants; competition for resources such as light and water; antimicrobial compounds that protect against pathogens; herbivorous defense mechanisms; mycorrhizal antagonism; and the ability of parasitic plants to extract nutrients from hosts. Examples include black walnut trees, which produce juglone; eucalyptus, which releases allelopathic chemicals; garlic mustard, which disrupts mycorrhizal relationships; cacti, which uses physical and chemical defenses; and dodder, which parasitizes host plants (Heisey 1990).

6.4.2.2 Induced Systemic Resistance (ISR)

Certain marine microorganisms can induce defense mechanisms in plants by colonizing the rhizosphere or phyllosphere and activating host signaling pathways. This induced systemic resistance enhances the plant's ability to withstand a wide range of diseases and can persist for long periods (Romera et al. 2019). The ISR in plants enhances their defense against pathogens and pests and is activated by beneficial rhizobacteria. ISR begins with the recognition of these bacteria, triggering a signal transduction pathway involving molecules such as jasmonic acid (JA) and ethylene (ET). This leads to systemic signaling throughout the plant, priming it for a quicker and stronger response to pathogen attacks. While ISR does not directly activate defense genes, it primes the plant to produce defensive enzymes and antimicrobial compounds and strengthens cell walls upon attack. The key components of the ISR include beneficial rhizobacteria such as *Pseudomonas fluorescens*, JA, ET, and priming of the plant's defense system. Examples of applications include the use of bacteria such as *P. fluorescens* for crop protection against pathogens and the reduction of pesticide use through biological control (Van Wees et al. 2008).

6.4.2.3 Biocontrol Products

Researchers are working to create commercially viable marine microbe-based biocontrol products for use in agriculture. These products are safe for the environment, nontoxic to people, and useful to beneficial creatures. They may contain live microbial strains or their metabolites that specifically target plant diseases for several agricultural diseases, and biocontrol formulations are being developed to provide farmers with alternative methods to chemical pesticides (Usta 2013). Biocontrol products, also known as biological control agents, are natural or modified organisms used to manage plant pests and diseases. These agents include beneficial bacteria, fungi, insects, or nematodes, which work through mechanisms such as predation, parasitism, competition, and the induction of plant defense responses. Predatory insects and mites feed directly on pest organisms, whereas parasitic organisms lay their eggs in or on pest insects, eventually killing the host (Glare et al. 2012). Microbes produce antibiotics or toxic metabolites to inhibit or kill plant pathogens, while certain fungi can directly attack and break down pathogenic fungi. Beneficial microbes compete with pathogens for nutrients and space, preventing pathogen establishment. Some biocontrol agents induce enhanced plant defenses, providing long-lasting protection against pathogens. Examples of biocontrol products include *Bacillus thuringiensis* (Bt), *Trichoderma* spp., *Beauveria bassiana*, and *Pseudomonas fluorescens*, which target specific pests and pathogens through various mechanisms (Harman et al. 2004). These biocontrol products offer environmentally friendly alternatives to chemical pesticides for managing plant pests and diseases.

6.4.2.4 Biological Soil Amendments

The application of marine microorganisms to organic fertilizers or soil amendments might increase soil health and reduce the occurrence of soil-borne illnesses. These microbial additions have the ability to increase the diversity and activity of microorganisms in the soil, fostering a competitive ecosystem that prevents the spread of dangerous infections. To further reduce the prevalence of sickness, certain marine bacteria can also break down the cell walls of dangerous fungi or create enzymes that break down fungal poisons (Shu et al. 2022). Biologically modified soil amendments are natural or organic compounds added to soil to increase its physical properties, sterility, and biological activity, increasing plant growth and health. They work through various mechanisms, such as nutrient enrichment, soil structure improvement, enhanced microbial activity, disease suppression, enhanced water retention and management, and enhanced root growth and development. For example, organic matter decomposition releases essential nutrients, organic amendments improve the soil structure, microbial inoculants increase nutrient cycling, and mycorrhizal fungi increase nutrient uptake (Kibblewhite et al. 2008). Common biological soil amendments include compost, manure, biochar, microbial inoculants such as *Rhizobium*, and mycorrhizal fungi. In comparison with soil, which is rich in nutrients and microbial activity, biochar improves soil porosity, manure enhances soil fertility, and microbial inoculants such as *Rhizobium* fix atmospheric nitrogen in legumes (Altieri and Nicholls 2003).

These amendments have been applied in agriculture, landscaping, and horticulture to improve soil health, fertility, and plant growth. By using these natural materials, farmers can reduce the need for synthetic fertilizers, improve the soil structure, increase water retention, and promote healthier plant growth through symbiotic relationships and disease suppression mechanisms.

6.4.2.5 Biological Seed Treatments

Introducing beneficial marine microorganisms into seeds before planting can help create a healthy rhizosphere microbiome and shield seedlings from soil-borne infections. Microbial inoculants can be used to coat seeds, or microbial solutions can be applied to seedling roots to aid in germination. This early inoculation aids in the formation of a protective microbial community surrounding the growing roots, avoiding the colonization of pathogens and promoting plant health from the earliest stages of growth (Cardarelli et al. 2022). Biological seed treatments involve the application of beneficial microorganisms or natural substances to seeds before they are planted to improve their health, enhance germination, and protect against diseases and pests. These treatments work through mechanisms such as disease suppression, induced resistance, enhanced nutrient availability, growth promotion, and stress tolerance, and examples include *Trichoderma*, *Bacillus*, *Rhizobium*, and mycorrhizal fungi (Van Loon 2007). *Trichoderma* produces enzymes and antibiotics that suppress fungal pathogens and induce systemic resistance, whereas *Bacillus* produces antimicrobial compounds, solubilizes phosphate, and increases nutrient availability. *Rhizobium* fixes atmospheric nitrogen in symbiosis with legume roots, providing essential nitrogen to the plant, and mycorrhizal fungi form symbiotic associations with plant roots, increasing nutrient and water uptake, particularly phosphorus. These treatments are applied to seeds of various crops, such as corn, wheat, and vegetables, to protect against soil-borne diseases, improve plant vigor, enhance germination, protect against pathogens, improve nutrient uptake, increase nitrogen fixation, strengthen crop yield, and improve stress tolerance (Kloepper and Ryu 2006). Biological germination is a sustainable and environmentally friendly way to improve plant health and productivity in agriculture.

6.4.2.6 Synthesis of Lytic Enzymes

Certain marine microorganisms generate enzymes that break down pathogen cell walls. The structural elements of bacterial and fungal cell walls can be broken down by enzymes such as chitinases, glucanases, and proteases, which results in parasite death (Gajera and Vakharia 2012). The synthesis of lytic enzymes in plants is vital for defense against pathogens, as these enzymes break down the cell walls of invading microorganisms. Examples of lytic enzymes include chitinases, glucanases, and proteases, which target specific components of pathogen cell walls. Plants have pattern recognition receptors (PRRs) that identify pathogens, triggering defense responses such as the synthesis of lytic enzymes (van Loon et al. 2006). The activation of the mitogen-activated protein kinase (MAPK) pathway and the production of phytohormones such as salicylic acid (SA) and jasmonic acid (JA) regulate the expression of genes encoding lytic enzymes. Specific transcription factors, such as proteins, regulate the expression of genes and enzyme synthesis, which leads to the formation of lytic enzymes that are transported to infection sites. Chitinases break down chitin in fungal cell walls, glucanases degrade β-glucans, and proteases disrupt pathogen cellular structures. Synergistic effects occur when lytic enzymes work together, enhancing antimicrobial effects (Jashni et al. 2015).

6.4.2.7 Integrated Pest Management (IPM)

The efficiency and sustainability of disease management techniques can be improved by integrating marine microbe-based disease control techniques into integrated pest management (IPM) programs. By integrating cultural, biological, and chemical management strategies, growers can limit their dependency on synthetic pesticides while maintaining effective disease control and ecological balance (Dara 2019). Integrated pest management (IPM) is a comprehensive approach to pest control that combines various strategies to effectively manage pests while minimizing damage to plants and the environment. The mechanisms of IPM include monitoring pest populations, using preventive cultural practices such as crop rotation and sanitation, employing biological controls such as natural enemies and habitat management, utilizing mechanical and physical controls such as barriers and traps, judiciously using chemical controls such as selective and botanical pesticides, enhancing host plant resistance through genetic and induced resistance, and implementing behavioral controls such as pheromones and repellents. Examples of IPM applications include managing corn rootworms through cultural practices, biological control with entomopathogenic nematodes, and targeted chemical control based on monitoring. Scientists are now investigating novel microbial strains, formulations, and application strategies to maximize the effectiveness of marine microorganisms in agricultural settings to control plant diseases. Innovation in this area is being driven by the growing demand for sustainable and environmentally friendly alternatives to chemical pesticides, as knowledge of the effects of these products on human health and the environment has developed (Fig. 6.3).

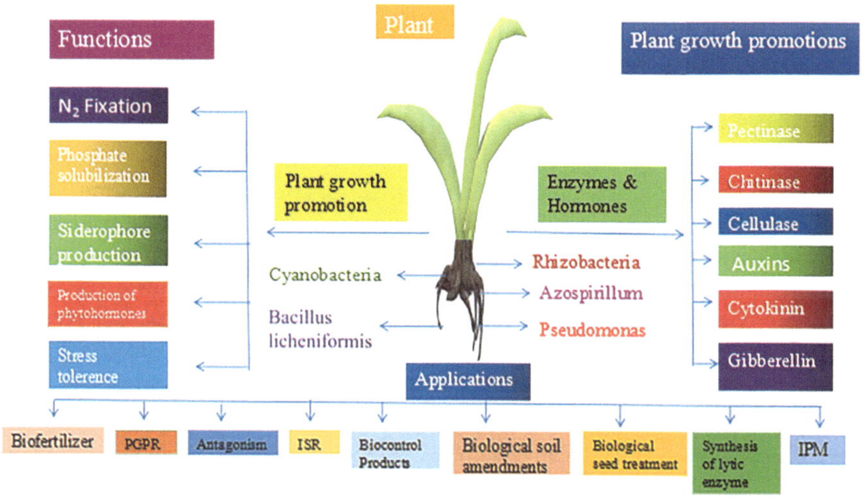

Fig. 6.3 Different marine microbes involved in plant growth promotion and applications

6.5 Applications of Marine Microbes in Agriculture

Bioprospecting marine microbes can improve agricultural productivity and fight plant diseases, which is promising for sustainable agriculture. Identifying, isolating, describing, and utilizing advantageous microorganisms in agricultural systems are all agricultural applications.

6.5.1 Biofertilizers

Marine microorganisms, such as diazotrophs and phosphate-solubilizing bacteria, play crucial roles in the cycling of nutrients in the environment. They are being explored as potential solutions for sustainable agriculture through the development of biofertilizers, which can increase crop productivity and soil health and contribute to climate change mitigation. These biofertilizers can increase nutrient availability, stimulate plant growth, control plant pathogens, and enhance soil health through increased microbial diversity and organic matter decomposition (Ortega Pérez et al. 2023). Applications include horticulture and rice farming; saline agriculture techniques such as smart delivery systems and Internet of Things (IoT) integration; urban and vertical farming; and climate-resilient crop production (Ayaz et al. 2019). Marine microorganisms are useful tools in sustainable agriculture because of their ability to persist in extreme environments and generate a wide range of bioactive compounds (Aswani et al. 2020). Farmers can reduce their need for artificial fertilizers while enhancing soil health and plant uptake of nutrients by inoculating crops with these beneficial marine bacteria.

6.5.2 Biostimulants

Plant growth and development can be stimulated by bioactive compounds produced by marine microorganisms, such as phytohormones and secondary metabolites. These substances might help plants proliferate their roots, become more resilient to stress, and absorb nutrients more effectively. To support sustainable farming practices, biostimulants generated from marine microbes provide an eco-friendly substitute for synthetic growth regulators (De Vasconcelos and Chaves 2019). Biostimulants are microorganisms that are used to increase plant growth, resilience, and yield. Marine microbes, including bacteria, fungi, and microalgae, can promote plant growth (Verma et al. 2019a, b). These microbes function as plant growth promoters through mechanisms such as solubilizing nutrients, fixing nitrogen, producing phytohormones, increasing stress tolerance, and suppressing diseases. Marine microbes can be applied to plants through soil amendments, seed treatments, foliar sprays, or hydroponic systems. The benefits of using marine microbes as biostimulants include increased plant growth, increased yield, sustainable agricultural practices, improved soil health, and stress resilience (Rai et al. 2021). Continued research and development are essential to fully harness their potential and integrate them into mainstream agricultural practices.

6.5.3 Biocontrol Agents

Marine bacteria produce antimicrobial substances that prevent plant diseases from growing. Scientists have discovered biocontrol agents that use these innate defensive systems to manage viral, bacterial, and fungal diseases in crops. These agents can be combined to create biopesticides or applied as seed treatments, reducing the need for chemical fungicides and antibiotics (De Vasconcelos and Chaves 2019). Marine microbes such as *Pseudomonas*, *Bacillus*, *Streptomyces*, and *Trichoderma* are effective biocontrol agents because they produce antibiotics, enzymes, and other compounds that inhibit pathogen growth. These agents are applied in agriculture through seed treatment, soil amendment, foliar application, and root drenching to protect plants from diseases caused by soil-borne and foliar pathogens (Babbal and Khasa 2017).

6.5.4 Bioremediation

Certain marine bacteria can detoxify and break down environmental toxins in addition to encouraging plant development and managing illnesses. The application of microbial consortia sourced from marine environments in bioremediation has demonstrated potential in addressing soil and water contamination due to organic pollutants, heavy metals, and agrochemicals. The use of marine microbes aids in the rehabilitation of damaged ecosystems and supports sustainable agricultural practices (Bala et al. 2022). Bioremediation involves the use of living organisms, particularly marine microbes, to remove or neutralize pollutants in the environment. Marine microbes aid in plant bioremediation by degrading pollutants through enzymatic breakdown and cometabolism. They can also bioaccumulate heavy metals and pollutants, transform inorganic pollutants, and enhance phytoremediation (Husain et al. 2022). Marine microbes produce biosurfactants to increase the efficiency of bioremediation by emulsifying and solubilizing pollutants. Specific marine microbes, such as *Alcanivorax* spp., *Marinobacter* spp., *Pseudomonas* spp., *Vibrio* spp., and *Halomonas* spp., play vital roles in bioremediation, hydrocarbon degradation, and biosurfactant production. Strategies for applying marine microbes in bioremediation include bioaugmentation, biostimulation, phytoremediation, and bioreactor systems to optimize pollutant degradation (Gidudu and Chirwa 2022). By assisting in the detoxification of pollutants and promoting healthier plant growth, marine microbes play crucial roles in restoring polluted environments and creating sustainable ecosystems.

6.5.5 Climate Change Adaptation

Marine bacteria, archaea, fungi, and protists are essential for plants to adapt to climate change. They help plants withstand stressors by enhancing stress tolerance through symbiotic relationships, producing stress-related hormones, assisting in nutrient acquisition and utilization, promoting disease resistance, modifying the environment, facilitating metabolic and genetic adaptations, and mitigating the effects of climate change (Kour et al. 2024). Marine microbes can also increase nutrient availability through nitrogen fixation and phosphate solubilization, improve disease resistance through biocontrol mechanisms, and modify the environment by forming biofilms and improving the soil structure. They can also facilitate genetic adaptations through horizontal gene transfer and produce metabolites that increase plant growth and stress resistance (Saeed et al. 2021; Nagpal et al. 2022).

6.6 Challenges and Future Directions

Marine microbes can promote plant development and manage diseases, but there are still obstacles to overcome. Subsequent studies should address these challenges, seize newly available prospects, and investigate their distinct biological characteristics.

6.6.1 Identification and Isolation

Identifying beneficial marine microorganisms with particular characteristics important for promoting the growth of plants and controlling disease is one of the challenges. Because of the scope and variety of marine ecosystems, identifying microbial strains with the right traits might be challenging. Enhanced techniques for sorting and separating marine microorganisms, along with cutting-edge genomics technology, can increase the effectiveness of bioprospecting endeavors.

6.6.2 Characterization and Optimization

To comprehend their modes of action and maximize their effectiveness in agricultural settings, comprehensive characterization and optimization are needed once potential microbial candidates have been identified. This entails researching the genetic composition, metabolic processes, and ecological interactions of marine microbes in addition to refining formulation techniques and culture conditions for realistic uses.

6.6.3 Field Efficacy and Scalability

There are issues with efficacy, consistency, and scalability when moving from lab research to field applications. Under a variety of environmental circumstances and agricultural techniques, such as varying soil types, temperatures, and cropping systems, marine microorganisms must exhibit dependable performance.

6.6.4 Commercialization and Regulatory Approval

The commercialization of microbial products for use in agriculture requires a discussion of complicated regulatory frameworks, safety evaluations, and consumer acceptability. For small-scale innovators, registering and labeling biofertilizers, biopesticides, and biostimulants made from marine microbes is a time-consuming and expensive process that hinders their ability to enter the market.

6.7 Future Directions

Integration of Multiomics Approaches The functional capacities of marine microbes and future research paths will integrate multiple techniques, such as transcriptomics, proteomics, metabolomics, and genomics. Researchers have identified new microbial candidates for agricultural applications and clarified the molecular mechanisms underlying plant–microbe interactions by integrating omics data with bioinformatic analysis and systems biology techniques.

Microbiome-Based Strategies Developing strategies to improve plant health and productivity requires an understanding of the intricate relationships that exist within the plant microbiome, particularly those between marine and native soil bacteria. Future studies might concentrate on utilizing the complementary qualities of various microbial communities, including marine and terrestrial microbes, for disease prevention and sustainable agricultural production.

Bioremediation and Ecosystem Services Future research in bioprospecting may look at the wider range of ecosystem services that marine microbes offer, such as the bioremediation of damaged soils and water bodies, in addition to their ability to promote plant growth and prevent disease. Bioprospecting marine microbes can improve agriculture by increasing plant growth, improving soil health, and reducing chemical inputs. However, challenges in identification, characterization, and field application require interdisciplinary collaboration, innovative technologies, and strategic investments. Future directions include omics-based approaches, synthetic biology, microbiome engineering, and ecosystem services.

6.8 Conclusion

Marine microbes offer a promising avenue in agricultural research for enhancing plant growth promotion, disease control, and sustainability in farming systems. They can be used in biofertilizers, biostimulants, biocontrol agents, bioremediation agents, and adaptations to abiotic stress. By leveraging their natural capabilities, farmers can improve soil fertility and nutrient uptake efficiency, suppress plant diseases, remediate environmental pollution, and increase resilience to climate-related stressors. Furthermore, marine microbes offer a natural solution for disease control. Many of these microorganisms produce antimicrobial compounds that protect plants from various pathogens, including bacteria, fungi, and viruses. This biocontrol potential reduces the need for chemical pesticides, promoting a more sustainable and environmentally friendly approach to agriculture. The use of marine microbes can also help manage soil health by promoting beneficial microbial communities and suppressing harmful microbial communities, leading to healthier and more productive soils. The integration of marine microbial solutions into agricultural practices could revolutionize crop management by increasing crop resilience to environmental stresses and improving yield quality and quantity. However, to fully realize these benefits, extensive research is needed to isolate, characterize, and optimize the use of these microbes under various agricultural conditions. Field trials and studies are essential for understanding the interactions between marine microbes and plants, ensuring their efficacy and safety. Future directions may involve integrating multiomics approaches, synthetic biology, genetic engineering, and microbiome-based approaches and exploring the ecosystem services provided by marine microbes. This approach can unlock the full potential of marine microbes for sustainable and resilient agricultural systems.

References

Ahmed E, Holmström SJ (2014) Siderophores in environmental research: roles and applications. Microb Biotechnol 7(3):196–208

Alamgir ANM, Alamgir ANM (2018) Secondary metabolites: secondary metabolic products consisting of C and H; C, H, and O; N, S, and P elements; and O/N heterocycles. In: Therapeutic use of medicinal plants and their extracts: volume 2: phytochemistry and bioactive compounds. Springer International Publishing, Cham, pp 165–309

Altaf MM, Imran M, Abulreesh HH, Khan MS, Ahmad I (2018) Diversity and applications of Penicillium spp. in plant-growth promotion. In: New and future developments in microbial biotechnology and bioengineering. Elsevier, pp 261–276

Altieri MA, Nicholls CI (2003) Soil fertility management and insect pests: harmonizing soil and plant health in agroecosystems. Soil Tillage Res 72(2):203–211

Aswani R, Jishma P, Radhakrishnan EK (2020) Endophytic bacteria from the medicinal plants and their potential applications. In: Microbial endophytes. Woodhead Publishing, pp 15–36

Avilala J, Golla N (2019) Antibacterial and antiviral properties of silver nanoparticles synthesized by marine actinomycetes. Int J Pharm Sci Res 10(3):1223–1228

Avinash TS, Rai RV (2014) Antifungal activity of plant growth promoting rhizobacteria against *Fusarium oxysporum* and Phoma sp. of cucurbitaceae. In: Microbial diversity and biotechnology in food security. Springer India, New Delhi, pp 257–264

Avis TJ, Gravel V, Antoun H, Tweddell RJ (2008) Multifaceted beneficial effects of rhizosphere microorganisms on plant health and productivity. Soil Biol Biochem 40(7):1733–1740

Ayaz M, Ammad-Uddin M, Sharif Z, Mansour A, Aggoune EHM (2019) Internet-of-things (IoT)-based smart agriculture: toward making the fields talk. IEEE access 7:129551–129583

Babbal A, Khasa YP (2017) Microbes as biocontrol agents. In: Probiotics and plant health. Singapore, Springer Singapore, pp 507–552

Bacon CW, White JF (2016) Functions, mechanisms and regulation of endophytic and epiphytic microbial communities of plants. Symbiosis 68:87–98

Bala S, Garg D, Thirumalesh BV, Sharma M, Sridhar K, Inbaraj BS, Tripathi M (2022) Recent strategies for bioremediation of emerging pollutants: a review for a green and sustainable environment. Toxics 10(8):484

Bashan Y, De-Bashan LE (2010) How the plant growth-promoting bacterium Azospirillum promotes plant growth—a critical assessment. Adv Agron 108:77–136

Beattie AJ, Hay M, Magnusson B, de Nys R, Smeathers J, Vincent JF (2011) Ecology and bioprospecting. Austral Ecol 36(3):341–356

Bensch K, Groenewald JZ, Braun U, Dijksterhuis J, de Jesús Yáñez-Morales M, Crous PW (2015) Common but different: the expanding realm of Cladosporium. Stud Mycol 82(1):23–74

Berde CV, Bramhachari PV, Berde VB (2022) Plant growth-promoting diazotrophs: current research and advancements. In: Trends of applied microbiology for sustainable economy, pp 207–223
Beringer JE, Johnston AW (1984) The significance of symbiotic nitrogen fixation in plant production. Crit Rev Plant Sci 1(4):269–286
Bernardet JF, Bowman JP (2006) The genus flavobacterium. The prokaryotes 7:481–531
Bhattacharyya PN, Jha DK (2012) Plant growth-promoting rhizobacteria (PGPR): emergence in agriculture. World J Microbiol Biotechnol 28:1327–1350
Billah M, Khan M, Bano A, Hassan TU, Munir A, Gurmani AR (2019) Phosphorus and phosphate solubilizing bacteria: keys for sustainable agriculture. Geomicrobiol J 36(10):904–916
Branchu P, Canette A, Medina Fernandez S, Mounier J, Meylheuc T, Briandet R, Grimaud R, Naitali M (2017) Impact of temperature on Marinobacter hydrocarbonoclasticus SP17 morphology and biofilm structure during growth on alkanes. Microbiology 163(5):669–677
Burris RH, Wilson PW (1945) Biological nitrogen fixation. Annu Rev Biochem 14(1):685–708
Calderone RA, Fonzi WA (2001) Virulence factors of *Candida albicans*. Trends Microbiol 9(7):327–335
Cardarelli M, Woo SL, Rouphael Y, Colla G (2022) Seed treatments with microorganisms can have a biostimulant effect by influencing germination and seedling growth of crops. Plants 11(3):259
Chandler JW (2015) Auxin biosynthesis. In: Amino acids in higher plants. CAB International, Wallingford, pp 340–361
Chaves MM, Flexas J, Pinheiro C (2009) Photosynthesis under drought and salt stress: regulatory mechanisms from whole plant to cell. Ann Bot 103(4):551–560
Chen J, Ullah C, Giddings Vassão D, Reichelt M, Gershenzon J, Hammerbacher A (2021) Sclerotinia sclerotiorum infection triggers changes in primary and secondary metabolism in *Arabidopsis thaliana*. Phytopathology 111(3):559–569
Dara SK (2019) The new integrated pest management paradigm for the modern age. J Integr Pest Manage 10(1):12
Das S, Lyla PS, Khan SA (2006) Marine microbial diversity and ecology: importance and future perspectives. Curr Sci:1325–1335
De Vasconcelos ACF, Chaves LHG (2019) Biostimulants and their role in improving plant growth under abiotic stresses. In: Biostimulants in plant science. IntechOpen, pp 3–16
Desale P, Patel B, Singh S, Malhotra A, Nawani N (2014) Plant growth promoting properties of Halobacillus sp. and Halomonas sp. in presence of salinity and heavy metals. J Basic Microbiol 54(8):781–791
Eid AM, Fouda A, Abdel-Rahman MA, Salem SS, Elsaied A, Oelmüller R, Hijri M, Bhowmik A, Elkelish A, Hassan SED (2021) Harnessing bacterial endophytes for promotion of plant growth and biotechnological applications: an overview. Plants 10(5):935
El Enshasy HA, Ambehabati KK, Hanapi SZ, Dailin DJ, Elsayed EA, Sukmawati D, Malek RA (2020) Trichoderma spp.: a unique fungal biofactory for healthy plant growth. In: Rhizosphere microbes: soil and plant functions. Singapore, Springer Singapore, pp 573–592
Emami-Karvani Z, Chitsaz-Esfahani Z (2021) Phosphorus solubilization: mechanisms, recent advancement and future challenge. In: Soil microbiomes for sustainable agriculture: functional annotation. Springer International Publishing, Cham, pp 85–131
Fatollahi P, Ghasemi M, Yazdian F, Sadeghi A (2021) Ectoine production in bioreactor by Halomonas elongata DSM2581: using MWCNT and Fe-nanoparticle. Biotechnol Prog 37(1):e3073
Flora Y, Rabha P, Shinde A, Jha P, Jobby R (2021) Nonsymbiotic bacteria for soil nitrogen fortification. Sustain Agric Rev 52:417–435
Francis, J.A., Mazur, W. and Bhadeshia, H.K.D.H., 2006. Review type IV cracking in ferritic power plant steels. Materials Science and Technology, 22(12), pp.1387–1395

Francis IM, Stes E, Zhang Y, Rangel D, Audenaert K, Vereecke D (2016) Mining the genome of Rhodococcus fascians, a plant growth-promoting bacterium gone astray. New Biotechnol 33(5):706–717
Fukami J, Cerezini P, Hungria M (2018) Azospirillum: benefits that go far beyond biological nitrogen fixation. AMB Express 8(1):73
Gajera HP, Vakharia DN (2012) Production of lytic enzymes by Trichoderma isolates during in vitro antagonism with *aspergillus Niger*, the causal agent of collar rot of peanut. Braz J Microbiol 43:43–52
Gidudu B, Chirwa EM (2022) Diversity of marine biosurfactant-producing bacteria and their role in the degradation of Heterogenous petrochemical hydrocarbons. In: Marine Surfactants. CRC Press, pp 163–183
Glare T, Caradus J, Gelernter W, Jackson T, Keyhani N, Köhl J, Marrone P, Morin L, Stewart A (2012) Have biopesticides come of age? Trends Biotechnol 30(5):250–258
Goodfellow M, Williams ST (1983) Ecology of actinomycetes. Ann Rev Microbiol 37(1):189–216
Guo S, Wang P, Wang X, Zou M, Liu C, Hao J (2020) Microalgae as biofertilizer in modern agriculture. In: Microalgae biotechnology for food, health and high value products. Springer Singapore, Singapore, pp 397–411
Gutiérrez-Mañero FJ, Ramos-Solano B, Probanza AN, Mehouachi J, Tadeo R, F. and Talon, M. (2001) The plant-growth-promoting rhizobacteria Bacillus pumilus and *bacillus licheniformis* produce high amounts of physiologically active gibberellins. Physiol Plant 111(2):206–211
Haas D, Keel C (2003) Regulation of antibiotic production in root-colonizing pseudomonas spp. and relevance for biological control of plant disease. Annu Rev Phytopathol 41(1):117–153
Hajek A, Eilenberg J (2018) Microbial antagonists combating plant pathogens and plant parasitic nematodes. In: Natural enemies: an introduction to biological control, pp 308–324
Harman GE, Howell CR, Viterbo A, Chet I, Lorito M (2004) Trichoderma species—opportunistic, avirulent plant symbionts. Nat Rev Microbiol 2(1):43–56
Hedden P, Kamiya Y (1997) Gibberellin biosynthesis: enzymes, genes and their regulation. Annu Rev Plant Biol 48(1):431–460
Heisey RM (1990) Allelopathic and herbicidal effects of extracts from tree of heaven (*Ailanthus altissima*). Am J Bot 77(5):662–670
Hung R, Rutgers SL (2016) Applications of aspergillus in plant growth promotion. In: New and future developments in microbial biotechnology and bioengineering. Elsevier, pp 223–227
Husain R, Vikram N, Yadav G, Kumar D, Pandey S, Patel M, Khan NA, Hussain T (2022) Microbial bioremediation of heavy metals by marine bacteria. In: Development in wastewater treatment research and processes. Elsevier, pp 177–203
Hyde KD, Jones EG, Leaño E, Pointing SB, Poonyth AD, Vrijmoed LL (1998) Role of fungi in marine ecosystems. Biodivers Conserv 7:1147–1161
Jashni MK, Dols IH, Iida Y, Boeren S, Beenen HG, Mehrabi R, Collemare J, de Wit PJ (2015) Synergistic action of a metalloprotease and a serine protease from *Fusarium oxysporum* f. sp. lycopersici cleaves chitin-binding tomato chitinases, reduces their antifungal activity, and enhances fungal virulence. Mol Plant-Microbe Interact 28(9):996–1008
Jha B, Gontia I, Hartmann A (2012) The roots of the halophyte Salicornia brachiata are a source of new halotolerant diazotrophic bacteria with plant growth-promoting potential. Plant Soil 356:265–277
Jones D, Keddie RM (2006) The genus arthrobacter. Prokaryotes 3:945–960
Khan MS, Gao J, Chen X, Zhang M, Yang F, Du Y, Moe TS, Munir I, Xue J, Zhang X (2020) The endophytic bacteria bacillus velezensis Lle-9, isolated from Lilium leucanthum, harbors antifungal activity and plant growth-promoting effects. J Microbiol Biotechnol 30(5):668
Kibblewhite MG, Ritz K, Swift MJ (2008) Soil health in agricultural systems. Philos Trans R Soc B Biol Sci 363(1492):685–701
Kloepper JW, Ryu CM (2006) Bacterial endophytes as elicitors of induced systemic resistance. In: Microbial root endophytes, pp 33–52

Kour D, Ahluwalia KK, Ramniwas S, Kumar S, Rustagi S, Singh S, Rai AK, Yadav AN, Ahluwalia AS (2024) Environment and climate change: influence on biodiversity, present scenario, and future prospect. J ApplBiol Biotechnol 12(2):1–12

Kshetri L, Naseem F, Pandey P (2019) Role of Serratia sp. as biocontrol agent and plant growth stimulator, with prospects of biotic stress management in plant. In: Plant growth promoting Rhizobacteria for sustainable stress management: volume 2: Rhizobacteria in biotic stress management. Springer Singapore, Singapore, pp 169–200

Kuhad RC, Singh S, Lata and Singh A (2011) Phosphate-solubilizing microorganisms. In: Bioaugmentation, biostimulation and biocontrol, pp 65–84

Kumar G, Teli B, Mukherjee A, Bajpai R, Sarma BK (2019) Secondary metabolites from cyanobacteria: a potential source for plant growth promotion and disease management. In: Secondary metabolites of plant growth promoting Rhizomicroorganisms: discovery and applications, pp 239–252

Kumbar B, Mahmood R, Nagesha SN, Nagaraja MS, Prashant DG, Kerima OZ, Karosiya A, Chavan M (2019) Field application of *Bacillus subtilis* isolates for controlling late blight disease of potato caused by *Phytophthora infestans*. Biocatal Agric Biotechnol 22:101366

Kurtzman CP, Fell JW (2006) Yeast systematics and phylogeny—implications of molecular identification methods for studies in ecology. In: Biodiversity and ecophysiology of yeasts. Springer Berlin Heidelberg, Berlin, Heidelberg, pp 11–30

Li W, Liu XM (2021) Systematic investigations on iron cycling in phosphorus/siderophore systems: synergism or antagonism? Appl Geochem 124:104796

Li JH, Laurent S, Konde V, Bedu S, Zhang CC (2003) An increase in the level of 2-oxoglutarate promotes heterocyst development in the cyanobacterium anabaena sp. strain PCC 7120. Microbiology 149(11):3257–3263

Li N, Wang W, Bitas V, Subbarao K, Liu X, Kang S (2018) Volatile compounds emitted by diverse Verticillium species enhance plant growth by manipulating auxin signaling. Mol Plant-Microbe Interact 31(10):1021–1031

Liu C, Wu Y, Li L, Ma Y, Shao Z (2007) Thalassospira xiamenensis sp. nov. and Thalassospira profundimaris sp. nov. Int J Syst Evol Microbiol 57(2):316–320

Lugtenberg B, Kamilova F (2009) Plant-growth-promoting rhizobacteria. Ann Rev Microbiol 63:541–556

Lugtenberg BJ, Kravchenko LV, Simons M (1999) Tomato seed and root exudate sugars: composition, utilization by pseudomonas biocontrol strains and role in rhizosphere colonization. Environ Microbiol 1(5):439–446

Lugtenberg BJ, Malfanova N, Kamilova F, Berg G (2013) Plant growth promotion by microbes. Molecular Microbial Ecology of the Rhizosphere 1:559–573

Mabrouk Y, Hemissi I, Salem IB, Mejri S, Saidi M, Belhadj O (2018) Potential of rhizobia in improving nitrogen fixation and yields of legumes. Symbiosis 107(73495):1–16

Mahapatra S, Yadav R, Ramakrishna W (2022) *Bacillus subtilis* impact on plant growth, soil health and environment: Dr. Jekyll and Mr. Hyde. J Appl Microbiol 132(5):3543–3562

Midhun SJ, Arun D (2023) Methods employed to study bacterial association in fish. In: Recent advances in aquaculture microbial technology. Academic Press, pp 51–64

Mishra A, Rajput S, Gupta PS, Goyal V, Singh S, Sharma S, Shukla S, Singh A, Shukla K, Varma A (2021) Role of cyanobacteria in Rhizospheric nitrogen fixation. In: Soil Nitrogen Ecology. Springer International Publishing, Cham, pp 497–519

Mok MC (2019) Cytokinins and plant development—an overview. Cytokinins:155–166

Moya P, Pedemonte D, Amengual S, Franco ME, Sisterna MN (2016) Antagonism and modes of action of Chaetomium globosum species group, potential biocontrol agent of barley foliar diseases. Boletín de la Sociedad Argentina de Botánica 51(4):569–578

Nagpal S, Mandahal KS, Kumawat KC, Sharma P (2022) Beneficial Rhizobacteria unveiling plant fitness under climate change. In: Plant stress Mitigators: action and application. Singapore, Springer Nature Singapore, pp 281–321

Negi, S, Barry, AN, Friedland, N, Sudasinghe, N, Subramanian, S, Pieris, S, Holguin, FO, Dungan, B, Schaub, T and Sayre, R, (2016) Impact of nitrogen limitation on biomass, photosynthesis, and lipid accumulation in Chlorella sorokiniana. Journal of applied phycology, 28(2), pp.803–812

Negi S, Kumar P, Kumar J, Singh A, Dubey RC (2022) Indigenous nitrogen fixing microbes engineer rhizosphere and enhance nutrient availability and plant growth. In: Rhizosphere engineering. Academic Press, pp 19–43

Neilands JB (1995) Siderophores: structure and function of microbial iron transport compounds. J Biol Chem 270(45):26723–26726

Orozco-Mosqueda MDC, Kumar A, Fadiji AE, Babalola OO, Puopolo G, Santoyo G (2023) Agroecological management of the gray mold fungus Botrytis cinerea by plant growth-promoting bacteria. Plants 12(3):637

Ortega Pérez R, Nieto García JC, Gallegos-Cedillo VM, Domene Ruiz MÁ, Santos Hernández M, Nájera C, Miralles Mellado I, Diánez Martínez F (2023) Biofertilizers enriched with PGPB improve soil fertility and the productivity of an intensive tomato crop. Agronomy 13(9):2286

Oyedele AO, Ezaka E, Taiwo LB (2024) Microbial biosynthesis of the classical phytohormones by plant growth-promoting microorganisms in plants. In: Microbial biostimulants for plant growth and abiotic stress amelioration. Elsevier, pp 345–366

Parrilli E, Tedesco P, Fondi M, Tutino ML, Giudice AL, de Pascale D, Fani R (2021) The art of adapting to extreme environments: the model system Pseudoalteromonas. Phys Life Rev 36:137–161

Pereira PR, Freitas CS, Paschoalin VM (2021) *Saccharomyces cerevisiae* biomass as a source of next-generation food preservatives: evaluating potential proteins as a source of antimicrobial peptides. Compr Rev Food Sci Food Saf 20(5):4450–4479

Prakash J, Egamberdieva D, Arora NK (2022) A novel bacillus safensis-based formulation along with mycorrhiza inoculation for controlling alternaria alternata and simultaneously improving growth, nutrient uptake, and steviol glycosides in Stevia rebaudiana under field conditions. Plants 11(14):1857

Premachandra D, Hudek L, Brau L (2016) Bacterial modes of action for enhancing of plant growth. J Biotechnol Biomater 6(3):1–8

Rai A, Rai PK, Singh S (2017) Exploiting beneficial traits of plant-associated fluorescent pseudomonads for plant health. In: Agro-environmental sustainability: volume 1: managing crop health, pp 19–41

Rai N, Rai SP, Sarma BK (2021) Prospects for abiotic stress tolerance in crops utilizing phyto-and biostimulants. Front Sustain Food Syst 5:754853

Ramke R, Jeyaraman A (2022) Microbial interaction with metals and metalloids. In: Microbes and microbial biotechnology for green remediation. Elsevier, pp 29–46

Rashid M, Khalil S, Ayub N, Alam S, Latif F (2004) Organic acids production and phosphate solubilization by phosphate solubilizing microorganisms (PSM) under in vitro conditions. Pak J Biol Sci 7(2):187–196

Rawat P, Das S, Shankhdhar D, Shankhdhar SC (2021) Phosphate-solubilizing microorganisms: mechanism and their role in phosphate solubilization and uptake. J Soil Sci Plant Nutr 21(1):49–68

Romera FJ, García MJ, Lucena C, Martínez-Medina A, Aparicio MA, Ramos J, Alcántara E, Angulo M, Pérez-Vicente R (2019) Induced systemic resistance (ISR) and Fe deficiency responses in dicot plants. Front Plant Sci 10:287

Saeed Q, Xiukang W, Haider FU, Kučerik J, Mumtaz MZ, Holatko J, Naseem M, Kintl A, Ejaz M, Naveed M, Brtnicky M (2021) Rhizosphere bacteria in plant growth promotion, biocontrol, and bioremediation of contaminated sites: a comprehensive review of effects and mechanisms. Int J Mol Sci 22(19):10529

Salimi F, Hamedi J (2021) Biofertilizers: microbes for agricultural productivity. In: Soil microbiomes for sustainable agriculture: functional annotation, pp 407–469

Samson RA, Visagie CM, Houbraken J, Hong SB, Hubka V, Klaassen CH, Perrone G, Seifert KA, Susca A, Tanney JB, Varga J (2014) Phylogeny, identification and nomenclature of the genus aspergillus. Stud Mycol 78(1):141–173

Sansinenea E (2019) Bacillus spp.: as plant growth-promoting bacteria. In: Secondary metabolites of plant growth promoting rhizomicroorganisms: discovery and applications, pp 225–237

Sharma R, Chauhan A, Shirkot CK (2015) Characterization of plant growth promoting bacillus strains and their potential as crop protectants against Phytophthora capsici in tomato. Biol Agric Hortic 31(4):230–244

Shu X, He J, Zhou Z, Xia L, Hu Y, Zhang Y, Zhang Y, Luo Y, Chu H, Liu W, Yuan S (2022) Organic amendments enhance soil microbial diversity, microbial functionality and crop yields: a meta-analysis. Sci Total Environ 829:154627

Silva LID, Pereira MC, Carvalho AMXD, Buttrós VH, Pasqual M, Dória J (2023) Phosphorus-solubilizing microorganisms: a key to sustainable agriculture. Agriculture 13(2):462

Simón-Mateo CARMEN, Depuydt S, DE Oliveira Manes CL, Cnudde F, Holsters M, Goethals K, Vereecke D (2006) The phytopathogen Rhodococcus fascians breaks apical dominance and activates axillary meristems by inducing plant genes involved in hormone metabolism. Mol Plant Pathol 7(2):103–112

Sreenivasulu N, Sopory SK, Kishor PK (2007) Deciphering the regulatory mechanisms of abiotic stress tolerance in plants by genomic approaches. Gene 388(1–2):1–13

Stewart A, Hill R (2014) Applications of Trichoderma in plant growth promotion. In: Biotechnology and biology of Trichoderma. Elsevier, pp 415–428

Svoboda T, Thon MR, Strauss J (2021) The role of plant hormones in the interaction of Colletotrichum species with their host plants. Int J Mol Sci 22(22):12454

Swarnalakshmi K, Senthilkumar M, Ramakrishnan B (2016) Endophytic actinobacteria: nitrogen fixation, phytohormone production, and antibiosis. In: Plant growth promoting Actinobacteria: a new avenue for enhancing the productivity and soil fertility of grain legumes, pp 123–145

Tariq A, Ahmed A (2022) Auxins interkingdom signaling molecules. In: Plant hormones: recent advances, new perspectives and applications, vol 25, p 3

Thomma BP (2003) Alternaria spp.: from general saprophyte to specific parasite. Mol Plant Pathol 4(4):225–236

Timofeeva A, Galyamova M, Sedykh S (2022a) Prospects for using phosphate-solubilizing microorganisms as natural fertilizers in agriculture. Plants 11(16):2119

Timofeeva AM, Galyamova MR, Sedykh SE (2022b) Bacterial siderophores: classification, biosynthesis, perspectives of use in agriculture. Plants 11(22):3065

Timofeeva AM, Galyamova MR, Sedykh SE (2023) Plant growth-promoting soil bacteria: nitrogen fixation, phosphate solubilization, siderophore production, and other biological activities. Plants 12(24):4074

Tyśkiewicz R, Nowak A, Ozimek E, Jaroszuk-Ściseł J (2022) Trichoderma: the current status of its application in agriculture for the biocontrol of fungal phytopathogens and stimulation of plant growth. Int J Mol Sci 23(4):2329

Usta C (2013) Microorganisms in biological pest control—a review (bacterial toxin application and effect of environmental factors). Curr Prog Biol Res 13:287–317

Van Loon LC (2007) Plant responses to plant growth-promoting rhizobacteria. In: New perspectives and approaches in plant growth-promoting Rhizobacteria research. Springer Netherlands, Dordrecht, pp 243–254

van Loon LC, Rep M, Pieterse CM (2006) Significance of inducible defense-related proteins in infected plants. Annu Rev Phytopathol 44:135–162

Van Wees SC, Van der Ent S, Pieterse CM (2008) Plant immune responses triggered by beneficial microbes. Curr Opin Plant Biol 11(4):443–448

Verma V, Ravindran P, Kumar PP (2016) Plant hormone-mediated regulation of stress responses. BMC Plant Biol 16:1–10

Verma M, Mishra J, Arora NK (2019a) Plant growth-promoting rhizobacteria: diversity and applications. In: Environmental biotechnology: for sustainable future, pp 129–173

Verma DK, Pandey AK, Mohapatra B, Srivastava S, Kumar V, Talukdar D, Yulianto R, Zuan ATK, Jobanputra AH, Asthir B (2019b) Plant growth-promoting rhizobacteria: an eco-friendly approach for sustainable agriculture and improved crop production. In: Microbiology for sustainable agriculture, soil health, and environmental protection. Apple Academic Press, pp 3–80

Vreeland RH, Litchfield CD, Martin EL, Elliot E (1980) Halomonas elongata, a new genus and species of extremely salt-tolerant bacteria. Int J Syst Evol Microbiol 30(2):485–495

Waghmode, BD, Kambale, AS, Navhale, VC, Chendge, PD and Mahadkar, UV (2017) Effect of plant population and fertilizer doses on yield attributes, yield and economics of summer groundnut. International Journal of Current Microbiology and Applied Science, 6(11), pp.2670–2675

Waghmode S, Suryavanshi M, Sharma D, Satpute SK (2020) Planococcus species–an imminent resource to explore biosurfactant and bioactive metabolites for industrial applications. Front Bioeng Biotechnol 8:996

Wagner L, Stielow B, Hoffmann K, Petkovits T, Papp T, Vágvölgyi C, De Hoog GS, Verkley G, Voigt K (2013) A comprehensive molecular phylogeny of the Mortierellales (Mortierellomycotina) based on nuclear ribosomal DNA. Pers-Mol Phylogeny Evol Fungi 30(1):77–93

Wang Y, Lin X, Zhang S, Sun W, Ma S, Zhao ZK (2016) Cloning and evaluation of different constitutive promoters in the oleaginous yeast Rhodosporidium toruloides. Yeast 33(3):99–106

Watve MG, Tickoo R, Jog MM, Bhole BD (2001) How many antibiotics are produced by the genus Streptomyces? Arch Microbiol 176:386–390

Weiss C, Eisner T (1998) Partnerships for value-added through bioprospecting. Technol Soc 20(4):481–498

Wong C (2024) Scientists discover first algae that can fix nitrogen—thanks to a tiny cell structure. Nature 628(8009):702–702

Yadav AN, Saxena AK (2018) Biodiversity and biotechnological applications of halophilic microbes for sustainable agriculture. J Appl Biol Biotechnol 6(1):48–55

Yanes ML, Bajsa N (2016) Fluorescent pseudomonas: a natural resource from soil to enhance crop growth and health. In: Microbial models: from environmental to industrial sustainability, pp 323–349

Zalar P, Gostinčar C, De Hoog GS, Uršič V, Sudhadham M, Gunde-Cimerman N (2008) Redefinition of Aureobasidium pullulans and its varieties. Stud Mycol 61(1):21–38

Chapter 7
Harnessing Plant-Associated Microorganisms for Increasing Agricultural Productivity and Biocontrol Potential in Harsh Environments

Fatma Aly Ahmed and Doaa S. Zaid

Abstract Plant-associated microorganisms play pivotal roles in enhancing agricultural productivity and resilience, particularly in harsh and marginal environments. These beneficial microbes—including rhizobacteria, endophytes, and mycorrhizal fungi—form symbiotic relationships with host plants, facilitating nutrient acquisition, improving soil structure, and increasing plant tolerance to abiotic stresses such as drought, salinity, and extreme temperatures. Their ability to modulate phytohormone levels, fix atmospheric nitrogen, solubilize phosphates, and produce siderophores enhances plant growth even under suboptimal conditions. Moreover, these microorganisms exhibit biocontrol potential by suppressing phytopathogens through mechanisms such as competitive exclusion, the production of antimicrobial compounds, and the induction of systemic resistance in host plants. In recent years, research has focused on identifying and engineering microbial consortia tailored to specific crops and stress conditions. Inoculants based on plant growth-promoting rhizobacteria (PGPR) and arbuscular mycorrhizal fungi (AMF) have demonstrated significant success in improving crop yield and health in arid, saline, and degraded soils. These microbial strategies offer sustainable and eco-friendly alternatives to chemical fertilizers and pesticides, aligning with the goals of climate-smart agriculture. This review explores the diversity and functional mechanisms of plant-associated microbes, emphasizing their applications in enhancing agricultural productivity and disease resistance under environmental stress. It also highlights the challenges in large-scale applications, such as microbial survival, host compatibility, and field efficacy, and discusses recent advances in microbial formulation and delivery systems. Harnessing the full potential of these microorganisms holds promise for ensuring food security and sustainable agriculture in the face of climate change and land degradation.

F. Aly Ahmed · D. S. Zaid (✉)
Department of Medicinal and Aromatic Plants, Desert Research Center, Cairo, Egypt

V. Gopikrishnan et al. (eds.), *Harnessing Microbial Resources for Sustainable Agriculture*, Microorganisms for Sustainability 61,
https://doi.org/10.1007/978-981-95-4817-0_7

Keywords Eco-friendly · Plant-associated microorganisms · Nitrogen-fixing bacteria · Biocontrol potential

7.1 Introduction

In light of the current and projected impacts of human-induced global climate change, with a steady increase in inputs of anthropogenic stressors and the salinization of soils and water, agricultural environments offer unique opportunities to explore interactions with Earth's biota. This is illustrated by the reliance on plant-associated microorganisms for increasing crop productivity and biocontrol (Loiko and Islam 2024). Arid and semiarid environments continue to present valuable opportunities to study and utilize plant-associated microorganisms that can promote crop growth, alleviate plant water stress, and function as biological control agents (Ben Gaied et al. 2024).

Microorganisms residing in or on plant components, including bacteria, fungi, and other microbes, develop multifaceted interactions with their host plants. These microorganisms can strongly support plant development, nutrient uptake, and stress adaptability, especially in arid and saline environments with persistent plant stress (Yang et al. 2009). For example, nitrogen-fixing microorganisms can increase crop yields by increasing nitrogen availability, especially in regions where fertilizers are either unavailable or environmentally undesirable (Vessey 2003). Similarly, fungi that live in relationships with plant roots can considerably increase a plant's water and nutrient uptake in saline soils by improving its ability to overcome osmotic stress (Aroca et al. 2013).

Additionally, microbial communities represent natural biocontrol agents that offer an environmentally beneficial alternative to chemical pesticides. This phenomenon is crucial in dry regions, where water stress fosters conditions that encourage the growth of diseases (Compant et al. 2005). The increasing use of plant growth-promoting bacteria (PGPB) on farms is now considered an eco-friendly way to increase a plant's drought tolerance, which is crucial as water scarcity increases worldwide (Bashan et al. 2014). The next chapter explores the diversity of plant-associated microorganisms and some of their mechanisms for assisting plants in adapting to harsh environmental conditions.

7.2 Mechanisms of Plant Growth Improvement via Plant-Associated Microorganisms in Dry Environments

Dry areas are characterized by water scarcity, elevated temperatures, and nutritional deprivation, rendering plant survival highly challenging in those habitats (Ingrao et al. 2023). However, microbes such as PGPR have several mechanisms to assist and improve plant growth under such challenging environmental conditions. These mechanisms include phytohormone synthesis, nitrogen fixation, phosphate solubilization, exopolysaccharide (EPS) production, volatile organic compound (VOC) emission, and plant stress response modulation (Bashan et al. 2014; Lotfi et al. 2022).

First and foremost, PGPR facilitate plant growth via the manufacture of phytohormones, including auxins, cytokinins, and gibberellins, which promote cell division and differentiation (Lotfi et al. 2022). PGPR-produced auxins, such as those of *Azospirillum brasilense*, encourage root elongation and improve lateral root branching. This increases the ability of plants to sift through the soil more deeply, absorb nutrients and water more effectively, and eventually increase their resilience to drought (Yang et al. 2009). Gibberellins activate root elongation, seed germination, and flowering, whereas cytokinins delay leaf aging. Together, these hormones promote growth at all stages, especially in harsh environments (Javid et al. 2011). Nitrogen fixation by PGPR provides essential nutrients in nutrient-deficient soils. Species such as *Rhizobium* benefit legumes, whereas *Azospirillum* and *Herbaspirillum* support nonleguminous plants, improving plant development and yields (Jaiswal et al. 2021; Kumar et al. 2019).

Phosphorus insufficiency, which is common in sandy soils, results from an increase in pH, where phosphorus is impermeable to plants (Khan et al. 2023). Phosphate-solubilizing bacteria (PSB), such as *Pseudomonas fluorescens*, release organic acids that dissolve phosphate and turn it into a form that can be easily absorbed by plants. This process results in significant development of plants and overall agricultural productivity (Abbasi 2023). Studies have demonstrated that inoculating plants with PSB strains accelerates and improves plant growth in dry environments. Furthermore, treating plants with two different strains, PSB and PGPR, considerably improves plant development, particularly under extreme conditions (Kangi et al. 2024). In addition to their role in nutrient acquisition, various PGPR also increase plant drought tolerance by producing exopolysaccharides (EPSs). EPSs are high-molecular-weight polymers released by bacteria, increasing their water-holding capacity and thus contributing to better hydration of the plant (Prihatna and Yan 2024). EPS produced by *Pseudomonas putida* significantly improved the soil moisture-holding capacity and root growth of plants under drought conditions. EPSs play a role in shielding plant roots from drying out and optimizing the efficiency of water usage in plants, supporting their resilience in states of limited water availability (Benmrid et al. 2024; Lee et al. 2024).

Another mechanism involves the release of volatile organic compounds (VOCs) by PGPRs. They function as chemical messengers to activate systemic resistance and stimulate the expression of plant defense genes (Tahir et al. 2017). VOCs stimulate the synthesis of plant protective enzymes and antioxidant elements, promoting plant responses to drought, salinity stress, and pathogen attack by initiating the expression of genes involved in these pathways (Fincheira and Quiroz 2018). VOCs produced by *Bacillus amyloliquefaciens,* such as acetoin and 2,3-butanediol, have been shown to increase plant tolerance to various stresses. These compounds also induce the primary response system to oxidative stresses by activating antioxidant defense mechanisms and facilitating stress adaptation (Wang et al. 2022). Other studies revealed the role of VOCs in the induction of vigorous root growth and architecture, which contributes to the plant's adaptation to osmotic stress (salinity or drought) and/or pathogen conditions (Vocciante et al. 2022). PGPRs help plants manage drought stress by regulating the synthesis of essential stress hormones such as abscisic acid, which aids in water conservation, and ethylene, whose excess is reduced to prevent growth inhibition (Lim and Kim 2013). ABA levels improve plant water usage efficiency and stomatal regulation (Grahovac et al. 2023).

7.3 Role of Nitrogen-Fixing Bacteria in Enhancing Crop Productivity Under Stress Conditions

Nitrogen (N) is a critical macronutrient for plant biosynthetic processes such as amino acid, protein, and nucleic acid formation. While nitrogen gas (N_2) accounts for approximately 78% of the Earth's atmosphere, it is relatively nonutilizable for most plants because of its stable molecular structure. To be usable, N_2 must be reduced through biological nitrogen fixation, a process that is supported by specialized nitrogen-fixing bacteria (NFB) (Pandey et al. 2024). This process is a highly energy-demanding operation and is traditionally more active under favorable conditions than under stress (Pandey et al. 2024; Wang et al. 2024). The interactivity between NFB, such as *Rhizobium,* and leguminous plants is crucial for improving plant development and soil richness (Goyal and Habtewold 2023). However, environmental stresses such as salinity, drought, and heavy mineral pollution can severely disrupt these symbiotic interactions, reducing nutrient accessibility and productivity and ultimately hindering plant development (Bellabarba et al. 2023).

NFB are either symbiotic or free-living microorganisms that increase soil richness by increasing nitrogen availability. *Rhizobium* is a symbiotic NFB that forms nodules on the roots of leguminous plants (Goyal and Habtewold 2023). The plant provides its nodule bacteria with carbon for their sustenance, and in return, the bacteria fix nitrogen to the plant (Hungria et al. 2006). Free-living NFB, such as *Azotobacter* and *Azospirillum* species, fix nitrogen independently either in the soil or attached to plant roots. These bacteria can "symbiosis" with nonleguminous plants, regulating their growth and increasing yield (Pandey et al. 2024). This contribution of NFB is valuable under conditions of stress and adversity. Biotic and abiotic stressors often reduce nutrient absorption and assimilation in plants, restricting their growth and decreasing their productivity (Bellabarba et al. 2023). NFB alleviates these adverse effects in several ways, improving the resilience of target plants and the yields of crops (Rutten and Poole 2019).

Plants experiencing salinity stress suffer from ionic and osmotic stress, which hinders their growth and metabolic activities. NFB reduces salinity stress by producing exopolysaccharides that bind sodium ions and reduce their toxicity (Bellabarba et al. 2023). They also enhance plant antioxidant mechanisms, counteracting the oxidative damage caused by salt stress. *Azospirillum brasilense* introduced into wheat plants promoted growth under saline conditions through increasing proline levels and increasing antioxidant enzyme activity (Omer et al. 2022).

The presence of heavy metals in the soil might have phytotoxic effects, inducing oxidative stress in plant cells and impairing their physiological functions (Mansoor et al. 2023). The metal-chelating ability of certain NFB reduces the toxicity of heavy minerals by promoting the expression of plant genes related to metal tolerance. *Rhizobium leguminosarum* PT298 alleviated cadmium toxicity by controlling the expression of phytochelatin synthase genes regulated by metallothionein proteins (Zulfiqar et al. 2022). Stress-responsive genes in NFB have been overexpressed through genetic engineering to increase nitrogen fixation performance (Rodríguez-Salazar et al. 2009). *Rhizobium etli* upregulates the trehalose-6-phosphate synthase (*tps*) gene, which improves drought tolerance and strengthens its symbiotic relationship with its host bean (Suárez et al. 2008).

7.4 Root Fungi and Their Impact on Water and Nutrient Uptake in Plants Grown in Saline Soils

Root fungi, especially arbuscular mycorrhizal fungi (AMF), are critical factors for plant survival in harsh environments. They increase water and nutrient uptake by stimulating root growth and increasing plant tolerance to salinity stress (Huang et al. 2023). AMF increase the exchange of key elements, including phosphorus (P), nitrogen (N), and potassium (K), while decreasing the interchange of sodium (Na), which helps maintain the ionic balance in saline soils. AMF modify the primary root structure, promoting elongation and increasing biomass, thereby expanding the surface area for water and essential nutrient absorption (Noshad et al. 2022). AMF inoculation has been demonstrated to considerably improve growth factors, including shoot development, root biomass, chlorophyll content, and overall plant output (Kang et al. 2023). AMF help minimize the side effects of salt by stimulating antioxidant defense mechanisms and osmotic adjustment, both of which are essential for plant health under stress. While AMF provide significant benefits in salty soils, their effectiveness can differ on the basis of the fungal species and plant type involved (Ortas et al. 2021).

When nutrients are deficient in the soil, AMF secrete organic acids and enzymes that help improve nutrient availability. The secreted organic acids enhance nutrient absorption and mobilization within the plant and influence metabolic processes to ensure maximum utilization of the absorbed elements. In addition, they prepare the plant to receive other symbiotic organisms that support it in adapting to difficult environmental conditions. This symbiotic interaction is essential to plant development and the ecological balance of the soil (Andrino et al. 2021). AMF improve phosphorus acquisition by releasing organic acids that dissolve phosphorus and increase its accessibility to plants (Wu et al. 2018). AMF synthesizes citric and oxalic acids, which aid in metal solubility and nutrient recruitment. These acids contribute to both plant nutrition and the chemical composition of the soil (Gadd 1999). AMF may accelerate the degradation of organic matter, allowing direct nitrogen uptake from complex organic sources, which is critical for plant development (Van Schöll et al. 2006). AMF secretions, along with plant secondary metabolites, can interact to modify organic acid exudation, which in turn affects rhizosphere nutrition dynamics and microbiome structure (Rahmat and Soheilikhah 2024).

Some root-associated fungi play important roles in regulating plant hormones, helping plants adapt to salinity stress. These microorganisms stimulate the production of phytohormones such as auxins, gibberellins, and cytokinins, which in turn increase the ability of plants to tolerate harsh environmental conditions. This process works by improving nutrient consumption and increasing water utilization efficiency, providing plants with effective mechanisms to cope with stress (Monika et al. 2022; Roy et al. 2021).

Research indicates that mycorrhizal fungi not only increase hormonal signaling but also increase plant adaptive resilience through other mechanisms, such as improved ionic homeostasis and oxidative defense (Aizaz et al. 2023). The diversity of regulatory mechanisms used by mycorrhizal fungi to mitigate salinity stress offers promising avenues for crop improvement and could greatly benefit agricultural practices, especially in saline environments (Huang et al. 2023). For example, *Acremonium alternatum* and *Piriformospora indica* fungal roots play vital roles in regulating the sodium–potassium equilibrium in plants, as they contribute to increasing potassium uptake, reducing sodium accumulation, and regulating hormone signaling, particularly ABA, which is needed for stress adaptation to harsh environmental situations (Berková et al. 2024). In addition, studies have shown that inoculating crops such as rice and wheat with AMF under salinity stress improves growth and productivity, indicating their potential as potent support biofertilizers, especially under difficult environmental situations (Huang et al. 2023).

7.5 Using Microorganisms in Dry Climates as Biological Control Agents to Control Fungal Diseases

Extreme conditions in dry areas induce microorganisms to produce bioactive compounds with antifungal properties as a defense mechanism. The extraction and utilization of these compounds as biological control agents introduces a sustainable approach to managing fungal diseases in dry climates as a conventional alternative to chemical fungicides, which pose ecological risks or lose efficacy over time (Eckstien et al. 2024). Microbial biocontrol agents, which include bacteria such as *Bacillus* spp. and *Pseudomonas* spp., as well as fungi such as *Trichoderma* spp., effectively suppress plant pathogens (Bonaterra et al. 2022). These microorganisms combat fungi by producing antifungal compounds, competing for resources, and engaging in direct parasitism. Dry conditions often augment the creation of these bioactive compounds, increasing microbial resilience to extreme conditions. In addition, their ability to colonize and protect roots makes them invaluable for sustaining plant survival and productivity (Librizzi et al. 2022). However, environmental considerations such as soil temperature and moisture can influence their effectiveness, and designed application strategies are needed to optimize their impact (Mawar et al. 2021).

The risk of fungal diseases arises from plant stress under dry conditions. For example, fungal diseases such as those caused by *Fusarium* and *Rhizoctonia* are permanent threats, as low soil moisture weakens plant defense, increasing the vulnerability of plants when conditions become favorable. Biological control agents alleviate these risks by optimizing the soil microbiome and enhancing plant defenses (Mawar et al. 2019). For example, *Trichoderma* species produce enzymes that can devastate fungal cell walls and release metabolites that trigger systemic resistance in plants (Lodha and Mawar 2010). Similarly, bacterial agents such as *Bacillus subtilis* produce antibiotics and volatile compounds that inhibit fungal growth. In addition to controlling pathogens, these microorganisms improve plant growth by facilitating nutrient availability, making them multifunctional tools in sustainable agriculture. These bioactive compounds have a promising future due to their sustainability and effectiveness. The complementarity of these agents with organic amendments, such as compost, has greatly improved their perseverance and efficiency under barren conditions (Wend et al. 2024).

Despite the promising future that these bioactive compounds have and their potential, challenges remain in the widespread adoption of biological control agents in dry climates. Many factors affect the performance and efficacy of these bioactive compounds, such as inconsistent performance due to environmental variability, limited commercial formulations, and regulatory hurdles that hinder their broader application (Mawar et al. 2021). The great advances in molecular tools and formulation technologies have helped overcome these issues by improving the stability and specificity of these agents under harsh conditions. The possibility of combining multiple strains or species, such as *Trichoderma* with beneficial bacteria, has also shown synergistic effects, increasing their robustness (Palmieri et al. 2022). It is essential to focus on region-specific trials, improved delivery systems, and integration into holistic pest management strategies to maximize their benefits. To unlock the full potential of microorganisms as eco-friendly solutions for fungal diseases in dry climates, more research and collaboration among stakeholders should be conducted (Al-Sadi 2016).

7.6 Interactions Between Microorganisms and Plants in Saline Soils and Their Role in Improving Agricultural Productivity

Meeting the necessities of the rapid increase in the human population is essential, especially in light of the growing demand for food. However, soil characteristics deteriorate significantly owing to the rigorous utilization of chemicals and various environmental difficulties, such as drought and salinity. Plants exposed to saline or arid soil conditions often present relatively low germination rates, poor root growth, reduced aerial biomass, and impaired photosynthetic efficiency. These factors combined lead to significant losses in crop productivity and significant economic damage (Li et al. 2022). Salinity and drought are important environmental difficulties that adversely affect crop growth and yield. Arid regions are estimated to occupy between 41% and 45% of the Earth's surface. Soil salinity arises from diverse causes, including a reduction in vegetation cover, which decreases evapotranspiration rates. Poor agricultural practices, such as the utilization of poor-quality irrigation water, inefficient irrigation practices, and excessive use of pesticides and fertilizers, also contribute to soil salinity. Additionally, natural factors such as the intrusion of saline groundwater into coastal areas and seawater intrusion lead to salt accumulation. Soil salinity is classified into two types on the basis of its origin. Primary salinity occurs due to natural processes, whereas secondary salinity arises from the improper management of natural resources (Daliakopoulos et al. 2016).

Plants can be divided into two main types according to their ability to survive in saline environments: halophytes and nonhalophytes (Green et al. 2017; Kosová et al. 2013). Halophytes are capable of tolerating salt and accumulating it within their tissues, whereas nonhalophytes represent the majority of plants, accounting for more than 98% of all species. Plants rely on reduced sodium uptake to maintain a low level of sodium ions (Na^+) in their bodies, which varies between 0.2 and 2.0 g/kg. Owing to this sensitivity, the presence of salts in the soil significantly and adversely affects these plants (Green et al. 2017; Naz et al. 2023). In response to salinity, halophyte plants exhibit distinct anatomical adaptations that vary depending on their capacity to accumulate, exclude, or excrete salts in their tissue. Halophytes can be divided into two main types: extreme halophytes, which include succulent plants characterized by their ability to reserve water in their vegetative organs, and intermediate halophytes, which have anatomical characteristics such as an expanded endodermis, the presence of aerenchyma in aerial tissues, and specialized cells adapted to moderate levels of salinity. These characteristics place them between those of halophytes and glycophytes (Grigore and Constantin 2010).

Salinity not only affects plants but also significantly affects the physical and chemical composition of the soil. It alters soil acidity (pH), reduces the accessibility of nutrients, decreases the soil mineral exchange capacity, and reduces organic carbon inputs, which alter the activity of soil microbial communities (Szoboszlay et al. 2019).

Plants do not function as isolated organisms in response to these stresses; rather, they depend closely on a network of mutualistic relationships with surrounding microorganisms. These microbes, especially PGPB, play fundamental roles in supporting plants and enhancing their adaptability to harsh environmental conditions. Therefore, to explore how plants respond to these challenges, it is necessary to study closely the impact of the accompanying microbes, which are an integral part of these vital relationships (Schlaeppi and Bulgarelli 2015). Soil microbial communities are shaped by several biotic and abiotic factors. Soil type is a key determinant in defining the structure and diversity of microorganisms. In addition, factors such as plant genetic makeup, growth stage, and agricultural practices also influence the composition of plant microbiomes. However, soil physical and chemical properties remain the primary factors determining the spectrum of microbial communities. Among these factors, pH is particularly significant; neutral soils tend to support greater bacterial diversity than acidic or alkaline soils do, such as those common in desert regions (Edwards et al. 2015).

Sustainable agriculture offers great promise for harnessing the potential of PGPB, which can effectively improve soil fertility and increase crop productivity. However, this technology faces major challenges, such as creating stable and effective commercial formulations, ensuring microbial survival and thriving under changing field conditions, and achieving compatibility with different plant species to minimize their impact. Consequently, it is essential to focus on studying the relationships between plants and their microbial partners, especially in environments suffering from salinity and drought (Gamalero et al. 2020).

7.7 Plant-Associated Microorganisms Are Strategic Partners in Sustainable Agriculture

Plants constitute the backbone of life on Earth, providing living organisms with food and oxygen and playing a vital role in maintaining environmental stability. However, plants face environmental challenges that affect their health and productivity, which harms the entire ecosystem (Chowdhury et al. 2023). In addition to these challenges, the importance of plant-associated microbes has emerged as an effective solution to support sustainable agriculture and increase the resistance of plants to environmental stresses (Schlaeppi and Bulgarelli 2015). These plant-associated microbes effectively increase the ability of plants to withstand harsh environments; they work to increase the absorption of nutrients, stimulate the production of plant hormones that promote growth, and improve plant tolerance to drought and salinity (Naeem et al. 2023). These multiple functions contribute to the sustainability of agricultural output and maintain its stability even under harsh environmental conditions (Lone et al. 2023). Additionally, they make valuable contributions to purifying polluted environments. These microorganisms purify soil and water from toxins via different methods, such as phytoextraction, which allows plants to absorb and break down pollutants present in the soil (Saeed et al. 2021). They also improve the effectiveness of roots in filtering water, which increases water purity and contributes to the stability of the overall ecosystem (Iqbal et al. 2023). Despite the enormous benefits of using microbes to support sustainable agriculture, careful control over their application is essential to prevent negative consequences. The introduction of nonnative microbial species can disrupt the natural ecological balance, which requires the implementation of these technologies within a framework that ensures environmental compatibility and prevents potential harm. Therefore, integrated plans must be developed that consider the environmental and ethical aspects of using these microbes in agriculture (Iqbal et al. 2023). Growth-promoting microbes offer exceptional opportunities to transform sustainable agricultural practices, support food security, and meet growing environmental challenges (Saeed et al. 2021). Advances in genomics and genetic engineering technologies have substantially improved the ability of these microbes to integrate more effectively into agricultural systems. Achieving these goals requires intensive collaboration among scientists, specialists, and policymakers to ensure that these technologies are applied in ways that support long-term sustainability. These efforts are directed toward achieving agriculture that relies on ecological balance and stimulates harmonious interactions between organisms and ecosystems (Ram et al. 2014).

References

Abbasi S (2023) Plant–microbe interactions ameliorate phosphate-mediated responses in the rhizosphere: a review. Front Plant Sci 14. https://doi.org/10.3389/FPLS.2023.1074279/PDF

Aizaz M, Ahmad W, Asaf S, Khan I, Saad Jan S, Salim Alamri S, Bilal S, Jan R, Kim KM, Al-Harrasi A (2023) Characterization of the seed biopriming, plant growth-promoting and salinity-ameliorating potential of halophilic fungi isolated from hypersaline habitats. Int J Mol Sci 24(5):4904–4904. https://doi.org/10.3390/IJMS24054904

Al-Sadi AM (2016) Epidemiology and management of fungal diseases in dry environments. Innovations in Dryland Agriculture:187–209. https://doi.org/10.1007/978-3-319-47928-6_7

Andrino A, Guggenberger G, Kernchen S, Mikutta R, Sauheitl L, Boy J (2021) Production of organic acids by arbuscular mycorrhizal fungi and their contribution in the mobilization of phosphorus bound to iron oxides. Front Plant Sci 12:661842. https://doi.org/10.3389/FPLS.2021.661842/BIBTEX

Aroca R, Ruiz-Lozano JM, Zamarreño ÁM, Paz JA, García-Mina JM, Pozo MJ, López-Ráez JA (2013) Arbuscular mycorrhizal symbiosis influences strigolactone production under salinity and alleviates salt stress in lettuce plants. J Plant Physiol 170(1):47–55. https://doi.org/10.1016/j.jplph.2012.08.020

Bashan Y, de-Bashan LE, Prabhu SR, Hernandez J-P (2014) Advances in plant growth-promoting bacterial inoculant technology: formulations and practical perspectives (1998–2013). Plant Soil 378(1–2):1–33. https://doi.org/10.1007/s11104-013-1956-x

Bellabarba A, Decorosi F, Fagorzi C, El Hadj Mimoune A, Buccioni A, Santoni M, Pacini GC, Bekki A, Azim K, Hafidi M, Mazzoncini M, Mengoni A, Pini F, Viti C (2023) Salt stress highlights the relevance of genotype × genotype interaction in the nitrogen-fixing symbiosis between *Sinorhizobium meliloti* and Alfalfa. Soil Syst 7(4). https://doi.org/10.3390/SOILSYSTEMS7040112

Ben Gaied R, Brígido C, Sbissi I, Tarhouni M (2024) Sustainable strategy to boost legume growth under salinity and drought stress in semiarid and arid regions. Soil Syst 8(3):84–84. https://doi.org/10.3390/SOILSYSTEMS8030084

Benmrid B, Ghoulam C, Ammar I, Nkir D, Saidi R, Staropoli A, Iacomino G, ELhajjami, E., Cheto, S., Geistlinger, J., Idbella, M., & Bargaz, A. (2024) Drought-tolerant rhizobacteria with predicted functional traits enhanced wheat growth and P uptake under moderate drought and low P-availability. Microbiol Res 285:127795–127795. https://doi.org/10.1016/J.MICRES.2024.127795

Berková V, Berka M, Štěpánková L, Kováč J, Auer S, Menšíková S, Ďurkovič J, Kopřiva S, Ludwig-Müller J, Brzobohatý B, Černý M (2024) The fungus *Acremonium alternatum* enhances salt stress tolerance by regulating host redox homeostasis and phytohormone signaling. Physiol Plant 176(3). https://doi.org/10.1111/PPL.14328

Bonaterra A, Badosa E, Daranas N, Francés J, Roselló G, Montesinos E (2022) Bacteria as biological control agents of plant diseases. Microorganisms 10(9). https://doi.org/10.3390/microorganisms10091759

Chowdhury PR, Medhi H, Bhattacharyya KG, Hussain CM (2023) Severe deterioration in food-energy-ecosystem nexus due to ongoing Russia-Ukraine war: A critical review. Sci Total Environ 902:166131. https://doi.org/10.1016/j.scitotenv.2023.166131

Compant S, Duffy B, Nowak J, Clément C, Barka EA (2005) Use of plant growth-promoting bacteria for biocontrol of plant diseases: principles, mechanisms of action, and future prospects. Appl Environ Microbiol 71(9):4951–4959. https://doi.org/10.1128/AEM.71.9.4951-4959.2005

Daliakopoulos IN, Tsanis IK, Koutroulis A, Kourgialas NN, Varouchakis AE, Karatzas GP, Ritsema CJ (2016) The threat of soil salinity: A European scale review. Sci Total Environ 573:727–739. https://doi.org/10.1016/j.scitotenv.2016.08.177

Eckstien D, Maximov N, Margolis N, Raanan H (2024) Toward sustainable biocontrol: inhibition of soil-borne fungi by microalgae from harsh environments. Front Microbiol 15. https://doi.org/10.3389/FMICB.2024.1433765

Edwards J, Johnson C, Santos-Medellín C, Lurie E, Podishetty NK, Bhatnagar S, Eisen JA, Sundaresan V (2015) Structure, variation, and assembly of the root-associated microbiomes of rice. Proc Natl Acad Sci 112(8). https://doi.org/10.1073/pnas.1414592112

Fincheira P, Quiroz A (2018) Microbial volatiles as plant growth inducers. Microbiol Res 208:63–75. https://doi.org/10.1016/J.MICRES.2018.01.002

Gadd GM (1999) Fungal production of citric and oxalic acid: importance in metal speciation, physiology, and biogeochemical processes. Adv Microb Physiol 41:47–92. https://doi.org/10.1016/s0065-2911(08)60165-4

Gamalero E, Bona E, Todeschini V, Lingua G (2020) Saline and arid soils: impact on bacteria, plants, and their interaction. Biology 9(6):116. https://doi.org/10.3390/biology9060116

Goyal RK, Habtewold JZ (2023) Evaluation of legume–rhizobial symbiotic interactions beyond nitrogen fixation that help the host survival and diversification in hostile environments. Microorganisms 11(6):1454–1454. https://doi.org/10.3390/MICROORGANISMS11061454

Grahovac J, Pajčin I, Vlajkov V (2023) Bacillus VOCs in the context of biological control. Antibiotics 12(3):581. https://doi.org/10.3390/ANTIBIOTICS12030581

Green TGA, Sancho LG, Pintado A, Saco D, Martín S, Arróniz-Crespo M, Angel Casermeiro M, de la Cruz Caravaca MT, Cameron S, Rozzi R (2017) Sodium chloride accumulation in glycophyte plants with cyanobacterial symbionts. AoB PLANTS 9(6). https://doi.org/10.1093/aobpla/plx053

Grigore MN, Constantin T (2010) A proposal for a new halophytes classification, based on integrative anatomy observations. Oltenia Studii Şi Comunicări Ştiinţele Naturii Tom 26(1) http://rbg-web2.rbge.org.uk/FE/fe.html

Huang S, Gill S, Ramzan M, Ahmad MZ, Danish S, Huang P, Al Obaid S, Alharbi SA (2023) Uncovering the impact of AM fungi on wheat nutrient uptake, ion homeostasis, oxidative stress, and antioxidant defense under salinity stress. Dental Sci Rep 13(1). https://doi.org/10.1038/S41598-023-35148-X

Hungria M, Franchini JC, Campo RJ, Crispino CC, Moraes JZ, Sibaldelli RNR, Mendes IC, Arihara J (2006) Nitrogen nutrition of soybean in Brazil: contributions of biological N_2 fixation and N fertilizer to grain yield. Can J Plant Sci 86(4):927–939. https://doi.org/10.4141/P05-098

Ingrao C, Strippoli R, Lagioia G, Huisingh D (2023) Water scarcity in agriculture: an overview of causes, impacts, and approaches for reducing the risks. Heliyon 9(8):e18507. https://doi.org/10.1016/j.heliyon.2023.e18507

Iqbal B, Li G, Alabbosh KF, Hussain H, Khan I, Tariq M, Javed Q, Naeem M, Ahmad N (2023) Advancing environmental sustainability through microbial reprogramming in growth improvement, stress alleviation, and phytoremediation. Plant Stress 10:100283. https://doi.org/10.1016/j.stress.2023.100283

Jaiswal SK, Mohammed M, Ibny FYI, Dakora FD (2021) Rhizobia as a source of plant growth-promoting molecules: potential applications and possible operational mechanisms. Front Sustain Food Syst 4. https://doi.org/10.3389/FSUFS.2020.619676/FULL

Javid MG, Sorooshzadeh A, Moradi F, Ali S, Sanavy MM, Allahdadi I (2011) The role of phytohormones in alleviating salt stress in crop plants. AJCS 5(6):726–734

Kang J, Yang W, Liu S, Xuan N, Shao Y, Geng Y, Afzal M, Zhang Y, Yue S, Mushtaq R, Chen G (2023) Arbuscular mycorrhizal fungi increased peanut (*Arachis hypogaea* L.) yield by changing the rhizosphere microbial community structure in saline–alkali soil. Front Microbiol 14. https://doi.org/10.3389/FMICB.2023.1303979

Kangi E, Brzostek ER, Bills RJ, Callister SJ, Zink EM, Kim YM, Larsen PE, Cumming JR (2024) A multiomic survey of black cottonwood tissues highlights coordinated transcriptomic and metabolomic mechanisms for plant adaptation to phosphorus deficiency. Front Plant Sci 15. https://doi.org/10.3389/FPLS.2024.1324608/PDF

Khan F, Siddique AB, Shabala S, Zhou M, Zhao C (2023) Phosphorus plays key roles in regulating plants' physiological responses to abiotic stresses. Plants 12(15). https://doi.org/10.3390/PLANTS12152861

Kosová K, Vítámvás P, Urban MO, Prášil IT (2013) Plant proteome responses to salinity stress—comparison of glycophytes and halophytes. Funct Plant Biol 40(9):775. https://doi.org/10.1071/FP12375

Kumar A, Meena VS, Roy P, Vandana, & Kumari, R. (2019) Role of rhizobia for sustainable agriculture: lab to land. Plant Growth Promoting Rhizobacteria Agric Sustain:129–149. https://doi.org/10.1007/978-981-13-7553-8_7

Lee S, Kim JA, Song J, Choe S, Jang G, Kim Y (2024) Plant growth-promoting rhizobacterium *bacillus megaterium* modulates the expression of antioxidant-related and drought-responsive genes to protect rice (*Oryza sativa* L.) from drought. Front Microbiol 15. https://doi.org/10.3389/FMICB.2024.1430546

Li W, Sun J, Zhang X, Ahmad N, Hou L, Zhao C, Pan J, Tian R, Wang X, Zhao S (2022) The mechanisms underlying salt resistance mediated by exogenous application of 24-epibrassinolide in peanut. Int J Mol Sci 23(12):6376. https://doi.org/10.3390/ijms23126376

Librizzi V, Malacrinò A, Nicosia MGLD, Barger N, Luzzatto-Knaan T, Pangallo S, Agosteo GE, Schena L (2022) Extracts from environmental strains of pseudomonas spp. effectively control fungal plant diseases. Plants 11(3):436–436. https://doi.org/10.3390/PLANTS11030436

Lim JH, Kim SD (2013) Induction of drought stress resistance by multifunctional PGPR *bacillus licheniformis* K11 in pepper. Plant Pathol J 29(2):201. https://doi.org/10.5423/PPJ.SI.02.2013.0021

Lodha S, Mawar R (2010) Efficacy of native biocontrol agents on soil microflora, dry root rot incidence and seed yield of rainfed arid crops. Indian Phytopathol 63(2):195–198. https://epubs.icar.org.in/index.php/IPPJ/article/view/11127

Loiko N, Islam MN (2024) Plant–soil microbial interaction: differential adaptations of beneficial vs. pathogenic bacterial and fungal communities to climate-induced drought and desiccation stresses. https://doi.org/10.20944/PREPRINTS202407.1261.V1

Lone R, Hassan N, Bashir B, Rohela GK, Malla NA (2023) Role of growth elicitors and microbes in stress management and sustainable production of sorghum. Plant Stress 9:100179. https://doi.org/10.1016/j.stress.2023.100179

Lotfi N, Soleimani A, Çakmakçı R, Vahdati K, Mohammadi P (2022) Characterization of plant growth-promoting rhizobacteria (PGPR) in Persian walnut associated with drought stress tolerance. Sci Rep 12(1). https://doi.org/10.1038/S41598-022-16852-6

Mansoor S, Ali A, Kour N, Bornhorst J, AlHarbi K, Rinklebe J, Abd El Moneim D, Ahmad P, Chung YS (2023) Heavy metal induced oxidative stress mitigation and ROS scavenging in plants. Plants 12(16):3003. https://doi.org/10.3390/PLANTS12163003

Mawar R, Tomer AS, Singh D (2019) Demonstration of efficacy of biocontrol agents in managing soil-borne diseases of various crops in arid region of India. Indian Phytopathol 72(4):699–703. https://doi.org/10.1007/S42360-018-0071-6

Mawar R, Manjunatha BL, Kumar S (2021) Commercialization, diffusion, and adoption of bioformulations for sustainable disease management in Indian arid agriculture: prospects and challenges. Circ Econ Sustain 1(4):1367–1385. https://doi.org/10.1007/S43615-021-00089-Y

Monika AK, Kumar N, Mamta NY, Kumar S, Arya SS (2022) Microbe-mediated amelioration of salinity stress in crops. Plant Stress Mitigators Action Applicat:429–450. https://doi.org/10.1007/978-981-16-7759-5_20

Naeem M, Zhao W, Ahmad N, Zhao L (2023) Beyond green and red: unlocking the genetic orchestration of tomato fruit color and pigmentation. Funct Integr Genomics 23(3):243. https://doi.org/10.1007/s10142-023-01162-5

Naz M, Ghani MI, Atif MJ, Raza MA, Bouzroud S, Afzal MR, Riaz M, Ali M, Tariq M, Fan X (2023) Sodium and abiotic stress tolerance in plants. In: Beneficial chemical elements of plants. Wiley, pp 307–330. https://doi.org/10.1002/9781119691419.ch13

Noshad MH, Chavoshi E, Mosaddeghi MR, Dorostkar V, Hosseini F (2022) Growth and nutrient uptake of haloxylon and atriplex as affected by mycorrhizal symbiosis under combined drought and salinity stresses. Can J Soil Sci 103(2):305–317. https://doi.org/10.1139/CJSS-2022-0069

Omer AM, Osman MS, Badawy AA (2022) Inoculation with *Azospirillum brasilense* and/or *Pseudomonas geniculata* reinforces flax (*Linum usitatissimum*) growth by improving physiological activities under saline soil conditions. Bot Stud 63(1):1–15. https://doi.org/10.1186/S40529-022-00345-W/FIGURES/6

Ortas I, Rafique M, Çekiç FÖ (2021) Do Mycorrhizal fungi enable plants to cope with abiotic stresses by overcoming the detrimental effects of salinity and improving drought tolerance, pp 391–428. https://doi.org/10.1007/978-3-030-51916-2_23

Palmieri D, Ianiri G, Del Grosso C, Barone G, De Curtis F, Castoria R, Lima G (2022) Advances and perspectives in the use of biocontrol agents against fungal plant diseases. Horticulturae 8(7). https://doi.org/10.3390/horticulturae8070577

Pandey A, Singh K, Singh AP (2024) Nitrogen forms and their availability-dependent root developmental adaptation in plants. In: Improving stress resilience in plants: physiological and biochemical basis and utilization in breeding. Academic Press, pp 97–117. https://doi.org/10.1016/B978-0-443-18927-2.00008-X

Prihatna C, Yan Q (2024) Exopolysaccharide is required by *Paraburkholderia phytofirmans* PsJN to confer drought-stress tolerance in pea. Front Microbiol 15. https://doi.org/10.3389/FMICB.2024.1442001

Rahmat S, Soheilikhah Z (2024) Arbuscular mycorrhizal fungi and plant secondary metabolism. In: Arbuscular Mycorrhizal fungi and higher plants. Springer Nature Singapore, Singapore, pp 99–121. https://doi.org/10.1007/978-981-99-8220-2_5

Ram P, Vivek K, Kumar SP (2014) Nanotechnology in sustainable agriculture: present concerns and future aspects. Afr J Biotechnol 13(6):705–713. https://doi.org/10.5897/AJBX2013.13554

Rodríguez-Salazar J, Suárez R, Caballero-Mellado J, Iturriaga G (2009) Trehalose accumulation in Azospirillum brasilense improves drought tolerance and biomass in maize plants. FEMS Microbiol Lett 296(1):52–59. https://doi.org/10.1111/J.1574-6968.2009.01614.X

Roy S, Chakraborty AP, Chakraborty R (2021) Understanding the potential of root microbiome influencing salt-tolerance in plants and mechanisms involved at the transcriptional and translational level. Physiol Plant 173(4):1657–1681. https://doi.org/10.1111/PPL.13570

Rutten PJ, Poole PS (2019) Oxygen regulatory mechanisms of nitrogen fixation in rhizobia. Adv Microb Physiol 75:325–389. https://doi.org/10.1016/BS.AMPBS.2019.08.001

Saeed Q, Xiukang W, Haider FU, Kučerik J, Mumtaz MZ, Holatko J, Naseem M, Kintl A, Ejaz M, Naveed M, Brtnicky M, Mustafa A (2021) Rhizosphere bacteria in plant growth promotion, biocontrol, and bioremediation of contaminated sites: A comprehensive review of effects and mechanisms. Int J Mol Sci 22(19):10529. https://doi.org/10.3390/ijms221910529

Schlaeppi K, Bulgarelli D (2015) The plant microbiome at work. Mol Plant-Microbe Interact 28(3):212–217. https://doi.org/10.1094/MPMI-10-14-0334-FI

Suárez R, Wong A, Ramírez M, Barraza A, Orozco MDC, Cevallos MA, Lara M, Hernández G, Iturriaga G (2008) Improvement of drought tolerance and grain yield in common bean by overexpressing trehalose-6-phosphate synthase in rhizobia. Molecular Plant–Microbe Interactions: MPMI 21(7):958–966. https://doi.org/10.1094/MPMI-21-7-0958

Szoboszlay M, Näther A, Liu B, Carrillo A, Castellanos T, Smalla K, Jia Z, Tebbe CC (2019) Contrasting microbial community responses to salinization and straw amendment in a semiarid bare soil and its wheat rhizosphere. Sci Rep 9(1):9795. https://doi.org/10.1038/s41598-019-46070-6

Tahir HAS, Gu Q, Wu H, Raza W, Safdar A, Huang Z, Rajer FU, Gao X (2017) Effect of volatile compounds produced by *Ralstonia solanacearum* on plant growth promoting and systemic resistance inducing potential of bacillus volatiles. BMC Plant Biol 17(1). https://doi.org/10.1186/s12870-017-1083-6

Van Schöll L, Hoffland E, Van Breemen N (2006) Organic anion exudation by ectomycorrhizal fungi and *Pinus sylvestris* in response to nutrient deficiencies. New Phytol 170(1):153–163. https://doi.org/10.1111/J.1469-8137.2006.01649.X

Vessey JK (2003) Plant growth promoting rhizobacteria as biofertilizers. Plant Soil 255(2):571–586. https://doi.org/10.1023/A:1026037216893

Vocciante M, Grifoni M, Fusini D, Petruzzelli G, Franchi E (2022) The role of plant growth-promoting rhizobacteria (PGPR) in mitigating plant's environmental stresses. Appl Sci 12(3):1231. https://doi.org/10.3390/APP12031231

Wang SY, Herrera-Balandrano DD, Wang YX, Shi XC, Chen X, Jin Y, Liu FQ, Laborda P (2022) Biocontrol ability of the bacillus amyloliquefaciens group, B. *Amyloliquefaciens*, *B. Velezensis*, *B. Nakamurai*, and *B. Siamensis*, for the management of fungal postharvest diseases: A review. J Agric Food Chem 70(22):6591–6616. https://doi.org/10.1021/acs.jafc.2c01745

Wang Q, Li S, Li J, Huang D (2024) The utilization and roles of nitrogen in plants. Forests 15(7):1191–1191. https://doi.org/10.3390/F15071191

Wend K, Zorrilla L, Freimoser FM, Gallet A (2024) Microbial pesticides—challenges and future perspectives for testing and safety assessment with respect to human health. Environ Health Glob Access Sci Source 23(1):1–29. https://doi.org/10.1186/S12940-024-01090-2/TABLES/3

Wu L, Kobayashi Y, Wasaki J, Koyama H (2018) Organic acid excretion from roots: a plant mechanism for enhancing phosphorus acquisition, enhancing aluminum tolerance, and recruiting beneficial rhizobacteria. Soil Sci Plant Nutr 64(6):697–704. https://doi.org/10.1080/00380768.2018.1537093

Yang J, Kloepper JW, Ryu C-M (2009) Rhizosphere bacteria help plants tolerate abiotic stress. Trends Plant Sci 14(1):1–4. https://doi.org/10.1016/j.tplants.2008.10.004

Zulfiqar U, Jiang W, Xiukang W, Hussain S, Ahmad M, Maqsood MF, Ali N, Ishfaq M, Kaleem M, Haider FU, Farooq N, Naveed M, Kucerik J, Brtnicky M, Mustafa A (2022) Cadmium phytotoxicity, tolerance, and advanced remediation approaches in agricultural soils; A comprehensive review. Front Plant Sci 13:773815. https://doi.org/10.3389/FPLS.2022.773815/BIBTEX

Chapter 8
Chaetomium: Goldmine for Sustainable Agriculture

Kasem Soytong and Jiaojiao Song

Abstract *Chaetomium* species are saprophytic fungi that are valuable resources for sustainable agriculture. It is found all over the world because it is durable and survives in living organisms in different climates. *Chaetomium* species are mostly ascomycete saprophytic fungi that survive well under stressful environmental conditions. This area has become one of the gold mines for sustainable agricultural production to maintain a balanced ecological system. The specific, potent isolates of *Chaetomium* were found to protect economic plants through disease, insect, and nematode control and to strongly produce cellulase to degrade organic materials in the fermentation process to increase soil nutrient fertility. All the discovered strains are recommended for acute and dermal toxicity, screening for agrochemical resistance and durability in different climates, including acidic or alkaline conditions, and growing in a wide range of temperature regimes.

Research has revealed that *Chaetomium* species can be developed as biological fungicides, biological insecticides, biological nematicides, and natural active metabolites for plant disease control as well as biosensors for soil revitalization. These bioproducts can be used for sustainable agricultural development and maintain biological diversity in surrounding environments. Research on microbial products of *Chaetomium* species and other fungi for plant disease control has been conducted since 1989. After several years, 22 effective strains of *Chaetomium* were discovered and patented as broad-spectrum microbial fungicides, noted as Patent No. 6266, International Code: AO 1 N 25/12, and then registered as Ketomium® microbial fungicides in Thailand, Laos, Cambodia, and Vietnam and as microbial fertilizers in P. R. China. *Chaetomium* as a microbial fungicide and microbial fertilizer can be applied in combination with integrated pest management (IPM). It has been shown

K. Soytong (✉) · J. Song
King Mongkut's Institute of Technology Ladkrabang (KMITL), Prince of Chumphon Campus, Chumphon, Thailand

Research Institute of Modern Organic Agriculture (RIMOA), King Mongkut's Institute of Technology Ladkrabang (KMITL), Bangkok, Thailand

Association of Agricultural Technology in Southeast Asia (AATSEA), Bangkok, Thailand

V. Gopikrishnan et al. (eds.), *Harnessing Microbial Resources for Sustainable Agriculture*, Microorganisms for Sustainability 61,
https://doi.org/10.1007/978-981-95-4817-0_8

to have pathogen and disease suppressive, curative, and protective effects to control plant diseases. It is strongly recommended to prove acute and dermal toxicity, agrochemical resistance screening, acidity or alkaline conditions, and temperature regimes before developing microbial fungicides and microbial fertilizers. *Chaetomium* spp. are distributed worldwide and survive well under stress conditions. It serves as a natural goldmine for sustainability to maintain and revitalize the environment and preserve the natural and ecological balance in nature. *Chaetomium* species are broad habitats in soil, water, and endophytes with various biological properties of antagonists of phytopathogens, some insects, plant parasitic nematodes, and human pathogens, as well as the production of active metabolites, enzymes, etc.

Keywords *Chaetomium* · Economic plants · Soil nutrition fertility · Soil revitalization

8.1 Introduction to *Chaetomium*

Chaetomium species were first reported in 1817 by Gustav Kunze. The type sample is *Chaetomium globosum*. The species described as perithecia is subglobose, with membranous fruiting bodies surrounded by terminal and lateral hairs; inside the perithecium contains subglobose asci with 8 ascospores per ascus, is pigmented, is smooth-walled, is one-celled, varies in shape, and domains into a lemon shape with a single germ pore. *Chaetomium* is a fungus belonging to Pyrenomycetes, Hymenoascomycetidae II (Xylaria type of centrum), and Ascomycotina, kingdom of fungi (Whiteside 1957; Seth 1970; Webster 1980; Von Arx et al. 1984). Chaetomiales (Ames 1963; Seth 1970) and Chaetomiaceae, Xylariales (Alexopoulos et al. 1979; Quimio 1984). Over 400 species of *Chaetomium* spp., Zopf (1881) described 10 species in the monographs. Chivers (1915) reported 28 species, Ames (1963) reported 85 species, and Von Arx et al. (1986) described 92 species in a monograph. *C. cochliodes* has been reported to produce cochliodinol, which can be used in the textile industry (Brewer et al. 1968).

8.2 Record of *Chaetomium* Species in Thailand

Udagawa and Horie (1973) first recorded some species, for example, *Ch. apiculatum* Lodha, *Ch. variosporum* Udagawa et Horie, and *Ch. venezuelense* Ames found in paddy soils for the first time in Thailand. *Ch. gracile* Udagawa was recorded by Cannon (1986), and *Ch. strumarium* Rai, Tewari & Mukerji was found by Udagawa et al. (1979).

Chaetomium species were further recorded by Soytong (1990), who reported 15 species found in fertile soil and animal dung as follows: *Ch. globosum* Kunze, *Ch. cupreum* Ames, *Ch. lucknowense*, *Ch. ampullare* Chivers, *Ch. bostrychodes* Zopf, *Ch. cochliodes* Palliser, *Ch. aureum* Chivers, *Ch. deceptivum* Malloch & Benny, *Ch.*

hamadae (Udagawa) v. Arx, *Ch. homopilatum* Omvik, *Ch. longicollum*, *Ch. malaysiense*, v. Arx, *Ch. megasporum* Sorgel, *Ch. seminudum* Ames, and *Ch. vitellinum* Carter (Figs. 8.1, 8.2, 8.3, 8.4, 8.5, 8.6, 8.7, 8.8, and 8.9). These species were first recorded in Thailand, except for *Ch. globosum*. Consequently, Petcharat and Soytong (1991) reported the six species of *Chaetomium* found in soils under the canopy of para rubber (*Hevea brasiliensis*) as *Ch. cupeum*, *Ch. globosum*, *Ch. lucknowense*, *Ch. aureum*, *Ch. fusiforme*, and *Ch. gracile*. Moreover, Pornsuriya et al. (2008) reported nine species in pineapple-planted soil as follows: *Ch. cochliodes*, *Ch. cupreum*, *Ch. bostrychodes*, *Ch. aureum*, *Ch. gracile*, *Ch. carinthiacum*, *Ch. flavigenum*, *Ch. Succineum*, and *Ch. perlucidum*, which was also found in soil planted with pineapple in Phatthalung and Rayong Provinces, Thailand. *Chaetomium siamense* sp. nov. was first discovered by Pornsuriya et al. (2011). The list of some *Chaetomium* species found in Thailand is presented in Tables 8.1 and 8.2.

Chaetomium siamense sp. nov. was first discovered by Pornsuriya et al. (2011) and was isolated from soil planted with pineapple in Thailand. The identification was confirmed by ITS rDNA to support *Ch. siamense* as a new species. It was found to produce chaetoviridin A, chaetoviridin F, cochliodone D, and chrysophanol (Fig. 8.10).

Species description: *Chaetomium siamense* sp. nov. Pornsuriya *et al.* (2011), MycoBank MB 506801. The colonies on PDA and CMA grew moderately, and after 14 days, they became red because of red pigment exudates. Ascomata superficial, globose, or ovate, 87–148 × 107–177 μm. Ascomatal hairs apically circulate or coiled, lateral hairs straight, finely verrucose. Asci clavate, 9–14 × 28–40 μm.

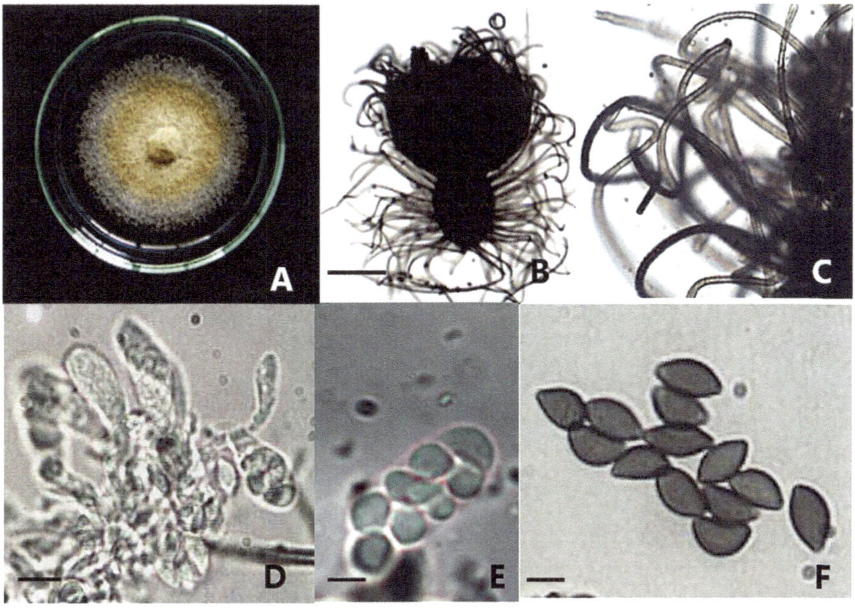

Fig. 8.1 *Chaetomium aureum*. (**a**) Ten-day-old culture on PDA, (**b**) ascomata, (**c**) ascomatal hairs, (**d**) young asci, (**e**) 8 ascospores in an ascus, (**f**) ascospores. Bar. B=100 μm, C,D,E,F=10μm. (Source: Pornsuriya et al. 2008)

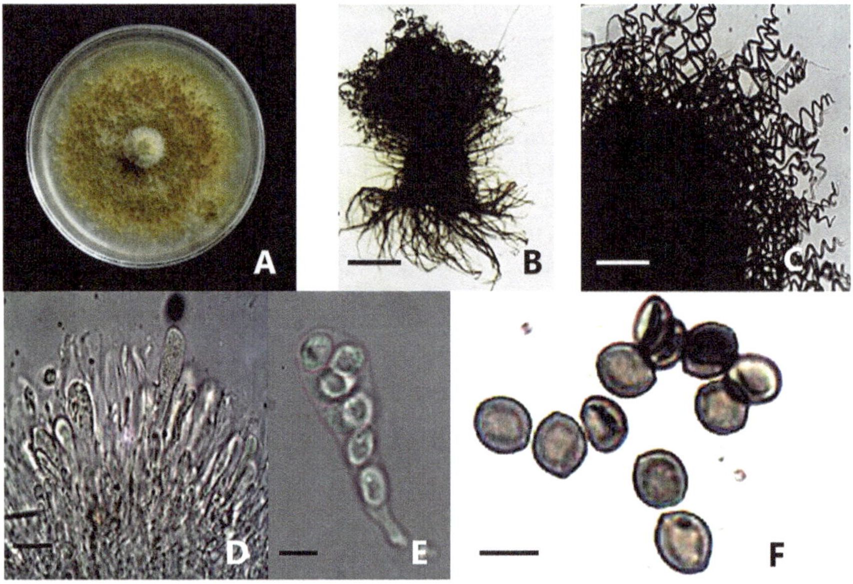

Fig. 8.2 *Chaetomium bostrychodes*. (**a**) Ten-day-old culture on PDA, (**b**) ascomata, (**c**) ascomatal hairs, (**d**) young asci, (**e**) 8 ascospores in an ascus, (**f**) ascospores. Bar. B=200 μm, C,D,E,F=10 μm. (Source: Pornsuriya et al. 2008)

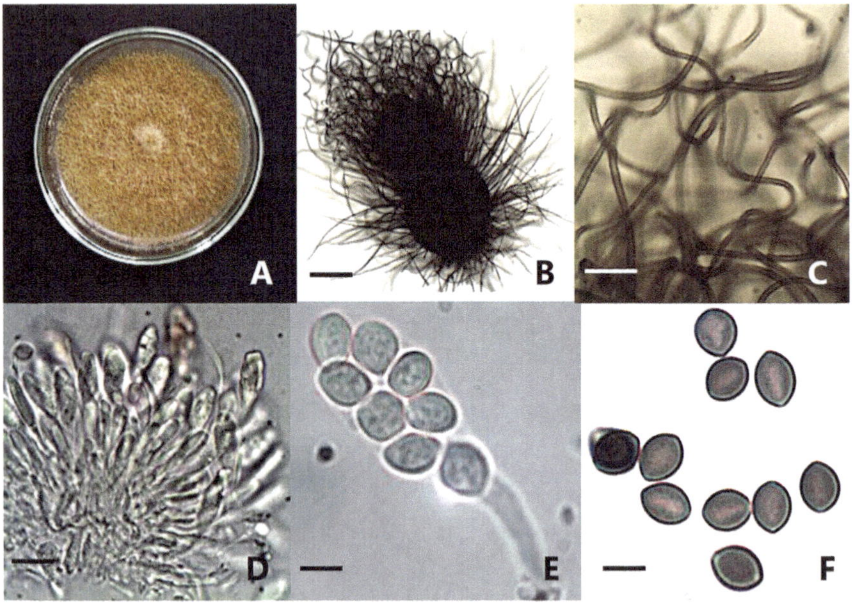

Fig. 8.3 *Chaetomium cochliodes*. (**a**) Ten-day-old culture on PDA, (**b**) ascomata, (**c**) ascomatal hairs, (**d**) young asci, (**e**) 8 ascospores in an ascus, (**f**) ascospores. Bar. B=100 μm, C,D,E,F=10 μm. (Source: Pornsuriya et al. 2008)

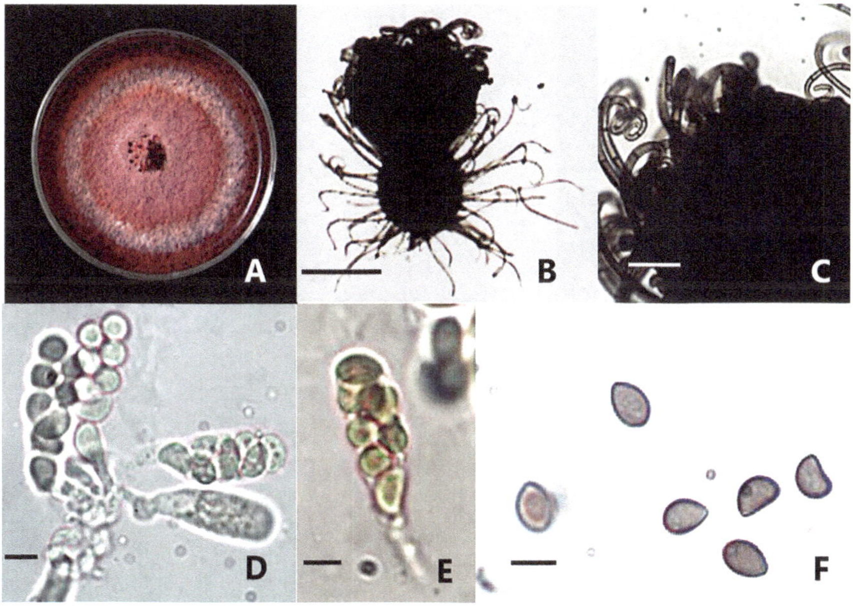

Fig. 8.4 Chaetomium cupreum. (**a**) Ten-day-old culture on PDA, (**b**) ascomata, (**c**) ascomatal hairs, (**d**) young asci, (**e**) 8 ascospores in an ascus, (**f**) ascospores. Bar. B=100 μm, C,D,E,F=10 μm. (Source: Pornsuriya et al. 2008)

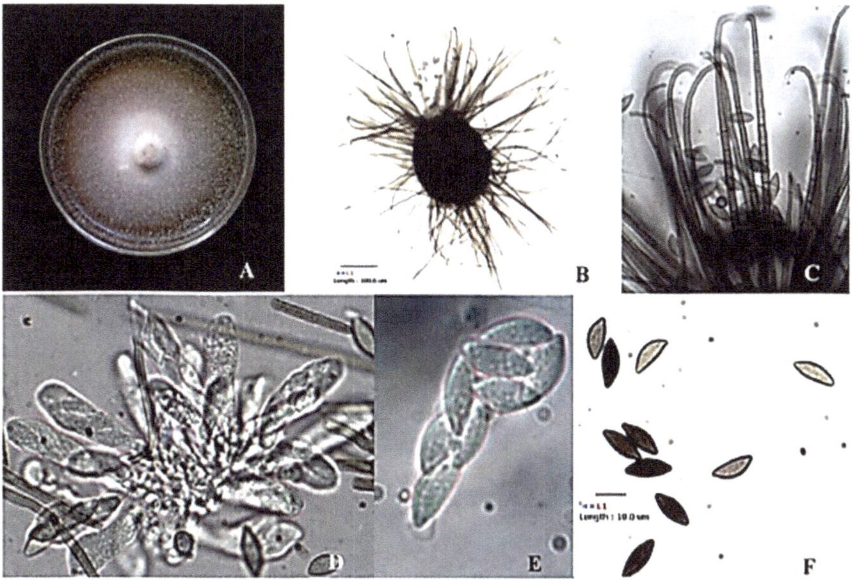

Fig. 8.5 Chaetomium gracile. (**a**) Ten-day-old culture on PDA, (**b**) ascomata, (**c**) ascomatal hairs, (**d**) young asci, (**e**) 8 ascospores in an ascus, (**f**) ascospores. Bar. B=100 μm, C,D,E,F=10 μm. (Source: Pornsuriya et al. 2008)

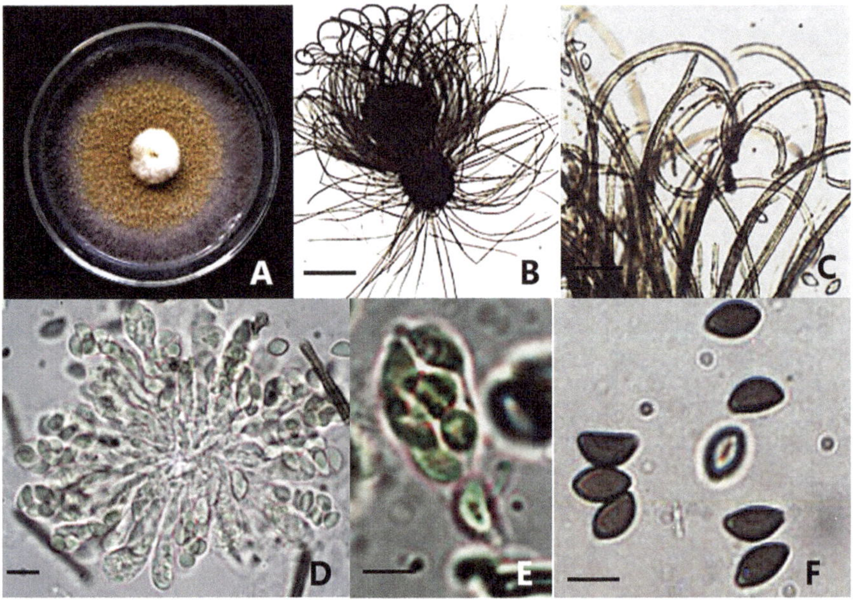

Fig. 8.6 *Chaetomium carinthiacum*. (**a**) Ten-day-old culture on PDA, (**b**) ascomata, (**c**) ascomatal hairs, (**d**) young asci, (**e**) 8 ascospores in an ascus, (**f**) ascospores. Bar. B=100 μm, C,D,E,F=10 μm (Source: Pornsuriya et al. 2008)

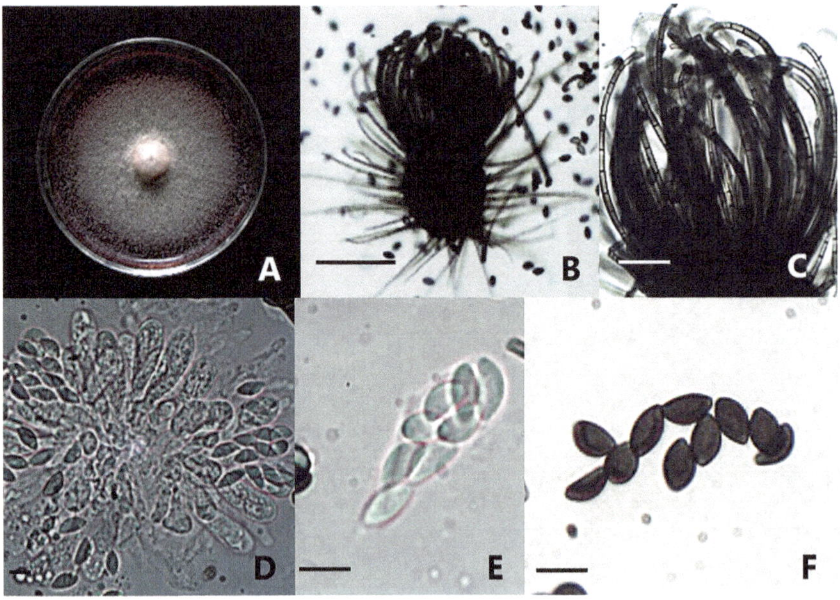

Fig. 8.7 *Chaetomium flavigenum*. (**a**) Ten-day-old culture on PDA, (**b**) ascomata, (**c**) ascomatal hairs, (**d**) young asci, (**e**) 8 ascospores in an ascus, (**f**) ascospores. Bar. B=100 μm, C,D,E,F=10 μm. (Source: Pornsuriya et al. 2008)

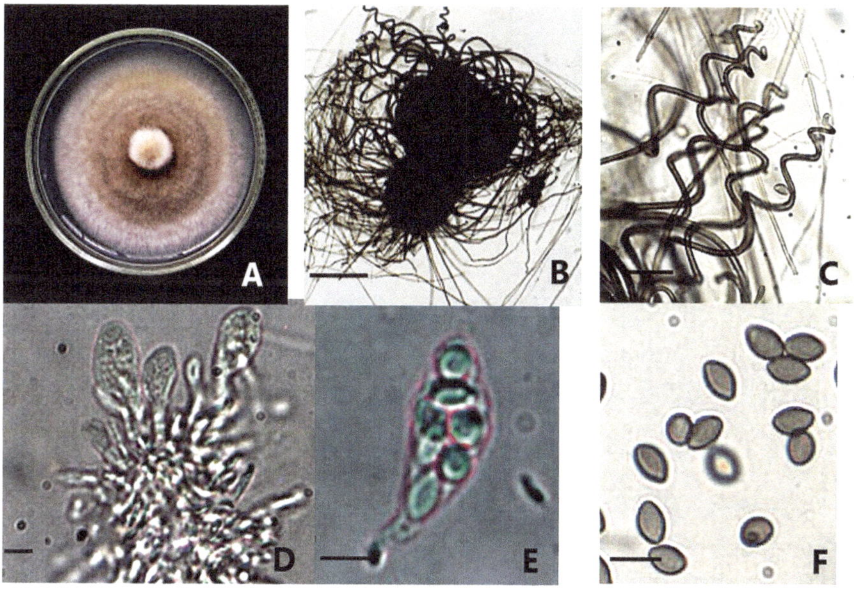

Fig. 8.8 *Chaetomium perlucidum*. (**a**) Ten-day-old culture on PDA, (**b**) ascomata, (**c**) ascomatal hairs, (**d**) young asci, (**e**) 8 ascospores in an ascus, (**f**) ascospores. Bar. B=100 μm, C,D,E,F=10 μm. (Source: Pornsuriya et al. 2008)

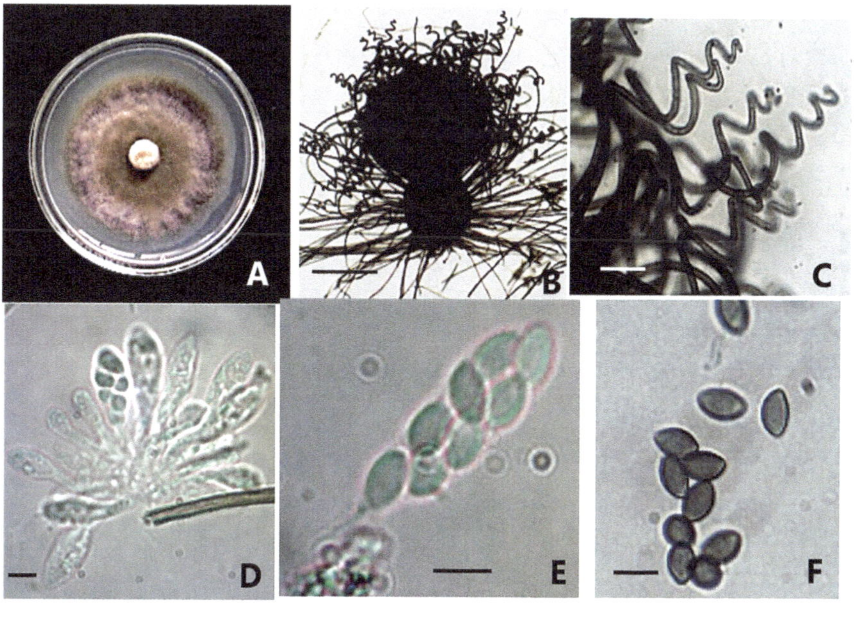

Fig. 8.9 *Chaetomium succineum*. (**a**) Ten-day-old culture on PDA, (**b**) ascomata, (**c**) ascomatal hairs, (**d**) young asci, (**e**) 8 ascospores in an ascus, (**f**) ascospores. Bar. B=100 μm, C,D,E,F=10 μm. (Source: Pornsuriya et al. 2008)

Table 8.1 Description of some *Chaetomium species* in Thailand

Figure	Species	Colony	Ascocarp	Asci	Ascospore	Terminal, hairs
1	*Chaetomium aureum*	Red pigment	Ovate, 78–142 × 90–180 μm	Clavate	Elliptical, 5–7 × 8–12 μm	Arcuate, septate
2	*Chaetomium bostrychodes*	Purple pigment exudate	Ovate, 190–349 × 272–419 μm	Clavate	Lemoniform, 7–9 × 8–11 μm	Spirally coiled
3	*Chaetomium cochliode*	Purple pigment	Ovate, 107–143 × 122–209 μm	Clavate	Lemon-shaped, 4–5 × 7–10 μm	Irregularly sinuous ones.
4	*Chaetomium cupreum*	Red pigment	Ovate, 79–142 × 94–151 μm	Clavate	Reniform, 4–6 × 6–10 μm	Arcuate, apically circinate or coiled
5	*Chaetomium gracile*	Yellow pigment	Ovate (75–161 × 110–202 μm)	Clavate	Fusiform, 5–6 × 12–17 μm	Arcuate
6	*Chaetomium carinthiacum*	Brown	Ovate, 96–146 × 101–153 μm	Clavate	Apical germ pore, elliptical, 4–6 × 8–11 μm	Irregularly sinuous with rough
7	*Chaetomium flavigenum*	Orange pigment	Ovate, 92–134 × 113–190 μm	Clavate	Fusiform, 4–6 × 6–11 μm	Arcuate
8	*Chaetomium perlucidum*	Graylish to black	Ovate, 91–145 × 120–190 μm	Clavate	Elliptical, 3–5 × 6–9 μm	Undulate and irregularly sinuous
9	*Chaetomium succineum*	Dark gray to black	Ovate, 107–143 × 122–209 μm	Clavate	Elliptical, 4–5 × 7–10 μm	Loosely hairs
10	*Chaetomium siamense*	Red pigment	Globose or ovate, 87–148 × 107–177 μm	Clavate	Fusiform,4–5 × 11–13 μm, with two apical germ pores	Straight, finely verrucose,

Source: Pornsuriya et al. (2008)

Table 8.2 List of *Chaetomium* species in Thailand

Species	References
Ch. floriforme Gené & Guarro	Gene and Guarro (1966)
Ch. apiculatum Lodha	Udagawa and Horie (1973)
Ch. variosporum Udagawa et Horie	Udagawa and Horie (1973)
Ch. venezuelense Ames	Udagawa and Horie (1973)
Ch. Ampullare Chivers	Soytong (1991)
Ch. aureum Chivers	Soytong (1991) and Petcharat and Soytong (1991)
Ch. bostrychodes Zopf	Soytong (1991)
Ch. cochliodes Palliser	Soytong (1991)
Ch. cupreum Ames	Soytong (1991) and Petcharat and Soytong (1991)
Ch. deceptivum Malloch & Benny	Soytong (1991)
Ch. gracile Udagawa	Petcharat and Soytong (1991)
Ch. vitellinum Carter	Soytong (1991)
Ch. fusiforme Chivers	Petcharat and Soytong (1991)
Ch. globosum Kunze	Soytong (1991), Petcharat and Soytong (1991), Somrithipol (2004), and Somrithipol et al. (2004)
Ch. hamadae (Udagawa) V. Arx	Soytong (1991)
Ch. homopilatum Omvik	Soytong (1991)
Ch. longicolleum Krezm & Badura	Soytong (1991)
Ch. lucknowense Rai & Tewari	Soytong (1991) and Petcharat and Soytong (1991)
Ch. malaysiense V. Arx	Soytong (1991)
Ch. megasporum Sorgel	Soytong (1991)
Ch. seminudum Ames	Soytong (1991)
Ch. thermophilum La Touche	Somrithipol (2004)
Ch. tortile Bainier	Somrithipol et al. (2004)
Ch. indicum Corda	Somrithipol et al. (2004)

Source: Pornsuriya et al. (2008)

Ascospores are fusiform, 4.3–5 × 11.1–13.5 μm with two apical germ pores (Fig. 8.11).

8.3 The Potential Biological Activities of Chaetomium Species

Chaetomium is a unique saprophytic fungus belonging to the Ascomycetes distributed worldwide in forestry soils, animal dungs, dead organic materials, marine, desert, and symbiotic endophytes in plants and marine algae. Many mycologists have identified

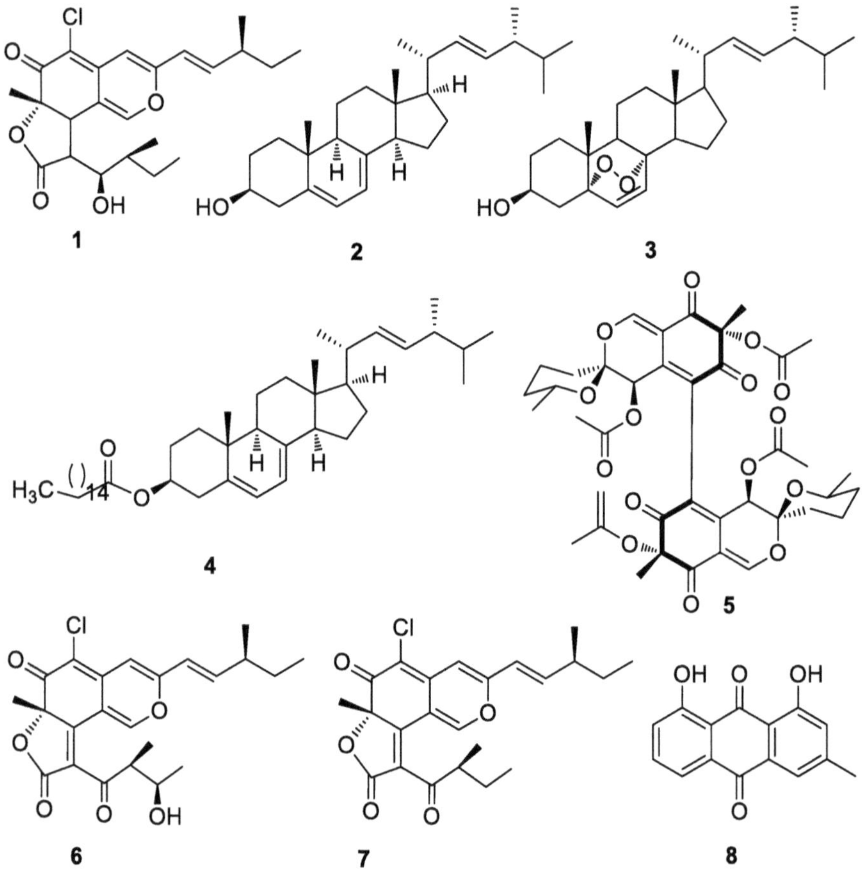

Fig. 8.10 Metabolites from *Chaetomium siamense*. 1, chaetoviridin G; 2, ergosterol; 3, 24(R)-5α,8α-epidioxyergosta-6-22-diene-3β-ol; 4, ergosterylplamitate; 5, cochliodone D; 6, chaetoviridin A; 7, chaetoviridin F; 8, chrysophanol (Pornsuriya et al. 2011)

their metabolites for anticancers, their antibacterial, antifungal, and enzymatic activities, their antagonistic activity against phytopathogens, and their actions as insecticides, nematicides, etc. This species is an interesting life member with biological diversity in nature and a wide range of activities for ecological balance (Table 8.3).

8.3.1 Enzymatic Activity

Chaetomium has been found to be a cellulolytic and saprophytic fungus in animal dung and organic material in fertile soil (Von Arx and Van der Aa 1987). It is reported to produce cellulase to macerate organic materials (Agarwal et al. 1963) and is good for biodegradation for composting. *Ch. globosum* produces a strong cellulase to degrade cellulosic materials during fermentation (Umikalsom et al. 1998) and expresses antagonistic activity against phytopathogens (Dhingra et al.

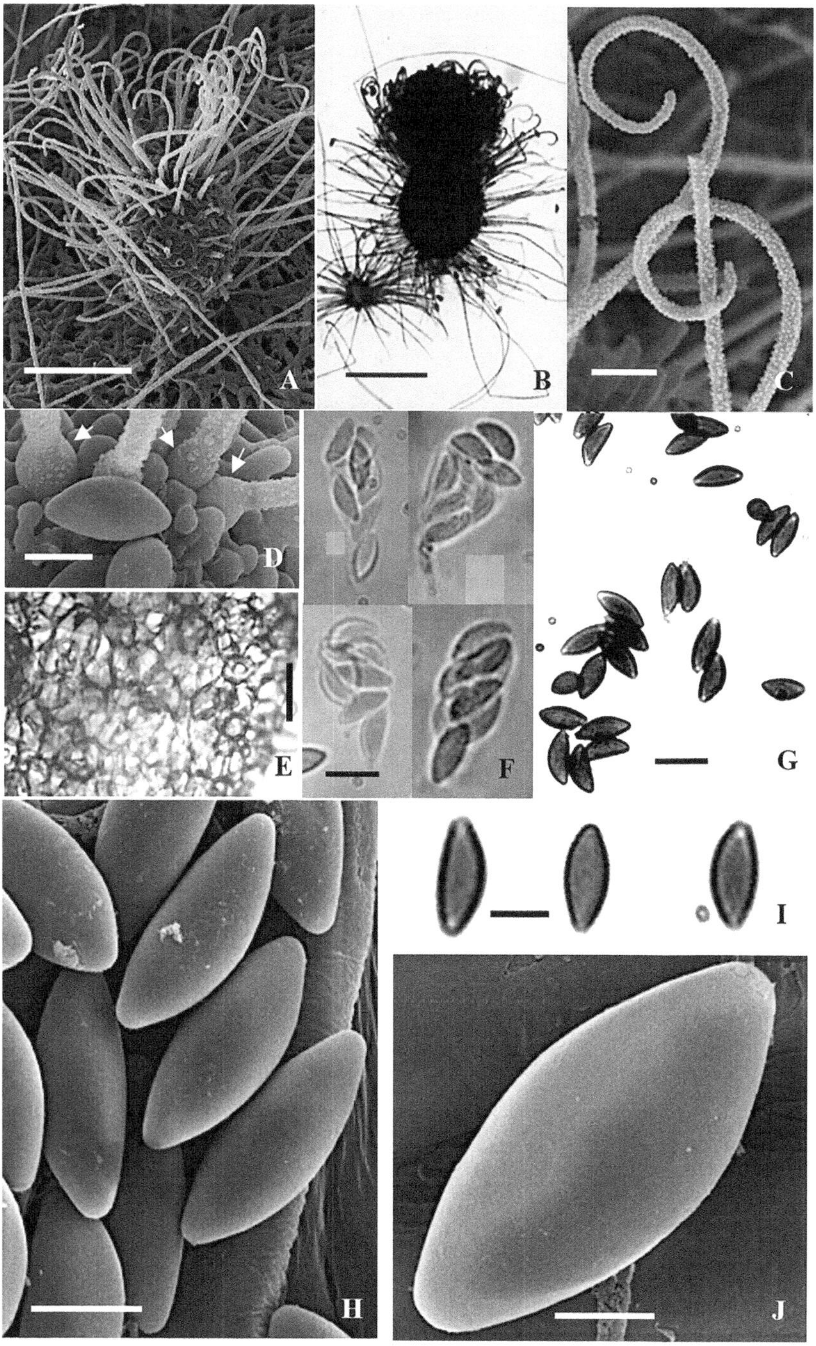

Fig. 8.11 Characteristics of *Chaetomium siamense* under a scanning electron microscope (SEM). (**a**) Ascomata. (**b**) Ascomata. (**c-d**) Ascomatal hairs (arrow). (**e**) Ascomata. (**f**) Asci. (**g**) Ascospores. (**h**) Ascospores. (**i**) Ascospores (arrow). (**j**) Ascospore. Bars A, B=50 μm. C, D, E, F=10 μm. G, H=5 μm. I=2 μm. (Pornsuriya et al. 2011)

Table 8.3 Biological activities of *Chaetomium* species

Species	Metabolites	Activity	References
Chaetomium subaffine	Chaetoglobosin A and prochaetoglobosins	Biological activity	Oikawa et al. (1933)
Chaetomium	Antagonism	*Helminthosporium victoriae*	Tveit and Moore (1954).
Chaetomium	Antagonism	*Fusarium* blight in oat	Tveit and Wood (1955)
Chaetomium globosum	Cellulase	Biological activity	Agarwal et al. (1963)
Chaetomium globosum	Antagonism	Seedling blight of corn	Chang and Kommedahl (1968)
Chaetomium globosum	Antagonism	*Venturia inaequalis*	Boudreau and Andrews (1987)
Chaetomium cupreum	Antagonism	*Pyricularia oryzae*, rice blast	Soytong (1989)
Chaetomium spp.	Chetomin	Ovine ill-thrift	Brewer et al. (1972)
Chaetomium globosum	Chaetoglobosin A and B	Cytotoxic	Sekita et al. (1973)
Chaetomium trilaterale	Oosporein	Biological activity	Cole et al. (1974)
Chaetomium coarctatum	Coarctation	Biological activity	Burrows et al. (1975)
Chaetomium globosum	Antibiosis	Corn root infection	Kommedahl and Mew (1975)
Chaetomium cellulolyticum	Enzymatic	Thermotolerant and cellulolytic	Chahal and Hawsworth (1976)
Chaetomium globosum	Chaetoglobosin C, D, E, and F	Cytotoxic	Sekita et al. (1976)
Chaetomium globosum	Chaetoglobosins, G. J.	Cytotoxic	Sekita et al. (1977)
Chaetomium udagawae	Sterigmatocystin	Biological activity	Udagawa et al. (1979)
Chaetomium thielavioideum	Chaetochromin	Mycotoxin	Sekita et al. (1980)
Chaetomium spp.	Chaetoglobosins A–J	Cytotoxic	Sekita et al. (1981)
Chaetomium globosum	Chetomin	Antimalarial	Kikuchi et al. (1981)
Chaetomium globosum	19-*O*-Acetylchaetoglobosin B and D,	Biological activity	Probst and Tamm (1981)
Chaetomium globosum	Antagonism	Apple scab pathogen	Heye and Andrews (1983)

(continued)

Table 8.3 (continued)

Species	Metabolites	Activity	References
Chaetomium spp.	Neocochliodinol and Isocochliodinol	Biological activity	Sekita (1983)
Chaetomium spp.	Chaetoglobosinsand	Cytotoxic	Sekita et al. (1983)
Chaetomium globosum	Antagonism	apple scab. *Venturia chaeque*,	Cullen and Andrews (1984)
Chaetomium globosum	Antagonism	Apple scab, and *Venturia inaequalis*	Cullen and Andrews (1984)
Chaetomium species	Cochliodinol	Biological activity	Brewer et al. (1968)
Chaetomium gracile	Chaetochromins B, C, and D	Biological activity	Koyama and Natori (1987)
Chaetomium spp.	Antagonism	*Alternaria raphanin* and *A. brassicicola*	Vannacci and Harman (1987)
Chaetomium spp.	Chetracin A, chaetocins B and C	Biological activity	Saito et al. (1988)
Chaetomium globosum	Antagonism	Sugarbeet and cotton damping-off	Walter and Gindrat (1988)
Chaetomium globosum var. *flavo-viridae*	Azaphilones	Biological activity	Takahashi et al. (1990)
Chaetomium globosum	Antagonism	*Pythium ultimum* damping-off of sugar- beet	Di Pietro et al. (1991)
Chaetomium cupreum	Antagonism	*Pyricularia oryzae*, *rice blast*	Soytong (1992a)
Chaetomium biofungicide	Antagonism	*F. oxysporum* f.sp. *lycopersici*	Soytong (1992b)
Chaetomium globosum	Antagonism	*Pythium ultimum* and damping-off	Di Pietro et al. (1992)
Chaetomium cochlioides	Tetrahydrofurans	Biological activity	Abraham and Arfmann (1992)
Chaetomium	Antagonism	Insecticidal	Imamura et al. (1993)
Chaetomium	Competition	*Botrytis cinerea*	Kohl et al. (1995)
Chaetomium globosum	Prenisatin (5-(3-methyl-2-butenyl)indole-2,3-dione)	Antifungal	Breinholt et al. (1996)
Chaetomium biofungicide	Antibiosis	*Phytophthora palmivora*	Prechaprome and Soytong (1997)
Chaetomium biofungicide	Antagonism	*Phytophthora*	Usuwan and Soytong (1999)
Chaetomium brasiliense	Chaetochalasin A	Biological activity	Oh et al. (1998)
Chaetomium globosum	*Cellulase*	Biological activity	Umikalsom et al. (1998)

(continued)

Table 8.3 (continued)

Species	Metabolites	Activity	References
Chaetomium biofungicide	Antibiosis	Black pepper root and basal stem rot	Sodsa-art and Soytong et al. (1999)
Chaetomium biofungicide	Antagonism	*Phytophthora*	Soytong et al. (1999)
Chaetomium globosum	PF1138A and B	Protease inhibitors	Tabata et al. (2000)
Chaetomium atrobrunneum	Chaetoatrosin A	Chitin synthase II inhibitor	Hwang et al. (2000)
Chaetomium biofungicide	Antagonism	Board-spectrum biofungicide	Soytong et al. (2001)
Marine *Chaetomium olivaceum*	Miliacin	Biological activity	Smetanina et al. (2001)
Chaetomium globosum	Anthraquinone-chromanone	Antimycobacterial	Kanokmedhakul et al. (2002))
Chaetomium quadrangulatum	Chromones	Monoamine oxidase	Fujimoto et al. (2002)
Chaetomium sp.	Avipin	Plant-parasitic nematodes	Nitao et al. (2002)
Chaetomium globosum	Antagonism	*Diaporthe phaseolorum* f. sp. *meridionalis*	Dhingra et al. (2003)
Chaetomium subspirale	Oxaspirodion	Inhibitor of inducible TNF-α expression	Rether et al. (2004)
Chaetomium globosum		*Cochliobolus sativus*	Agarwal et al. (2004)
Chaetomium seminudum	Immunomodulatory constituents	Biological activity	Fujimoto et al. (2004)
Chaetomium globosum	Chaetoglobosins Q, R, and T	Biological activity	Jiao et al. (2004)
Chaetomium biofungicide	Antagonism	*Hyophorbe lagenicaulis*	Soytong et al. (2005)
Chaetomium biofungicide	Antibiosis	Phytopathogenic fungi	Shternshis and Soytong et al. (2005)
Chaetomium biofungicide	Antagonism	Late blight of potato	Soytong and Ratancherdchai (2005)
Chaetomium globosum (endophyte)	Globosumones A–C	Cytotoxic orsellinic acid esters	Bashyal et al. (2005)
Chaetomium globosum	Chaetoviridins	Antifungal	Park et al. (2005)
Chaetomium sp.	Antibiotics	Antifungal	Kobayashi et al. (2005)

(continued)

Table 8.3 (continued)

Species	Metabolites	Activity	References
Chaetomium cupreum	Azaphilones	Antifungal	Kanokmedhakul et al. (2006)
Chaetomium spp.	Antagonism	Phytopathogenic fungi	Tomilova and Shternshis (2006)
Chaetomium chiversii	Isocoumarins	Sonoran desert plant-associated with fungus	Wijeratne et al. (2006b)
Chaetomium globosum	Dihydroxanthenone	Anticancer	Wijeratne et al. (2006a)
Endophytic *Chaetomium globosum*	Chaetopyranin	Biological activity	Wang et al. (2006)
Endophytic *Chaetomium*	Chaetominine	Biological activity	Jiao et al. (2006)
Chaetomium globosum	Chaetoglobosin U	Biological activity	Ding et al. (2006)
Chaetomium cochliodes	Chaetocochins A–C	Biological activity	Li et al. (2006a)
Chaetomium indicum	Chaetoindicins A–C	Biological activity	Li et al. (2006b)
Endosymbiotic *Chaetomium*	Chaetocyclinones	Endosymbiotic f	Lösgen et al. (2007)
Chaetomium sp.	Musanahol	Biological activity	Marwah et al. (2007)
Endophytic Ch. globosum from marine red alga	Benzaldehydes	Biological activity	Wang et al. (2007)
Chaetomium globosum	Chemokine receptor CCR-5 inhibitor	Biological activities	Yang et al. (2006)
Chaetomium cochliodes	Bis-spiro-azaphilones and azaphilones	Antimicrobial	Pholkerd et al. (2008)
Chaetomium sp.	Chaetominedione,	Tyrosine kinase inhibitor	Abdel-Lateff (2008)
Chaetomium brasiliense	Secondary metabolites	Biological activity	Li et al. (2008)
Marine fish-Chaetomium spp.	Chaetomugilins A–C	Cytotoxic	Yamada et al. (2008)
Chaetomium brasiliense	Depsidones	Antimalarial, cytotoxic	Khumkomkat et al. (2009)
Endophytic *Chaetomium globosum*	Polyhydroxylated steroids	Biological activity	Qin et al. (2009)
Marine fish-*Chaetomium globosum*	Chaetomugilins A and D	Biological activities	Yamada et al. (2009)

(continued)

Table 8.3 (continued)

Species	Metabolites	Activity	References
Chaetomium elatum	Cytochalasans	Cytotoxic	Thohinung et al. (2010)
Chaetomium biofungicide	Antibiosis	Citrus root rot	Kean et al. (2010)
Chaetomium biofungicide	Antibiosis	*Rigidoporus microporus*	Kaewchai and Soytong (2010)
Chaetomium aureus	Chromones	Biological activity	Li et al. (2010)
Ch. globosum	Cytoglobosins A–G and cytochalasans	Marine-derived endophytic	Cui et al. (2010)
Endophyte *Chaetomium globosum*	Chaetoglobosins	Cytotoxic	Zhang et al. (2010)
Ch longirostre	Azaphilones	Cytotoxic, antimalarial	Panthama et al. (2011)
Chaetomium siamense	Chaetoviridin	Antifungal	Pornsuriya et al. (2011)
Marine fish-derived *Ch globosum*	Azaphilone	Bioactivity	Yamada et al. (2011)
Endophytic *Chaetromium globosum*	Chaetoglocins A–D	Biological activity	Ge et al. (2011)
Chaetomium globosum	Chaetoglobosin A	Nematicidal activity, *Meloidogyne incognita*	Hu et al. (2013)
Chaetomium globosum	Azaphilones	Built environment	McMullin et al. (2013)
Chaetomium lucknowense and *Chaetomium cochiliodes*	Antagonism	*Ganoderma boninense*	Soytong (2014a)
Chaetomium	Antagonism	*Phytophthora nicotianae*	Hung et al. (2015a)
Chaetomium	Antagonism	*Phytophthora palmivora*	Hung et al. (2015b)
Chaetomium	Antagonism	Orchid anthracnose	Vannak et al. (2015)
Endophytic *Chaetomium globosum*	Metabolites	Antiphytopathogen, multitarget acetylcholinesterase inhibitory, and antioxidant	Li et al. (2016)
Chaetomium spp.	Antifungal	*Fusarium* wilt of tea	Phong et al. (2016)

(continued)

Table 8.3 (continued)

Species	Metabolites	Activity	References
Endophyte *Chaetomium globosum*	Metabolites	Negatively affects both above- and belowground herbivores in cotton	Zhou et al. (2016)
Chaetomium biofungicide	Antagonism	*Colletotrichum gloeosporioides*	Vilavong and Soytong (2016)
Chaetomium globosum	Cytochalasan	Extreme environment	Wang et al. (2017)
Ch. elatum	Nanomatebolites	*Phytophthora*	Thongkham et al. (2018)
Ch. brasiliense	Antagonism	*Pythium*	Udompongsuk et al. (2018)
Ch. brasiliense	Nanometabolites	*Dreschlera* sp.	Vareeket et al. (2018)
Aquatic endophyte *Chaetomium globosum*	Bioactive compounds	Bioactivity	Ruan et al. (2018)
Enodophytic *Chaetomium globosum*	Nematicidal metabolites	Nematicidal	Khan et al. (2019)
Endophytic *Chaetomium globosum*	Xanthoquinodins	Bioactivity	Tantapakul et al. (2020)
Chaetomium sp.	Nanoactive metabolites	Rice blast pathogen	Song et al. (2020a)
Endophytic *Chaetomium globosum*	Xanthoquinodins	Biological activity	Tantapakul et al. (2020)
Chaetomium sp.	Nanoactive metabolites	Rice blast pathogen	Song et al. (2020a)
Chaetomium biofungicide	Active metabolites	Broad spectrum biofungicide	Soytong et al. (2001)
Chaetomium cochliodes	Active metabolites	Durian root rot and *Phytophthora palmivora*	Tongon and Soytong (2021)
Chaetomium Brasiliense (endophyte)	Active metabolites	Cytotoxic and antibacterial	Promgool et al. (2021)
Chaetomium cupreum	Nanoactive metabolites	*Plant immunity*	Tongon and Soytong (2021)
Chaetomium	Nanoactive metabolites	*Plant immunity*	Tongon and Soytong (2022)
Chaetomium brasilense (endophyte)	Nanoactive metabolites	Stimulation of rice	Song et al. (2022)
Chaetomium brasiliense	Nanoactive metabolites	Plant immunity	Vareeket et al. (2023)

2003; Agarwal et al. 2004). *Chaetomium* spp. strongly compete with other microorganisms to suppress phytopathogenic microbes in the soil. It is known as a strictly saprophytic coprophilous fungus that is found in plant debris, animal debris, and high soil fertility (Soytong 1990). Some species of *Chaetomium* produce cellulase and ligninase (Soytong et al. 2001). It is a member species that produces biodecomposers.

8.3.2 Potential to Control Pathogens

Chaetomium globosum and *Chaetomium cochliodes* were first reported to control seed-borne pathogens, for example, *Alternaria, Fusarium, Helminthosporium,* and *Rhizoctonia* (Tveit and Moore 1954). *Ch. globosum* decreased the disease prevalence of oat seedlings infested with *Helminthosporium victoriae* and protected the infection of corn seeds coated with *Ch. globosum* (Chang and Kommedahl 1968). *Ch. globosum* has been reported to control *Venturia inaequalis*, which causes apple scab disease (Heye and Andrews 1983). *Ch. globosum has been* reported to control *Phomopsis sojae* as a soybean pathogen (Manandhar et al. 1986). The specific isolate of *Ch. globosum* was reported for rice blast control caused by *Pyricularia oryzae* (Soytong and Quimio 1989b). Furthermore, *Chaetomium cupreum* was first recorded to control rice blast (Soytong 1992a).

Chaetomium species are known to be distributed in fertile soil, artificial dung, and sediments that are found to survive under stress conditions. *Chaetomium* species are reported to release active metabolites for inhibited phytopathogens and antimicrobial activities (Soytong et al. 2001). *Ch. globosum* is recorded to release chetomin, which suppresses *Pythium ultimum* (Di Pietro et al. 1992) and produces prenisatin, which inhibits *Botrytis cinerea* (Breinholt et al.,1996).

Our research revealed that *Ch. globosum* releases echinulin and chaetomanone, which express antituberculosis agents (Kanokmedhakul et al. 2002), inhibit *Pyricularia oryzae* (rice blast) (Soytong and Quimio 1989b; Soytong 1992a), and control *Fusarium* tomato wilt (Soytong 1990; Sibounnavong et al. 2012). Moreover *Ch. cupreum* is reported to release rotiorinols A–C, (–)-rotiorin, and rubrorotiorin, which inhibit *Candida albicans* (human disease) (Kanokmedhakul et al. 2006).

Chaetomium siamense sp. nov. (MycoBank MB 506801) is a new species isolated from soil planted with pineapple and found in Thailand. It has been shown to produce chaetoviridin G, chaetoviridin A, chaetoviridin F, cochliodone D, and chrysophanol, which exhibit some antimicrobial activity (Pornsuriya et al. 2011). It can antagonize *Phytophthora* spp., *F. oxysporum* f sp. *lycopersici*, and *Colletotrichum* spp., which can cause plant diseases. *Chaetomium cochliodes* VTh01 and CTh05 produce chaetochalasin A and chaetoviridine F, which express antimalaria (*Plasmodium falciparum*). Chaetochalasin A, cochliodones C, and chaetoviridins E and F actively suppressed *M. tuberculosis*. Chaetoviridins E and F are antimicrobial against BC1, NCI-H187, KB, and other cell lines (Pholkerd et al. 2008). These strains of *Ch. cochliodes* VTh01 and CTh05 have been formulated as microbial fungicides that effectively control *D. oryzae* and *Ganoderma boninense* (Soytong 2014a, b). *Chaetomium brasiliense* produces mollicellins B, C, and E, which suppress *P. falciparum*. Mollicellin K inhibited *C. albicans* and *M. tuberculosis*. Mollicellins H, J, and K-N have antimicrobial effects on BC1, NCI-H187, and KB (Khumkomkhet et al. 2009). These active metabolites from *Ch. brasiliense* also actively inhibited *Fusarium solani,* causing wilt in tomato (Tongon and Soytong 2016). *Chaetomium elatum* ChE01 releases prochaetoglobosin III; prochaetoglobosin III; chaetoglobosins B, C, D, F, and G; chaetoglobosin V; and isochaetoglobosin D, which inhibits cytotoxicity in cholangiocarcinoma cell lines and human breast cancer (Thohinung et al. 2010). It is formulated to be a microbial fungicide that can decrease diseases, including *Fusarium* tomato wilt (Soytong 2015), anthracnose of banana plants (Soytong et al. 2019), and rice blast (Song et al. 2020b). Microbial fungicides developed from *Ch. cupreum* effectively controls white rot in rubber trees (Soytong and Kaewchai 2014). With this, *Ch. cupreum* RY202 was found to produce sochromophilonol, ochrephilonol, and clearanol B, which are cytotoxic to the NCI-H187 and KB cell lines (Panthama et al. 2014).

8.3.3 Endophytic Chaetomium

Endophytic *Chaetomium* species and their biological activities are reported to produce more than 142 active metabolites (Fatima et al. 2016). *Ch. The Globosum* endophyte is isolated from plant cells of *Gingko biloba,* which releases chaetoglobosins A, C, D, E, G, and R to suppress *Rhizopus stolonifer* and *Coniella diplodiella*, causing grape rot (Li et al. 2016). *Ch. fusiforme, Ch. globosum*, and *Ch. madrasense* are endophytic fungi that grow in healthy plants (Elkhateeb et al. 2021). These endophytic fungi are plant-associated fungi that appear inside the plant and are found in the intercellular spaces of plant cells with no symptoms. The endophyte *Chaetomium* lives in plant cells and produces active metabolites related to phytochemical biosynthesis to stimulate immunity or resistance to abiotic and biotic stresses in plants (Wen et al. 2002).

Our research investigation revealed that the endophytic *Ch. The Globosum* strain 7 s^{-1} is found inside vascular plant cells of palm trees (Song et al. 2015) and rice (Leewijit et al. 2016). The endophytic *Ch. The Globosum* strain 7 s^{-1}, which is isolated from *Rhapis cochinchinensis* (Palm tree), produces chetomin and chaetocochin C, which are antimicrobial agents against gram-positive bacteria, inhibit a normal cell line, and are cytotoxic to NCI-H187, MCF-7, and KB (Tantapakul et al. 2020). Endophytic *Chaetomium brasiliense* is found in the plant cells of rice and can release active metabolites to inhibit its cytotoxic and antibacterial activities (Promgool et al. 2021). We found that both endophytes of *Ch. globosum* 7 s^{-1} and *Ch. brasiliense*, which effectively controls anthracnose disease (*Colletotrichum* sp.), wilt (*Fusarium* sp.), and root rot (*Phytophthora* sp.).

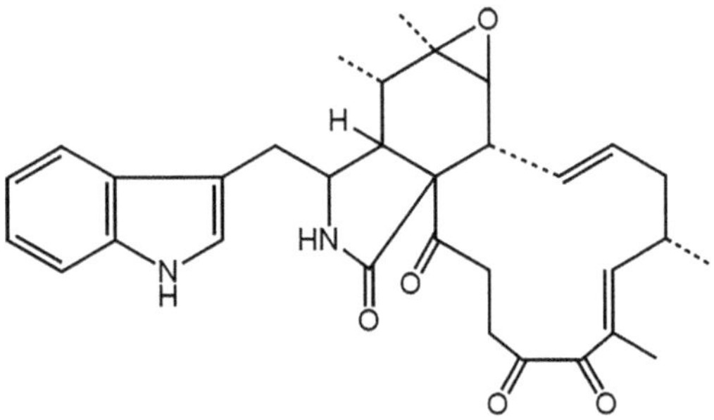

Fig. 8.12 Chaetoglobosin C produced by *Chaetomium globosum*

8.4 Application of *Chaetomium* in Sustainable Agriculture

The selected species of *Chaetomium* is a powerful enzymatic and antagonistic fungus used to control plant pathogens that can be found throughout almost the entire world. The active metabolites of *Chaetomium* mostly exhibit antimicrobial activity against phytopathogens. Research has revealed that the crude metabolite from *Ch. globosum* inhibited the spore production of phytopathogens, for example, *F. oxysporum* f. sp. *lycopersici*, *C. gloeosporioides* (citrus anthracnose), *P. palmivora* (black pepper rot), and *P. parasitica* (citrus rot). *Ch. globosum* has been shown to release chaetoglobosin C, which inhibits several phytopathogens, for example, *F. oxysporum*, *C. dematium*, *C. gloeosporioides*, *Rhizoctonia solani*, *Sclerotium rolfsii*, *Magnaporthe oryzae*, *P. cactorum*, *P. palmivora*, *P. parasitica* (Soytong et al. 2001), and chaetoviridins A and inhibits *P. ultimum* and *M. oryzae* (Park et al. 2005). Rotiorinol pure compounds from *Ch. cupreum* exhibited antimicrobial activity (Kanokmedhakul et al. 2006). *Ch. globosum* was found to release chaetoglobosin C (Fig. 8.12), which inhibited *P. palmivora* (root rot of durian).

The crude metabolites of *Ch. Cochliodes were* found to inhibit *Drechslera oryzae* (leaf blight) at an ED_{50} of 66 μg/mL, resulting in the formation of abnormal spores of *D. oryzae* and subsequent loss of pathogenicity. (Vareeket et al. 2023). Crude metabolites derived from *Ch. globosum Ch. lucknowense* and *Ch. cupreum* reportedly inhibits *F. oxysporum*, causing tea wilt (Phong et al. 2016). The crude extract of *Ch. lucknowense*, *Ch. globosum*, and *Ch. cupreum* was found to suppress *P. nicotianae* (citrus root rot) at ED_{50} values between 2.6 and 4.6 μg/mL (Hung et al. 2015b). Rotiorinol produced from *Ch. cupreum was* found to inhibit *Rigidophorus microporus* (rot of oil palm), for which the ED_{50} was 26 μg/l (Kaewchai and Soytong 2010).

Chaetomium is a powerful saprophytic fungus that belongs to the largest species of Ascomycetes (Von Arx et al. 1986; Soytong and Quimio 1989a; Rodríguez et al. 2002). It is a potential biological control agent against various phytopathogens (Soytong and Quimio 1989b; Park et al. 2005). *Chaetomium* controls plant pathogenic fungi and bacteria through antibiosis, mycoparasitism, and nutrient competition in combination (Zhang and Yang 2007) and can be grown under stress conditions. *Ch. globosum* and *Ch.* Cochliodes inhibit *Helminthosporium victoriae* to protect oat seeds (Tviet and Moor 1954) and produce metabolites to control the damping-off of sugar beet (*Pythium ultimum*) (Di Pietro et al. 1991). *Chaetomium* also inhibits more phytopathogens, for example, *Alternaria brassicicola* (Vannacci and Harman 1987), *Pythium ultimum* (Di Pietro et al. 1991), *Botrytis cinerea* (Kohl et al. 1995), and *Rhizoctonia solani* (Walter and Gindrat 1988). *Ch. cupreum* antagonizes *Colletotrichum* sp. and *Phomopsis* sp. (Manandhar et al. 1986) and *Venturia inaequalis* (apple scab) (Cullen et al. 1984).

Since 1989, antagonistic *Chaetomium* species have been found to control several phytopathogens (Soytong et al. 2001). *Ch. cupreum* and *Ch. globosum* inhibit *C. lunata* and *M. oryzae* (Soytong 1989, 1992a). Ch. *globosum was* found to inhibit *F. oxysporum* f. sp. *lycopersici* and *Pseudomonas solanacearum* (Soytong 1991). Different synergistic species and isolates of the *Chaetomium* microbial fungicide have been formulated in pellet, powder, and liquid forms (Soytong et al. 2001; Kaewchai et al. 2009). It has been registered as a biofertilizer in China and a biofungicide in Thailand, Laos, Vietnam, and Cambodia. *Chaetomium* microbial fungicides have successfully controlled rot disease in bottle palm in the field (Soytong et al. 2005) and root rot in black pepper, strawberry, citrus, and sugar beet (Soytong et al. 2001; Tomilova and Shternshis 2006). *Chaetomium* mycofungicide also successfully inhibits *Rhizoctonia solani* (potato rot), *Didymella applanata* (raspberry spur blight) (Shternshis et al. 2005; Tomilova and Shternshis 2006), and other phytopathogens, for example, *F. oxysporum* f. sp. *lycopersici*, *P. palmivora* (Prechaprome and Soytong 1997), and *P. parasitica* (Usuwan et al. 1999; Soytong 1999; Sodsa-art and Soytong 1999).

In 2020, all the identified species of Chaetomium were tested for synergistic growth, for example, *Ch. brasiliense*, *Ch. cupreum*, *Ch. cochlioildes*, *Ch. longirostre*, *Ch. elatum*, *Ch. globosum*, *Ch. Siamense*, and *Ch. lucknowense*, which is a novel synergistic strain that combines broad-spectrum microbial fungicides for plant disease control. It has been successfully applied to control several diseases, for example, *Phytophthora* rot of durian, citrus, black pepper, leaf blight of durian, anthracnose, rice blast, sheath blight, and wilt diseases.

8.5 *Chaetomium* as a Microbial Nematicide and Microbial Insecticide

The wide range of potential uses of *Chaetomium* for agriculture has led to the identification of other isolates for other purposes, such as nematode and insect control. Various studies have been carried out on *Chaetomium* species. The endophyte *Chaetomium* produces active metabolites to control root-knot nematodes (*Meloidogyne incognita*) and soybean cyst nematodes (*Heterodera glycines* and *Globodera pallida*). The active metabolites of *Chaetomium* species have been reported to control *Aphis gossypii* (cotton aphids) and *Spodoptera exigua* (beet armyworms). The active metabolites of *Ch. globosum* suppressed the hatching eggs of root-knot nematodes and controlled *H. glycines*. Avipin is reported to be an active compound derived from *Ch. globosum* and has antagonistic effects on root-knot nematodes (Nitao et al. 2002).

In 2012, *Ch. globosum* is patented in China for its nematicidal activity, which produces metabolites for nematode control, namely, patent No. CN102925369A by Chinese scientists. It is claimed that chaetoglobosin A is produced by *Ch. globosum* NK106 (CGMCC6716), which was applied at 300 mg/mL to kill the root-knot nematode because of its high toxicity to the juvenile stage of *M. incognita*, resulted in a 63% reduction in nematode eggs and 90% lethality.

Chaetoglobosin A is reported to be produced by *Ch. globosum* NK102, which has a nematicidal effect on the control of *M. incognita. Ch. globosum* NK102 has repellent effects on the secondary stage of juveniles. Chaetoglobosin A strongly promoted a nematicidal effect in the secondary stage, with 99.8% mortality at 300 μg/mL and an LC_{50} of 77.0 μg/mL. Chaetoglobosin A can prevent the second stage of nematodes from penetrating host plants, and the number of nematode eggs decreases (Hu et al. 2013).

Endophytic *Ch. globosum* TAMU 520 derived from cotton is reported to be systemically colonized in cotton plant tissue by treated seeds. It reduces root-knot disease and decreases the reproduction of female nematodes (Zhou et al. 2016). *Ch. globosum* can infect the egg of a cyst nematode (*Globodera pallida*), which was reported in the United States and was associated with the greatest reduction in potato disease caused by *G. pallida* (76%) (Kooliyattil et al. 2017). Other reports showed endophytic *Ch. Globosum* strain YSC5, which showed a nematicidal effect by releasing the active compound flavipin, and Chaetoglobosin A & B, which expressed strong nematicidal effects of 83%, 87%, and 91%, respectively, and caused mortality in *Meloidogyne javanica* (second-stage juvenile stage) at a concentration of 200 μg/mL, for which the LC_{50} values were 107, 99, and 88 μg/mL, respectively, after treatment for 72 h. *Ch. globosum* YSC5 was also reported to contain 4,5,6-trihydroxy-7-methylphthalide and 3-methoxyepicoccone, which inhibited *M. javanica* at 75 and 78%, respectively, with LC_{50} values of 131 and 124 μg/mL, respectively (Khan et al. 2019). *Ch.* The presence of the *Globosum* endophyte had a negative effect on the fecundity of beet armyworms and cotton aphids, which colonized the plants *treated* with *Ch. globosum.* Lately, beet armyworms have a smooth head capsule (Zhou et al. 2016). Moreover, *Chaetomium cupreum* has also been found to digest ruminant nutrition (Suphalucksana and Soytong 2006).

8.6 Future Remarks

Chaetomium species are mostly ascomycete saprophytic fungi that can be found worldwide and survive well under stressful environmental conditions. This area has become one of the gold mines for sustainable agricultural production to reserve an ecologically balanced system. The specific, potent isolates of Chaetomium were found to protect economic plants through disease protection and insect and nematode control and to strongly produce cellulase to degrade organic materials in the fermentation process to increase soil nutrient fertility. All the discovered strains are recommended for acute and dermal toxicity; screening for agrochemical resistance and durability in different climates, including acidic or alkaline conditions; and growth in a wide range of temperature regimes.

References

Abdel-Lateff A (2008) Chaetominedione, a new tyrosine kinase inhibitor isolated from the algicolous marine fungus *Chaetomium* sp. Tetrahedron Lett 49:6398–6400

Abraham WR, Arfmann HA (1992) Rearranged tetrahydrofurans from *Chaetomium cochlioides*. Phytochemistry 31:2405–2408

Agarwal PN, Sahgal D, Verma G, Verma R (1963) Decomposition of cellulose by fungus *Chaetomium globosum*: studies on enzyme activity. Indian J Exp Biol 1:46

Agarwal R, Tewari AK, Srivastava KD, Singh DV (2004) Role of antibiosis in the biological control of spot blotch (*Cochliobolus sativus*) of wheat by *Chaetomium globosum*. Mycopathologia 157:369–377

Alexopoulos CJ, Ulos CJ, Mims CW (1979) Introductory mycology, 4th edn. Wiley. 632 pp

Ames LM (1963) A monograph of the Chaetomiaceae. US Army Res Dev, Ser 2. 125 pp

Bashyal BP, Wijeratne EMK, Faeth SH, Gunatilaka AAL (2005) Globosumones A-C, cytotoxic orsellinic acid esters from the Sonoran Desert endophytic fungus *Chaetomium globosum*. J Nat Prod 68:724–728

Boudreau MA, Andrews JH (1987) Factors influencing antagonism of *Chaetomium globosum* to *Venturia inequalis:* a case study in failed biocontrol. Phytopathology 77:1470–1475

Breinholt J, Demuth H, Heide M, Jensen G, Moller I, Nielsen R et al (1996) Prenisatin (5-(3-methyl-2-butenyl)indole-2,3-dione) is an antifungal isatin derivative from *Chaetomium globosum*. Acta Chem Scand 50:443–445

Brewer D, Jerram WA, Taylor A (1968) The production of cochliodinol and a related metabolite by *Chaetomium* species. Can J Microbiol 14:861–866

Brewer D, Duncan JM, Jerram WA, Leach CK, Safe S, Taylor A, Christensen CM (1972) Ovine ill-thrift in Nova Scotia. 5. The production and toxicology of chetomin, a metabolite of *Chaetomium* spp. Can J Microbiol 18:1129–1137

Burrows BF, Turner WB, Walker ERH (1975) 8-Ethylidene-7,8-dihydro-4 methoxypyrano[4,3-*b*]pyran-2,5-dione (Coarctatin), a metabolite of *Chaetomium coarctatum*. J Chem Soc Perkin Trans I:999–1000

Cannon PE (1986) A revision of *Achaetomium, Achaetomiella Subramantula,* and some similar species of *Chaetomium*. Trans Br Mycol Soc 87:45–76

Chahal DS, Hawsworth DL (1976) *Chaetomium cellulolyticum*, a new thermotolerant and cellulolytic *Chaetomium* I. Isolation, description and growth rate. Mycologia 68:600–610

Chang I, Kommedahl T (1968) Biological control of seedling blight of corn by coating kernels with antagonistic microorganisms. Phytopathology 58:1395–1401

Chivers AH (1915) A monograph of the genera *Chaetomium* and *Ascotricha*. Mem Torrey Bot Club 14:155–240

Cole RJ, Kirksey JW, Gutter HG, Davis EE (1974) Toxic effects of oosporein from *Chaetomium trilaterale*. J Agric Food Chem 22:517–520

Cui CM, Li XM, Li CS, Proksch P, Wang BG (2010) Cytoglobosins A-G, cytochalasans from a marine-derived endophytic fungus, *Chaetomium globosum* QEN-14. J Nat Prod 73:729–733

Cullen D, Andrews JH (1984) Evidence for the role of antibiosis in the antagonism of *Chaetomium globosum* to the apple scab pathogen, *Venturia inaequalis*. Can J Bot 62:1819–1823

Cullen D, Berbee FM, Andrews JA (1984) *Chaetomium globosum* antagonizes the apple scab pathogen. *Venturia chaeque*, under field conditions. Can J Bot 62:1814–1818

Dhingra OD, Mizubuti ESG, Santana FM (2003) *Chaetomium globosum* for reducing primary inoculum of *Diaporthe phaseolorum* f. sp. *meridionalis* in soil- surface soybean stubble in field conditions. Biol Control 26:302–310

Di Pietro A, Kung R, Gutrella M, Schwinn FJ (1991) Parameters influencing the efficacy of *Chaetomium globosum* in controlling *Pythium ultimum* damping-off of sugar- beet. J Plant Dis Prot 98:565–573

Di Pietro A, Gut-Rella M, Pachlatko JP, Schwinn FJ (1992) Role of antibiotics produced by *Chaetomium globosum* in biocontrol of *Pythium ultimum*, a causal agent of damping-off. Phytopathology 82:131–135

Ding G, Song YC, Chen JR, Xu C, Ge HM, Wang XT et al (2006) Chaetoglobosin U, a cytochalasan alkaloid from the endophyte *Chaetomium globosum* IFB-E019. J Nat Prod 69:302–304

Elkhateeb WA, Kolaibe AGAL, Elnahas MO, Daba GM (2021) Highlights on Chaetomium morphology, secondary metabolites and biological activates. J Pharm Pharmacol Res 4(1). https://doi.org/10.31579/2693-7247/030

Fatima N, Mukhtae U, Ul-Haq I, Qazi MA, Jadoon M, Ahmed S (2016) Biological evaluation of Endophytic fungus *Chaetomium* sp. NF15 of *Justicia adhatoda* L.: a potential candidate for drug discovery. Jundishapur J Microbiol 9(6):e29978. https://doi.org/10.5812/jjm.29978

Fujimoto H, Nozawa M, Okuyama E, Ishibashi M (2002) Five new chromones possessing monoamine oxidase inhibitory activity from an ascomycete, *Chaetomium quadrangulatum*. Chem Pharm Bull 50:330–336

Fujimoto H, Sumino M, Okuyama E, Ishibashi M (2004) Immunomodulatory constituents from an ascomycete, *Chaetomium seminudum*. J Nat Prod 67:98–102

Ge HM, Zhang Q, Xu SH, Guo ZK, Song YC, Huang WY et al (2011) Chaetoglocins A-D. Four new metabolites from the endophytic fungus *Chaetomium globosum*. Planta Med 77:277–280

Gene J, Guarro J (1966) A new Chaetomium from Thailand. Myclo Res 100(8):1005–1009

Heye CC, Andrews JH (1983) Antagonism of *Athelia bombacina* and *Chaetomium globosum* to the apple scab pathogen, *Venturia inaequalis*. Phytopathology 73:650–654

Hu Y, Zhang WP, Zhang P, Ruan WB, Zhu XD (2013) Nematicidal activity of Chaetoglobosin a Poduced by *Chaetomium globosum* NK102 against *Meloidogyne incognita*. J Agric Food Chem 61:41–46

Hung PN, Pongnak W, Soytong K, Supattra P (2015a) Efficacy of *Chaetomium* species as biological control agents against *Phytophthora nicotianae* root rot in citrus. Mycobiology 43:288–296

Hung PN, Pongnak W, Soytong K, Supatta P (2015b) Biological control of *Phytophthora palmivora* causing root rot of pomelo using *Chaetomium* spp. Mycobiology 43:63–70

Hwang EI, Yun BS, Kim YK, Kwon BM, Kim HG, Lee HB et al (2000) Chaetoatrosin a, a novel chitin synthase II inhibitor produced by *Chaetomium atrobrunneum* F449. J Antibiot 53:248–255

Imamura K, Gomi S, Yaguchi T, Moryama C, Iwata M (1993) Novel insecticide PF1093 and its manufacture with *Chaetomium*. Jpn. Kokai Tokkyo Koho. 8 pp

Jiao W, Feng Y, Blunt JW, Cole ALJ, Munro MHG (2004) Chaetoglobosins Q, R, and T, three further new metabolites from *Chaetomium globosum*. J Nat Prod 67:1722–1725

Jiao RH, Xu S, Liu JY, Ge HM, Ding H, Xu C et al (2006) Chaetominine is a cytotoxic alkaloid produced by the endophyte *Chaetomium* sp. IFB-E015. Org Lett 8:5709–5712

Kaewchai S, Soytong K (2010) Application of biofungicides against *Rigidoporu microporus* causing white root disease of rubber trees. J Agric Technol 6:349–363

Kaewchai S, Soytong K, Hyde KD (2009) Mycofungicides and fungal biofertilizers. Fungal Divers 38:25–50

Kanokmedhakul S, Kanokmedhakul K, Phonkerd N, Soytong K, Kongsaeree P, Suksamrarn A (2002) Antimycobacterial antraquinone-chromanone compound and diketopiperazine alkaloid from the fungus *Chaetomium globosum* KMITL-N0802. Planta Med 68:834–836

Kanokmedhakul SK, Nasomjai P, Loungsysouphanh S, Soytong K, Isobe M, Kongsaeree K, Prabpai S, Suksamran A (2006) Antifungal Azaphilones from the fungus, *Chaetomium cupreum* CC3003. J Nat Prod 69:891–895

Kean S, Soytong K, To-anun, C (2010) Application of biofungicide to control citrus root rot under field condition in Cambodia. J Agric Technol 6:219–230

Khan B, Yan W, Wei S, Wang ZY, Zhao SS, Cao LL, Rajput NA, Ye Y (2019) Nematicidal metabolites from endophytic fungus *Chaetomium globosum* YSC5. doi:https://doi.org/10.1093/femsle/fnz169

Khumkomkat P, Kanokmedhakul S, Kanokmedhakul K, Hahnvajanawong C, Soytong K (2009) Antimalarial and cytotoxic depsidones from the fungus *Chaetomium brasiliense*. J Nat Prod 72:1487–1491

Khumkomkhet P, Kanokmedhakul S, Kanokmedhakul K, Hahnvajanawong C, Soytong K (2009) Antimalarial and cytotoxic depsidones from the fungus *Chaetomium brasiliense*. J Nat Prod 72:1487–1491

Kikuchi T, Kodata S, Nakamura K, Nishi A, Taga T, Kaji T (1981) Dethio-tetra (methylthio) chetomin, a new antimicrobial metabolite of *Chaetomium globosum* KUNZE ex FR. Structure and partial synthesis from chetomin. Chem Pharm Bull 30:3846–3848

Kobayashi M, Kanasaki R, Sato I, Abe F, Nitta K, Ezaki M et al (2005) FR207944, an antifungal antibiotic from *Chaetomium* sp. no. 217 I. Taxonomy, fermentation, and biological properties. Biosci Biotechnol Biochem 69:515–521

Kohl JW, Molhoek HL, van der Plas CH, Fokkema HJ (1995) Effect of *Ulocladium atrum* and other antagonists on sporulation of *Botrytis cinerea* on dead lily leaves exposed to field condition. Phytopathology 85:393–400

Kommedahl T, Mew IC (1975) Biocontrol of corn root infection in the field by seed treatment with antagonists. Phytopathology 65:296–300

Kooliyattil R, Dandua LM, Knudsen GR (2017) Prospecting fungal parasites of the potato cyst nematode *Globodera pallida* using a rapid screening technique. J Basic Microbiol 57:386–392

Koyama K, Natori S (1987) Chaetochromins B, C, and D; *bis*(naphtho-γ-pyrone) derivatives from *Chaetomium gracile*. Chem Pharm Bull 35:578–584

Leewijit T, Pongnak W, Soytong K, Poeaim S (2016) Isolation of soil and Endophytic Fungi from Rice (*Oryza sativa* L.). Int J Agric Technol 12(7.1)

Li GY, Li BG, Yang T, Yan JF, Liu GY, Zhang GL (2006a) Chaetocochins A-C, epipolythiodioxopiperazines from *Chaetomium cochliodes*. J Nat Prod 69:1374–1376

Li GY, Li BG, Yang T, Liu GY, Zhang GL (2006b) Chaetoindicins A-C, three isoquinoline alkaloids from the fungus *Chaetomium indicum*. Org Lett 8:3613–3615

Li GY, Li BG, Yang T, Liu GY, Zhang GL (2008) Secondary metabolites from the fungus *Chaetomium brasiliense*. Helv Chim Acta 91:124–129

Li LM, Zou Q, Li GY (2010) Chromones from an ascomycete, *Chaetomium aureus*. Chin Chem Lett 21:1203–1205

Li W, Yang X, Yang Y, Duang R, Chen G, Li X, Li Q, Qin S, Li S, Zhao L, Ding Z (2016) Anti-phytopathogen, multitarget acetylcholinesterase inhibitory and antioxidant activities of metabolites from endophytic *Chaetomium globosum*. Nat Prod Res 30:2616–2619

Lösgen S, Schlörke O, Meindl K, Herbst-Irmer R, Zeeck A (2007) Structure and biosynthesis of chaetocyclinones, new polyketides produced by an endosymbiotic fungus. Eur J Org Chem 2007:2191–2196

Manandhar JB, Thapliyal N, Cavanaugh KJ, Sinclair JB (1986) Interaction between pathogenic and saprobic fungi isolated from soybean roots and seeds. Mycopathologia 98:69–75

Marwah RG, Fatope MO, Deadman ML, Al-Maqbali YM, Husband J (2007) Musanahol: a new aureonitol-related metabolite from a *Chaetomium* sp. Tetrahedron 63:8174–8180

McMullin DR, Sumarah MW, Blackwell BA, Miller JD (2013) New azaphilones from *Chaetomium globosum* isolated from the built environment. Tetrahedron Lett 54:568–572

Nitao JK, Meyer SLF, Oliver JE, Schmidt WF, Chitwood DJ (2002) Isolation of avipin, a fungus compound antagonistic to plant-parasitic nematodes. Nematology 4:55–63

Oh H, Swenson DC, Gloer JB, Wicklow DT, Dowd PF (1998) Chaetochalasin A: a new bioactive metabolite from *Chaetomium brasiliense*. Tetrahedron Lett 39:7633–7636

Oikawa H, Murakami Y, Ichihara A (1933) 20-Ketoreductase activity of chaetoglobosin a and prochaetoglobosins in a cell-free system of *Chaetomium subaffine* and the isolation of new chaetoglobosins. Biosci Biotechnol Biochem 57:628–631

Panthama N, Kanokmedhakul S, Kanokmedhakul K, Soytong K (2011) Cytotoxic and antimalarial Azaphilones from *Chaetomium longirostre*. J Nat Prod 74:2395–2399

Panthama, N., Kanokmedhakul, S., Kanokmedhakul, K. and K. Soytong (2014). Chemical constituents from the fungus *Chaetomium cupreum* RY202. Arch Pharm Res DOI https://doi.org/10.1007/s12272-014-0418-1 1–6 p

Park JH, Choi GJ, Jang KS, Lim HK, Kim HT, Cho KY, Kim JC (2005) Antifungal activity against plant pathogenic fungi of chaetoviridins isolated from *Chaetomium globosum*. FEMS Microbiol Lett 252:309–313

Petcharat V, Soytong K (1991) *Chaetomium* in soil under para rubber. Songklanakarin J Sci Technol 13:129–132

Pholkerd N, Kanokmedhakul S, Kanokmedhakul K, Soytong K, Prabpai S, Kongsearee P (2008) Bis-spiro-azaphilones and azaphilones from the fungi *Chaetomium cochliodes* VTh01 and C. Cochliodes CTh05. Tetrahedron 64:9636–9645

Phong NH, Pongnak W, Soytong K (2016) Antifungal activities of *Chaetomium* spp. against *Fusarium* wilt of tea. Plant Prot Sci 52:10–17

Pornsuriya C, Lin FC, Kanokmedhakul S, Soytong K (2008) New record of *Chaetomium* species isolated from soil under pineapple plantation in Thailand. J Agric Technol 4:91–103

Pornsuriya C, Soytong K, Poeaim S, Kanokmedhakul S, Khumkomkhet P, Lin F-C, Wang HK, Hyde KD (2011) *Chaetomium siamense* sp. nov., a soil isolate from Thailand, produces a new chaetoviridin. G Mycotaxon 115:19–27

Prechaprome S, Soytong K (1997) Integrated biological control of durian stem and root rot caused by *Phytophthora palmivora*. Proceedings of the first international symposium on biopesticides. Thailand, October 27–31, 1996:228–237

Probst A, Tamm C (1981) 19-*O*-acetylchaetoglobosin B and 19-*O*-acetylchaetoglobosin D, two new metabolites of *Chaetomium globosum*. Helv Chim Acta 64:2056–2064

Promgool T, Kanokmedhakul K, Leewijit T, Song JJ, Soytong K, Yahuafai J, Kudera T, Kokoska L, Kanokmedhakul S (2021) Cytotoxic and antibacterial depsidoines from endophytic fungus, *Chaetomium brasiliense* isolated from Thai rice. Nat Prod Res. Accepted in October 2021, ISSN: (Print) (Online) Journal homepage: https://www.tandfonline.com/loi/gnpl20. https://doi.org/10.1080/14786419.2021.1999947

Qin JC, Gao JM, Zhang YM, Yang SX, Bai MS, Ma YT et al (2009) Polyhydroxylated steroids from an endophytic fungus, *Chaetomium globosum* ZY-22 isolated from *Ginkgo biloba*. Steroids 74:786–790

Quimio TH (1984) Advanced. Mycology. Lecture syllabus. Department of Plant Pathology. University of the Philippines at Los Banos, College, Laguna, 228 pp

Rether J, Erkel G, Anke T, Sterner O (2004) Oxaspirodion, a new inhibitor of inducible TNF-α expression from the ascomycete *Chaetomium subspirale*. Production, isolation and structure elucidation. J Antibiot 57:493–495

Rodríguez K, Stchigel A, Guarro J (2002) Three new species of *Chaetomium* from soil. Mycologia 94:116–126

Ruan BH, Yu ZF, Yang XQ, Yang YB, Hu M, Zhang ZX, Zhou QY, Zhou H, Ding ZT (2018) New bioactive compounds from aquatic endophyte *Chaetomium globosum*. Nat Prod Res 32:1050–1055

Saito T, Suzuki Y, Koyama K, Natori S, Iitaka Y, Kinoshita T (1988) Chetracin A and chaetocins B and C, three new epipolythiodioxopiperazines from *Chaetomium* spp. Chem Pharm Bull 36:1942–1956

Sekita S (1983) Isocochliodinol and neocochliodinol, bis(3-indolyl)-benzoquinones from *Chaetomium* spp. Chem Pharm Bull 31:2998–3001

Sekita S, Yoshihira K, Natori S (1973) Structures of chaetoglobosin a and B, cytotoxic metabolites of *Chaetomium globosum*. Tetrahedron Lett 23:2109–2112

Sekita S, Yoshihira K, Natori S (1976) Structures of chaetoglobosin C, D, E and F, cytotoxic indol-3-yl-[13]cytochalasans from *Chaetomium globosum*. Tetrahedron Lett 17:1351–1354

Sekita S, Yoshihira K, Natori S, Chaetoglobosins GJ (1977) Cytotoxic indol-3-yl[13] cytochalasans from *Chaetomium globosum*. Tetrahedron Lett 32:2771–2774

Sekita S, Yoshihira K, Natori S (1980) Chaetochromin, a bis(naphthodihydropyran-4-one) mycotoxin from *Chaetomium thielavioideum*: application of ^{13}C-^{1}H long-rang coupling to the structure elucidation. Chem Pharm Bull 28:2428–2435

Sekita S, Yoshihira K, Natori S, Udagawa SI, Sakabe F, Kurata H et al (1981) Chaetoglobosins, cytotoxic 10-(indol-3-yl)-[13]cytochalasans from *Chaetomium* spp. I. Production, isolation and some cytological effects of chaetoglobosins A-J. Chem Pharm Bull 30:1609–1617

Sekita S, Yoshihira K, Natori S (1983) Chaetoglobosins, cytotoxic 10-(indol-3-yl) [13]cytochalasans from *Chaetomium* spp. IV. ^{13}C-nuclear magnetic resonance spectra and their application to a biosynthetic study. Chem Pharm Bull 31:490–498

Seth HK (1970) A monograph of the genus *Chaetomium*. Nova Hedwigia 37:1–133

Shternshis M, Tomilova O, Shpatova T, Soytong K (2005) Evaluation of the effects of Ketomium mycofungicide on Siberian isolates of phytopathogenic fungi. J Agric Technol 1:247–253

Sibounnavong P, Sibounnavong PS, Kanokmedhakul S, Soytong K (2012) Antifungal activities of *Chaetomium brasilense* CB01 and *Chaetomium cupreum* CC03 against *F. oxysporum* f. sp. *lycopersici* race 2. Int J Agric Technol 8(3):1029–1038

Smetanina OF, Denisenko VA, Pivkin MV, Khudyakova YV, Gerasimenko AV, Popov DY et al (2001) 3β-Methoxyolean-18-ene (miliacin) from the marine fungus *Chaetomium olivaceum*. Russ Chem Bull 50:2463–2465

Sodsa-art P, Soytong K (1999) Biological control of black pepper root and basal stem rot in the field. Proceeding of the symposium on biological control in tropics. MARDI Training Centre, March 18–19, 1999, Malaysia, 68–70 pp

Song JJ, Wattanachai P, Kasem S (2015) Biological activity of endophytic fungi associated with palm trees. Int J Agric Technol 11(2):567–579

Song JJ, Soytong K, Kanokmedhakul S, Kanokmedhakul K, Poeaim S (2020a) Natural product of nanoparticles constructed from *Chaetomium* spp. to control rice blast disease caused by *Magnaporthe oryzae*. Int J Agric Biol 23:1013–1020

Song JJ, Soytong K, Kanokmedhakul S, Kanokmedhakul K, Poeaim S (2020b) Antifungal activity of microbial nanoparticles derived from *Chaetomium* spp. against *Magnaporthe oryzae* causing rice blast. Plant Prot Sci 56:180–190

Song JJ, Soytong K, Thianrat L (2022) Nanofibers derived from endophytic *Chaetomium* brasilense for growth stimulation of rice. Int J Agric Technol 18(2):843–856

Somrithipol S (2004) Coprophilous fungi. In: Thai fungal diversity (eds. E.B.G. Jones, M. Tanticharoen and K.D. Hyde). BIOTEC, Thailand: 119–128

Somrithipol S, Hywel-Jones N L, Jones E B G (2004) Seed fungi. In: Thai fungal diversity (eds. E.B.G. Jones, M. Tanticharoen and K.D. Hyde). BIOTEC, Thailand:129–140

Soytong K (1989) Antagonism of *Chaetomium cupreum* to *Pyricularia oryzae*: a case study to biocontrol of a rice blast disease. Thai Phytopathology 9:28–33

Soytong K (1990) Biological control of tomato wilt disease caused by *Fusarium oxysporum* f. sp. *lycopersici* using a new biofungicide. In: A paper presented at the international conference on biotechnology and environmental science. August, 21–24, 1990. Chulabhorn Research Institute

Soytong K (1991) Species of *Chaetomium* in Thailand soils. Thai Phytopathology 11:86–94

Soytong K (1992a) Antagonism of *Chaetomium cupreum* to *Pyricularia oryzae*. J Plant Prot Tropics 9:17–23

Soytong K (1992b) Biological control of tomato wilt caused by *Fusarium oxysporum* f.sp. *lycopersici* using *Chaetomium cupreum*. Kasetsart J (Nat Sci) 26:310–313

Soytong K (2014a) Biological control of stem rots of oil palms caused by *Ganoderma boninense* using *Chaetomium lucknowense* and *Chaetomium cochliodes*. J Agric Technol 10:87–92

Soytong K (2014b) Bioformulation of *Chaetomium cochliodes* for controlling brown leaf spot of rice. Int J Agric Technol 10(2):321–337

Soytong K (2015) Testing bioformulation of *Chaetomium elatum* ChE01to control Fusarium wilt of tomato. Int J Agric Technol 11(4):975–996

Soytong K, Kaewchai S (2014) Biological control of white root of rubber trees using *Chaetomium cupreum*. Int J Agric Technol 10(1):93–103

Soytong K, Quimio TH (1989a) A taxonomic study on the Philippine species of *Chaetomium*. Philippine Agriculturist 72(1):59–72

Soytong K, Quimio TH (1989b) Antagonism of *Chaetomium globosum* to the rice blast pathogen, *Pyricularia oryzae*. Kasetsart J (Nat Sci) 23:198–203

Soytong K, Ratancherdchai K (2005) Application of mycofungicide to control late blight of potato. J Agric Technol 1:19–32

Soytong, K., Usuwan, P., Kanokmedhakul, S., Knokmedhakul, K., Kukongviriyapan, V. and M. Isobe. (1999). Integrated biological control of *Phytophthora* rot of sweet orange using mycofungicides in Thailand. Proceeding of the 5th international conference on plant protection in the tropics. 15–18 March 1999. Malaysia. 329–331 pp

Soytong K, Kanokmedhakul S, Kukongviriyapa V, Isobe M (2001) Application of *Chaetomium* species (Ketomium®) as a new broad spectrum biological fungicide for plant disease control: a review article. Fungal Divers 7:1–15

Soytong K, Pongnak W, Kasiolarn H (2005) Biological control of *Thielaviopsis* bud rot of *Hyophorbe lagenicaulis* in the field. J Agric Technol 1:235–245

Soytong K, Suthammarit S, Kanokmedhakul S (2019) Evaluation of bioformulation from *Chaetomium elatum* ChE01 to control banana anthracnose caused by *Colletotrichum musae*. Int J Agric Technol 15(4):665–674

Suphalucksana W, Soytong K (2006) Significance of *Chaetomium cupreum* for ruminant nutrition improvement through biodegradation. Int J Agric Technol 2(2):155–163

Tabata Y, Miiko N, Yaguchi T, Hatsu M, Ishii S, Imai S (2000) PF1138A and B, novel trans-epoxysuccinate-type cysteine protease inhibitors produced by *Chaetomium globosum*. Meiji Seika Kenkyu Nenpo 39:55–64

Takahashi M, Koyama K, Natori S (1990) Four new azaphilones from *Chaetomium globosum* var. *flavo-viridae*. Chem Pharm Bull 38:625–628

Tantapakul C, Promgool T, Kanokmedhakul K, Soytong K, Song J, Hadsadee S, Jungsuttiwong S, Kanokmedhakul S (2020) Bioactive xanthoquinodins and epipolythiodioxopiperazines from *Chaetomium globosum* 7 s^{-1}, an endophytic fungus isolated from Rhapis cochinchinensis (lour.) Mart. Nat Prod Res 34(4):494–502

Thohinung S, Kanokmedhakul S, Kanokmedhakul K, Kukongviriyapan V, Tusskorn O, Soytong K (2010) Cytotoxic 10-(indol-3-yl)-[13] cytochalasans from the fungus *Chaetomium elatum* ChE01. Arch Pharm Res 33:1135–1141

Thongkham D, Soytong K, Kanokmedhakul S, Kanokmedhakul K (2018) Nanoparticles derived from *Chaetomium elatum* against *Phytophthora* rot of durian. Int J Agric Technol 14(7):2115–2124

Tomilova OG, Shternshis MV (2006) The effect of a preparation from *Chaetomium* fungi on the growth of phytopathogenic fungi. Appl Biochem Microbiol 42:76–80

Tongon R, Soytong K (2016) Fungal metabolites from *Chaetomium brasiliense* to inhibit *Fusarium solani*. Int J Agric Technol 12(7.1)

Tongon R, Soytong K (2021) Application of *Chaetomium cochliodes* CTh02 to against durian root rot cause by *Phytophthora palmivora* RT01. Int J Agric Technol 17(2):753–766

Tongon R, Soytong K (2022) Nanoparticles derived from active metabolites of *Chaetomium cupreum* CC3003 against *Phytophthora* rot of durian. Int J Agric Biol. Issn Print: 1560–8530; ISSN Online: 1814–9596 21–0142/2022/27–1–19–27. https://doi.org/10.17957/IJAB/15.1894. https://www.fspublishers.org

Tveit M, Moore MB (1954) Isolates of *Chaetomium* that protect oats from *Helminthosporium victoriae*. Phytopathology 44:686–689

Tveit M, Wood RKS (1955) The control of *Fusarium* blight in oat seedlings with antagonistic species of *Chaetomium*. Ann Appl Biol 43:538–552
Udagawa S, Horie Y (1973) Three *Chaetomium* species from Thailand soil. Reports of the Tottori Mycological Institute 10:429–435
Udagawa S, Muro IT, Kurata H, Sekita S, Yoshihira K, Natori S (1979) *Chaetomium udagawae*: a new producer of sterigmatocystin. Trans Mycol Soc Jpn 20:475–480
Udompongsuk M, Soytong K, Kanokmedhakul S, Kanokmedhakul K (2018) In vitro efficacy of *Chaetomium brasiliense* against *Pythium* spp. causing root rot disease of tangerine. Int J Agric Technol 14(7):2181–2190
Umikalsom MS, Ariff AB, Hassan MA, Karim MIA (1998) Kinetics of cellulose production by *Chaetomium globosum* at different levels of dissolved oxygen tension using oil palm empty fruit bunch fiber as substrate. World J Microbiol Biotechnol 14:491–498
Usuwan P, Soytong K, Kanokmedhakul S, Kanokmedhakul K, Kukongviriyapan V, Isobe M (1999) Integrated biological control of Phytophthora rot of sweet orange using mycofungicides in Thailand. Proc. of the 5th International Conference on Plant Protection in the Tropics 329–331
Vannacci G, Harman GE (1987) Biocontrol of seed-borne *Alternaria raphani* and *A. brassicicola.* Can J Microbiol 33:850–856
Vannak S, Sarayut P, Soytong K (2015) Antifungal activities of endophytic fungi isolated from orchids against *Colletotrichum* sp. caused anthracnose in orchids. Int J Agric Technol 8(11):1949–1961
Vareeket R, Soytong K, Kanokmedhakul S, Kanokmedhakul K (2018) Nanoparticles from *Cheatomium brasiliense* against brown spot of rice. Int J Agric Technol 14(7):2207–2214
Vareeket R, Song JJ, Soytong K (2023) New discovery of natural product nanoparticles constructed from active metabolites from *Chaetomium brasiliense* for immunity to brown spot of rice var. RD 47. Int J Agric Technol 19(1):291–300
Vilavong S, Soytong K (2016) Application of bioformulation of *Chaetomium* for biocontrol of *Colletotrichum gloeosporioides* causing coffee anthracnose in Arabica variety in Laos. J Agric Technol 12:1239–1245
Von Arx JA, Van der Aa HA (1987) Spororminula tenerifae gen. et sp. nov. Trans Br Mycol Soc 89:117–120
Von Arx JA, Dreyfuss M, Muller E (1984) A revolution of *Chaetomium* and the Chaetomiaceae. Persoonia 12:169–179
Von Arx JA, Guarro J, Figuers MJ (1986) The ascomycete genus *Chaetomium.* Nova Hedwigia 84:1–162
Walter D, Gindrat D (1988) Biological control of damping-off of sugarbeet and cotton with *Chaetomium globosum* or a fluorescent *Pseudomonas* sp. Can J Microbiol 34:631–637
Wang S, Li XM, Teuscher F, Li DL, Diesel A, Ebel R et al (2006) Chaetopyranin, a benzaldehyde derivative, and other related metabolites from *Chaetomium globosum*, an endophytic fungus derived from the marine red alga *Polysiphonia urceolata*. J Nat Prod 69:1622–1625
Wang S, Zhang Y, Li XM, Wang BG (2007) Benzaldehydes from endophytic fungus *Chaetomium globosum* separated from marine red alga *Polysiphonia urceolata.* Haiyang Yu Huzhao 38:131–135
Wang XY, Yan X, Fang MJ, Wu Z, Wang D, Qiu YK (2017) Two new cytochalasin derivatives from *Chaetomium globosum* SNSHI-5, a fungus derived from extreme environment. Nat Prod Res 31:1669–1675
Webster J (1980) Introduction to fungi, 2nd edn. Cambridge University Press. 669 pp
Wen J, Okyere SK, Wang S, Wang J, Xie L, Ran Y, Hu Y (2002) Endophytic fungi: an effective alternative source of plant-derived bioactive compounds for pharmacological studies. J Fungi (Basel) 8(2):205. https://doi.org/10.3390/jof8020205
Whiteside WC (1957) Perithecial initials of *Chaetomium*. Mycologia 49:420–425
Wijeratne EMK, Turbyville TJ, Fritz A, Whitesell L, Gunatilaka AAL (2006a) A new dihydroxanthenone from a plant-associated strain of the fungus *Chaetomium globosum* demonstrates anticancer activity. Bioorg Med Chem 14:7917–7923

Wijeratne EMK, Paranagama PA, Gunatilaka AAL (2006b) Five new isocoumarins from Sonoran desert plant-associated fungal strains *Paraphaeosphaeria quadriseptata* and *Chaetomium chiversii.* Tetrahedron 62:8439–8446

Yamada, T., Doi, M., Shigeta, H., Muroga, Y., Hosoe, S., Numata, A., *et al.* (2008). Absolute stereostructures of the cytotoxic metabolites chaetomugilins A–C produced by *Chaetomium* species separated from a marine fish. Tetrahedron Lett, 49:4192-4195

Yamada T, Muroga Y, Tanaka R (2009) New azaphilones, secochaetomugilins a and D, produced by a marine-fish-derived *Chaetomium globosum.* Mar Drugs 7:249–257

Yamada T, Muroga Y, Jinno M, Kajimoto T, Usami Y, Numata A (2011) New class azaphilone produced by a marine fish-derived *Chaetomium globosum.* The stereochemistry and biological activities. Bioorg Med Chem 19:4106–4113

Yang SW, Mierzwa R, Terrcciano J, Patel M, Gullo V, Wagner N (2006) Chemokine receptor CCR-5 inhibitors produced by *Chaetomium globosum.* J Nat Prod 69:1025–1028

Zhang H, Yang Q (2007) Expressed sequence tag-based identification of genes associated with the biocontrol agent *Chaetomium cupreum.* Appl Microbiol Biotechnol 74:650–658

Zhang J, Ge HM, Jiao RH, Li J, Peng H, Wang YR et al (2010) Cytotoxic chaetoglobosins from the endophyte *Chaetomium globosum.* Planta Med 76:1910–1914

Zhou W, Starr JL, Krumm JL, Sword GA (2016) The fungal endophyte *Chaetomium globosum* negatively affects both above- and belowground herbivores in cotton. FEMS Microbiol Ecol 92(10), fiw158. https://doi.org/10.1093/femsec/fiw158

Zopf W (1881) Zur Entwicklungsgeschichte der scomyceten. Chaetomium Nova Acta 42:199–292. 14–20

Chapter 9
Microbial Bioagents as Tools for Eliminating Toxigenic Fungi

Noran A. Al-Ansary, Wafaa Mohamed Haggag, and Medhat K. Ali

Abstract A significant amount of information about the potential application of helpful microbes to manage plant diseases has been produced by laboratory and field trial experiments conducted in recent decades. Mycotoxigenic fungal diseases have received particular attention because of their direct negative impact on crop productivity and the possibility of mycotoxin formation, which is harmful to both human and animal health. The development and registration of new biopesticides have been encouraged by new legislative attempts to limit the use of current commercial chemical pesticides. Current methods for the preharvest control of mycotoxigenic fungi and their corresponding toxins are covered in this chapter. We will concentrate on the various ways in which the most commonly researched biological control agents work. Additionally, a thorough analysis of their capacity to inhibit the production of mycotoxins will be included.

Keywords Agriculture · Effect · Food · Mycotoxin · Probiotic bacteria · Toxigenic fjungi

9.1 Introduction

A significant portion of the calories that individuals throughout the world consume on a daily basis come from cereals. To prevent global food insecurity, food production must increase by 50–70% over the next 30 years because of the growing worldwide population (FAO 2009). Food insecurity is a severe threat to emerging nations, notably those in sub-Saharan Africa, where the number of hungry people is predicted to rise in the coming years (Sasson 2012). Furthermore, it is impossible to

N. A. Al-Ansary (✉) · W. M. Haggag
Plant Pathology Department, National Research Centre, Giza, Egypt

M. K. Ali
Plant Pathology Department, Faculty of Agriculture, Ain Shams University, Cairo, Egypt

V. Gopikrishnan et al. (eds.), *Harnessing Microbial Resources for Sustainable Agriculture*, Microorganisms for Sustainability 61,
https://doi.org/10.1007/978-981-95-4817-0_9

overstate the impact of plant diseases. One of the greatest risks to the availability and safety of food worldwide is the effect that fungus-related diseases and novel forms of existing pathogens have on crops that are crucial to agriculture. The damage caused by diseases to our most significant crops is estimated to be sufficient to feed 8.5% of the world's population (Fisher et al. 2012). Therefore, effective management of these plant diseases is essential for achieving the goal of feeding the world's expanding population. Toxigenic plant pathogens are a significant group of plant pathogens that produce mycotoxins, secondary metabolites with unrelated chemical structures and biological properties that have a wide range of toxic effects on humans and livestock. As a result, these pathogens threaten not only food security but also food safety (Bennett and Klich 2003; Zain 2011; De Ruyck et al. 2015).

Numerous tactics, including the development of resistant cultivars, the application of sensible crop rotation plans, and the use of chemical management, can be used to manage plant diseases. The European Union (EU Directive 2009/128/EC) has encouraged research on alternative and environmentally friendly solutions, such as integrated pest management and the use of biological control agents (BCAs) or bioelicitors, in response to the detrimental effects of plant protection products on the environment and the health of humans and animals. In plant pathology, biological control, or bioelicitors, is the use of microorganisms to stop colonization and/or prevent the spread of dangerous plant infections (Pal et al. 2006; Haggag et al. 2017). In this chapter, BCAs are described as beneficial bacteria that can protect plants by resisting plant diseases (Agrios 1969; Weller 1988; Punja and Utkhede 2003; El-Tarabily and Sivasithamparam 2006).

Fusarium, *Alternaria*, *Claviceps*, *Stachybotrys*, and *Aspergillus* are the most researched plant pathogenic genera that produce mycotoxins (Bennett and Klich 2003; Yazar and Omurtag 2008; Ferrigo et al. 2016; Pitt and Miller 2017). Cereals, nuts, beans, sugarcane, and sugar beet are among the many commodities that these genera infect in the field (e.g., *Fusarium*, *Alternaria*, and *Claviceps* spp.) and/or in storage (e.g., *Aspergillus* spp.). *Aspergillus flavus* infects groundnuts and maize, *Fusarium graminearum* is the most common pathogen in both wheat and maize, and *Fusarium verticillioides* contaminates maize. Research has focused less on other mycotoxigenic plant diseases, such as *Alternaria alternata*, *Claviceps purpurea*, and other members of the genus *Aspergillus* (such as *A. carbonarius*, *A. niger*, and *A. parasiticus*) and *Fusarium* (such as *F. avenaceum*, *F. acuminatum*, and *F. proliferatum*).

There are many reports of obtaining and testing a large number of potentially useful fungal and bacterial species and strains (Gao et al. 2016; Zubair et al. 2021; Ayaz et al. 2023). However, their results are too unpredictable and depend on many biotic and abiotic factors. Antitoxic pathogens are mostly undemanding microorganisms with naturally occurring ecological connections with the objects of interest; for example, they counteract major phytopathogens, are utilizable as environmentally healthy components, decompose, and feed experimental chickens and pigs. However, when applied to toxigenic fungi infecting grain, these microorganisms may cause unpredictable, uncharacteristic, abnormal toxic effects that are only undemandingly influenced by the environment and organism settings or the absence of symptoms. Therefore, objects and tools to differentiate the cooperation or resistance of good microorganisms with toxigenic phytopathogens that cause frank or covert abnormal or toxic symptoms are urgently needed. This chapter discusses the biological control of contaminated grain mycobiota by fungal, bacterial, and yeast communities and integrates partner-based, community-level, and biopesticide methods with mycotoxin inhibition via antimicrobial consortium production, with a general overview of microorganisms manipulating toxigenic fungal growth and toxic properties (Corbu et al. 2023).

Total toxicity control of grain and food products is an important problem for pharmaceutical and agricultural chemistry, microbiology, and biotechnology. The motivation for the biosynthesis of mycotoxins is not well known, but their ecological role is likely linked to their ability to control population dynamics, adapt to new habitats, and withstand the stress of other populations. Mycotoxins are likely defensive compounds allowing fungi to eliminate their opponents. Understanding the invasion and evolution of their interaction with toxic fungi is a complex scientific problem. Agriculture is already of practical interest in limiting large mycotoxin outbreaks and controlling invasions and the spread of toxigenic fungal populations, including biological management. Two main directions of agronomy and mycology research are defined: studying the development and functioning of beneficial organisms displaced in habitats of toxigenic species that are able to eliminate them directly or indirectly, using antimicrobial metabolites, restricting sources, or disrupting self-regulation systems via chemical, ecological, and integrated approaches.

9.2 Toxigenic Fungi: Identification, Characteristics, and Risks

The possibility of fungal contamination generally extends from field to storage, distribution, and consumption of a product. In addition to the deterioration of the quality of the product, some of these fungi are capable of producing toxic substances, so-called mycotoxins, which pose serious risks to human and animal health. Currently, more than 400 mycotoxins or closely related chemical substances have been tested for their potential toxicity. Only approximately 40 out of approximately 400 mycotoxins are known and can be detected with the methods available for public use. These chemicals are considered mutagenic, teratogenic, neurotoxic, and carcinogenic agents. Common mycotoxigenic fungal species can occur anywhere under the right moisture content and storage climate, in which case their impact as spoilage organisms is the greatest determinant of the microbiological safety of the final food products.

The taxonomy of toxigenic fungi is highly variable and complex, but the majority of these fungi are able to produce dangerous mycotoxins (aflatoxins, fumonisins, trichothecenes, zearalenones, etc.) (Haggag et al. 2023, 2024). The following genera are most frequently isolated from air, substrata, crops, and foods contaminated by toxigenic fungi: *Aspergillus niger*, *Alternaria*, *Aspergillus flavus*, *Aspergillus parasiticus*, *Cladosporium*, *Fusarium*, *Mortierella alpina*, *Penicillium chrysogenum*, *Penicillium funiculosum*, *Penicillium griseofulvum*, *Penicillium expansum*, *Rhizopus stolonifer*, and *Trichoderma viride. Aspergillus flavus* and *Aspergillus parasiticus* are potential producers of aflatoxins, the most toxic substances known to date. *Aspergillus niger*, *Penicillium chrysogenum*, and *Penicillium funiculosum* produce ochratoxins, citrinin, and other serious harmful mycotoxins, such as roquefortines and PR toxins.

The prevalence of toxigenic fungi results in the contamination of food products by mycotoxins, which pose significant risks to human and animal health. Even low levels of toxicogenic mycotoxins lead to severe organ damage, several types of cancer, and immunological disorders. Toxicological studies have shown that the target organs of FB1 and AFB1 differ, that is, AFB1 is the main target for primary immune organs and the liver, whereas FB1 affects the kidney. However, it is important to note the synergistic interaction between *Fusarium moniliforme* and *Aspergillus flavus* fungi, which, when present, resulted in CPE when no CPE with only FB1/AFB1, was observed. This synergistic action has been observed in the liver, kidney, and broiler chick pancreas. Moreover, prolonged frequent exposure to low but constant levels of various mycotoxins may induce additional toxic effects, which were not identified in experiments where a single mycotoxin was administered.

Numerous agricultural crops include mycotoxins, which are produced during and/or after plant colonization under specific environmental circumstances (Bennett and Klich 2003; Marin et al. 2013). Numerous organisms can experience a variety of harmful effects from short-term or long-term exposure to mycotoxins (Voss et al. 2007; Zain 2011; Marin et al. 2013; De Ruyck et al. 2015; Ferrigo et al. 2016). These fungal toxins frequently cause harm to plants (phytotoxins), vertebrates, and invertebrates (mycotoxins). These contaminants from nature have major negative economic impacts on the global economy and impede international trade because borders are rejected when mycotoxin concentrations are above the upper limit of what is allowed. Many research organizations do not consider the economic significance of mycotoxin production by these toxigenic plant diseases when examining biological control methods. The impact of BCAs on mycotoxin generation is frequently disregarded in favor of focusing only on their fungicidal or fungistatic properties.

9.3 Cultural Practices

Control of *Aspergillus* infection and, thereby, mycotoxin production in the field presently relies on agronomic practices, indirect interventions, and chemical treatments. However, none of these methods are fully effective (Al-Ansary et al. 2022).

The direct avoidance of ear contact with water by in-place drying would be virtually impossible. However, sustainable agricultural practices and irrigation through the root zone rather than the furrow may altogether avoid this problem. Fungicidal treatments or engineered microbial bioagents may provide promising, environmentally benign, acceptable, and conventional alternatives to conventional fungicide application. These methods could be integrated into the current system to provide a comprehensive control package suitable for different plant diseases of different crops.

Water stress strongly affects *A. flavus* infection and subsequent mycotoxin production. Late-season drought and residual soil moisture are strongly associated with elevated levels of *A. flavus* proliferation and aflatoxins. Flooding or wetting of fields is common with furrow or check-basin irrigation to provide sufficient surface moisture for initial infection and mycotoxin production by *A. flavus*.

Different methods of culture, such as irrigation systems for decreasing plant stress, soil tillage for improving water and air relationships in the root zone, and the use of resistant cultivars against infection by phytopathogenic fungi and mycotoxin production, are agronomically and environmentally relatively benign approaches to minimize disease development. Nevertheless, only the development of ecologically safe and sustainable agricultural practices based on sound ecological principles can ultimately minimize risks.

9.4 Chemical Fungicides

The main components of cereal culture provide humans with protein, carbohydrates, and various types of vitamins. The main organ involved in increasing the storage period of this product is an inflorescence with grain, which requires protection from fungal infestation. Microbial infections can occur in the field if the weather becomes cooler and there is a decrease in air pressure during pollination and, to a greater extent, indoors when water from grain infestation leads to the growth of saprophytes. The quality of the stored material is improved by its chemical processing via fungicides, which have toxic and mutagenic effects on the human body. Currently, the future of the battle against toxigenic wood odors is hopelessly linked to the biological containment of these microorganisms.

Fungicides are a major component of the economic and successful production of cereals. Although they are useful in achieving pest control in the formation of the first fruits, they do not guarantee a harmless environment. For example, their excessive use can lead to contamination of the product (bread) by harmful reaction impurities, such as carcinogenic nitroso compounds. Chemically synthesized fungicides lead to environmental pollution if they are left in the soil. Among the main disinfective components of current fungicides used to control toxigenic fungi are triazoles and strobilurins. Triazoles are synthetic chemicals with multiple biological activities. They contain substances with antifungal characteristics such as tebuconazole, paclobutrazol, metconazole, difenoconazole, etc. These substances are used to control the mold of almost all crops and populations of predatory fungi, including those with which the production of fusarium mycotoxins is associated.

Therefore, to limit these harmful effects, new possibilities are being sought for biological pest control agents (biofungicides). Because microorganisms can limit the viability of a wide range of their host populations, they apply their biocontrol capabilities to fight the infection of cereal plants by pathogenic wood molds.

9.5 The Role of Microbial Bioagents in Toxigenic Fungi Elimination

The efficiency of using microbial bioagents to suppress toxigenic molds found in different types of plant-derived food and animal feed. These molds were obtained from plants grown in a controlled greenhouse environment. The main goal of the research is to identify potential biocontrol agents that could be used to effectively control and regulate these harmful fungi. Recent scientific reports have presented significant evidence suggesting that the utilization of antagonistic microorganisms can play a crucial role in achieving robust and efficient biological control of toxigenic fungi. These antagonists, through their diverse array of modes of action, have demonstrated remarkable potential in mitigating and counteracting the deleterious effects caused by these mycotoxigenic fungi. To reduce plant disease, they may interact with pathogens directly or indirectly through one or more processes (Farzand et al. 2019a, b; Ayaz et al. 2021). In the rhizosphere, BCA competes with pathogens for resources and space. By inhibiting pathogen growth or metabolic activity, it also prevents pathogen pathogenicity through the use of a variety of compounds, including volatiles, lipopeptides, biosurfactants, bacteriocins, and enzymes that have antimicrobial properties (Babalola 2010). BCA may interfere with pathogens' ability to sense quorum through the inhibition of signal molecule production that initiates infections, such as the production of QS inhibitors that can degrade QS signal molecules such as chitinases, pectinases, and lactonases. These inhibitors reduce the symptoms of plant diseases by preventing pathogen invasion (Saeki et al. 2020). The preharvest application of these microorganisms has emerged as an extremely promising strategy, as it has the ability to substantially reduce the severity of diseases induced by mycotoxigenic fungi. This approach holds particular significance considering the high concentrations of these antagonists that can be achieved within the environment and fruits. These extraordinary and highly valuable properties of biocontrol agents indicate their immense potential for utilization in critical agricultural field conditions. By effectively deploying these agents, it is possible to notably decrease or even completely minimize the severity of disease outbreaks caused by mycotoxigenic fungi. Therefore, the incorporation of biocontrol agents in disease management practices can significantly contribute to safeguarding and preserving crop yields. Moreover, the successful control of mycotoxigenic fungi through the utilization of biocontrol agents also relies on various factors, including the intricate structure and composition of the population of mycotoxigenic fungi present in the environment, as well as the prevailing environmental conditions. In conclusion, investigations conducted in recent scientific reports aimed to thoroughly evaluate the efficacy of microbial bioagent antagonism in suppressing toxigenic molds obtained from food and feeds of plant origin. The findings from this research shed light on the immense potential of biocontrol agents in mitigating the severity of diseases caused by mycotoxigenic fungi. These agents, when employed strategically and judiciously, hold great promise for reducing crop losses and ensuring food safety in critical field conditions. However, it is crucial to consider the intricate interplay among biocontrol agents, mycotoxigenic fungi, and environmental factors when designing and implementing effective disease management strategies.

The elimination of toxigenic fungi from food, plants, and the environment is difficult, and the control of toxigenic mold is usually performed with fungicides. The importance of reducing fungal spoilage and maintaining good-quality final products has increased interest in the development of alternative methods to fungicides that pose a risk to human health. Microbial antagonists represent a potential alternative to the use of chemical fungicides because of the rarity of the development of resistance to biological control agents and their reentry interval, as well as the short heightened risk period for early biting compared with most chemical pesticides. The continued development of microbial antagonists with improved protection, durability, and within-field efficacy provides a new opportunity to achieve environmentally sound and sustainable control of toxigenic plant pathogens.

9.5.1 Definition and Types of Microbial Bioagents

The use of microbial control agents in agriculture has gained attention because of their potential to reduce the reliance on synthetic pesticides, minimize environmental impacts, and promote sustainable farming practices. In recent years, interest in developing microbial products specifically for large-scale plantations, with a focus on cereals, has increased. Cereals, such as wheat, rice, and maize, are staple crops that play a vital role in global food security. However, they are also susceptible to various diseases and pests that can significantly impact crop yield and quality. When a new microbial product for cereals is created, several factors need to be considered. First, the product must possess good grain quality, as it directly affects the overall crop yield and market value. Additionally, the active phase of the product should be short, ensuring that it effectively combats the targeted pathogens or pests within a specific timeframe. Finally, the product should not react with various seed treatment methods, ensuring compatibility and ease of application. Promising advancements have been made in the development of microbial products for cereals. For example, *Trichodex Ocris*, derived from *Trichoderma gamsii*, has shown promising results in field trials. This product has demonstrated the ability to effectively combat pathogens and pests in selected cereal plants. Our research and development activities also revolve around creating original and effective active components for plant protection. To achieve this goal, we explored a range of potential mycoparasitic strains and formulated them into suitable formulations for cereal use. Fungi, in particular, have proven to be effective bioagents for combating plant diseases. Given their wide prevalence, most biological preparations containing fungi serve as active agents against fungal pathogens. At present, the *Pseudotsuga genevuensis* variety is being utilized for the elimination of *Fusarium* fungi, as it inhibits the growth of their mycelium. Similarly, the active agents present in *Trichoderma harzianum* fungal agents have shown inhibitory effects on the radial growth stages of *Fusarium*. Although there are currently approximately 25 *Trichoderma* active agents available on the Polish market, none of them are specifically intended for cereals. According to the European and Mediterranean Plant Protection Organization, microbial control agents are classified into four main groups: fungi, bacteria, viruses, and other biological entities, such as protists and nematodes. Microbial bioagents are living microorganisms that have the potential to benefit plants when applied directly or through seed inoculation or environmental treatments. This approach offers an alternative to conventional agricultural chemicals, promoting environmentally friendly and sustainable farming practices. In conclusion, the development of microbial products for large-scale cereal plantations is a complex task that requires careful consideration of various factors. However, promising advancements have been made, and ongoing research aims to create effective active components tailored specifically for cereal protection. By harnessing the power of fungi and other microbial control agents, we can work toward sustainable agricultural practices and mitigate the impact of plant diseases and pests on cereal crops.

9.5.2 Fungi as Tools for Toxigenic Fungus Management

Large amounts of bioactive substances produced by beneficial fungi can be utilized as agrochemicals to protect crops. Since several beneficial fungi have been shown to limit the growth of different plant pathogens, there has been much interest in the development of fungal strains as BCAs for plant diseases (Thambugala et al. 2020; Zubair et al. 2021). Among the fungal taxa most widely employed as BCAs to combat bacterial and fungal plant diseases are *Trichoderma*, *Aspergillus*, and *Penicillium*. There have also been reports of fungi with antagonistic effects against a range of diseases, including *Gliocladium* and *Saccharomyces* (Farzand et al. 2019a, b; Zimmermann and Marqués-gálvez 2021). The majority of the various microorganisms known as fungal endophytes reside in the interior tissues of plants asymptomatically. Through mutualism and, in rare instances, parasitism, endophytic fungi engage in complex and intimate interactions with their hosts (Alblooshi et al. 2021; Poveda 2021). Owing to their ability to generate a wide range of physiologically active and structurally diverse secondary metabolites, they are essential for defending their hosts against pathogenic pathogens and pests (Farzand et al. 2019a, b; Zubair et al. 2021). The identification of new secondary metabolites with a variety of possible agricultural uses is facilitated by endophytic fungi (Premjanu and Jayanthy 2012). Endophytic fungi contain a wide range of metabolites with different chemical structures, such as terpenoids, alkaloids, steroids, peptides, isocoumarins, benzopyranones, and quinones. A strong chemical basis for the creation of agrochemicals with potential antibacterial, antifungal, herbicidal, nematicidal, insecticidal, and other agricultural applications was established by the identification of these metabolites (Contreras-Cornejo et al. 2016; Risoli et al. 2022). As a result, a wide range of fungi engage in endophytic and rhizospheric interactions with plants.

Numerous fungi, particularly *Trichoderma*, are well known for their broad-spectrum antagonistic actions against a variety of phytopathogens. Trichoderma is a common genus of fungi that exist in soil as mycoparasites, saprotrophs, and plant symbionts.

Filamentous fungi belonging to this genus have been the subject of much research and are used as biocontrol agents in agriculture to combat plant diseases (Djonovi'c et al. 2007; Degani and Dor 2021). To guarantee effective plant disease management, the direct and indirect control capabilities of BCAs against phytopathogens have been extensively researched throughout the past 10 years. Through parasitism and the production of insecticidal compounds, repelling metabolites, and antifeedant chemicals, *Trichoderma* has been shown to directly reduce insect pests. The indirect methods include parasitizing insect symbiotic bacteria, enticing natural enemies, and promoting plant defensive systems (Leonetti et al. 2017). *Trichoderma* is therefore a possible future alternative for the promotion of sustainable agriculture because of its effectiveness in combating plant diseases and insect pests. Over 80% of plant diseases are strongly hostile to this fungus (Thambugala et al. 2020; Degani and Dor 2021). By interfering with mycelial growth, *T. aggressivum f. europaeum* TAET1 demonstrated potent antagonistic activity against *Botrytis cinerea*, *Sclerotinia sclerotiorum*, and *Mycosphaerella melonis*. Furthermore, it completely prevents the germination of *S. sclerotiorum* sclerotia (Sánchez-Montesinos et al. 2021; Lahlali et al. 2022). It has been demonstrated that *Trichoderma asperellum* TaspHu1, which was isolated from *Juglans mandshurica* Maxim. rhizosphere soils in China, increases tomato seedling resistance to *Alternaria* alternata-induced leaf spot disease (Thambugala et al. 2020). Through improved plant immunity, *Trichoderma* has been demonstrated to play a major role in indirectly suppressing plant diseases in the rhizosphere. For sustainable agriculture, *Trichoderma* spp. can therefore be used in conjunction with several other alternative pest management strategies (Thambugala et al. 2020; Degani and Dor 2021).

9.5.3 Bacteria as a Tool for Toxigenic Fungus Management

The contribution of beneficial bacteria to crop disease resistance and plant growth has been the subject of numerous studies. Many bacteria from different genera, including *Bacillus*, *Paenibacillus*, *Agrobacterium*, *Bradyrhizobium*, *Acinetobacter*, *Azospirillum*, *Azotobacter*, *Pseudomonas*, *Rhizobium*, and *Streptomyces*, have been identified as biocontrol agents for managing a variety of diseases in main crops at this time (Massawe et al. 2018; Ayaz et al. 2021). Significant plant diseases caused by phytopathogens—primarily bacteria, fungi, and nematodes—pose a substantial threat to the long-term health and productivity of crops (Tahir et al. 2017). Although inoculation with *Bacillus* spp. suppresses pathogen growth and reduces disease severity in many infected crops, crops are severely impacted when infected by pathogenic bacteria such as *Pseudomonas savastanoi*, *Xanthomonas axonopodis*, and *Ralstonia solanacearum* (Tahir et al. 2017; Massawe et al. 2018).

By forming plant–microbial interactions in the rhizosphere, beneficial bacteria produce biofilms and secondary compounds, including fengycin, bacillomycin, iturin, and surfactin, that lower the population of plant pathogens (Farzand et al. 2019a, b; Ayaz et al. 2022). By producing extracellular enzymes such as chitosanase, protease, glucanase, and cellulase, *Bacillus* species adhere to mycelial cell walls and deform hyphae (da Silva et al. 2020; Zubair et al. 2021). Fengycin, iturin, pumilacidin, mixirin, and surfactin are examples of lipopeptides, which are antifungal peptides that work against harmful fungi in rhizospheres (Gao et al. 2016; Ayaz et al. 2022). A variety of bacteria, especially *Bacillus*, *Pseudomonas*, and *Burkholderia* species, suppress nematodes in a range of plants by altering their behavior, including feeding and reproduction (Gao et al. 2016; Cao et al. 2019). Root-knot nematode infestation was successfully managed via biological treatment with *Bacillus* isolates, according to earlier research. Additionally, they have been shown to lower nematode populations in the soil and roots that are affected (Ayaz et al. 2022; Ali et al. 2023). Additionally, *Bacillus* species increase the activity of defense-related enzymes such as phenylalanine ammonia-lyase (PAL), peroxidase, and polyphenol oxidase, as well as root exudates and modifications such as polysaccharides and amino acids (Farzand et al. 2019a, b; Jacob et al. 2020).

The biocontrol bacteria *Bacillus* and *Pseudomonas* spp. are efficient against a variety of phytopathogens in important crops. *Bacillus velezensis* isolates contain many biosynthesis-related gene clusters that are involved in the manufacture of secondary metabolites (Li et al. 2019; Ayaz et al. 2022). The development of novel medications may benefit from the utilization of these chemically diverse bioactive metabolites. To guarantee sustainable agriculture, new research on the production of bioactive substances, chemical compositions, intriguing bioactive gene clusters, and biological uses of *B. velezensis* and allied *Bacillus* species would be useful in controlling plant diseases (Katz and Baltz 2016; Zubair et al. 2021). The antibacterial chemicals produced by *B. velezensis* have a wide range of applications in the control of plant diseases. Surfactin, iturin, and fengycin are lipopeptides that *B. velezensis* utilizes to induce antagonistic effects against *Fusarium oxysporum* and *R. solanacearum* (Wu et al. 2014; Farzand et al. 2019a, b). In an industrial setting, *B. velezensis* QST713 is employed as a biocontrol agent to prevent the growth of green mold disease-causing *Trichoderma aggressivum* f. *europium* and to safeguard the edible mushroom *Agaricus bisporus* in a compost micromodel (Pandin et al. 2018). They appear to rely mostly on antibiosis and the emergence of systemic resistance across a range of plant species to carry out their activities (Molinari and Leonetti 2019; Ayaz et al. 2022). Furthermore, *B. subtilis* plays a significant role in the management of various agricultural diseases. One possible biocontrol agent for watermelon *Fusarium* wilt is *B. subtilis* IBFCBF-4. The *B. subtilis* R31 strain, which was isolated from *Dendrobium orchid* leaves, is a significant example of an exogenous endophyte that prevents banana *Fusarium* wilt in China (Zhu et al. 2020;

Li et al. 2021). To effectively control diverse crop diseases, much research should be conducted to create effective formulation technologies and examine different natural products with the assessment of their antimicrobial activity both in vitro and in the field.

9.5.4 Mechanisms of Action

Recently, the practical application of microbial bioagents as tools for reducing the contents of harmful fungi and their metabolites has received increasing attention in scientific research. As a result, data have been accumulated, which are well specified for the mechanisms of action of bioagents. Beneficial microbial bioagents carry out various functions because of their specific and highly effective nonspecific mechanisms. The metabolic exometabolites of nonpathogenic microbial strains suppress competitive microflora via the main principles of pathogenic nonpathogenic interrelations: metabolic competition, antibiosis, mycoparasitism, and stimulation or enhancement of plant defense. To counteract the pathogen, BCAs typically use multiple modes of action, that is, the existence of one dominating mode of action does not preclude the others.

1. Antibiosis includes the synthesis of secondary metabolites that can inhibit growth, reduce virulence, or eradicate pathogenic fungi, including antibiotics (Nagórska et al. 2007; Wang et al. 2007; Couillerot et al. 2009), lytic enzymes (El-Tarabily and Sivasithamparam 2006), and other proteins (Daguerre et al. 2014).
2. Competition takes place when two or more fungi compete for the same vital nutrients needed for growth and development (Benítez et al. 2004; Alabouvette et al. 2009). Exclusion by sharing a niche is another form of competition (Abbas et al. 2006a, b; Ehrlich 2014).
3. Mycoparasitism or hyperparasitism occurs when direct parasitic attack of one fungus by another results in the eventual death of the host pathogen (Barnett 1963; Howell 2003; Kim and Vujanovic 2016).
4. Beneficial microorganisms colonizing a plant can increase resistance to plant diseases by inducing systemic or local defense responses (Shoresh et al. 2010; Lugtenberg et al. 2016).

9.5.4.1 Antibiosis

Several fungal BCAs, including *Trichoderma* spp. and *Clonostachys* spp. (Weller 1988; Punja and Utkhede 2003; Benítez et al. 2004; Xue et al. 2009); bacterial BCAs, including *Bacillus* spp., *Pseudomonas* spp., *Streptomyces* spp., and *Lactobacillus* spp. (Gupta et al. 2001; Wang et al. 2007; Zhao et al. 2014); and yeast BCAs, including *Cryptococcus* spp., *Kluyveromyces* spp., and *Saccharomyces* spp.

(El-Tarabily and Sivasithamparam 2006; Etcheverry et al. 2009); are known to produce a variety of antibiotics, enzymes, and other antifungal compounds that have detrimental effects on plant pathogens. Each of these BCAs contains a variety of metabolites that target distinct pathogen structures, thereby reducing or eliminating pathogen proliferation.

(a) *Hydrolysis of fungal cell walls by enzymes*

Chitin and glucan polymers make up the majority of the complex structure that is the fungal cell wall. Molecules that disrupt this cell wall have been identified for a number of BCAs. *Trichoderma* spp. create linear oligopeptides called peptidaibols, which block beta-glucan synthase and prevent the pathogen from repairing its cell wall (Vinale et al. 2008). *A. flavus* colony color was altered by culture filtrates of a *T. harzianum* isolate, which clearly affected growth. *A. flavus* morphological changes were observed under a microscope, as evidenced by aberrant vesicle formation and a variety of aberrant conidial heads that reflected cell wall deformation (Calistru et al. 1997). Several extracellular enzymes, including lipolytic, cellulolytic, pectinolytic, amylolytic, and proteolytic enzymes, have also been shown to be produced; however, this inhibition is directly linked to the medium's supply of carbon (glucose or sucrose) or nitrogen (L-alanine or another) (Calistru et al. 1997).

(b) *Metabolite production that impacts the fungal membrane*

Several BCAs have been reported to produce antifungal metabolites that disrupt membrane structures. Lipopeptides, which disrupt the membrane and its sterols, are the most significant class (Romero et al. 2007). These lipopeptides work well against a number of toxigenic fungal genera, including *Aspergillus* and *Fusarium* species.

Two *Bacillus* species (P1 and P11) demonstrated strong antifungal activity (Nagórska et al. 2007) against *A. flavus* (Veras et al. 2016) when two antimicrobial lipopeptides, iturin and surfactin, were present. Similarly, the ability of *B. subtilis* BS119m to generate a significant amount of surfactin allows it to completely inhibit *A. flavus* growth (Mohammadipour et al. 2009). Iturin levels and the incidence of *Fusarium* disease were found to be inversely correlated by Crane et al., who tracked iturins produced by *B. amyloliquefaciens* in wheat grown in both greenhouse and field settings (Crane et al. 2013). Another lipopeptide that was isolated from *Bacillus subtilis* IB culture, fengycin, inhibited *F. graminearum* (Wang et al. 2007).

(c) *Synthesis of antifungal substances with antibiotic properties unrelated to effects on cell walls and membranes*

A number of fungicidal compounds have been identified, in addition to antibiotics, which have been characterized as effective allies in the fight against bacterial contamination. *P. fluorescens* produces the polyketide molecule 2,4-diacetylphloroglucinol (DAPG), which has drawn special attention because of its broad-spectrum efficacy against a variety of fungal infections (Shanahan et al. 1992; Mavrodi et al. 2001; Maurhofer et al. 2004; Brazelton et al. 2008). This molecule has been recovered from the rhizosphere of many crops (De La Fuente et al. 2006) after it was first obtained from the sugar beet rhizosphere of *Pseudomonas* spp. strain F113 (Shanahan et al. 1992). DAPG has been demonstrated to exhibit antifungal properties against *Alternaria* and *Fusarium* species (Müller et al. 2016).

9.5.5 Competition

While competition for nutrients occurs when two or more microorganisms vie for the same source of macro- and micronutrients needed for growth and the production of secondary metabolites, competition for niches, also known as competitive exclusion, occurs when another microorganism restricts access to a pathogen's habitat on the plant or seeds (Pal et al. 2006).

A. flavus control is among the most well-known and promising instances of competition for ecological niches and nutrients (Ehrlich 2014). However, there have also been reports of competition between other mycotoxigenic infections, including *F. pseudograminearum* through nutritional competition (Lakhesar et al. 2010), as well as *F. culmorum* and *F. graminearum* (Dawson et al. 2004). Atoxigenic *A. flavus* strains have been shown to be effective BCAs for controlling toxigenic strains of *A. flavus* in maize (Abbas et al. 2006a, b; Dorner 2009, 2010; Atehnkeng et al. 2014), cottonseed (Cotty 1989, 1990; Cotty and Bhatnagar 1994), and different kinds of nuts (Horn and Dorner 2009; Alaniz Zanon et al. 2013; Alaniz Zanon et al. 2016; Hulikunte Mallikarjunaiah et al. 2017). Depending on the endemic region, various strains of atoxigenic *A. flavus* are now being used, and a combination of strains is occasionally employed in the field. Coinoculating corn kernels with GFP-labeled AF70 and wild-type AF36 has recently provided in situ confirmation of this competitive exclusion idea. After being visualized under UV light, the results revealed that the kernels coinfected with both fungi and the control kernel inoculated with only GFP-labeled AF70 presented a population differential (up to 82% reduction). Additionally, measurements of aflatoxins (AFs) revealed a 73% decrease in aflatoxins in comparison with the control (Hruska et al. 2014).

However, *A. flavus* produces other chemicals that are harmful in addition to AFs. Another mycotoxin produced by some strains of *A. flavus*, notably atoxigenic strains, is cyclopiazonic acid (CPA), which mostly affects the liver and muscles of animals (Chang and Ehrlich 2011; Uka et al. 2017). For example, the commercially registered BCA AF36 has been shown to produce CPA in cottonseeds even if it is effective against the toxic *A. flavus*. For the same crops that were previously described, researchers screened and tested novel strains that did not produce both toxins (Abbas et al. 2011; Mauro et al. 2015; Zhou et al. 2015). Because *A. parasiticus* is less virulent and does not predominate in soil, such as *A. flavus*, it is less common to test atoxigenic strains of *A. flavus* against other fungi that produce AFs (Alaniz Zanon et al. 2013).

When treated prior to the development of a disease, *Trichoderma* and *Clonostachys* species also compete for nutrients and niches (Punja and Utkhede 2003; Verma et al. 2007). When there is little iron available in the environment, *Trichoderma* species, particularly *T. harzianum*, create siderophores, which are low-molecular-mass ferric-iron-specific chelators (Daguerre et al. 2014). *Trichoderma* spp. are able to compete for iron, which is necessary for the growth of numerous plant diseases, since siderophores chelate the oxidized ferric ions (Fe + 3), making them available as an iron source (Altomare et al. 1999; Benítez et al. 2004; Vinale et al. 2008).

9.5.6 Mycoparasitism

A direct parasitic interaction between a fungus and another fungal host is known as mycoparasitism (Benítez et al. 2004). The parasitic fungus can break down and infiltrate host cells via mycoparasitic interactions, which are mediated by a specific gene involved in the synthesis of several metabolites, primarily chitinases, glucanases, and proteases (Howell 2003; Benítez et al. 2004; Harman et al. 2004). This tactic is used by a variety of BCAs to compete with various mycotoxigenic diseases, particularly *Fusarium* spp. Among these, *Trichoderma* spp. are common mycoparasitic BCAs that are naturally found in plants and soil (Howell 2003; Punja and Utkhede 2003; Druzhinina et al. 2011).

In addition to competing for niches and nutrients, fungi are mostly biotrophic, engage in mycoparasitic interactions with other living fungi, increase plant resistance both locally and systemically, and secrete secondary antifungal compounds (Howell 2003; Verma et al. 2007). When *Trichoderma* spp. and *Fusarium* interact, certain chitinase-encoding genes are upregulated (Druzhinina et al. 2011; Matarese et al. 2012). In an in vitro test, *T. viride* showed strong antagonistic effects against *F. verticillioides*, as evidenced by the fungal radial expansion being suppressed by 46% after 6 days and by 90% after 14 days (Yates et al. 1999).

Through mycoparasitism, *T. harzianum* outperformed *F. verticillioides* on rice, and interactions revealed reciprocal antagonism (Sempere and Santamarina 2009). Since the cryoscopy electron microscopy data in this study confirmed the presence of mycoparasitism, the author proposed that certain metabolites, including chitinases, ß-1,3 glucanases, and enzymes that break down cell walls, were involved in this mechanism. Similar findings and outcomes were reported in a previous study that used the same experimental setting and the same BCA on rice but against *Alternaria alternata* (Sempere and Santamarina 2007). When *Trichoderma* spp.-produced chitinases break down the fungal cell wall, another class of enzymes known as exochitinases is released, and the pathogen is eradicated (Benítez et al. 2004). The majority of tests have used *Trichoderma* spp. as a BCA against *F. graminearum* in wheat (Calistru et al. 1997; Dal Bello et al. 2002; Dawson et al. 2004; Inch and Gilbert 2007; Schoneberg et al. 2015).

In a field assessment, the T-22 strain decreased *F. graminearum* perithecial production by 70% (77) (Inch and Gilbert 2007).

Using mycoparasitism strategies, *Sphaerodes* spp. have recently been identified as possible biocontrol agents against *Fusarium* spp. with encouraging outcomes. *Sphaerodes mycoparasitica* is a species that was isolated from wheat and asparagus fields alongside *Fusarium* spp. (Vujanovic and Goh 2009). It has demonstrated the ability to restrict *Fusarium* infection in both the 3-ADON and 15-ADON chemotypes as well as to restrict DON synthesis in both plants and in vivo (Vujanovic and Goh 2009, 2012). According to Palumbo et al. (2007), *P. fluorescens* (strains JP2034 and JP2175) produces chitinase and antifungal metabolites for bacterial BCAs, which inhibits the growth of *A. flavus* and *F. verticillioides*.

9.5.7 Indirect Through the Plant

It has been well documented that employing plant growth-promoting rhizobacteria to increase systemic plant resistance leads to efficient defense against a wide range of diseases (Heil and Bostock 2002; Choudhary et al. 2007; Pieterse et al. 2014). Numerous plant growth regulators, including gibberellins, cytokinins, and indole acetic acid, which disrupt plant signaling, are known to be produced by *P. fluorescens* (Mantelin and Touraine 2004). Furthermore, it generates enzymes, volatile chemicals, and antibiotics (Alimi et al. 2012). When certain pathogens, such as *F. verticillioides* M1, are present, *P. fluorescens* MPp4 produces indole-3-acetic acid, which in turn increases its antagonistic action (Hernández-Rodríguez et al. 2008). When *F. graminearum* infects barley, *P. fluorescens* CHA0 stops carbon diversion and decreases plant biomass (Henkes et al. 2011). Khan et al. (2006) reported the antagonistic activity of *P. fluorescens* MKB158 against *F. culmorum;* nevertheless, the authors noted that the antagonistic activity is mediated indirectly by the plant's systemic resistance. In addition to producing lytic enzymes, *Lysobacter enzymogenes* strain C3 also produces resistance in wheat against *F. graminearum*, which is another way to exert its biocontrol impact (Jochum et al. 2006). The existence of some fungal elicitors was confirmed by the effective decrease in the pathogen following heat treatment of C3 broth cultures to inactivate the bacterial cells and lytic enzymes.

In addition to rhizobacteria, the fungus *T. harzianum* colonizes plant roots, increases nutrient availability, stimulates plant development, and strengthens resistance to fungal infections (Benítez et al. 2004; Harman et al. 2004; Vinale et al. 2008). The use of *Trichoderma* spp. against *F. verticillioides* (Danielsen and Funck 1999), *F. graminearum* (Schoneberg et al. 2015), and *A. flavus* (Aiyaz et al. 2015) has been extensively studied. According to reports, *T. harzianum* inhibits *F. verticillioides* in maize by triggering the ethylene and jasmonate signaling pathways, which results in systemic resistance (Ferrigo et al. 2014). Since they have been shown to invade plant tissue away from roots, cause transcriptomic changes in plants, and shield plants from disease, new species of *Trichoderma* (*T. stromaticum, T. amazonicum, T. evansii, T. martiale, T. taxi,* and *T. theobromicola*) have recently been categorized as true endophytes (Mukherjee et al. 2012).

Colonization is another strategy used to increase plant resistance. To reduce plant diseases and mycotoxins in crops (Waller et al., 2005; Deshmukh and Kogel, 2007; Mousa et al., 2015; Rabiey et al., 2015; Rabiey and Shaw, 2016), much research is being conducted to identify endophytic bacteria that colonize plants without causing damage to them (Porras-Alfaro and Bayman, 2011) . By triggering plant defense responses, endophytes can improve plant growth and fitness while providing resistance against biotic and abiotic stressors. It should be mentioned, therefore, that in specific stages of their life cycles or under particular environmental circumstances, some of them are harmful to the plant (Porras-Alfaro and Bayman, 2011) . By improving nutrient intake from the soil, certain endophytes, such as *Piriformospora indica*, a cultivable root fungal endophyte that belongs to the order Sebacinales in the Basidiomycota, work to strengthen the host immune system against a variety of fungal diseases (Varma et al., 1999; Sherameti et al., 2005) .

9.6 Advantages and Limitations

The application of bioagents to suppress toxin-initiating fungal growth and secondary spread of fungal contamination, especially by *A. flavus* and *A. parasiticus*, has been perhaps the most intensively developed method (*Aspergillus* section *Nidulantes*). This is because they are not only widespread but also the most common causative agents of aflatoxin formation in many crops. Currently, hydrophilic antiaflatoxins, which are strains of nonaflatoxigenic *A. flavus* isolated from fuzzy and sclerotial mutants, are dominant in practical use. They are easily applied with the help of efficient and convenient techniques, including chemical-initiated events, the coating of crops with biomass, and various biopreparations. The main immunization effect of the nonaflatoxigenic fungus is associated with both the ease of colonization and the competition of the animal agent with the pathogenic fungus for crop resources, as well as the production of multiple agriculturally active secondary metabolites with antifungal biological activity. These bioagents have shown tremendous potential in mitigating fungal contamination and controlling the outbreak of toxins in various crops. Additionally, hydrophilic antiaflatoxins have emerged as the most effective strains capable of inhibiting the growth and spread of the highly prevalent *A. flavus* and *A. parasiticus*. The widespread distribution of these fungi has made them a significant threat to crop production, emphasizing the urgent need for effective solutions. Through the use of fuzzy and sclerotial mutants, nonaflatoxigenic *A. flavus* strains have been successfully isolated, providing a promising avenue for combatting aflatoxin formation. The implementation of efficient application techniques, such as chemical-initiated events and the coating of crops with biomass, offers practical and convenient means to utilize these bioagents. Moreover, various biopreparations have proven to be effective in maximizing their impact. The key to the success of these nonaflatoxigenic bioagents lies in their ability to colonize the crop easily and outcompete the pathogenic fungus for vital resources. Additionally, these bioagents produce multiple agriculturally active secondary metabolites, which possess potent antifungal properties. This dual approach of colonization and competition, coupled with the release of antifungal metabolites, provides a robust immunization effect against aflatoxigenic fungi. The constant advancement and optimization of these bioagents hold great promise in revolutionizing crop protection strategies, ensuring a safe and abundant food supply for future generations.

As one of the most promising and innovative methods for effectively suppressing toxigenic activity in food and feed components, the antiaflatoxigenic effect of microbial agents has been achieved with great success. By utilizing various biocontrol techniques not only to combat fungal growth itself but also to employ other bioagents, such as bacteria, fungi, and nonpathogenic parasitic insects (nematodes), we can maximize the efficacy of these interventions. Moreover, we explore the limitations and potential applications of phytobiotics, which serve as exciting new sources of antitoxinogenic and phytogenetic activity promoters. These groundbreaking ideas present new avenues for addressing food safety and enhancing the overall quality of our agricultural practices.

9.7 Future Directions and Research Opportunities

Among the numerous selection criteria for biocontrol agents, commonly tested surrogates for mycotoxin production might indicate biocontrol efficacy. Thus far, the results of research aimed at creating naturally based preparations of biologically active substances suitable for suppressing the life activities of toxigenic fungi suggest the possibility of developing commercial agents based on microorganisms. Some of these offer advantages over chemical agents in terms of simplicity of use and potential safety. Furthermore, industry will likely progress with attempts to create more effective bioagents for decontamination of cereals with mycotoxins to offer additional choices for producers of feed and grain.

Research has focused on creating preparations of biologically active bacterial and yeast cells that can suppress the growth of toxigenic fungi, and daily support uses show the ability of microbial agents to limit the contamination of grains with fungi and decrease the content of mycotoxins. There is potential for the future expansion of commercial bioagents that might be effective in the field under a wide range of environmental conditions. Currently, promising microbial bioagents are proposed for inclusion in an integrated protection system against toxigenic fungi and meet current requirements for chemical agents. The key issue is the demonstration of the efficiency of application criteria for the preparation of microorganisms.

9.8 Conclusion and Key Takeaways

Microorganisms, including bacteria, yeasts, and filamentous fungi, can be easily isolated from plants and should have a good chance of interacting with endophytic microorganisms. Moreover, they may play the first role in plant organs, from being the first ones to colonize the plant to elimination by plant defense mechanisms. This would have considerable advantages for the application model since the inoculation of axenic seeds, which need to meet more stringent requirements, may be bypassed.

This type of tool may be used for seed treatment or application to the soil in large-scale planting areas. Additionally, these tools can be applied under controlled conditions for the production of safe plants for seed or vegetative propagation. For future commercial implementation, the construction of microbial consortia, which mix effective phyllosphere and rhizosphere biocontrol agents, will be crucial. The study of the possible differential behavior of these microorganisms under seed or vegetative propagation conditions is essential.

The rapid growth of the global population has led to increased demand for food, which in turn affects the amount of toxic and low-quality food produced, which may cause pathogen diseases. Thus, diseases related to mycotoxin exposure pose a major risk to human and animal health. Therefore, my partner and I propose the use of an innovative combination of various bacterial probiotic strains (competitively) to eliminate toxigenic fungi, the major reason for the presence of mycotoxins in different crops. Owing to the resistance of these strains, they are thought to be the basis for reducing the mycotoxin content in a variety of products. The practical enforcement of probiotic products on the market and the concept of using bacteria and their effects are similar to our bioagent control bacteria products that are currently commercially available on the market for chemical, biological, and ecological regulation with up to weekly application, allowing for longer durations of protective effects in various climatic, soil, and social conditions for a variety of trials.

References

Abbas H, Zablotowicz R, Bruns HA et al (2006a) Biocontrol of aflatoxin in corn by inoculation with nonaflatoxigenic *Aspergillus flavus* isolates. Biocontrol Sci Tech 16:437–449

Abbas H, Zablotowicz R, Bruns HA, Abel CA (2006b) Biocontrol of aflatoxin in corn by inoculation with nonaflatoxigenic *Aspergillus flavus* isolates. Biocontrol Sci Tech 16(5):437–449

Abbas HK, Zablotowicz RM, Horn BW et al (2011) Comparison of major biocontrol strains of nonaflatoxigenic *aspergillus flavus* for the reduction of aflatoxins and cyclopiazonic acid in maize. Food Addit Contam Chem Anal Control Expo Risk Assess 28:198–208

Agrios GN (1969) Control of plant diseases. Plant Pathol 28:173–206

Aiyaz M, Divakara ST, Nayaka SC et al (2015) Application of beneficial rhizospheric microbes for the mitigation of seed-borne mycotoxigenic fungal infection and mycotoxins in maize. Biocontrol Sci Tech 25:1105–1119

Alabouvette C, Olivain C, Migheli Q, Steinberg C (2009) Microbiological control of soil-borne phytopathogenic fungi with special emphasis on wilt-inducing *Fusarium oxysporum*. New Phytol 184(3):529–544

Alaniz Zanon MS, Chiotta ML, Giaj-Merlera G et al (2013) Evaluation of potential biocontrol agents for aflatoxin in Argentinean peanuts. Int J Food Microbiol 162:220–225

Alaniz Zanon MS, Barros GG, Chulze SN (2016) Nonaflatoxigenic *Aspergillus flavus* as potential biocontrol agents to reduce aflatoxin contamination in peanuts harvested in northern Argentina. Int J Food Microbiol 231:63–68

Al-Ansary NA, Haggag WM, Ali MK (2022) Evaluation of antifungal activity of some natural essential oils against fungal pathogens associated with maize grains. Int J Agric Technol 18(5):1897–1916

Alblooshi AA, Purayil GP, Saeed EE, Ramadan GA, Tariq S, Altaee AS, El-Tarabily KA, AbuQamar SF (2021) Biocontrol potential of Endophytic Actinobacteria against *Fusarium solani*, the causal agent of sudden decline syndrome on date palm in the UAE. J Fungi 8:8

Ali Q, Yu C, Wang Y, Sheng T, Zhao X (2023) High killing rate of nematode and promotion of rice growth by synthetic volatiles from bacillus strains due to enhanced oxidative stress response. Physiol Plant 175:e13868

Alimi M, Javad M, Darzi MT (2012) Characterization and application of microbial antagonists for control of Fusarium head blight of wheat caused by *Fusarium graminearum* using single and mixture strain of antagonistic bacteria on resistant and susceptible cultivars. Afr J Microbiol Res 6:326–334

Altomare C, Norvell WA, Björkman T et al (1999) Solubilization of phosphates and micronutrients by the plant-growth- promoting and biocontrol fungus *Trichoderma harzianum* Rifai 1295-22. Appl Environ Microbiol 65:2926–2933

Atehnkeng J, Ojiambo PS, Cotty PJ et al (2014) Field efficacy of a mixture of atoxigenic *Aspergillus flavus* link: FR vegetative compatibility groups in preventing aflatoxin contamination in maize (*Zea mays* L.). Biol Control 72:62–70

Ayaz M, Ali Q, Farzand A, Khan AR, Ling H, Gao X (2021) Nematicidal volatiles from *bacillus atrophaeus* GBSC56 promote growth and stimulate induced systemic resistance in tomato against Meloidogyne incognita. Int J Mol Sci 22:5049

Ayaz M, Ali Q, Jiang Q, Wang R, Wang Z, Mu G, Khan SA, Khan AR, Manghwar H, Wu H et al (2022) Salt tolerant bacillus strains improve plant growth traits and regulation of Phytohormones in wheat under salinity stress. Plants 11:2769

Ayaz M, Li CH, Ali Q, Zhao W, Chi YK, Shafiq M, Ali F, Yu XY, Yu Q, Zhao JT, Yu JW (2023) Bacterial and fungal biocontrol agents for plant disease protection: journey from lab to field, current status, challenges, and global perspectives. Molecules 28(18):6735–6759

Babalola OO (2010) Beneficial bacteria of agricultural importance. Biotechnol Lett 32:1559–1570

Barnett HL (1963) The nature of Mycoparasitism by fungi. Ann Rev Microbiol 17:1–14

Benítez T, Rincón AM, Limón MC, Codon AC (2004) Biocontrol mechanisms of *Trichoderma* strains. Int Microbiol 7(4):249–260

Bennett JW, Klich M (2003) Mycotoxins M. Mycotoxins. Clin Microbiol Rev 16(16):497–516

Brazelton JN, Pfeufer EE, Sweat TA et al (2008) 2,4-Diacetylphloroglucinol alters plant root development. Mol Plant-Microbe Interact 21:1349–1358

Calistru C, McLean M, Berjak P (1997) *In vitro* studies on the potential for biological control of *Aspergillus flavus* and *Fusarium moniliforme* by Trichoderma species. A study of the production of extracellular metabolites by Trichoderma species. Mycopathologia 137:115–124

Cao H, Jiao Y, Yin N, Li Y, Ling J, Mao Z, Yang Y, Xie B (2019) Analysis of the activity and biological control efficacy of the *Bacillus subtilis* strain Bs-1 against *Meloidogyne incognita*. Crop Prot 122:125–135

Chang P-K, Ehrlich KC (2011) Cyclopiazonic acid biosynthesis by *Aspergillus flavus*. Toxin Rev 30:79–89

Choudhary DK, Prakash A, Johri BN (2007) Induced systemic resistance (ISR) in plants: mechanism of action. Indian J Microbiol 47:289–297

Contreras-Cornejo HA, Macías-Rodríguez L, Del-Val E, Larsen J (2016) Ecological functions of *Trichoderma* spp. and their secondary metabolites in the rhizosphere: interactions with plants. FEMS Microbiol Ecol 92:fiw036

Corbu VM, Gheorghe-Barbu I, Dumbravă AŞ, Vrâncianu CO, Şesan TE (2023) Current insights in fungal importance—a comprehensive review. Microorganisms 11(6):1384–1436

Cotty PJ (1989) Virulence and cultural characteristics of two *Aspergillus flavus* strains pathogenic on cotton. Phytopathology 79:808–814

Cotty PJ (1990) Effect of Atoxigenic strains of *Aspergillus flavus* on aflatoxin contamination of developing cottonseed. Plant Dis 74:233–235

Cotty PJ, Bhatnagar D (1994) Variability among atoxigenic *aspergillus flavus* strains in ability to prevent aflatoxin contamination and production of aflatoxin biosynthetic pathway enzymes. Appl Environ Microbiol 60:2248–2251

Couillerot O, Prigent-Combaret C, Caballero-Mellado J, Moënne-Loccoz Y (2009) *Pseudomonas fluorescens* and closely related fluorescent pseudomonads as biocontrol agents of soil-borne phytopathogens. Lett Appl Microbiol 48:505–512

Crane JM, Gibson DM, Vaughan RH et al (2013) Iturin levels on wheat spikes linked to biological control of Fusarium head blight by *Bacillus amyloliquefaciens*. Phytopathology 103:146–155
da Silva LR, Inglis MCV, Moraes MCB, Magalhães DM, Sifuentes DN, Martins I, de Mello SCM (2020) Morphological and protein alterations in *Sclerotinia sclerotiorum* (lib.) de Bary after exposure to volatile organic compounds of *Trichoderma* spp. Biol Control 147:104279
Daguerre Y, Siegel K, Edel-Hermann V, Steinberg C (2014) Fungal proteins and genes associated with biocontrol mechanisms of soil-borne pathogens: a review. Fungal Biol Rev 28(4):97–125
Dal Bello GM, Mónaco CI, Simón MR (2002) Biological control of seedling bright of wheat caused by *Fusarium graminearum* with beneficial rhizosphere microorganisms. World J Microbiol Biotechnol 18:627–636
Danielsen S, Funck JD (1999) Fungal endophytes from stalks of tropical maize and grasses: isolation, identification, and screening for antagonism against *Fusarium verticillioides* in maize stalks. Biocontrol Sci Tech 9:545–553
Dawson WAJM, Jestoi M, Rizzo A et al (2004) Field evaluation of fungal competitors of *Fusarium culmorum* and *F. graminearum*, causal agents of ear blight of winter wheat, for the control of mycotoxin production in grain. Biocontrol Sci Tech 14:783–799
De La Fuente L, Mavrodi DV, Landa BB et al (2006) phlD-based genetic diversity and detection of genotypes of 2,4- diacetylphloroglucinol-producing *Pseudomonas fluorescens*. FEMS Microbiol Ecol 56:64–78
De Ruyck K, De Boevre M, Huybrechts I, De Saeger S (2015) Dietary mycotoxins, coexposure, and carcinogenesis in humans: short review. Mutat Res Rev Mutat Res 766:32–41
Degani O, Dor S (2021) Biological control of Trichoderma to protect sensitive maize hybrids against late wilt disease in the field. J. Fungi 7:315
Deshmukh SD, Kogel K-H (2007) Piriformospora indica protects barley from root rot caused by Fusarium graminearum. Journal of Plant Disease Protocol. 114(6):263–268
Djonovi'c S, Vittone G, Mendoza-Herrera A, Kenerley CM (2007) Enhanced biocontrol activity of *Trichoderma virens* Transformants constitutively Coexpressing _-1,3- and _-1,6-Glucanase genes. Mol Plant Pathol 8:469–480
Dorner JW (2009) Biological control of aflatoxin contamination in corn using a nontoxigenic strain of *Aspergillus flavus*. J Food Prot 72:801–804
Dorner JW (2010) Efficacy of a biopesticide for control of aflatoxins in corn. J Food Prot 73:495–499
Druzhinina IS, Seidl-Seiboth V, Herrera-Estrella A et al (2011) *Trichoderma*: the genomics of opportunistic success. Nat Rev Microbiol 9:749–759
Ehrlich KC (2014) Nonaflatoxigenic *Aspergillus flavus* to prevent aflatoxin contamination in crops: advantages and limitations. Front Microbiol 5:1–9
El-Tarabily KA, Sivasithamparam K (2006) Potential of yeasts as biocontrol agents of soil-borne fungal plant pathogens and as plant growth promoters. Mycoscience 47:25–35
Etcheverry MG, Scandolara A, Nesci A et al (2009) Biological interactions to select biocontrol agents against toxigenic strains of *Aspergillus flavus* and *Fusarium verticillioides* from maize. Mycopathologia 167:287–295
FAO (2009) How to feed the world in 2050. Available from http://www.fao.org/fileadmin/templates/wsfs/docs/expert_paper/How_to_Feed_the_World_in_2050
Farzand A, Moosa A, Zubair M, Khan AR, Massawe VC, Tahir HAS, Sheikh TMM, Ayaz M, Gao X (2019a) Suppression of *Sclerotinia sclerotiorum* by the induction of systemic resistance and regulation of antioxidant pathways in tomato using Fengycin produced by *Bacillus amyloliquefaciens* FZB42. Biomolecules 9:613
Farzand A, Moosa A, Zubair M, Rashid Khan A, Hanif A, Tahir HAS, Gao X (2019b) Marker-assisted detection and LC–MS analysis of antimicrobial compounds in different bacillus strains and their antifungal effects on *Sclerotinia sclerotiorum*. Biol Control 133:91–102
Ferrigo D, Raiola A, Piccolo E et al (2014) *Trichoderma harzianum* T22 induces in maize systemic resistance against Fusarium verticillioides. J Plant Pathol 96:133–142

Ferrigo D, Raiola A, Causin R (2016) Fusarium toxins in cereals: occurrence, legislation, factors promoting the appearance and their management. Molecules 21:627

Fisher M, Henk D, Briggs C, Brownstein J, Madoff L, McCraw S, Gurr S (2012) Emerging fungal threats to animal, plant and ecosystem health. Nature 484(7393):186–194

Gao H, Qi G, Yin R, Zhang H, Li C, Zhao X (2016) *Bacillus cereus* strain S2 shows high Nematicidal activity against Meloidogyne incognita by producing sphingosine. Sci Rep 6:28756

Gupta CP, Dubey RC, Kang SC et al (2001) Antibiosis-mediated necrotrophic effect of *Pseudomonas* GRC 2 against two fungal plant pathogens. Curr Sci 81:91–94

Haggag WM, Tawfik MM, Abouziena HF, Abd El Wahed MSA, Ali RR (2017) Enhancing wheat production under arid climate stresses using bio-elicitors l. Gesunde Pflanzen 69(3):149–158

Haggag WM, Diab MM, Al-Ansary NA, Ali MK (2023) Activity of five natural essential oils against fungal contamination in fresh and stored maize grains. Int J Agric Technol 19(6):2427–2448

Haggag, Wafaa M.; Mohamed M. Diab; Noran A. Al-Ansary; Mohamed I. M. Ibrahim; Abd El-Nasser A. Khattab;·Mosaad A. Abdel-Wahhab, Medhat K. Ali (2024). Molecular identification and management of mycotoxigenic fungi in stored corn grains. Cereal Res Commun, 2024, 4, p. 1631-1644. https://doi.org/10.1007/s42976-024-00502-w

Harman GE, Howell CR, Viterbo A et al (2004) *Trichoderma* species - opportunistic, avirulent plant symbionts. Nat Rev Microbiol 2:43–56

Heil M, Bostock RM (2002) Induced systemic resistance (ISR) against pathogens in the context of induced plant defenses. Ann Bot 89:503–512

Henkes GJ, Jousset A, Bonkowski M et al (2011) *Pseudomonas fluorescens* CHA0 maintains carbon delivery to *Fusarium graminearum*-infected roots and prevents reduction in biomass of barley shoots through systemic interactions. J Exp Bot 62:4337–4344

Hernández-Rodríguez A, Heydrich-Pérez M, Acebo-Guerrero Y et al (2008) Antagonistic activity of native Cuban rhizobacteria against *Fusarium verticillioides* (Sacc.) Nirenb. In maize (*Zea mays* L.). Appl Soil Ecol 39:180–186

Horn BW, Dorner JW (2009) Effects of nontoxigenic *aspergillus flavus* and *A. Parasiticus* on aflatoxin contamination in wounded peanut seeds inoculated with agricultural soil containing natural fungal populations. Biocontrol Sci Tech 19:249–262

Howell CR (2003) Mechanisms employed by *Trichoderma* species in the biological control of plant diseases: the history and evolution of current concepts. Plant Dis 87:4–10

Hruska Z, Rajasekaran K, Yao H et al (2014) Coinoculation of aflatoxigenic and nonaflatoxigenic strains of *Aspergillus flavus* was used to study fungal invasion, colonization, and competition in maize kernels. Front Microbiol 5:122

Hulikunte Mallikarjunaiah N, Jayapala N, Puttaswamy H et al (2017) Characterization of non aflatoxigenic strains of *aspergillus flavus* as potential biocontrol agent for the management of aflatoxin contamination in groundnut. Microb Pathog 102:21–28

Inch S, Gilbert J (2007) Effect of *Trichoderma harzianum* on perithecial production of *Gibberella zeae* on wheat straw. Biocontrol Sci Tech 17:635–646

Jacob J, Krishnan GV, Thankappan D, Amma DKBNS (2020) Endophytic bacterial strains induced systemic resistance in agricultural important crop plants. In: Microbial endophytes. Elsevier, Amsterdam, pp 75–105

Jochum CC, Osborne LE, Yuen GY (2006) Biological control of *Fusarium* head blight with *Lysobacter enzymogenes* strain C3. Biol Control 39:336–344

Katz L, Baltz RH (2016) Natural product discovery: past, present, and future. J Ind Microbiol Biotechnol 43:155–176

Khan MR, Fischer S, Egan D et al (2006) Biological control of *Fusarium* seedling blight disease of wheat and barley. Phytopathology 69:386–394

Kim SH, Vujanovic V (2016) Relationships between mycoparasite lifestyles and biocontrol behaviors against *Fusarium* spp. and mycotoxin production. Appl Microbiol Biotechnol 100:5257–5272

Lahlali R, Ezrari S, Radouane N, Kenfaoui J, Esmaeel Q, El Hamss H, Belabess Z, Barka EA (2022) Biological control of plant pathogens: a global perspective. Microorganisms 10:596

Lakhesar DPS, Backhouse D, Kristiansen P (2010) Nutritional constraints on displacement of *Fusarium pseudograminearum* from cereal straw by antagonists. Biol Control 55:241–247

Leonetti P, Zonno MC, Molinari S, Altomare C (2017) Induction of SA-Signaling pathway and ethylene biosynthesis in *Trichoderma harzianum* treated tomato plants after infection of the root-knot nematode *Meloidogyne incognita*. Plant Cell Rep 36:621–631

Li Y, Héloir M-C, Zhang X, Geissler M, Trouvelot S, Jacquens L, Henkel M, Su X, Fang X, Wang Q et al (2019) Surfactin and Fengycin contribute to the protection of a *Bacillus subtilis* strain against grape downy mildew by both direct effects and Defense stimulation. Mol Plant Pathol 20:1037–1050

Li C, Cheng P, Zheng L, Li Y, Chen Y, Wen S, Yu G (2021) Comparative genomics analysis of two Banana Fusarium wilt biocontrol endophytes *Bacillus subtilis* R31 and TR21 provides insights into their differences in Phytobeneficial traits. Genomics 113:900–909

Lugtenberg BJJ, Caradus JR, Johnson LJ (2016) Fungal endophytes for sustainable crop production. FEMS Microbiol Ecol 92:1–17

Mantelin S, Touraine B (2004) Plant growth-promoting bacteria and nitrate availability: impacts on root development and nitrate uptake. J Exp Bot 55:27–34

Marin S, Ramos AJ, Cano-Sancho G, Sanchis V (2013) Mycotoxins: occurrence, toxicology, and exposure assessment. Food Chem Toxicol 60(1):218–237

Massawe VC, Hanif A, Farzand A, Mburu DK, Ochola SO, Wu L, Tahir HAS, Gu Q, Wu H, Gao X (2018) Volatile compounds of Endophytic bacillus spp. have biocontrol activity against Sclerotinia sclerotiorum. Phytopathology 108:1373–1385

Matarese F, Sarrocco S, Gruber S et al (2012) Biocontrol of Fusarium head blight: interactions between *Trichoderma* and mycotoxigenic Fusarium. Microbiology 158:98–106

Maurhofer M, Baehler E, Notz R et al (2004) Cross talk between 2,4-Diacetylphloroglucinolproducing biocontrol pseudomonads on wheat roots. Appl Environ Microbiol 70:1990–1998

Mauro A, Battilani P, Cotty PJ (2015) Atoxigenic *Aspergillus flavus* endemic to Italy for biocontrol of aflatoxins in maize. BioControl 60:125–134

Mavrodi OV, McSpadden Gardener BB, Mavrodi DV et al (2001) Genetic diversity of phlD from 2,4-Diacetylphloroglucinol-producing fluorescent pseudomonas spp. Phytopathology 91:35–43

Mohammadipour M, Mousivand M, Salehi Jouzani G et al (2009) Molecular and biochemical characterization of Iranian surfactin-producing *Bacillus subtilis* isolates and evaluation of their biocontrol potential against *Aspergillus flavus* and *Colletotrichum gloeosporioides*. Can J Microbiol 55:395–404

Molinari S, Leonetti P (2019) Bio-control agents activate plant immune response and prime susceptible tomato against root-knot nematodes. PLoS One 14:e0213230

Mousa WK, Shearer CR, Limay-Rios V, Zhou T, Raizada MN (2015) Bacterial endophytes from wild maize suppress Fusarium graminearum in modern maize and inhibit mycotoxin accumulation. Frontiers in Plant Science. 6:805

Mukherjee M, Mukherjee PK, Horwitz BA et al (2012) *Trichoderma*-plant–pathogen interactions: advances in genetics of biological control. Indian J Microbiol 52:522–529

Müller T, Behrendt U, Ruppel S et al (2016) Fluorescent pseudomonads in the Phyllosphere of wheat: potential antagonists against fungal Phytopathogens. Curr Microbiol 72:383–389

Nagórska K, Bikowski M, Obuchowski M (2007) The multicellular behavior and production of a wide variety of toxic substances support the use of *Bacillus subtilis* as a powerful biocontrol agent. Acta Biochim Pol 54:495–508

Pal KK, Spadden M, Gardener B (2006) Biological control of plant pathogens. Plant Heal Instr, pp 1–25

Palumbo JD, O'Keeffe TL, Abbas HK (2007) Isolation of maize soil and rhizosphere bacteria with antagonistic activity against *Aspergillus flavus* and *Fusarium verticillioides*. J Food Prot 70:1615–1621

Pandin C, Le Coq D, Deschamps J, Védie R, Rousseau T, Aymerich S, Briandet R (2018) Complete genome sequence of *bacillus velezensis* QST713: a biocontrol agent that protects *Agaricus bisporus* crops against the green Mold disease. J Biotechnol 278:10–19
Pieterse CMJ, Zamioudis C, Berendsen RL et al (2014) Induced systemic resistance by beneficial microbes. Annu Rev Phytopathol 52:347–375
Pitt JI, Miller JD (2017) A concise history of mycotoxin research. J Agric Food Chem 65(33):7021–7033
Porras-Alfaro A, Bayman P (2011) Hidden fungi, emergent properties: Endophytes and microbiomes. Annual Review of Phytopathology. 49(1):291–315
Poveda J (2021) Trichoderma as biocontrol agent against pests: new uses for a Mycoparasite. Biol Control 159:104634
Premjanu N, Jayanthy C (2012) Endophytic fungi a repository of bioactive compounds-a review. Int J Institutional Pharm Life Sci 2:135–162
Punja ZK, Utkhede RS (2003) Using fungi and yeasts to manage vegetable crop diseases. Trends Biotechnol 21:400–407
Rabiey M, Ullah I, Shaw MW (2015) The endophytic fungus Piriformospora indica protects wheat from fusarium crown rot disease in simulated UK autumn conditions. Plant Pathology. 64(5):1029–1040
Rabiey M, Shaw MW (2016) Piriformospora indica reduces fusarium head blight disease severity and mycotoxin DON contamination in wheat under UK weather conditions. Plant Pathology. 65(6):940–952
Risoli S, Cotrozzi L, Sarrocco S, Nuzzaci M, Pellegrini E, Vitti A (2022) *Trichoderma*-induced resistance to *Botrytis cinerea* in Solanum species: a meta-analysis. Plants 11:180
Romero D, de Vicente A, Rakotoaly RH et al (2007) The iturin and fengycin families of lipopeptides are key factors in the antagonism of *Bacillus subtilis* toward *Podosphaera fusca*. Mol Plant Microbe Interact 20:430–440
Saeki EK, Kobayashi RKT, Nakazato G (2020) Quorum sensing system: target to control the spread of bacterial infections. Microb Pathog 142:104068
Sánchez-Montesinos B, Santos M, Moreno-Gavíra A, Marín-Rodulfo T, Gea FJ, Diánez F (2021) Biological control of fungal diseases by *Trichoderma aggressivum* f. *europaeum* and its compatibility with fungicides. J. Fungi 7:598
Sasson A (2012) Food security for Africa: an urgent global challenge. Agric Food Sec 1(1):2
Schoneberg A, Musa T, Voegele RT et al (2015) The potential of antagonistic fungi for the control of *Fusarium graminearum and Fusarium crookwellense* varies depending on the experimental approach. J Appl Microbiol 118:1165–1179
Sempere F, Santamarina MP (2007) *In vitro* biocontrol analysis of *Alternaria alternata* (Fr.) Keissler under different environmental conditions. Mycopathologia 163:183–190
Sempere F, Santamarina MP (2009) Antagonistic interactions between the fungal rice pathogen *Fusarium verticillioides* (Sacc.) Nirenberg and *Trichoderma harzianum* Rifai. Ann Microbiol 59:259–266
Shanahan P, O'Sullivan DJ, Simpson P et al (1992) Isolation of 2,4-diacetylphloroglucinol from a fluorescent pseudomonad and investigation of physiological parameters influencing its production. Appl Environ Microbiol 58:353–358
Sherameti I, Shahollari B, Venus Y, Altschmied L, Varma A, Oelmüller R (2005) The endophytic fungus Piriformospora indica stimulates the expression of nitrate reductase and the starch-degrading enzyme glucan water dikinase in tobacco and Arabidopsis roots through a homeodomain transcription factor that binds to a conserved motif in. The Journal of Biological Chemistry. 280(28): 26241–26247
Shoresh M, Harman GE, Mastouri F (2010) Induced systemic resistance and plant responses to fungal biocontrol agents. Annu Rev Phytopathol 48:21–43
Tahir HAS, Gu Q, Wu H, Raza W, Safdar A, Huang Z, Rajer FU, Gao X (2017) Effects of volatile compounds produced by *Ralstonia solanacearum* on the growth-promoting and systemic resistance-inducing potential of Bacillus volatiles. BMC Plant Biol 17:133

Thambugala KM, Daranagama DA, Phillips AJL, Kannangara SD, Promputtha I (2020) Fungi vs. fungi in biocontrol: an overview of fungal antagonists applied against fungal plant pathogens. Front Cell Infect Microbiol 10:604923

Uka V, Moore GG, Arroyo-Manzanares N et al (2017) Unraveling the diversity of the cyclopiazonic acid family of mycotoxins in *aspergillus flavus* by UHPLC triple-TOF HRMS. Toxins 2017(9):35

Veras FF, Correa APF, Welke JE et al (2016) Inhibition of mycotoxin-producing fungi by bacillus strains isolated from fish intestines. Int J Food Microbiol 238:23–32

Varma A, Verma S, Sudha, Sahay N, Bütehorn B, Franken P (1999) Piriformospora indica, a cultivable plantgrowth-promoting root endophyte. Applied and Environmental Microbiology. 65(6):2741–2744

Verma M, Brar SK, Tyagi RD et al (2007) Antagonistic fungi, *Trichoderma* spp.: panoply of biological control. Biochem Eng J 37:1–20

Vinale F, Sivasithamparam K, Ghisalberti EL et al (2008) *Trichoderma*-plant–pathogen interactions. Soil Biol Biochem 40:1–10

Voss KA, Smith GW, Haschek WM (2007) Fumonisins: Toxicokinetics, mechanism of action and toxicity. Anim Feed Sci Technol 137:299–325

Vujanovic V, Goh YK (2009) *Sphaerodes mycoparasitica* sp. *nov*., a new biotrophic mycoparasite *of Fusarium avenaceum, F. Graminearum* and *F. Oxysporum*. Mycol Res 113:1172–1180

Vujanovic V, Goh YK (2012) Qpcr quantification of *sphaerodes mycoparasitica* biotrophic mycoparasite interaction with *Fusarium graminearum*: *in vitro* and *in planta* assays. Arch Microbiol 194:707–717

Waller F, Achatz B, Baltruschat H, Fodor J, Becker K, Fischer M, Heier T, Hückelhoven R, Neumann C, Wettstein DV, Franken P, Kogel K-H (2005) The endophytic fungus Piriformospora indica reprograms barley to salt-stress tolerance, disease resistance, and higher yield. Proceedings of the National Academy of Sciences. 102(38):13386–13391

Wang J, Liu J, Chen H, Yao J (2007) Characterization of the *Fusarium graminearum* inhibitory lipopeptide from *Bacillus subtilis* IB. Appl Microbiol Biotechnol 76:889–894

Weller DM (1988) Biological control of soilborne plant pathogens in the rhizosphere with bacteria. Annu Rev Phytopathol 26:379–407

Wu L, Wu H, Chen L, Xie S, Zang H, Borriss R, Gao X (2014) Bacilysin from *bacillus amyloliquefaciens* FZB42 has specific bactericidal activity against harmful algal bloom species. Appl Environ Microbiol 80:7512–7520

Xue AG, Voldeng HD, Savard ME et al (2009) Biological control of fusarium head blight of wheat with *Clonostachys rosea* strain ACM941. Can J Plant Pathol 31:169–179

Yates IE, Meredith F, Smart W et al (1999) *Trichoderma viride* suppresses fumonisin B1 production by *Fusarium moniliforme*. J Food Prot 62:1326–1332

Yazar S, Omurtag GZ (2008) Fumonisins, trichothecenes and zearalenone in cereals. Int J Mol Sci 9(11):2062–2090

Zain ME (2011) Impact of mycotoxins on humans and animals. J Saudi Chem Soc 15:129–144

Zhao Y, Selvaraj JN, Xing F et al (2014) Antagonistic action of *Bacillus subtilis* strain SG6 on *Fusarium graminearum*. PLoS One 9:1–11

Zhou L, Wei D, Selvaraj JN et al (2015) A strain of *aspergillus flavus* from China shows potential as a biocontrol agent for aflatoxin contamination. Biocontrol Sci Tech 25:583–592

Zhu J, Tan T, Shen A, Yang X, Yu Y, Gao C, Li Z, Cheng Y, Chen J, Guo L (2020) Biocontrol potential of *Bacillus subtilis* IBFCBF-4 against the Fusarium wilt of watermelon. J Plant Pathol 102:433–441

Zimmermann SD, Marqués-gálvez JE (2021) Endophytic fungi: from Symbiosis to secondary metabolite communications or vice versa? Front Plant Sci 12:791033

Zubair M, Farzand A, Mumtaz F, Khan AR, Sheikh TMM, Haider MS, Yu C, Wang Y, Ayaz M, Gu Q, Gao X, Wu H (2021) Novel genetic dysregulations and oxidative damage in *Fusarium graminearum* induced by plant Defense eliciting psychrophilic *Bacillus atrophaeus* Ts1. Int J Mol Sci 22(22):12094

Chapter 10
Remediating Heavy Metals from the Environment via Secondary Interactions Between Plants and Microorganisms

Samah Mostafa El-Sayed, Amr Said Mohamed, and Azza Abdel-Hamid Mohamed Mazher

Abstract When heavy metals (HMs) contaminate food, they pose a serious risk to both human health and the environment. This is due to their inability to decompose naturally and the presence of organic pollutants, which can persist in their harmful form for extended durations. HMs such as zinc (Zn), cobalt (Co), nickel (Ni), and iron (Fe) play notable roles in plant metabolic activities. However, if these metals exceed a plant's needs, they can become toxic. Cadmium, lead, and zinc are superfluous minerals that the plant does not necessitate in any manner. The phytoremediation strategy was selected as the principal method because of its cost-effectiveness, minimal labor requirements, ability to maintain soil attributes, and potential to increase soil quality and fertility. Certain plants have the ability to absorb and store significant quantities of heavy minerals without displaying any signs of toxicity. These plants are commonly referred to as "hyperaccumulators." Nevertheless, continuous exposure to these metals might injure certain plants, underscoring the crucial role of microorganisms in safeguarding these plants, increasing their effectiveness, and enhancing their ability to tolerate elevated concentrations of heavy metals. The objective of this chapter is to gather and record as much information as possible from current research on the serious problem of heavy metal contamination in the environment, along with the importance of using plant and biological approaches to address it, which has attracted the attention of experts.

Keywords Heavy metals · Hyperaccumulators · Pollution · Microorganisms · Isoenzymes · Siderophores · Surfactants · Biopolymers

---

S. M. El-Sayed (✉) · A. A.-H. M. Mazher
Ornamental Plants and Woody Trees Department, Agricultural and Biological Research Institute, National Research Centre, Giza, Egypt

A. S. Mohamed
Botanical Gardens Research Department, Horticulture Research Institute of Agricultural Research Center (ARC), Giza, Egypt

V. Gopikrishnan et al. (eds.), *Harnessing Microbial Resources for Sustainable Agriculture*, Microorganisms for Sustainability 61,
https://doi.org/10.1007/978-981-95-4817-0_10

10.1 Introduction

The primary factor contributing to the contamination of the environment with organic and inorganic (metal) substances caused by human activities is the rapid progress of various industries, agriculture, and technology. The continuous buildup of these pollutants in the environment leads to a disruption in many elements of the ecosystem, all of which seep into the soil, resulting in detrimental effects on its properties and overall quality. The natural microbial diversity of soil is depleted, which results in a loss of vegetation cover and reduced agricultural output and quality. Heavy metals create substantial vulnerability to the environment and human health due to their continuous buildup and multiple sources (Rajendran et al. 2021; Emenike et al. 2018). In addition to being caused by human activity, these substances also occur spontaneously in the crust of the Earth. They are released through the environment and can be transported from one location to another due to increasingly severe climate change events, thereby exacerbating pollution levels. Moreover, the risk associated with heavy metals arises from their nondegradability, in contrast to organic compounds, which can be broken down through natural biological processes, thereby changing their structure and reducing their harmful impacts (Rajendran et al. 2021).

Heavy metals are recognized as elements with a density greater than 5 g/cm^3. These minerals are classified as either necessary or not depending on their necessity for the plant and their participation in the plant's vital functions. Plants require certain amounts of essential metals, such as iron, copper, cobalt, zinc, nickel, and manganese, for various metabolic processes, such as pigment biosynthesis, ion homeostasis, gene regulation, respiration, enzyme activity, sugar metabolism, photosynthesis, and nitrogen fixation. However, if these metals exceed the plant's requirements, they can accumulate in the tissues and become toxic (Tiwari and Lata 2018; Bashir et al. 2016). Nonessential elements, even in small amounts, are poisonous and do not participate in any of the plant's metabolic activities. The elements mentioned are cadmium, lead, mercury, and arsenic (Ashraf et al. 2019).

Various physical methods, including excavation, disposal, and soil washing, have been used to address the issue of soil contamination (Chakrabartty and Harun-Or-Rashid 2021). Additionally, chemical methods such as stabilization, solidification, electrokinetics, and chemical precipitation have also been utilized (Jiang et al. 2022b). While these technologies may be efficient at removing pollutants, they pose a risk to the environment by causing soil degradation and the eradication of important soil microorganisms. Consequently, it diminishes the appropriateness of agricultural soil for subsequent utilization. Additionally, this procedure is costly and labor intensive, making it unsuitable for soil with deep contaminant dispersion. Consequently, numerous scientists have suggested the use of phytoremediation techniques to remediate contaminated soil, either independently or in combination with specific microbes, to eliminate, transform, or restrict the dispersion of these toxic substances present in the soil. There are also environmentally advantageous methods that not only preserve the environment but also enhance its quality.

Hence, this chapter aims to collect and record the detrimental consequences of heavy metals, the process of phytoremediation, and the data presented by numerous studies that have investigated the capacity of microorganisms to ameliorate the effectiveness of plants in safeguarding the environment, decreasing heavy metal levels, or mitigating negative effects.

10.2 Plant Growth in Soils Polluted with HMs and Its Negative Consequences

Heavy metals in the soil have a direct or indirect effect on plant growth. The absorption of heavy elements through the root area and their subsequent transfer to other areas of the plant directly impact the plant.

The accumulation of heavy metals in soil has detrimental effects on plant morphology, physical characteristics, and chemical composition. Heavy metals have several noticeable effects, including stunted growth rates, chlorosis (yellowing of leaves), necrosis (cell death), leaf rolling, changes in stomatal action, decreased water potential, cation efflux (movement of positively charged ions out of cells), alterations in membrane functions, inhibition of photosynthesis and respiration, changes in metabolism, and effects on the activities of important enzymes (Sharma and Dubey 2007; Dubey 2011). Aquaporin impairment has been shown to govern water responses, which are among the initial plant reactions to HMs. In their study, Chatterjee and Chatterjee (2000) reported that elevated amounts of chromium (Cr) in cauliflower led to a decrease in the water potential and transpiration rate while simultaneously increasing the diffusive resistance and relative water content.

Heavy metals have a high level of toxicity. When the concentration of these metals in the cytosol of plants exceeds the normal level, phytotoxicity occurs. This phytotoxicity hampers the processes of photosynthesis, cell respiration, and nitrogen metabolism. Additionally, it induces oxidative stress, which negatively impacts the growth and development of plants (Maksymiec et al. 2007). Heavy metals (HMs) affect the functions of PSI and PSII, alter stomatal function, and decrease photosynthesis (Appenroth 2010). Respiration is often diminished by high levels of toxic heavy metals (Losch 2004); nevertheless, several plant species experience an increase in respiration at extremely low concentrations (Appenroth 2010). Respiratory sensitivity to chromium has been identified in *Pisum sativum* (Bishnoi et al. 1993). Furthermore, As and Al decreased the activity of nitrogen absorption enzymes, including nitrate reductase, nitrite reductase, and glutamine synthetase, in rice seedlings (Dubey 2011).

The growth of a plant is indirectly affected by altering the characteristics of the soil in which it resides. On the basis of specific research findings, the accumulation of heavy metals in the soil can potentially lead to a shift in pH. For example, the accumulation of cadmium and aluminum leads to a decrease in soil pH, resulting in acidity (Liu et al. 2013). Conversely, the accumulation of nickel and chromium minerals causes an increase in soil pH, leading to alkalinity (Zhu et al. 2017).

Modifying the pH of the soil has a dual impact on nutrients and microbial activity, leading to a decline in soil fertility. This is because microbial activity plays a crucial role in breaking down organic compounds in the soil and promoting fertility (Li et al. 2018). In addition, certain heavy metals compete with and hinder the absorption of essential nutrients necessary for plant growth. Lead and zinc hinder the absorption of magnesium and iron, leading to symptoms of deficiency in plants, even though these nutrients are present in the soil (Das et al. 2019; Kabata-Pendias and Mukherjee 2007).

The data shown in Table 10.1 depict the potential hazards associated with the exposure of some crops to heavy metals. Research suggests that *Triticum aestivum* L. and *Zea mays* plants have reduced growth rates when cultivated in soils contaminated with nickel (Ni) and zinc (Zn). Furthermore, these plants presented decreased levels of photosynthetic pigments, proteins, sugars, and relative water content. Nevertheless, the levels of proline and lipid peroxidation (MDA) and the activities of the isoenzymes CAT and SOD increased, as reported by Shahzad et al. (2024) and Anbuganesan et al. (2024), respectively. Ni induces a decrease in the weight of cauliflower plants; water use efficiency (WUE), expressed as a percentage; photosynthetic pigments; transpiration rates; and nutrient element contents, including nitrogen (N), phosphorus (P), potassium (K), magnesium (Mg), and calcium (Ca) contents (Khan et al. 2024). The presence of cadmium pollution resulted in a decrease in sorghum plant biomass and nutrient component content; additionally, it caused changes in soil characteristics, including a decrease in accessible potassium (K), phosphorus (P), and alkaline hydrolysis of nitrogen, as well as an increase in pH and DTPA-Cd levels (Zhang et al. 2024). In their study, Direk et al. (2024) reported that cultivating wheat plants in soil contaminated with heavy metals resulted in a 32% decrease in the relative growth rate when exposed to As, an 18% decrease when exposed to Cd, and a 24% decrease when both metals were present. Additionally, the contaminated soil led to reductions in photosynthetic pigments, stomatal conductance, transpiration, relative water content (RWC), and the carbon assimilation rate. The following substances and enzymes are involved: proline, H_2O_2, MDA, CAT, SOD, NOX, APX, and GST isoenzymes. A different study revealed that various factors related to plant growth and function, such as the photosynthetic rate, transpiration rate, water use efficiency (WUE), stomatal conductance, vapor pressure deficit, relative water content (RWC), photosynthetic pigments, and the root/shoot ratio, decreased in *Helianthus annuus* plants that were exposed to soil contaminated with lead at a concentration of 500 mg kg^{-1}. On the other hand, the activities of certain enzymes, including superoxide dismutase (SOD), peroxidase (POD), ascorbate peroxidase (APX), and catalase (CAT), are increased in these plants (Ayub et al. 2024). *Cucurbita pepo* experiences growth issues when cultivated in contaminated media containing 200 μM Cu and Cd (Acila et al. 2024).

Table 10.1 Some recent examples that illustrate the toxic effects of heavy metals on plants

Plant species	Heavy metal	Concentration range	Effects	References
Zea mays	Ni	150 $ml.l^{-1}$	Vegetative growth parameters, total chlorophyll, and protein, RWC (%) were decreased. EC (%), SOD, phenol, MDA, and proline were increased	Anbuganesan et al. (2024)
	Zn	300 $ml.l^{-1}$		
Cauliflower	Ni	60 mg. kg^{-1} soil	Plant fresh and dry matter, photosynthetic pigments, nutrient content, WUE (%), and stomata conductance were decreased. CAT, POD, APX, H_2O_2, MDA, proline were increased	Khan et al. (2024)
Triticum aestivum	Ni	100 mg. kg^{-1} soil	Vegetative growth indicators, protein, proline, sugar content, and nutrient elements content were decreased.	Shahzad et al. (2024)
	Zn	100 mg. kg^{-1} soil		
Sorghum	Cd	100–700 $ml.l^{-1}$	Biomass and nutrient element content in plant were decreased. Cadmium accumulation was increased	Zhang et al. (2024)
Triticum aestivum	Cd	100 μM	RGR, photosynthetic pigments, stomata conductance, transpiration, RWC, and carbon assimilation rate were decreased. Proline, H_2O_2, MDA, the activity of CAT, SOD, NOX, APX, GST isoenzymes,	Direk et al. (2024)
	As	100 μM		
Helianthus annuus	Pb	500 mg. kg^{-1} soil	photosynthetic rate, transpiration rate, WUE (%), stomata conductance, vapor pressure deficit, RWC (%), photosynthetic pigments, and root/shoot ratio were decreased. SOD, POD, APX, and CAT isoenzymes activity were increased.	Ayub et al. (2024)
Cucurbita pepo L.	Cd Cu	200 μM	Vegetative growth parameters and activity of SOD isoenzyme were decreased. The activity of glutathione S-transferase was increased.	Acila et al. (2024)

The accumulation of pollution in the soil has detrimental effects on crop quality, development, and production. Additionally, it poses a risk to human health and can lead to various diseases when these contaminated products are consumed. Consequently, it is necessary to remediate this contaminated environment to maintain a pristine environment devoid of pollutants that are safe for human well-being and life.

10.3 Phytoremediation of Soil Contaminated with Heavy Metals

Multiple chemical and physical methods exist for treating contaminated soil and purifying it from heavy metals, all of which are renowned for their rapid pollutant removal capabilities. Chemical methods involve processes such as binding the elements in the soil and stiffening them with binders or collecting them through electrical charges (Jiang et al. 2022b). On the other hand, the physical method involves digging, removing the soil, and relocating it to a different location (Chakrabartty and Harun-Or-Rashid 2021). These approaches are typically seen as expensive and demanding, necessitating significant time and effort while also diminishing soil quality (Ojuederie and Babalola 2017; Liu et al. 2020).

Phytoremediation is a sustainable and ecologically sound biological technique that effectively reduces or eliminates pollutants from soil while preserving its inherent characteristics. In contrast, it enhances soil characteristics while simultaneously enhancing visual appeal and the quality of the surrounding environment. Additionally, it is a cost-effective option that demands minimal effort or human resources. An inherent drawback of this procedure is its sluggishness, as it requires a significant amount of time, often spanning months or even years, to completely cleanse the soil. Additionally, plants must fulfill specific criteria to carry out their role and achieve maximum benefit. The specifications are succinctly outlined in the following points: (1) rapidly growing plants with high biomass production, (2) deep and broad root systems, (3) herbivore-repellent properties to prevent component transfer, and (4) ability to withstand and tolerate the detrimental effects of heavy metals. Certain plant species have the capacity to accumulate significantly high levels of heavy metals, surpassing other plants by up to 500 times. These plants are referred to as "hyperaccumulators" (Chaney et al. 1997). Another study indicated that hyperaccumulator plants have the capacity to accumulate heavy metals in their shoots at levels 100–1000 times greater than the plants' ability to withstand heavy metals (Rascio and Navari-Izzo 2011). Baker et al. (1994) and Raskin et al. (1994) established threshold amounts of heavy metals that classify a plant as hyperaccumulating. According to their findings, hyperaccumulating plants can accumulate more than 0.01% Cd; more than 0.1% Ni, Co, Cu, Cr, and Pb; and more than 1% Zn.

The technology of phytoremediation can be categorized into five specific approaches, as listed below:

1. Phytoextraction refers to the capacity of specific plants to collect heavy metals from the soil and store them in their aboveground parts. Phytoremediation is a technique that effectively eliminates pollutants from the soil without causing any harm to its inherent properties (Jacob et al. 2018; Oladoye et al. 2022).
2. Rhizostabilization is a technique employed by certain plants to restrict the movement of pollutants in the soil. This is achieved by adsorbing pollutants through the roots or by binding them to compounds secreted by the roots, which leads to a reduction in the mobility of heavy elements and prevents their transfer from one area to another (Ullah et al. 2015; DalCorso et al. 2019; Gavrilescu 2022).

3. Phytovolatilization is a process in which plants take in pollutants, convert them into less harmful or nontoxic substances, and then release them into the atmosphere by evaporation (Ali et al. 2013; Gavrilescu 2022; Oladoye et al. 2022).
4. Phytotransformation or phytodegradation is a technique that relies on the ability of plants to break down organic pollutants via specific enzymes. However, it is important to note that this method cannot break down heavy elements because they are not biodegradable. Nonetheless, it is still considered a valuable phytoremediation technique (Ali et al. 2013; Singh et al. 2018).
5. Phytofiltration is a method in which plants utilize their biomass to gather toxins present in water and liquid waste while also absorbing both organic and mineral pollutants (Rezania et al. 2016; Singh et al. 2018; Hakeem et al. 2020). Rhizofiltration refers to the accumulation of substances in the root, whereas blastofiltration refers to accumulation in early seedlings. On the other hand, caulofiltration is the term used when accumulation occurs in the shoot (Hakeem et al. 2020).

The domain of genetic engineering has immense potential for increasing the ability of plants to take up heavy metals. Guo et al. (2024) employed *Agrobacterium* strain GV3101 to transfer the SsIRT9 gene, which triggers the accumulation of cadmium in plants, from *Salix suchowensis* (willow trees) to *Nicotiana benthamiana* (tobacco plants). They reported that when these genetically modified plants were exposed to cadmium stress, the accumulation of cadmium increased 2–4.5-fold as a result of the ability of the gene to be expressed within the plant.

Ornamental herbaceous plants have a quick growth period, a large amount of plant material, the ability to adjust to different environments, a wide geographic range, and a low risk of toxicity. Moreover, these plants can provide economic advantages while simultaneously improving the visual appeal of the polluted surroundings. Additionally, as ornamental plants are rarely consumed, the risk of hazardous substances, especially heavy metals, entering the food chain is reduced (Deepika and Haritash 2023). Numerous researchers have investigated the propensity of certain ornamental plants to grow and absorb heavy metals, as mentioned in Table 10.2. *Plectranthus* sp. can absorb mercury (Hg), a related element in locations of gold mining and extraction, from shoots at concentrations ranging from 13.8 to 220.05 ppm (Muryani et al. 2024). *Silybum marianum* (L.) Gaerth was grown in soil contaminated with Cd at a concentration of 3.30 mg. kg^{-1} soil and accumulate Cd in roots (Papadimou et al. 2024). Hussain et al. (2024) noted that *Celosia argentea* L. plants are good accumulator plants for Cr, Cu, Pb, and Zn in leaves and roots. An investigation has been made to test the ability of some ornamental plants to grow in an environment contaminated with a heavy metal mixture consisting of Cd 10–20, Pb 200–750, Ni 100–150, and Zn 250–400 mg. The plant *Trifolium pratense* was highly effective at accumulating Cd, Zn, and Ni in both roots and leaves, as well as adapting to and increasing the production of biomass and photosynthetic pigments. *Plantago lanceolata* was also successful at collecting Cd and Zn in the leaves, whereas the *Amaranthus retroflexus* plant showed a high ability to absorb and accumulate Pb in the roots (Cakaj et al. 2024). *Spergularia rubra* accumulated Cd and

Table 10.2 Recent examples of some ornamental plants that accumulate heavy elements

Plant species	Heavy metal	Concentration range	Accumulator organ	References
Plectranthus sp.	Hg	13.80–220.05 ppm	Shoot	Muryani et al. (2024)
Silybum marianum (L.) Gaerth	Cd	3.30 mg. kg^{-1} soil	Root	Papadimou et al. (2024)
Celosia argentea L.	Cr	100–400 mg. l^{-1}	Leaves and root	Hussain et al. (2024)
	Cu			
	Pb			
	Zn			
Trifolium pretense *Plantago lanceolata* *Amaranthus retroflexus*	Cd	10–20 mg. kg^{-1} soil	Leaves and root	Cakaj et al. (2024)
	Pb	200–750 mg. kg^{-1} soil		
	Ni	100–150 mg. kg^{-1} soil		
	Zn	250–400 mg. kg^{-1} soil		
Spergularia rubra	Zn	2225.97 mg. kg^{-1} soil	Shoot	Medina-Díaz et al. (2024)
	Cd	4.74 mg. kg^{-1} soil		
Hydrangea macrophylla (Thunb.) Ser.	Cd	15–240 mg. kg^{-1} soil	Root	Chen et al. (2024)
Polygonum chinense L.	Cd	5–50 μg.l^{-1}	Root	Li et al. (2024)
Gladiolus grandifloras	Cu	80–100 μg.kg^{-1} soil	Root, shoot, flowers	Abdul Haseeb et al. (2024)
Chrysanthemum	As	80–100 μg.kg^{-1} soil		
Pteridium aquilinum	Cr	800–6400 μM	Leaves, rhizomes and adventitious roots	Eslava-Silva et al. (2023)

Zn in shoots when the Cd concentration was 4.74 and the Zn concentration was 2225.97 mg. kg^{-1} soil (Medina-Díaz et al. 2024). *Hydrangea macrophylla* (Thunb.) Ser. Cv. Classic Rood are grown and tolerated in Cd-contaminated soil at a concentration of 240 mg. kg^{-1} soil (Chen et al. 2024). *Polygonum chinense* L. accumulates Cd in roots when it is grown at concentrations ranging from 5 to 50 μg.L^{-1} (Li et al. 2024). Abdul Haseeb et al. (2024) reported that *Gladiolus grandifloras* and *Chrysanthemum* have the ability to accumulate Cu and As in all parts of both plants. Eslava-Silva et al. (2023) reported that the uptake capacity of sporophytes (leaves, rhizomes, and adventitious roots) of *Pteridium aquilinum* may be successfully used in the phytoremediation process.

When some plants are exposed to heavy elements at low concentrations, their growth is stimulated better than that of plants that are not exposed, which is known as "hormesis" (Mattson 2008; Xu et al. 2023). Some studies have demonstrated that cadmium (Cd) has an effect on *Dimorphotheca ecklonis* (El-Sayed et al. 2022) and *Picris divaricata* (Hu et al. 2023) plants.

Some researchers have warned against irrigating vegetables and crops with marginal water or planting them in contaminated soil, as some hyperaccumulating plants can grow and tolerate heavy elements, accumulating inside them and causing severe harm to humans when consumed. Cress, parsley, and rocket are examples of plants that may be irrigated with iron (Fe)-contaminated water that they deposit in their shoots (Nasircilar et al. 2024), and tapioca plants are also able to grow in soil contaminated with the heavy elements Cd, Cu, Mn, Ni, and Pb, which accumulate these elements in plant organs (Pugazhendhi et al. 2024). Flax (*Linum usitatissimum* L.) accumulates Cd in shoots when grown in contaminated soil (Kotoula et al. 2024).

10.4 Microorganism–Plant Conjugation in the HM Phytoremediation Process

Associations among hyperaccumulating plants and metal-tolerant microorganisms (mostly fungi, bacteria, and yeasts) play critical roles in bioremediation procedures, whereby microbes accumulate or mobilize heavy metals and then increase hyperaccumulating plant development and metal uptake ability (Glick 2010). As a result, the phytoremediation of heavy metals by microbes has garnered much interest worldwide. Furthermore, symbiotic soil microbes can increase plant tolerance to a variety of environmental challenges (Hadi and Fano 2010).

10.4.1 Fungi

Fungi are microorganisms that live in a symbiotic relationship with plants via their roots, acting as an extension of the root by penetrating and spreading their hyphae in the soil, increasing the area where nutrients and water may be absorbed from the soil. Hence, they improve the plant's ability to develop better, and the plant provides the carbon needed for its growth in the form of sugar; hence, they are biofertilizers or biostimulants for plant growth under diverse stress circumstances (Xavier and Boyetchko 2002). Mycorrhizae produce extraradical mycelia and intraradical hyphae, which penetrate the intercellular gaps and enter cortical root cells. In the event of reduced heavy metal absorption, fungal vacuoles play crucial roles in retention, binding, and immobilization. They regulate cytosolic metal ion concentrations and detoxify potentially harmful metal ions. Christie et al. (2004) noted that the fungal cell wall, specifically that of chitin and glomalin, is especially essential since it contains free amino, hydroxyl, carboxyl, and other functional groups (Gadd 1993). Anand et al. (2006) reported that mycorrhizae can grow spontaneously and play an important role in plants even in severe situations such as high temperatures and acidity. Darwish et al. (2017) reported that mycorrhizal fungi enhanced the growth of *Eucalyptus gomphocephala*, increasing the values of morphological traits, photosynthetic pigments, and carbohydrate contents. Elhindi et al. (2018) tested the growth of *Tagetes erecta* plants in media inoculated with *Glomus constrictum* and polluted with a mixture of heavy metals Zn, Cu, Mn, and Co and reported that the growth and flowering attributes, photosynthetic pigment contents, and nutrient element (N, P, and K) contents increased. Jiang et al. (2024) noted that the microbial population beneficial to the rhizosphere and plant biomass of *Sedum alfredii* was promoted in the presence of a phyllosphere microbiome. *Phomopsis columnaris* fungi enhance the ability of *Noccaea caerulescens* to accumulate Ni (Ważny et al. 2021). Pedroso et al. (2018) reported that *Glomus macrocarpum* and *Paraglomus occultum* fungus enhanced the morphological traits and chlorophyll content of *Acacia mangium* when it was grown in polluted media.

10.4.2 Bacteria

Microorganisms are fundamental components of natural ecosystems and play an important role in their rehabilitation. Indeed, plant–microbe relationships in metal-polluted soils are critical for plants to survive metal toxicity and grow in such hostile settings (Alves et al. 2022; Montreemuk et al. 2024). The rhizosphere is an important bioregion where many microorganisms accumulate (Glick 2012). The root zone is defined as the narrow space surrounding the roots of plants in which the effects on the roots are physically and chemically induced (Benizri and Kidd 2018). Mendes et al. (2013) and Novo et al. (2018) reported that because this location is rich in nutritional substances, it includes an estimated 1011 microbial cells per gram of soil, which include bacteria, fungi, nematodes, and certain invertebrates. These components are created from the diverse secretions of the roots, which account for the microbial variety (Goswami et al. 2016). Amino acids, sugars, organic acids, fatty acids, vitamins, and enzymes are examples of molecules with high nitrogen and carbon contents (Badri and Vivanco 2008; Novo et al. 2018). This is what beneficial bacteria want, so they increase plant growth by facilitating the conversion of nutrients in the soil into a form that the plant can easily absorb. These microorganisms also help plants withstand many types of environmental stress (Rashid et al. 2016; Novo et al. 2018); via these secretions, plants select bacterial strains with unique traits that allow them to grow in the presence of heavy metals in the soil, and these bacterial strains can accumulate and transform heavy metals into nontoxic forms (Mishra et al. 2017). Certain bacterial species reside in (1) the root area (rhizosphere) and (2) internally in the stem and leaf areas (phyllosphere) or (3) may colonize the entire plant, producing what is known as an endophytic association (Compant et al. 2010; Kong and Glick 2017).

Bacteria are able to accelerate the growth of various yield crops, allowing them to withstand and grow in heavy metal-contaminated soils, such as *Bacillus pseudomycoides* strain ARN7 with the plant *Zea mays* (Anbuganesan et al. 2024); *Bacillus mycoides* with the plant *Triticum aestivum* L. (Shahzad et al. 2024); *Enterococcus* sp. strains NVY-101, 102, 103, 105, and 107; and *Bacillus* sp. strains NSL-104 and 106 on sorghum plants (Zhang et al. 2024); *Bacillus cereus* with the plant *Triticum aestivum* L. (Direk et al. 2024); and *Pseudomonas fluorescens* and *P. gessardii* with the plant *Helianthus annuus* (Ayub et al. 2024).

Some bacterial strains can withstand the harmful effects of heavy metals present at high concentrations in the soil because they have the genes responsible for these effects (Silver and Phung 1996). It prepares for this by employing processes that either limit the solubility of metals to stabilize them or increase their solubility for removal (Gadd 2000). Bacteria stabilize metals through precipitation, alkalinity, and complex creation processes, reducing their transfer to plant aerial organs (Ma et al. 2016). These bacterial strains create biopolymers, such as exopolysaccharides, which interact with metals. This decreases its mobility and bioavailability in the soil, allowing the plant to carry out the phytostabilization process more efficiently. While some bacteria increase bioavailability through acidification, chelation, and protonation, they also lower the pH of heavy metal-polluted soil (Ma et al. 2016). Furthermore, some plants produce organic acids, siderophores, and surfactants such as fatty acids, lipoproteins, lipopeptides, and glycolipids, which increase the solubility and availability of heavy metals in the soil, allowing the plant to carry out the phytoextraction process (Sessitsch et al. 2013). Bacteria can thrive and function in heavy metal-contaminated soil by controlling permeability, limiting metal absorption, and enzymatically removing harmful effects. Some bacteria use the methylation process to decrease the harmful effects of elements, which involves the transfer of a methyl group to metals, a process known as biotransformation (Ramasamy et al. 2007). Some bacterial species, which have independent metabolisms, bioabsorb elements on their outer surface by producing chelating substances and peptides that form complexes with metals such as glutathione and metallothionein, as well as the presence of biopolymers that aid in the absorption process (Pokethitiyook and Poolpak 2016; Ayangbenro and Babalola 2018; Lal et al. 2018). Then, bioaccumulation takes place inside the cell cavity, and oxidation and reduction processes occur to isolate these metals (Hou et al. 2020; Fomina and Gadd 2014; Vijayaraghavan and Yun 2008). Previous studies have demonstrated the ability of bacterial activity to promote the growth and biomass of ornamental plants, hence increasing the ability of plants to remove heavy ions from the soil by increasing their antioxidant enzymatic activities, as reported by Abdollahi et al. (2020) for ornamental cabbage plants with *Rhizobium pusense* and *Bacillus cereus* bacteria, Thanh et al. (2023) for *Hibiscus rosa-sinensis* plants with *Acidithiobacillus thiooxidans* bacteria, and Li et al. (2022) for *Coreopsis grandiflora* plants with *Penicillium*, *Achromobacter*, and *Rhizobium* bacteria.

Although the interaction of bacteria with plants clearly clears the soil and eliminates the negative effects of heavy metals, the continued exposure of these bacterial cells to pollution for long periods of time may affect their activity in the soil, in addition to increasing their numbers, leading to competition for nutrients (Jiang et al. 2022a). Teng et al. (2022) reported that the use of carrier materials was required to maintain the stability of bacterial cells in terms of density and activity, as well as to make it easier to collect, redistribute, store, and protect them from adverse environmental conditions. Carriers can exist in either organic or inorganic forms (Jiang et al. 2022b). Organic carriers are excellent suppliers of additional nutrients and help bacterial cells eliminate heavy metals more quickly; these include organic compounds such as agar, carrageenan, sodium alginate, polyurethane, and polyvinyl alcohol (Bouabidi et al. 2018; Wang et al. 2021). The inorganic carriers include biochar, activated charcoal, and diatoms (El-Tohory et al. 2023). These compounds help to preserve soil porosity, increase the carbon content, and improve cation exchange and water retention (Kafle et al. 2022; Ghosh and Maiti 2020). However, Samoraj et al. (2023) warned against the overuse of biochar because some of its sources contain high levels of heavy metals and polycyclic aromatic hydrocarbons, which can harm the soil and cause some environmental risks.

10.4.3 Yeast

Yeast and plant interactions have drawn increasing attention recently because of their positive effects on plants. These interactions are sometimes referred to as possible biofertilizers. Yeasts are unicellular fungi that excel in breaking down organic compounds in soil and plants, making them very efficient decomposers in nature (Rhodes 2014). Many studies have described the possible advantages of yeast in phytoremediation, such as stimulating plant growth and breaking down contaminants. Yeast may produce phytohormones, amino acids, and vitamins, all of which promote plant growth (Csutak et al. 2017). A variety of beneficial yeasts survive in the root, enabling the plant to withstand biotic and abiotic stress (Freimoser et al. 2019). Yeasts produce siderophores, which serve to supply iron in an easily absorbable form for plants (Di Francesco et al. 2021). Siderophores also improve soil fertility and promote plant resistance to the harmful effects of heavy metals (Ramos-Garza et al. 2016). Surfactants, including mycolic acid, lipopeptides, lipoproteins, and humic compounds, are chemicals produced by yeasts. These compounds lower surface tension, allowing heavy metals, particularly those that are poorly soluble, to dissolve (Singh et al. 2020). Exopolysaccharides, for example, form compounds with heavy metals, leading them to precipitate and become immobilized (Rahbar et al. 2021). Moreover, because of their capacity to generate chelating chemicals, yeasts can remove heavy metals, such as metallothionein and glutathione, by converting them into a form that is simpler for plants to absorb (Rono et al. 2021). Yeasts synthesize vitamin B12, which promotes the production of mycorrhizal fungi, which aid in the absorption of poisons more effectively (Nafady et al. 2019).

In nature, yeast cells are common and proliferate quickly, making them an inexpensive and easily available source of high biomass. Numerous physiological and molecular pathways for the interaction between heavy metals and yeast cells result in heavy metal degradation, even in the face of environmental hardship (Ozturk et al. 2021; Sharma et al. 2021). The distinct cell wall structure of yeast and its features make bioremediation more effective. The yeast cell wall has a negative charge because certain functional groups (phosphates and/or carboxyl groups), negatively charged yeast cell walls, and positively charged heavy metal ions interact via electrostatic interactions (Stanila et al. 2015). Furthermore, the yeast cell wall contains a range of polysaccharides and glycoproteins, including glycans, chitin, chitosan, mannan, and phosphomannan. These macromolecules contain a range of metal-binding groups, including phosphates, amines, hydroxyl groups, imidazoles, sulfates, and sulfhydryl groups. Furthermore, yeast mycelia provide a wide surface area by entwining mycelia and producing floes, resulting in net-structured yeast. The broad surface area of yeast increases oxygen diffusion, resulting in efficient oxygen delivery (Hegde et al. 2018). Heavy metal removal is influenced by the dissolved oxygen level, which affects whether the metal is oxidized or reduced. As a result, when dissolved oxygen is present, the reactions of different types of metal ions change (Li et al. 2013). Some environmental factors affect yeast biosorption effectiveness; consequently, maintaining a pH between 5 and 6 is preferable (Chen et al. 2021). *Saccharomyces cerevisiae* flourishes at temperatures ranging from 15 to 40 °C (Massoud et al. 2019; Zinicovscaia et al. 2020). Qiao et al. (2023) reported that *Saccharomyces cerevisiae* improved glutathione transferase activity and glutathione metabolism in *Dianthus spiculifolius* Schur. cultivated in polluted media containing Cd and Pb. Table 10.3 shows the promoting effects of several microorganisms on accumulator plants cultivated in HM-polluted soil.

10.5 Conclusion and Future Prospects

Research has demonstrated the dangers associated with the accumulation of heavy elements in the environment as well as the different natural or human-caused sources from which these metals originate. Thus, these metals injure their ultimate beneficiary, man, by causing the degradation of environmental elements (soil, water, and air) and adversely affecting crop output. Numerous physical and chemical approaches have been used to solve this issue. These techniques are rapid fixes, but they have the drawbacks of being costly, labor intensive, and potentially damaging other things. As a result, scientists have turned to employing plants that either barely harm them or have the capacity to absorb and retain minerals without harm. However, these plants might not be able to carry out their functions as effectively as they did at first because of the repeated buildup of these minerals in the soil. They therefore require naturally occurring auxiliary mechanisms, which take the form of soil-dwelling microorganisms such as fungi, bacteria, and yeast. These microbes secrete certain compounds that either make these metals easy for plants to absorb or secrete other compounds that bind the metals in the soil and prevent them from being absorbed or transferred to other locations. Since some plants can absorb and store heavy metals without exhibiting any symptoms of poisoning and may

Table 10.3 Role of microorganisms in enhancing the tolerance of HM-accumulator plants

Plant species	Microorganisms	Category	Heavy metal	Signs of boosting tolerance	References
Tagetes erecta	*Glomus constrictum*	Fungi	Sewage water containing (10.53 $\mu g.g^{-1}$ Zn, 5.6 $\mu g.g^{-1}$ Cu, 31.3 $\mu g.g^{-1}$ Mn, 1.23 $\mu g.g^{-1}$ Mn.)	Enhancing growth and flowering attributes, photosynthetic pigments content, and nutrient elements (N, P, K) content	Elhindi et al. (2018)
Sedum alfredii	1. *Streptomyces* and *Nocardia* 2. *Cladosporium* and *Acremonium*	Phyllosphere microbiome (bacteria and fungi, respectively)	Cd 2.2–15.2 + Zn378.70–1135.80 mg. kg^{-1} soil	Increasing the microbial population beneficial to the plant, increasing plant biomass	Jiang et al. (2024)
Noccaea caerulescens	*Phomopsis columnaris*	Fungi	Ni 150 μM	Increasing anthocyanin pigments in shoot and Ni accumulation in shoot and root	Ważny et al. (2021)
Acacia mangium	*Glomus macrocarpum*, *Paraglomus occultum*	Fungi	7200 mg/kg Zn + 1140 mg/kg Cu + 480 mg/kg Pb + 72 mg/kg Cd	Increasing the values of morphological traits, chlorophyll content	Pedroso et al. (2018)
Eucalyptus gomphocephala	Mycorrhiza	Fungi	Cd 50–200 ppm	Increasing the values of morphological traits, photosynthetic pigments and carbohydrates content	Darwish et al. (2017)

Ornamental cabbage	*Bacillus cereus Rhizobium pusense*	Bacteria	Cd 100 + Pb 3500 μg. l^{-1}	Promoting plant growth, increasing plant biomass and absorption of Cd and Pb	Abdollahi et al. (2020)
Hibiscus rosa-sinensis	*Acidithiobacillus thiooxidans*	Bacteria	Cr 218+ Pb 308.65 + Mn 882.38 + Cd 46.25 mg. kg^{-1} soil	Increasing seed germination, vegetative growth parameters, protein, total chlorophyll, total carbohydrates, enhancing heavy metal accumulation	Thanh et al. (2023)
Coreopsis grandiflora	*Penicillium Achromobacter Rhizobium*	Bacteria	Cd 20 mg. kg^{-1} soil	Enhancing growth biomass, increasing SOD, POD, and CAT isoenzymes activity, increasing free amino acids	Li et al. (2022)
Dianthus spiculifolius Schur	*Saccharomyces cerevisiae*	Yeast	Cd 3 mM, Pb 80 mM	Increasing glutathione transferase activity and enhancing the glutathione metabolism pathway with genes encoding antioxidant enzymes	Qiao et al. (2023)

seriously injure humans, many researchers have advocated caution when edible plants are planted on contaminated lands.

In addition to interest in genetic modification, which increases the efficiency of the bioaccumulation process, from a future perspective, the relationship between plants and microbes requires more research and testing to determine the equivalent relationships between plants and microorganisms that increase the efficiency of removing heavy metals.

References

Abdollahi S, Golchin A, Shahryari F (2020) Lead and cadmium-resistant bacterial species isolated from heavy metal-contaminated soils show plant growth-promoting traits. Int Microbiol 23:625–640. https://doi.org/10.1007/s10123-020-00133-1

Abdul Haseeb M, Akhtar I, Javed A, Tehreem F, Imran U, Sanaullah (2024) Phytorehzial technique for phytoremediation of heavy metals using ornamentals plants. Asian J Res Crop Sci 9(1):96–102. https://doi.org/10.9734/AJRCS/2024/v9i1250

Acila S, Derouiche S, Allioui N (2024) Embryo growth alteration and oxidative stress responses in germinating *Cucurbita pepo* seeds exposed to cadmium and copper toxicity. Sci Rep 14:8608. https://doi.org/10.1038/s41598-024-58635-1

Ali H, Khan E, Sajad MA (2013) Phytoremediation of heavy metals-concepts and applications. Chemosphere 2013(91):869–881

Alves ARA, Yin Q, Oliveira RS, Silva EF, Novo LAB (2022) Plant growth-promoting bacteria in phytoremediation of metal polluted soils: current knowledge and future directions. Sci Total Environ 838:156435

Anand P, Isar J, Saran S, Saxena RK (2006) Bioaccumulation of copper by Trichoderma viride. Bioresour Technol 97:1018–1025. https://doi.org/10.1016/j.biortech.2005.04.046

Anbuganesan V, Vishnupradeep R, Bruno LB, Sharmila K, Freitas H, Rajkumar M (2024) Combined application of biochar and plant growth-promoting Rhizobacteria improves heavy metal and drought stress tolerance in *Zea mays*. Plants 13:1143. https://doi.org/10.3390/plants13081143

Appenroth KJ (2010) Definition of "heavy metals" and their role in biological systems. In: Sherameti I, Varma A (eds) Soil heavy metals, soil biol, vol 9. Springer, Berlin, pp 19–29

Ashraf S, Ali Q, Zahir ZA, Ashraf S, Asghar HN (2019) Phytoremediation: environmentally sustainable way for reclamation of heavy metal polluted soils. Ecotoxicol Environ Saf 174:714–727

Ayangbenro A, Babalola O (2018) Metal(loid) bioremediation: strategies employed by microbial polymers. Sustainability 10:3028. https://doi.org/10.3390/su10093028

Ayub A, Shabaan M, Malik M, Asghar HN, Zulfiqar U, Ejaz M, Alarjani KM, Al Farraj DA (2024) Synergistic application of Pseudomonas strains and compost mitigates lead (Pb) stress in sunflower (*Helianthus annuus* L.) via improved nutrient uptake, antioxidant defense and physiology. Ecotoxicol Environ Saf 274:116194. https://doi.org/10.1016/j.ecoenv.2024.116194

Badri DV, Vivanco JM (2008) Regulation and function of root exudates. Plant Cell Environ 32(6):666–681. https://doi.org/10.1111/j.1365-3040.2008.01926.x

Baker AJM, Mcgrath SP, Sidoli CMD, Reeves RD (1994) The possibility of in situ heavy metal decontamination of polluted soils using crops of metal accumulating plants. Resour Conserv Recycl 11:41–49. https://doi.org/10.1016/0921-3449(94)90077-9

Bashir K, Rasheed S, Kobayashi T, Seki M, Nishizawa NK (2016) Regulating subcellular metal homeostasis: the key to crop improvement. Front Plant Sci 7:1192

Benizri E, Kidd PS (2018) The role of the rhizosphere and microbes associated with hyperaccumulator plants in metal accumulation. In: van der Ent A, Echevarria G, Baker AJM, Morel,

JL (eds) Agromining: farming for metals. Issue November. 2017 Springer International Publishing, pp 157–188. https://doi.org/10.1007/978-3-319-61899-9_9

Bishnoi NR, Chugh LK, Sawhney SK (1993) Effect of chromium on photosynthesis, respiration and nitrogen fixation in pea (*Pisum sativum* L.) seedlings. J Plant Physiol 142:25–30

Bouabidi ZB, El-Naas MH, Zhang Z (2018) Immobilization of microbial cells for the biotreatment of wastewater: a review. Environ Chem Lett 17(1):241–257. https://doi.org/10.1007/s10311-018-0795-7

Cakaj A, Drzewiecka K, Hanc A, Lisiak-Zielinska M, Ciszewska L, Drapikowska M (2024) Plants as effective bioindicators for heavy metal pollution monitoring. Environ Res 256:119222. https://doi.org/10.1016/j.envres.2024.119222

Chakrabartty M, Harun-Or-Rashid GM (2021) Feasibility study of the soil remediation technologies in the natural environment. Am J Civ Eng 9:91–98

Chaney RL, Malik M, Li YM, Brown SL, Brewer EP, Angle JS, Baker AJM (1997) Phytoremediation of soil metals. Curr Opin Biotechnol 8:279–284

Chatterjee J, Chatterjee C (2000) Phytotoxicity of cobalt, chromium and copper in cauliflower. Environ Pollut 109:69–74

Chen X, Tian Z, Cheng H, Xu G, Zhou H (2021) Adsorption process and mechanism of heavy metal ions by different components of cells, using yeast (*Pichia pastoris*) and Cu2þ as biosorption models. RCS Adv 11:17080e17091. https://doi.org/10.1039/d0ra09744f

Chen H, Zhao H, Zhao B (2024) Exploring the remediation potential of *Hydrangea macrophylla* (Thunb.) Ser. In cadmium-contaminated soil by comparing cultivars and seedling age. Environ Technol Innov 33:103474. https://doi.org/10.1016/j.eti.2023.103474

Christie P, Li X, Chen B (2004) *Arbuscular mycorrhiza* can depress translocation of zinc to shoots of host plants in soils moderately polluted with zinc. Plant Soil 261:209–217

Compant S, Clément C, Sessitsch A (2010) Plant growth-promoting bacteria in the rhizo and endosphere of plants: their role, colonization, mechanisms involved and prospects for utilization. Soil Biol Biochem 42(5):669–678. https://doi.org/10.1016/j.soilbio.2009.11.024

Csutak O, Simon-Gruita A, Duta G (2017) Preliminary studies on yeast-plant systems with applications in phytoremediation. Sci Bull Ser F Biotechnol 21:183–189

DalCorso G, Fasani E, Manara A, Visioli G, Furini A (2019) Heavy metal pollutions: state of the art and innovation in phytoremediation. Int J Mol Sci 20:3412

Darwish MA, Shanan NT, Mazhar AAM, Abd El-Aziz NG, Mahgoub MH, El-Sayed SM (2017) Mitigating the harmful effect of cadmium on *Eucalyptus gomphocephala* by using mycorrhiza fungi. Biosci Res 14(3):455–463

Das P, Samantaray S, Rout GR (2019) Heavy metal-induced oxidative stress and protection by mycorrhizal fungi. In: Hossain MA, Hasanuzzaman MZ, Fujita S (eds) Oxidative damage to plants: antioxidant networks and signaling. Academic, pp 377–397

Deepika GP, Haritash AK (2023) Phytoremediation potential of ornamental plants for heavy metal removal from contaminated soil: a critical review. Hortic Environ Biotech:1–26. https://cdnsciencepub.com/doi/full/10.1139/er-2017-0022

Di Francesco A, Sciubba L, Bencivenni M, Marzadori C, Baraldi E (2021) Application of *Aureobasidium pullulans* in iron-poor soil. Can the production of siderophores improve iron bioavailability and yeast antagonistic activity? Ann Appl Biol:398e406. https://doi.org/10.1111/aab.12742

Direk A, Arikan-Abdulveli B, Ozfidan-Konakci C, Yildiztugay E, Uysal A (2024) Effects of *Bacillus cereus* on physiological and biochemical characteristics of wheat under arsenic and cadmium stress: a biological agent to reduce heavy metal stress. Plant Stress 12:100458. https://doi.org/10.1016/j.stress.2024.100458

Dubey RS (2011) Metal toxicity, oxidative stress and antioxidative defense system in plants. In: Gupta SD (ed) Reactive oxygen species and antioxidants in higher plants. CRC Press, Boca Raton, FL, pp 177–203

Elhindi KM, Al-Mana FA, El-Hendawy S, Al-Selwey WA, Elgorban AM (2018) *Arbuscular mycorrhizal* fungi mitigates heavy metal toxicity adverse effects in sewage water contaminated soil on *Tagetes erecta* L. Soil Sci Plant Nutr 64:662–668. https://doi.org/10.1080/0038076 8.2018.1490631

El-Sayed SM, Abd El-Aziz NG, Mazhar AAM (2022) Antioxidant isoenzymes, chemical constituents and growth parameters of cadmium-stressed *Dimorphotheca ecklonis* plant and affected by humic acid. Egypt J Chem 65(12):519–532. https://doi.org/10.21608/EJCHEM.2022.119441.5370

El-Tohory S, Zeng W, Huang J, Moussa MG, Dong L, Mohamed A, Khalifa O, Saleem MH, Zhran M, Salama MA, Wu J (2023) Effect of intercropping and biochar application on cadmium removal capacity by *Corchorus olitorius* and *Zea mays*. Environ Technol Innov 29:103033. https://doi.org/10.1016/j.eti.2023.103033

Emenike CU, Jayanthi B, Agamuthu P, Fauziah SH (2018) Biotransformation and removal of heavy metals: a review of phytoremediation and microbial remediation assessment on contaminated soil. Environ Rev 26:156–168

Eslava-Silva FdJ, Muñíz-Díaz de León ME, Jiménez-Estrada M (2023) *Pteridium aquilinum* (Dennstaedtiaceae), a novel hyperaccumulator species of hexavalent chromium. Appl Sci 5621:13. https://doi.org/10.3390/app13095621

Fomina M, Gadd GM (2014) Biosorption: current perspectives on concept, definition and application. Bioresour Technol 160:3–14. https://doi.org/10.1016/j.biortech.2013.12.102

Freimoser FM, Rueda-Mejia MP, Tilocca B, Migheli Q (2019) Biocontrol yeasts: mechanisms and applications. World J Microbiol Biotechnol 35(10):1–19. https://doi.org/10.1007/s11274-019-2728-4

Gadd GM (1993) Interactions of fungi with toxic metals. New Phytol 124:25–60

Gadd GM (2000) Bioremedial potential of microbial mechanisms of metal mobilization and immobilization. Curr Opin Biotechnol 11(3):271–279. https://doi.org/10.1016/S0958-1669(00)00095-1

Gavrilescu M (2022) Enhancing phytoremediation of soils polluted with heavy metals. Curr Opin Biotechnol 74:21–31

Ghosh D, Maiti SK (2020) Biochar assisted phytoremediation and biomass disposal in heavy metal contaminated mine soils: a review. Int J Phytoremediation 23:559–576. https://doi.org/10.108 0/15226514.2020.1840510

Glick BR (2010) Using soil bacteria to facilitate phytoremediation bacteria. Biotechnol Adv 28:367–374

Glick BR (2012) Plant growth-promoting bacteria: mechanisms and applications. Scientifica 2012:1–15. https://doi.org/10.6064/2012/963401

Goswami D, Thakker JN, Dhandhukia PC (2016) Portraying mechanics of plant growth promoting rhizobacteria (PGPR): a review. Cogent Food Agric 2(1):1127500. https://doi.org/10.108 0/23311932.2015.1127500

Guo N, Ling H, Yu R, Gao F, Cao Y, Tao J (2024) Expression of *Sailx suchowensis* SsIRT9 enhances cadmium accumulation and alters metal homeostasis in tobacco. J Hazard Mater 463:132958. https://doi.org/10.1016/j.jhazmat.2023.132958

Hadi F, Fano A (2010) Effect of diazotrophs (*Rhizobium* and *Azotobacter*) on growth of maize (*Zea mays* L.) and accumulation of lead (Pb) in different plant parts. Pak J Bot 42:4363–4370

Hakeem KR, Bhat RA, Qadri H (2020) Phytoremediation of heavy metals: an eco-friendly and sustainable approach. In: Hakeem KR, Bhat RA, Qadri H (eds) Bioremediation and biotechnology: sustainable approaches to pollution degradation. Springer International Publishing, Cham, pp 1–327. ISBN 978-3-030-35690-3

Hegde S, Lodge S, Trabold TA (2018) Characteristics of food processing wastes and their use in sustainable alcohol production. Renew Sustain Energ Rev 81:510–523. https://doi.org/10.1016/j.rser.2017.07.012

Hou D, O'Connor D, Igalavithana AD, Alessi DS, Luo J, Tsang DCW, Sparks DL, Yamauchi Y, Rinklebe J, Ok YS (2020) Metal contamination and bioremediation of agricultural soils for food safety and sustainability. Nat Rev Earth Environ 1(7):366–381. https://doi.org/10.1038/s43017-020-0061-y

Hu Z, Zhao C, Li Q, Feng Y, Zhang X, Lu Y, Ying R, Yin A, Ji W (2023) Heavy metals can affect plant morphology and limit plant growth and photosynthesis processes. Agronomy 13:2601. https://doi.org/10.3390/agronomy13102601

Hussain U, Afza R, Gul I, Sajad MA, Shah GM, Muhammad Z, Khan SM (2024) Phytoremediation of heavy metals spiked soil by *Celosia argentea* L.: effect on plant growth and metal stabilization. Environ Sci Pollut Res 31:15339–15347. https://doi.org/10.1007/s11356-024-32176-6

Jacob JM, Karthik C, Saratale RG, Kumar SS, Prabakar D, Kadirvelu K, Pugazhendhi A (2018) Biological approaches to tackle heavy metal pollution: a survey of literature. J Environ Manag 217:56–70

Jiang X, Yang Y, Feng S, Hu Y, Cao M, Luo J (2022a) Reactive effects of presowing magnetic field exposure on morphological characteristics and antioxidant ability of *Brassica juncea* in phytoextraction. Chemosphere 303:135046. https://doi.org/10.1016/j.chemosphere.2022.135046

Jiang Y, Yang F, Dai M, Ali I, Shen X, Hou X, Alhewairini SS, Peng C, Naz I (2022b) Application of microbial immobilization technology for remediation of Cr(VI) contamination: a review. Chemosphere 286:131721. https://doi.org/10.1016/j.chemosphere.2021.131721

Jiang Y, Luo J, Guo X, Qiao Y, Li Y, Zhang Y, Zhou R, Vaculík M, Li T (2024) Phyllosphere microbiome assists the hyperaccumulating plant in resisting heavy metal stress. J Environ Sci (In press). https://doi.org/10.1016/j.jes.2024.05.032

Kabata-Pendias A, Mukherjee AB (2007) Trace elements from soil to human. Springer

Kafle A, Timilsina A, Gautam A, Adhikari K, Bhattarai A, Aryal N (2022) Phytoremediation: mechanisms, plant selection and enhancement by natural and synthetic agents. Environ Adv 8:100203. https://doi.org/10.1016/j.envadv.2022.100203

Khan R, Sarwar JM, Shabaan M, Asghar HN, Zulfiqar U, Iftikhar I, Aijaz N, Haider FU, Chaudhary T, Soufan W (2024) Exploring the synergistic effects of indole acetic acid (IAA) and compost in the phytostabilization of nickel (Ni) in cauliflower rhizosphere. BMC Plant Biol 24:275. https://doi.org/10.1186/s12870-024-04920-0

Kong Z, Glick BR (2017) The role of plant growth-promoting bacteria in metal phytoremediation. In: Poole RK (ed) Advances in microbial physiology, vol 71. Academic, pp 97–132. https://doi.org/10.1016/bs.ampbs.2017.04.001

Kotoula D, Papazoglou EG, Economou G, Trigas P, Bouranis DL (2024) Phytoremediation potential of flax grown on multimetal contaminated soils: a field experiment. Plants 13:1541. https://doi.org/10.3390/plants13111541

Lal S, Ratna S, Said O, Ben, Kumar R (2018) Biosurfactant and exopolysaccharide-assisted rhizobacterial technique for the remediation of heavy metal contaminated soil: an advancement in metal phytoremediation technology. Environ Technol Innov 10:243–263. https://doi.org/10.1016/j.eti.2018.02.011

Li H, Shi A, Li M, Zhang X (2013) Effect of pH, temperature, dissolved oxygen, and flow rate of overlying water on heavy metals release from storm sewer sediments. J Chem 2013:434012. https://doi.org/10.1155/2013/434012

Li Y, Wang X, Li R, Liu H, Zhang W (2018) Heavy metal pollution in soils on a global scale: sources, risks and remediation techniques. Environ Skept Crit 7(2):32–43

Li X, Li B, Zheng Y, Luo L, Qin X, Yang Y, Xu J (2022) Physiological and rhizospheric response characteristics to cadmium of a newly identified cadmium accumulator *Coreopsis grandiflora* Hogg. (Asteraceae). Ecotoxicol Environ Saf 241:113739. https://doi.org/10.1016/j.ecoenv.2022.113739

Li N, Wu F, Zhang J, Yu Y (2024) Remediation of heavy metal cadmium polluted water by a variety of garden plant schemes. Desalin Water Treat 319:100485. https://doi.org/10.1016/j.dwt.2024.100485

Liu H, Probst A, Liao B, Wang J (2013) Metal contamination of soils and crops affected by the chelungpu fault in Taiwan. Sci Total Environ 463–464, 530–542

Liu S, Yang B, Liang Y, Xiao Y, Fang J (2020) Prospect of phytoremediation combined with other approaches for remediation of heavy metal-polluted soils. Environ Sci Pollut Res 27:16069–16085

Losch R (2004) Plant mitochondrial respiration under the influence of heavy metals. In: Prasad MNV (ed) Heavy metal stress in plants: biomolecules to ecosystem, 3rd edn. Springer, Berlin, pp 182–200

Ma Y, Oliveira RS, Freitas H, Zhang C (2016) Biochemical and molecular mechanisms of plant–microbe-metal interactions: relevance for phytoremediation. Front Plant Sci 7. https://doi.org/10.3389/fpls.2016.00918

Maksymiec W, Wójcik M, Krupa Z (2007) Variation in oxidative stress and photochemical activity in *Arabidopsis thaliana* leaves subjected to cadmium and excess copper in the presence or absence of jasmonate and ascorbate. Chemosphere 66:421–427

Massoud R, Rasoul M, Hamzehlou P, Khosravi-darani K (2019) Bioremediation of heavy metals in food industry: application of *Saccharomyces cerevisiae*. Electron J Biotechnol 37:56–60. https://doi.org/10.1016/j.ejbt.2018.11.003

Mattson MP (2008) Hormesis defined. Aging Res Rev 7(1):1–7. https://doi.org/10.1016/j.arr.2007.08.007. Epub 2007 Dec 5. PMID: 18162444; PMCID: PMC2248601.

Medina-Díaz HL, Lopez-Bellido FJ, Alonso-Azcarate J, Fernandez-Morales FJ, Rodriguez L (2024) A new hyperaccumulator plant (*Spergularia rubra*) for the decontamination of mine tailings through electrokinetic-assisted phytoextraction. Sci Total Environ 912:169543. https://doi.org/10.1016/j.scitotenv.2023.169543

Mendes R, Garbeva P, Raaijmakers JM (2013) The rhizosphere microbiome: significance of plant beneficial, plant pathogenic, and human pathogenic microorganisms. FEMS Microbiol Rev 37(5):634–663. https://doi.org/10.1111/1574-6976.12028

Mishra J, Singh R, Arora NK (2017) Alleviation of heavy metal stress in plants and remediation of soil by rhizosphere microorganisms. Front Microbiol 8:1–7. https://doi.org/10.3389/fmicb.2017.01706

Montreemuk J, Stewart TN, Prapagdee B (2024) Bacterial-assisted phytoremediation of heavy metals: concepts, current knowledge, and future directions. Environ Technol Innov 33:103488. https://doi.org/10.1016/j.eti.2023.103488

Muryani E, Sajidan, Budiastuti MS, Pranoto (2024) The capacity of *Plectranthus* sp. to perform phytoremediation on heavy metal mercury (Hg) in the tailing of gold processing industries. 6th IOP Conf Ser Earth Environ Sci 1339:012038. https://doi.org/10.1088/1755-1315/1339/1/012038

Nafady NA, Hashem M, Hassan EA, Ahmed HAM, Alamri SA (2019) The combined effect of *Arbuscular mycorrhizae* and plant-growth-promoting yeast improves sunflower defense against *Macrophomina phaseolina* diseases. Biol Control 138(April):104049. https://doi.org/10.1016/j.biocontrol.2019.104049

Nasircilar AG, Ulukapi K, Topcuoglu B, Kurubas S, Erkan M (2024) Salt and heavy metal stress responses and metal uptake potentials of some leafy vegetables. Agrosyst Geosci Environ 7:e20487. https://doi.org/10.1002/agg2.20487

Novo LAB, Castro PML, Alvarenga P, da Silva EF (2018) Plant growth-promoting rhizobacteria-assisted phytoremediation of mine soils. In: Biogeotechnologies for mine site rehabilitation, vol 1, pp 281–295. https://doi.org/10.1016/B978-0-12-812986-9.00016-6

Ojuederie OB, Babalola OO (2017) Microbial and plant-assisted bioremediation of heavy metal polluted environments: a review. Int J Environ Res Public Health 14:1504

Oladoye PO, Olowe OM, Asemoloye MD (2022) Phytoremediation technology and food security impacts of heavy metal contaminated soils: a review of literature. Chemosphere 288:132555

Ozturk M, Metin M, Altay V, Filippis LD, Unal BT, Khursheed A (2021) Molecular biology of cadmium toxicity in *Saccharomyces cerevisiae*. Biol Trace Elem Res 199(12):4832e4846

Papadimou SG, Barbayiannis N, Golia EE (2024) Preliminary investigation of the use of *Silybum marianum* (L.). Gaertn. as a Cd accumulator in contaminated Mediterranean soils: the relationships among cadmium (Cd) soil fractions and plant Cd content. Euro-Mediterr J Environ Integr 9:405–417. https://doi.org/10.1007/s41207-023-00430-x

Pedroso FD, Barbosa MV, dos Santos JV, Pinto FA, Siqueira JO, Carneiro MAC (2018) *Arbuscular mycorrhizal* fungi favor the initial growth of *Acacia mangium*, *Sorghum bicolor*, and *Urochloa brizantha* in soil contaminated with Zn, Cu, Pb, and Cd. Bull Environ Contam Toxicol 101:386–391. https://doi.org/10.1007/s00128-018-2405-6

Pokethitiyook P, Poolpak T (2016) Biosorption of heavy metal from aqueous solutions. In: Phytoremediation. Springer International Publishing, pp 113–141. https://doi.org/10.1007/978-3-319-40148-5_4

Pugazhendhi A, Govindasamy C, Sharma A (2024) Heavy metal accumulation in root and shoot tapioca plant biomass grown in agriculture land situated around the magnesite mine tailings. Environ Res 257:119287. https://doi.org/10.1016/j.envres.2024.119287

Qiao K, Wang Q, Liu X, Gong S, Wang J (2023) Cadmium/lead tolerance of six dianthus species and detoxification mechanism in *Dianthus spiculifolius*. Chemosphere 312:137258. https://doi.org/10.1016/j.chemosphere.2022.137258

Rahbar SY, Khosroushahi YA, Gargari PB (2021) Yeast exopolysaccharides and their physiological functions. Folia Microbiol 66(2):171–182. https://doi.org/10.1007/s12223-021-00856-2

Rajendran S, Priya TAK, Khoo KS, Hoang TKA, Ng HS, Munawaroh HSH, Karaman C, Orooji Y, Show PLA (2021) Critical review on various remediation approaches for heavy metal contaminants removal from contaminated soils. Chemosphere 287:132369

Ramasamy K, Kamaludeen, Banu SP (2007) Bioremediation of metals: microbial processes and techniques. In: Singh SN, Tripathi RD (eds) Environmental bioremediation technologies. Springer, Berlin Heidelberg, pp 173–187. https://doi.org/10.1007/978-3-540-34793-4_7

Ramos-Garza J, Bustamante-Brito R, Angeles de Paz G, Medina-Canales MG, VasquezMurrieta MS, Wang ET, Rodríguez-Tovar AV (2016) Isolation and characterization of yeasts associated with plants growing in heavy-metal-and arsenic-contaminated soils. Can J Microbiol 62(4):307–319. https://doi.org/10.1139/cjm-2015-0226

Rascio N, Navari-Izzo F (2011) Heavy metal hyperaccumulating plants: how and why do they do it? And what makes them so interesting? Plant Sci 180:169–181

Rashid MI, Mujawar LH, Shahzad T, Almeelbi T, Ismail IMI, Oves M (2016) Bacteria and fungi can contribute to nutrients bioavailability and aggregate formation in degraded soils. Microbiol Res 183:26–41. https://doi.org/10.1016/j.micres.2015.11.007

Raskin I, Nandu K. PBA, Dushenkov S, Salt DE (1994) Bioconcentration of heavy metals by plants. Curr Opin Biotechnol 5:285–290. https://doi.org/10.1016/0958-1669(94)90030-2

Rezania S, Taib SM, Md Din MF, Dahalan FA, Kamyab H (2016) Comprehensive review on phytotechnology: heavy metals removal by diverse aquatic plants species from wastewater. J Hazard Mater 318:587–599

Rhodes CJ (2014) Mycoremediation (bioremediation with fungi) growing mushrooms to clean the earth. Chem Speciat Bioavailab 26(3):196e198. https://doi.org/10.3184/095422914X14047407349335

Rono JK, Le Wang L, Wu XC, Cao HW, Zhao YN, Khan IU, Yang ZM (2021) Identification of a new function of metallothionein-like gene OsMT1e for cadmium detoxification and potential phytoremediation. Chemosphere 265. https://doi.org/10.1016/j.chemosphere.2020.129136

Samoraj M, Dmytryk A, Tuhy Ł, Zdunek A, Rusek P, Moustakas K, Chojnacka K (2023) Applicability of alfalfa and goldenrod residues after supercritical CO2 extraction to plant micronutrient biosorption and renewable energy production. Energy 262:125437. https://doi.org/10.1016/j.energy.2022.125437

Sessitsch A, Kuffner M, Kidd P, Vangronsveld J, Wenzel WW, Fallmann K, Puschenreiter M (2013) The role of plant-associated bacteria in the mobilization and phytoextraction of trace elements in contaminated soils. Soil Biol Biochem 60:182–194. https://doi.org/10.1016/j.soilbio.2013.01.012

Shahzad A, Aslam U, Ferdous S, Qin M, Siddique A, Billah M, Naeem M, Mahmood Z, Kayani S (2024) Combined effect of endophytic *Bacillus mycoides* and rock phosphate on the amelioration of heavy metal stress in wheat plants. BMC Plant Biology 24:125. https://doi.org/10.1186/s12870-024-04812-3

Sharma P, Dubey RS (2007) Involvement of oxidative stress and role of antioxidative defense system in growing rice seedlings exposed to toxic levels of aluminum. Plant Cell Rep 26:2027–2038

Sharma R, Jasrotia T, Sharma S, Sharma M, Kumar R (2021) Chemosphere sustainable removal of Ni(II) from waste water by freshly isolated fungal strains. Chem 282:130871

Silver S, Phung LT (1996) Bacterial heavy metal resistance: new surprises. Ann Rev Microbiol 50(1):753–789. https://doi.org/10.1146/annurev.micro.50.1.753

Singh J, Sharma D, Kumar G, Sharma NR (2018) Bioremediation: an eco-sustainable approach for restoration of contaminated sites. In: Singh J, Sharma D, Kumar G, Sharma NR (eds) Microbial bioprospecting for sustainable development. Springer, pp 1–397. ISBN 978-981-13-0052-3

Singh S, Kumar V, Singh S, Dhanjal DS, Datta S, Sharma D, Singh NK, Singh J (2020) Biosurfactant-based bioremediation. In: Bioremediation of pollutants. INC. https://doi.org/10.1016/b978-0-12-819025-8.00016-8

Stanila A, Mihaescu T, Odagiu A (2015) Studies regarding cytotoxicity of copper, lead and zinc ions solutions on living *Saccharomyces cerevisiae* cells. ProEnvironment 8:445e450

Teng Y, Li Z, Yu A, Guan W, Wang Z, Yu H, Zou L (2022) Phytoremediation of cadmium-contaminated soils by *Solanum nigrum* L. enhanced with biodegradable chelating agents. Environ Sci Pollut Res 29(37):56750–56759. https://doi.org/10.1007/s11356-022-19879-4

Thanh NC, Narayanan MN, Saravanan M, Chinnathambi A, Alahmadi TA, Brindhadevi K, Sharma A, Pugazhendhi A (2023) *Hibiscus rosa-sinensis* as a potential hyperaccumulator in metal contaminated magnesite mine tailings. Chemosphere 339:139737. https://doi.org/10.1016/j.chemosphere.2023.139738

Tiwari S, Lata C (2018) Heavy metal stress, signaling, and tolerance due to plant-associated microbes: an overview. Front Plant Sci 9:452

Ullah A, Heng S, Farooq M, Munis H, Fahad S, Yang X (2015) Phytoremediation of heavy metals assisted by plant growth promoting (PGP) bacteria: a review. Environ Exp Bot 117:28–40

Vijayaraghavan K, Yun YS (2008) Bacterial biosorbents and biosorption. Biotechnol Adv 26(3):266–291. https://doi.org/10.1016/j.biotechadv.2008.02.002

Wang Y, Li B, Li Y, Chen X (2021) Research progress on enhancing the performance of autotrophic nitrogen removal systems using microbial immobilization technology. Sci Total Environ 774:145136. https://doi.org/10.1016/j.scitotenv.2021.145136

Ważny R, Rozpądek P, Domka A, Jędrzejczyk RJ, Nosek M, Hubalewska-Mazgaj M, Lichtscheidl I, Kidd P, Turnau K (2021) The effect of endophytic fungi on growth and nickel accumulation in Noccaea hyperaccumulators. Sci Total Environ 10(768):144666. https://doi.org/10.1016/j.scitotenv.2020.144666

Xavier IJ, Boyetchko SM (2002) Arbuscular mycorrhizal fungi as biostimulants and bioprotectants of crops. Appl Mycol Biotechnol 2:311–340

Xu N, Sun Y, Wang Y, Cui Y, Jiang Y, Zhang C (2023) Hormesis effects in tomato plant growth and photosynthesis due to acephate exposure based on physiology and transcriptomic analysis. Pest Manag Sci 79:2029–2039

Zhang Y, Zhao S, Zhang R, Li BL, Li Y, Han H, Duan P, Chen Z (2024) Screening of plant growth-promoting rhizobacteria helps alleviate the joint toxicity of PVC+Cd pollution in sorghum plants. Environ Pollut 355:124201. https://doi.org/10.1016/j.envpol.2024.124201

Zhu F, Li F, Zhou H, Sun R (2017) Interactive effect of Cd and Pb on soil enzyme activities and microbial community structure in farmland soil. Environ Sci Pollut Res Int 24(11):10884–10894

Zinicovscaia I, Yushin N, Abdusamadzoda D, Grozdov D, Shvetsova M (2020) Efficient removal of metals from synthetic and real galvanic zinc containing effluents by brewer's yeast. Materials 13(3624):1–17

Chapter 11
Microbial Nanotechnology in Agricultural Applications

Jiaojiao Song and Kasem Soytong

Abstract Nanotechnology has been developed for plant disease management, and many kinds of organic nanoparticles have been developed to induce the immunity of plants, such as plant vaccines. Natural products from *Chaetomium lucknowense, Chaetomium brasiliense, Chaetomium cochliodes, Chaetomium cupreum, Chaetomium elatum, Chaetomium globosum, Chaetomium siamense, Emericella nidulans, Trichoderma harzianum*, and *Trichoderma hamatum* are reported to have antimicrobial activities against plant pathogens. In particular, *Ch. siamense* is a newly discovered and endophytic *Chaetomium* spp. that produces active metabolites. These active metabolites from *Chaetomium* spp., *E. nidulans, T. harzianum,* and *T. hamatum* are constructed as nanofibers for inducing plant disease immunity through phytoalexin production as nanoplant vaccines. The nanofibers are used as elicitors for the induction of immunity in plants that induce phytoalexin biosynthesis, for example, scopoletin and anthocyanidin against *Phytophthora* or *Pythium* rot and scoparone against *Phytophthora* or *Pythium rot*, capsidiol in chili against anthracnose, alpha-tomatine in tomato against *Fusarium* wilt, and sakuranetin and oryzalexin in rice plants against blast. The natural product nanoelicitors derived from *Chaetomium* spp., *Emericella* spp., and *Trichoderma* spp. help reduce disease incidence in plants.

Keywords Nanotechnology · *Chaetomium* · Nanofibers · Disease immunity

---

J. Song · K. Soytong (✉)
Research Institute of Modern Organic Agriculture (RIMOA), King Mongkut's Institute of Technology Ladkrabang (KMITL), Bangkok, Thailand

Association of Agricultural Technology in Southeast Asia (AATSEA), Bangkok, Thailand

King Mongkut's Institute of Technology Ladkrabang (KMITL), Prince of Chumphon Campus, Chumphon, Thailand

V. Gopikrishnan et al. (eds.), *Harnessing Microbial Resources for Sustainable Agriculture*, Microorganisms for Sustainability 61,
https://doi.org/10.1007/978-981-95-4817-0_11

11.1 Introduction

Nanotechnological applications have been developed for agricultural application and much advanced innovation (Li et al. 2011a, b). It is an alternative method of immunity and disease control for economic plants in various environmental conditions and involves the restructuring of organic materials to small particles (Soutter 2012). Many researchers are actively investigating organic nanoparticles for their biological activities (Elibol et al. 2003). The use of nanoparticles and nanofibers is an effective method for plant disease management (Rai and Ingle 2012). Researchers have investigated organic nanoparticles for sustainable agriculture (Pan et al. 2019). Agricultural development has attracted increasing attention in the field of nanotechnological application (Li et al. 2011a; b). Nanobiotechnology may overcome both biotic and abiotic factors involved in plant stress. Nanotechnology is known as a new scientific technique for constructing materials at the atomic level (Soutter 2012). Natural products from potent microorganisms, such as *Chaetomium* species, have been confirmed to have antifungal effects on plant disease control (Soytong et al. 2001). It is an alternative method for disease control because of its environmental stability. The construction of nanofibers loaded with active metabolites of the potent microorganism would be developed rather than toxic agrochemical pesticides.

Ditta (2012) reported that climate change causes urbanization and natural resources due to toxic agrochemical application, which results in the accumulation of toxic agrochemicals in cultivated soils and water, which urgently needs to be solved by decreasing the use of agrochemical pesticides, which are substitutes for other management methods. The proposed method of nanotechnology for sustainable agriculture is one of the strategies used to make natural products in nanoform for agriculture, but few studies have investigated this approach (Nguyen et al. 2012).

Nanotechnological research can lead to effective plant disease and insect management.

This new method can serve sustainable agriculture to increase quality of life and can be used to carefully mitigate either negative or positive impacts (Ditta 2012). Chemical pesticides, which are formulated at the nanoscale or microscale in the form of colloidal suspensions or powders, must be considered for their safety in the environment. Nanotechnological application in agriculture has become a novel industrial revolution for agricultural development (Soutter 2012). Potential innovations in nanotechnology are challenging in agricultural development (Ditta 2012). NPs are magic bullets that contain microbial substances from effective microorganisms, leading to easy and fast penetration of plant cells (Perlatti et al.). Nanopreparative, which is used as a synthetic pesticide, and chemical fertilizers, which constitute alternative methods, should be applied with appropriate applications.

The new strategy for nanotechnology in agriculture involves the construction of nanoproducts from microbial metabolites for controlling insect pests and pathogens for application in green agriculture and organic crop production, which helps to balance the previous ecosystem in the surrounding environment. Nanotechnological research for agriculture in general involves controlling insect pests and phytopathogens and applying them as plant growth regulators or nanofertilizers.

NPs derived from synthetic chemicals that are released into cultivated soil may cause toxicity to living organisms and affect biodiversity imbalance, which may ultimately accumulate in vegetables, fruits, and food, affecting human health (Morales-Díaz et al. 2017). Therefore, the safe and sustainable application of agricultural nanotechnology requires regulatory standards for nanomaterial application in water, soil, and plants (OECD 2014).

Nanoscience defines atoms, molecules, or macromolecules on a scale of 1—100 nm (EPA 2007). However, the modification of nanometric scales may occur in larger nanomaterials (Wang et al. 2013); thus, nanoscience has been redefined to expand the structures to 1000 nm (Mukhopadhyay 2014).

Nanotechnology has rapidly expanded in agricultural applications. The large quantities of synthetic nanotechnological materials are approximately 10,000 tons per year of TiO_2 and 1000 tons of AlO_x, CeO_2, FeO_x, and ZnO (Baby et al. 2010). The number of nanoparticles in ecosystems is expected to increase (Brillas and Martínez-Huitle 2015), and the surrounding environment is expected to contaminate agricultural soil, air, and water (Lellis et al. 2019).

Natural product nanoparticles derived from metabolites of effective microorganisms are being developed for green agriculture to protect our environments. The secondary metabolites produced in plants are low-molecular-weight compounds that act as phytoalexins that accumulate in plants after application, similar to plant phytoalexins (VanEtten et al. 1994). These high-molecular-weight metabolites are fine structures that can be used as elicitors to induce plant immunity. Soytong et al. (2014) reported that chaetomanone A from *Ch. lucknowense*, chaetoglobosin C derived from *Ch. Globosum*, and the trichotoxin A50 from *T. harzianum* PC01 induced immunity after infection in chili caused by *Colletotrichum capsici*, which is characterized by the presence of the capsidiol phytoalexin. I believe that chaetomanone A, chaetoglobosin C, and trichotoxin A50, which are natural product nanoelicitors, can induce phytoalexins in plants against diseases.

11.2 Nanofiber Construction from the Metabolites of Microorganisms for Plant Immunity

The natural product nanoparticles constructed from metabolites of *Chaetomium* spp. have been proven to be nanoelicitors of plant disease immunity. Scientists have started to possess the biological properties of nanobioproducts via agriculture (Salata 2004). The natural product metabolites of *Chaetomium* sp. were constructed through electrospun materials with metabolite-loaded nanoparticles of 241 nm in size (Dar and Soytong 2014).

The nanoproduct constructed from a metabolite of *Ch. cupreum* CC3003 successfully expresses plant disease immunity and controls *Curvularia lunata* causing

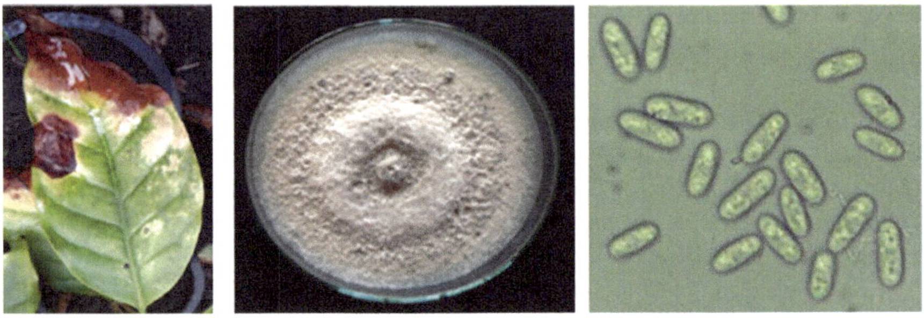

Fig. 11.1 Coffee anthracnose on a leaf infected by *Colletotrichum gloeosporioides* (**a**); pure culture on PDA (**b**); and conidia, 400X (**c**). (Source: Vilavong and Soytong (2016))

rice leaf spot in the field in Cambodia (Tann and Soytong 2016). Nanorrotiorinol application derived from a metabolite of *Ch. cupreum,* a nanotrichotoxin derived from a pure *Trichoderma harzianum* PC01, increased plant immunity and reduced coffee anthracnose caused by *C. gloeosporioides* by 46 and 43%, respectively, in Lao PDR (Vilavong and Soytong 2016). Further investigations of nanoproducts, namely, nano-CGE, nano-CGH, and nano-CGM, derived from metabolites of *Ch. globosum* that has been shown to control *C. lunata* (rice leaf spot) in Cambodia have been conducted. The ED_{50} values of nano-CGE, nano-CGH, and nano-CGM were 1.1, 1.2, and 1.9 μg/mL, respectively. The pathogen cells lost pathogenicity because the nanoparticles penetrated the pathogen cells with disruption, and the protoplasts emerged from the pathogen cells. These nanoparticles have been shown to increase plant height and tiller number after treatment for 60 days (Tann et al. 2011).

11.3 Nanoelicitor Application to Induce Immunity Against Anthracnose in Coffee

The pure compound of nanorotiorinol constructed from *Ch. cupreum* and nanotrichotoxin was constructed from *T. harzianum* PC01. Noriorinol treatment resulted in a reduction in the incidence of anthracnose caused by *C. gloeosporioides* by 46% (Fig. 11.1). Compared with the control, the nanotrichotoxin treatment reduced anthracnose to 42%. These nanomicrobial elicitors induce immunity through phytoalexes against coffee anthracnose (Vilavong and Soytong 2016).

11.4 Nanoelicitors Derived from *Chaetomium* spp. Against Blast Disease in Rice

Nanotechnological agriculture has become a new strategy for crop management, especially to protect against plant disease through phytoalexin production as part of plant immunity. The pure compounds and effective metabolites constructed to be

Table 11.1 The effects of nanoparticles derived from *Ch. cochliodes* isolate CTh05 against *M. oryzae*

Nano-particles	Concentration (ppm)	Colony diameter (cm)	Growth inhibition (%)	Number of spores (10^5)	Spore Inhibition (%)	ED_{50}(ppm)
Nano-CCoH	0	5.00^{a}	–	69.00^{a}	–	33.41
	3	4.96^{ab}	0.75^{e}	67.00^{ab}	2.78^{i}	
	5	4.88^{bc}	2.25^{de}	56.00^{bcd}	18.98^{f}	
	10	4.82^{cd}	3.50^{cd}	50.75^{cd}	26.63^{e}	
	15	4.76^{d}	4.75^{c}	45.00^{de}	34.87^{cd}	
Nano-CCoE	0	5.00^{a}	–	69.00^{a}	–	9.47
	3	4.93^{ab}	1.25^{e}	60.00^{abc}	13.23^{fg}	
	5	4.87^{bc}	2.50^{de}	49.00^{cd}	28.99^{de}	
	10	4.78^{d}	4.25^{cd}	36.00^{e}	47.99^{b}	
	15	4.58^{e}	8.25^{b}	22.00^{f}	68.26^{a}	
Nano-CCoM	0	5.00^{a}	–	69.00^{a}	–	16.51
	3	4.87^{bc}	2.50^{de}	64.00^{ab}	7.36^{gh}	
	5	4.77^{d}	3.75^{cd}	51.25^{cd}	26.00^{e}	
	10	4.56^{e}	8.75^{b}	44.00^{de}	36.40^{c}	
	15	3.88^{f}	22.25^{a}	36.00^{e}	47.97^{b}	
C.V.(%)		1.18	25.04	11.03	13.91	-

Source: Song et al. (2020a)
Means followed by a common letter are not significantly different by DMRT at $P \leq 0.05$

nanofibers are investigated from metabolites of *Ch. cochliodes,* which inhibited *Magnaporthe oryzae* (rice blast), as shown in Table 11.1. Nano-CcoE, Nano-CcoH, and Nano-CC0M suppressed sporulation at ED_{50} values of 9, 33, and 16 ppm, respectively. The blast of rice can decrease the disease incidence by 59% after treatment with nano-CcoM. The treatment with nano-CCoE and nano-CCo-H resulted in a decrease in rice blast disease incidence of 57 and 50%, respectively, and tricyclazole decreased the disease prevalence to 55% within 30 days, as shown in Table 11.2 (Song et al. 2020a).

The antifungal activity of natural product nanoparticles from *Ch. Brasiliense, Ch. elatum*, and *Ch. lucknowense* was found to control *M. oryzae* (blast of rice var. PSL 2). With this, the nanofibers from *Ch. elatum*, namely, nano-CEE, nano-CEH, and nano-CEM, resulted in the suppression of conidia production at ED_{50} values of 7, 16, and 8 µg/mL, respectively. Nano-CBE, nano-CBH, and nano-CBM constructed from *Ch. brasiliense* inhibited the conidia production of *M. oryzae* at ED_{50} values of 9, 6, and 13 μg/mL, respectively. Moreover, the evolution of nanofibers derived from *Ch.* Notably, nano-CLE, nano-CLH, and nano-CLM were found to suppress sporulation at ED_{50} values of 7, 10, and 5 µg/mL, respectively. The conidia of the blast pathogen treated with the nanofibers were broken, and the protoplasts inside were removed; then, pathogenicity was lost after the conidia were

Table 11.2 Effects of the nanoparticle *Chaetomium cochliodes* isolate CTh05 on plant height and disease severity in rice var RD57

Treatments	15 (d)	30 (d)	Disease reduction (%)
T1 (Non-inoculated control)	38.70[c]	54.50[c]	–
T2 (Inoculated with *M. oryzae*)	38.60[c]	53.75[c]	–
T3 (Nano-CCoH)	41.00[ab]	62.25[b]	50.2
T4 (Nano-CCoE)	40.50[b]	76.25[a]	57.5
T5 (Nano-CCoM)	43.00[a]	77.75[a]	59.8
T6 (Tricyclazole)	38.50[c]	75.00[a]	55.5

Source: Song et al. (2020a)
Means followed by a common letter are not significantly different by DMRT at $P \leq 0.05$

reinoculated into the host rice seedlings. The natural product nano-CBH from *Ch. Brasiliense*-treated rice plants expressed sakuranetin and oryzalexin B phytoalexin at Rf values of 0.05 and 0.28, respectively, compared with the phytoalexin standard. Nano-CBHs act as microbial elicitors of immunity in rice plants (Song et al. (2020b).

The induction of plant immunity by microbial nanofibers is an alternative method for disease protection. Phytoalexins in rice plants were first reported in rice plants infected with *Pyricularia oryzae* (Uehara 1958).

The phytoalexins discovered in plants infected by *M. oryzae*, which causes blast, are reported as 15 phytoalexins. Therefore, sakuranetin is a flavonoid and diterpenoid, for example, oryzalexins A–F, oryzalexin S, momilactones A and B, and phytocassanes A–E (Kodama et al. 1992; Dillon et al. 1997). Sakuranetin phytoalexin is expressed by the resistant variety of rice plants inoculated with *M. oryzae* for three days in young leaves, resulting in an HR (hypersensitive reaction) (Hasegawa et al. 2014).

The active metabolites were constructed from the metabolites of *Ch. elatum, Ch. lucknowense*, and *Ch. brasiliense* by electron spinning that yielded nano-CLM and nano-CEE nano-CBH, respectively. These microbial nanoparticles are applied to inhibit blast pathogens and induce phytoalexin accumulation in rice var. PSL2 to prevent *M. oryzae* from causing blast disease. Nano-CBH strongly promoted phytoalexin accumulation in rice leaves, which was confirmed via thin layer chromatography (TLC), and the Rf value of 0.28 was confirmed to be oryzalexin B compared with the standard value, and the Rf value of 0.08 was confirmed to be sakuranetin compared with the standard value (Fig. 11.2) (Song et al. 2020b).

These findings demonstrated that nano-*Ch. elatum* from hexane extract, nano-*Ch. elatum* from EtOAc extract, and nano-*Ch. elatum* from the MeOH extract inhibited spore production, with ED_{50} values of 16, 7, and 8 ppm, respectively. Nano-*Ch. brasiliense* from hexane extract, nano-*Ch. brasiliense* from EtOAc extract, and nano-*Ch. brasiliense* from the MeOH extract inhibited spore production at ED_{50} values of 6, 9, and 13 ppm. Nano-*Ch. lucknowense* from hexane extract, nano-*Ch. lucknowense* from EtOAc extract, and nano-*Ch. lucknowense* from the MeOH extract inhibited sporulation, with ED_{50} values of 10, 7, and 5 ppm, respectively (Table 11.3).

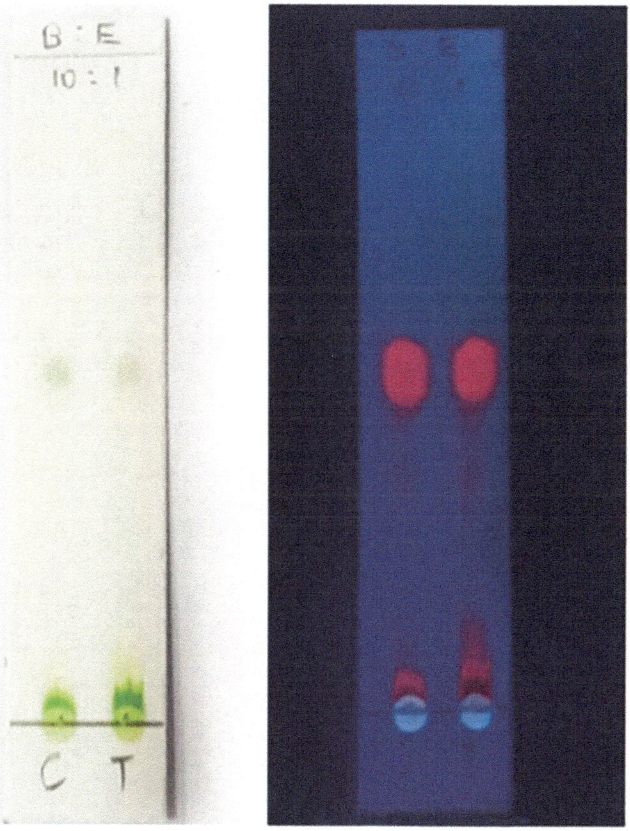

Fig. 11.2 Sakuranetin and oryzalexin B, which are phytoalexins with Rf values of 0.08 and 0.28, respectively, were detected via TLC in combination with benzene:ethyl acetate (10:1) under 365 nm UV light. (Song et al. 2020b)

11.5 Active Metabolite Nanoparticles of Endophytic *Chaetomium brasiliense* PT302 as Plant Growth Stimulants

Endophytic fungi isolated from rice plants include many fungi, including *Ch. cupreum, Ch. globosum,* and *Ch. brasilense* (Leewijit et al. 2016). The characteristics of the metabolites and nanofibers from the endophytic *Chaetomium brasilense* are shown in Fig. 11.3. The metabolites from endophytic *Ch. Brasiliense* at a concentration of 50 ppm stimulated the germination of both var. Supanburi 1 and Pathumtani 80 rice seeds. The results revealed that the germination of rice seeds coated with metabolites from the endophytic *Ch. brasiliense* was better than that of the non-treated control. These active metabolites were constructed to be nanofibers and tested for their ability to germinate rice seeds var. Supanburi 1 and Prathumthani 80

Table 11.3 Effects of nanoparticles derived from *Chaetomium brasiliense, Chaetomium lucknowense,* and *Chaetomium elatum* to inhibit the sporulation of *Magnaporthe oryzae*

		Nano-*C. elatum*			Nano-*C. brasiliense*			Nano-*C. lucknowense*		
Nano particles	Conc. (ppm)	No. of spores (10^5)^1	Spore inh. (%)^1	ED_{50} (ppm)	No. of spores (10^5)^1	Spore inh. (%)^{2,3}	ED_{50} (ppm)	No. of spores (10^5)^1	Spore inh. (%)^1	ED_{50} (ppm)
Hexane	0	32.50^{a}	–	16.70	17.25^{a}	–	6.86	40.50^{a}	–	10.72
	3	30.50^{ab}	5.69^{g}		14.50^{bc}	15.75^{e}		34.00^{b}	15.32^{j}	
	5	27.00^{bc}	16.45^{ef}		10.75^{d}	37.57^{d}		28.75^{cd}	28.53^{h}	
	10	22.50^{de}	30.63^{cd}		5.50^{f}	68.05^{b}		23.50^{ef}	41.81^{f}	
	15	15.75^{f}	51.85^{b}		3.50^{g}	79.74^{a}		14.50^{g}	64.18^{cd}	
EtOAc	0	32.50^{a}	–	7.89	17.25^{a}	–	9.76	40.50^{a}	–	7.01
	3	29.75^{ab}	7.91^{fg}		14.00^{bc}	18.47^{e}		31.25^{bc}	22.41^{i}	
	5	20.50^{e}	37.06^{c}		11.75^{d}	31.72^{d}		25.00^{def}	37.82^{fg}	
	10	15.50^{f}	52.73^{b}		7.50^{e}	56.36^{c}		15.00^{g}	63.00^{d}	
	15	6.50^{g}	80.46^{a}		4.50^{fg}	73.89^{ab}		10.50^{gh}	73.87^{ab}	
MeOH	0	32.50^{a}	–	8.66	17.25^{a}	–	13.42	40.50^{a}	–	5.24
	3	30.25^{ab}	6.80^{fg}		15.75^{ab}	8.34^{f}		26.50^{de}	34.25^{g}	
	5	25.00^{cd}	23.01^{de}		13.75^{c}	20.03^{e}		21.25^{f}	47.53^{e}	
	10	15.50^{f}	52.43^{b}		11.50^{d}	33.29^{d}		12.75^{gh}	68.65^{bc}	
	15	7.25^{g}	77.78^{a}		7.25^{e}	57.92^{c}		8.75^{h}	78.52^{a}	
CV%		12.45	19.01		10.23	8.14		11.75	7.75	

Source: Song et al. (2020b)

^{1,2,3} The average of four replications; means followed by a common letter are not significantly different by Duncan's multiple range test (DMRT) at $P = 0.01$; *inh.* inhibition, *conc.* concentration

Fig. 11.3 Characteristics of metabolites and nanofibers from endophytic *Chaetomium brasilense*, (**a**) crude hexane, ethyl acetate, and methanol and (**b**) nano-CBH, nano-CBE, and nano-CBM. (Source: Song et al. (2022a))

at 5 ppm, which resulted in better seed germination, number of tillers, and plant height than those of the control. Nanofibers constructed from *Ch. The brasiliense* endophyte stimulated the growth of rice at lower concentrations than did the crude extracts of *Ch. brasiliense* endophyte (Song et al. 2022a). Endophytic microorganisms live in plant tissue by releasing a volatile organic substance mixture that is lethal to fungal and bacterial plant pathogens. Fungal endophytes are associated with host plants that react to plant resistance. *Ch. The globosum* endophyte belongs to Ascomycetes and has been shown to produce antimicrobial metabolites in *Ginkgo* plants that act against plant pathogens (Qin et al. 2009; Li et al. 2011a; b).

The nanofibers yielded nano-CBD, nano-CBM, and nano-CBE, which stimulated the growth of the rice variety Pathumthani 80, whose seed germination, plant height, and root length after treatment with the nanofibers at all concentrations were greater than those of the control (Table 11.4).

11.6 Natural Product Nanoparticles Derived from *Emericella nidulans* EN for Rice Blast Disease Control

Emericella is an ascomycete fungus that is distributed worldwide. *Emericella rugulosa* has been reported to produce the tajixanthone compound (Moosophon et al. 2009), which inhibits *F. oxysporum* sp. *lycopersici* (wilt of tomato) and promotes plant immunity (Sibounnavong and Soytong 2011). *Magnaporthe oryzae* causes rice blast, which infects rice plants, leading to loss of yield (Wang and Valent 2009). The crude hexane metabolite of *Emericella nidulans* is sterigmatocystin and emericellin, whereas the ethyl acetate crude extract is specifically demethylsterigmatocystin. The ethyl acetate crude extract of *E. nidulans* suppressed the rice blast

Table 11.4 Nanofibers derived from *Chaetomium brasilense* for stimulating the growth of the rice variety Pathumthani 80

Nano-particles	Conc. (ppm)	Germination (%)	Plant height (mm)	Root length (mm)
Nano-CBH	0	67	9.79g[1]	10.77g[2]
	1	72	12.33ef	15.57c–e
	3	72.5	17.14ab	19.02a
	5	60	14.28c–e	18.97a
	7	50	13.08e	18.16ab
	10	42.5	9.95g	11.5fg
Nano-CBE	0	67	10.77fg	10.78
	1	67.5	13.97c–e	13.95de
	3	60	15.63b–d	14.54de
	5	70	18.30a	17.60a–c
	7	55	13.88c–e	13.58ef
	10	40	9.65g	10.34g
Nano-CBM	0	67	10.52fg	10.54g
	1	40	13.48de	15.10de
	3	55	15.73b–d	14.42de
	5	70	17.80ab	16.22b–d
	7	67.5	15.91bc	13.67ef
	10	42.5	9.76g	8.04h
	C.V. (%)		10.68	10.91

[1]Averages of four repeated experiments which followed by a common letter were not significantly differed

[2]Averages of four repeated experiments which followed by a common letter were not significantly differed by DMRT at P 0.01

pathogen *M.* oryzae, whose ED_{50} was 66 μg/mL. NPs constructed from crude ethyl acetate, crude methanol, and crude hexane from *E. nidulans* yielded nano-ENE, nano-ENH, and nano-ENM, which strongly inhibited the rice blast pathogen, with ED50 values of 4.2 μg/mL, 4.5 μg/mL, and 8.9 μg/mL, respectively. Nano-EN was used to treat inoculated rice plants, and sakuranetin was detected at R_f 0.09 (Song et al. 2022b) (Table 11.5).

These findings revealed that nanofibers derived from *E. nidulans* at low concentrations can control rice blast caused by *M. oryzae* better than can the crude metabolites of *E. nidulans.* Compared with crude metabolites, organic nanofibers are known to quickly penetrate plant tissues (Song et al. 2022b).

Nanofibers derived from this fungus can be applied at low concentrations, resulting in greater inhibition of blast pathogens than can crude metabolites at high concentrations. These nanofibers at 15 μg/mL elicited plant immunity. The investigation of phytoalexin revealed that the control did not detect a spot via thin-layer chromatography, whereas nano-EN and tricyclazole applications detected spots with Rf values of 0.09 at 3, 5, and 7 days after rice plants were inoculated with the blast pathogen, and sakuranetin (Salata, 2004)) was detected. The Rf of 0.09 under 365° UV light detected on a spot of TLC plate demonstrated sakuranetin from rice leaves inoculated with the rice blast pathogen (Mok (2009)).

Table 11.5 Effects of nanoparticles constructed from *Emericella nidulans* on the inhibition of *Magnaporthe oryzae*, which causes rice blast, for 14 days

Nanoparticles	Concentration (μg/mL)	Colony diameter (cm)	Growth inhibition (%)	Number of spores (10^5)	Spore inhibition (%)	Effective dose$_{50}$ (μg/mL)
From crude hexane	0	5.00^{a1}	–	36.00^{a}	–	8.85
	3	4.98^{ab}	0.25^{i}	28.00^{b}	22.05^{h}	
	5	4.92^{bc}	1.50^{hi}	22.75^{c}	36.50^{f}	
	10	4.82^{e}	3.50^{fg}	18.00^{d}	49.92^{e}	
	15	4.73^{f}	5.25^{de}	12.25^{ef}	66.17^{c}	
From crude ethyl acetate	0	5.00^{a}	–	36.00^{a}	–	4.26
	3	4.94^{abc}	1.00^{hi}	22.25^{c}	38.06^{f}	
	5	4.81^{e}	3.75^{ef}	15.50^{de}	56.88^{d}	
	10	4.28^{i}	14.25^{b}	10.00^{f}	72.27^{b}	
	15	4.04^{j}	19.00^{a}	3.50^{g}	90.31^{a}	
From crude methanol	0	5.00^{a}	–	36.00^{a}	–	4.49
	3	4.89^{cd}	2.00^{gh}	25.25^{bc}	29.54^{g}	
	5	4.83^{de}	2.50^{fgh}	13.00^{ef}	64.03^{c}	
	10	4.66^{g}	6.75^{d}	9.75^{f}	72.94^{b}	
	15	4.46^{h}	10.75^{c}	5.00^{g}	86.15^{a}	
Coefficient of variation (%)	–	0.92	18.27	10.93	4.55	–

Source: Song et al. (2022b)
Means followed by a common letter are not significantly different by Duncan's multiple range test at $P \leq 0.05$

11.7 Natural Product Nanoelicitors from *Chaetomium brasiliense*, *Chaetomium cochliodes*, and *Chaetomium elatum* for Immune Root Rot in Tangerine

Pythium aphanidermatum PN02 has been confirmed to be a pathogenic isolate that causes root rot in tangerines (*Citrus reticulata*) in wide-area plantations (Udompongsuk et al. 2018). However, root rot in citrus is a serious disease caused by Pythiaceous fungi, for example, *Phytophthora* spp. and *Pythium* spp. (Maseko and Coutinho 2002).

Chaetomium brasiliense, *Chaetomium cochliodes*, and *Chaetomium elatum* successfully inhibited the causal agent as microbial fungicides. Crude metabolites were further investigated for their ability to suppress the pathogen, with a concentration of 1000 ppm resulting in a high ability to inhibit pathogen mycelia and abnormal oogonia. Research on microbial nanotechnology has advanced the development of more effective microbial products, such as nanoelicitors, to induce immunity to control root rot in *Citrus*.

Nano-elicitors derived from metabolites of *Chaetomium brasiliense*, *Chaetomium cochliodes*, and *Chaetomium elatum*, which were used to inoculate citrus seedlings with *P. aphanidermatum* PN02, effectively inhibited pathogen propagation, with

ED50 values ranging from 2.69 to 12.88 ppm. These nanoelicitors have been shown to induce phytoalexin in infected seedlings, leading to recovery from root rot disease. All the tested nanoelicitors demonstrated a fluorescent spot at an Rf value of 0.6, which was a phytoalexin of scoparone under UV illumination at 254 nm. Scoparone was discovered for the first time in citrus plants infected by *Pythium aphanidermatum* (Udompongsuk et al. 2018). However, Afek and Sztejnberg (1989) reported the production of scoparone in two *Phytophthora*-resistant citrus varieties compared with susceptible varieties. The citrus infection of *Phytophthora citrophthora* after treatment with fosetyl-Al or phosphorus acid revealed that macrophylla, sour orange, and rough lemon varieties presented 2–4-fold higher concentrations of scoparone. On the other hand, Sulistyowati et al. (1990) reported that the phytoalexin 6,7-dimethoxycoumarin (DMC) is present in the stems and roots of citrus plants infected with *P. citrophthora.*

11.8 Natural Product Nanoelicitors to Recover Phytophthora Rot of Durian (*Durio zibethinus* L.)

Rots of durian (*Durio zibethinus* L.) have been infected by *Phytophthora palmivora*, which has invaded durian plantations for years. The pathogen has always developed into a resistant strain to chemical fungicides. The bioactive compounds from microorganisms are made by electron spinning to yield nanofibers that quickly penetrate plant cells and tissue to increase active metabolite stability and decrease loss from the surface of plants after application (Perlatti et al. 2013). This method is one of the techniques used to increase the efficacy of the application of natural products for plant disease control (Rai and Ingle 2012) in various formulations of liquid or wettable powder into plants (Ditta 2012).

Phytoalexins are related to defense mechanisms in plants (Ahuja et al. 2012), and they accumulate in healthy plants around infected tissues (Deverall 1982), resulting in hypersensitive reactions. However, abiotic conditions can elicit the plants to induce phytoalexins in many species of plants (Angelova et al. 2006; Yean et al. 2009), and biotic factors related to microorganisms also elicit phytoalexin accumulation in plants (Liu et al. 1995). In plants, phytoalexin production is stimulated by either abiotic or biotic stresses, which elicit phytoalexin to protect against plant disease. The production of phytoalexins in plants can delay pathogen infection (Glazebrook and Ausubel 1994). Nanoparticles constructed from *Ch. brasiliense* has been reported to effectively inhibit *P. palmivora* (durian rot) at an ED_{50} of 1.08 μg/L, and root rot disease decreased significantly (Tongon et al. 2018).

Microbial natural products from *Ch. cupreum* in the form of crude materials and nanoparticles have been shown to control *Phytophthora palmivora* (durian rot) and to induce phytoalexin accumulation in durian trees to prevent *Phytophthora* rot. The crude metabolites CC-E, CC-H, and CC-M inhibited *P. palmivora* at ED_{50} values of 60, 97, and 140 μg/L, respectively. The particle sizes of the natural product

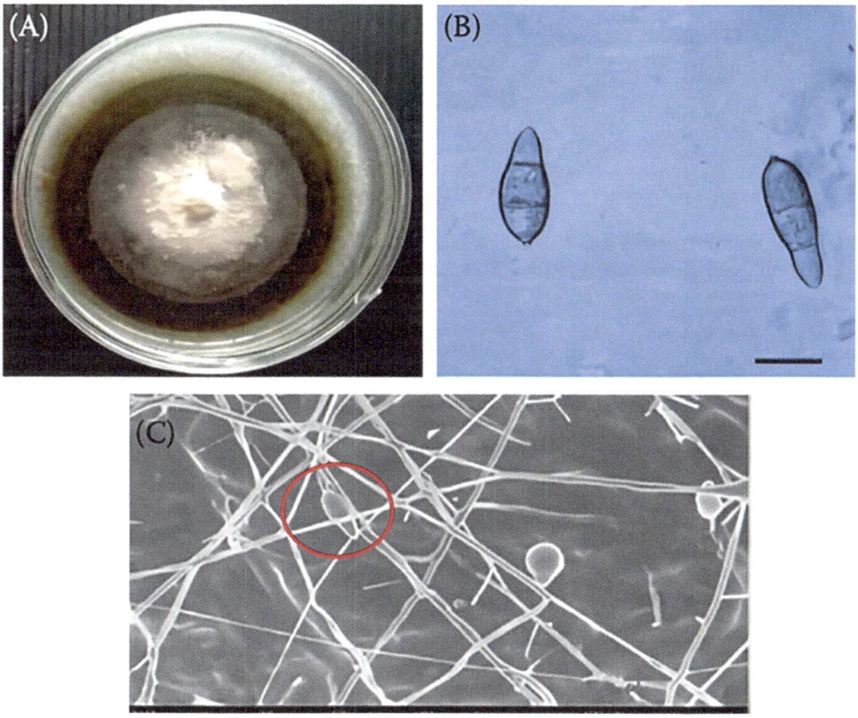

Fig. 11.4 *Magnaporthe oryzae*, (**a**) colony on media after 15 days, (**b**) conidia under a binocular microscope, and (**c**) conidia under SEM, scale bar = 10 μm. (Source: Song et al. (2022b))

nanofibers constructed from *Ch cupreum* were measured, and the results revealed that the sizes of the nano-CC-E, nano-CC-H, and nano-CC-M particles were an average of 534, 499, and 537 nm, respectively. Nano-CC-M, nano-CC-H, and nano-CC-E inhibited *P. palmivora* at ED_{50} values of 16, 13, and 11 μg/L, respectively. Durian seedlings inoculated with *P. palmivora* and treated with nano-CC-E presented phytoalexin accumulation in durian plants, especially in leaves, namely, scopoletin, which acts as an elicitor to induce plant immunity against rot disease in durian (Tongon and Soytong 2022). Scopoletin (6-methoxy-7-hydroxycoumarin) is produced by plants and acts as an antifungal metabolite against *Phytophthora* spp. in tobacco plants (Gnonlonfin et al. 2012) (Figs. 11.4, 11.5, 11.6, 11.7, 11.8, and 11.9) (Table 11.6).

Nanoelicitors derived from *Chaetomium* species tested as growth regulators on dwarf mandarin in Russia. The tangerine fruits treated with the nanoelicitor presented greater fruit weights (44.8 g), peel weights (20.75), juice contents (54.5%), and soluble dry substances (11%) than the control fruits did, as shown in Table 11.8 (Kunina et al. 2023) (Table 11.7).

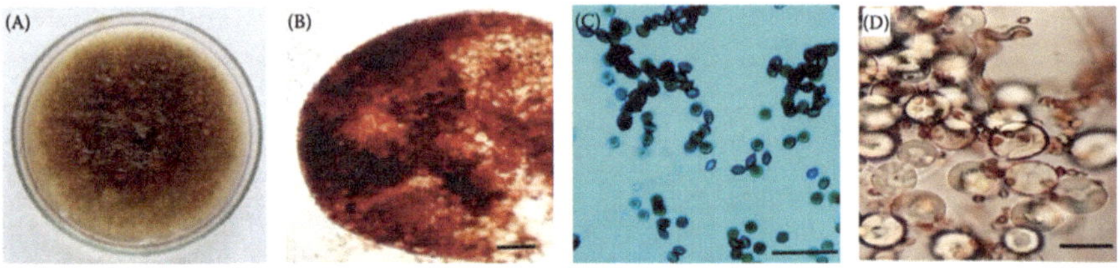

Fig. 11.5 Characteristic of *Emericella nidulans*, (**a**) colony on PDA, (**b**) ascomata, (**c**) ascospores, and (**d**) hulle cells, scale bars = 10 μm. (Source: Song et al. (2022b))

Fig. 11.6 Hexane crude extract, ethyl acetate crude extract, and methanol crude extract of *Emericella nidulans* (upper); nanofibers (middle) and nanofibers under a scanning electron microscope (lower part). Nano-ENH, nano-ENE, and nano-ENM. (Source: Song et al. (2022b))

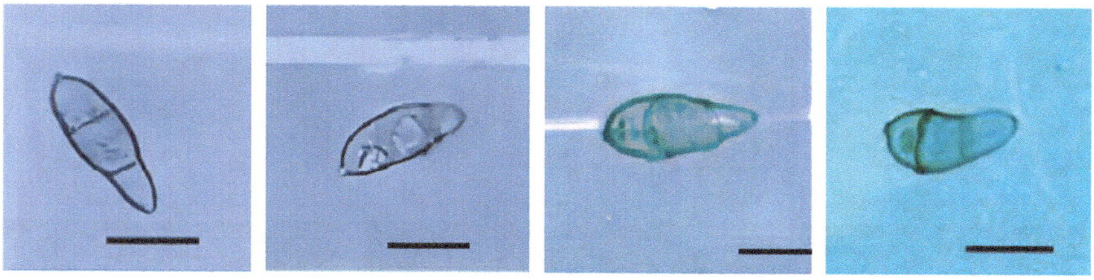

Fig. 11.7 The nanofibers of *Emericella nidulans* caused abnormal conidia of *Magnaporthe oryzae* after treatment with nano-ENH, nano-ENE, or nano-ENM; scale bars = 10 μm. (Source: Song et al. (2022b))

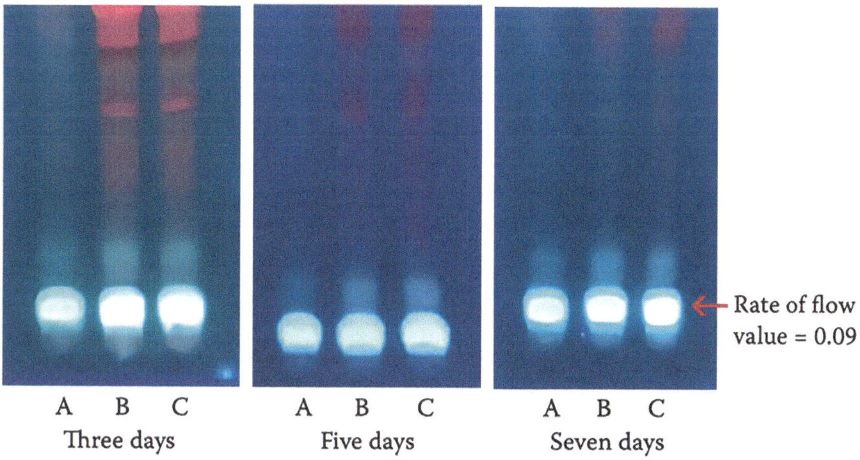

Fig. 11.8 Sakuranetin after treatment of nanofibers from *E. nidulans* and tricyclazole. A, standard of sakuranetin; B, nanofibers; C, tricyclazole. (Source: Song et al. (2022b))

In conclusion, the nanoelicitor acts as a plant growth regulator on the fruits of dwarf mandarin, which presented higher growth parameters than did the nontreated control. Nanoelicitor treatment of mandarin fruits led to a thicker peel, which is related to a defense mechanism that protects plants from pathogens. Treated nanoelicitors accelerate the ripening of fruits (Kunina et al. 2023).

11.9 Nanofibers from *Chaetomium brasiliense* for Plant Immunity

Song et al. (2020b) reported that natural product nanoparticles constructed from *Ch. brasiliense* expressed antifungal effects on *M. oryzae*, causing the blast pathogen rice var. PSL 2. The nano-CBM, nano-CBE, and nano-CBH were constructed from *C. brasiliense*, which was inoculated with rice var. RD47 with *Drechslera oryzae* at

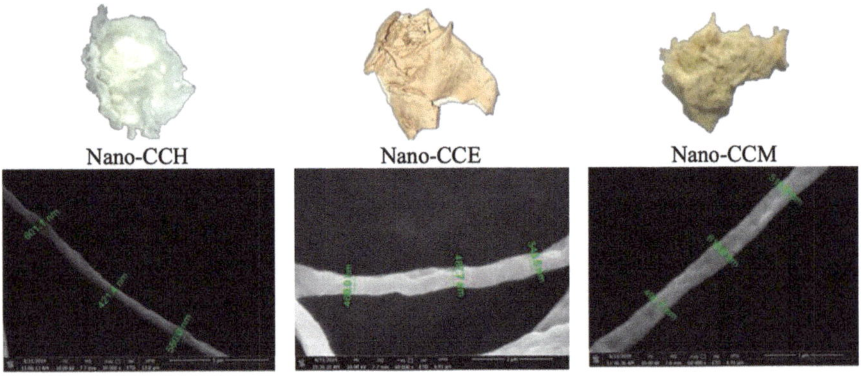

Fig. 11.9 Characteristics of nanofibers constructed from *Chaetomium cupreum* CC3003

Table 11.6 Effects of a nanoelicitor derived from *Chaetomium cupreum* CC3003 on *Phytophthora palmivora*

Metabolites	Concentration (mg.kg^{-1})	Colony diameter (cm)^{2,3}	Growth inhibition (%)^{2,3}	ED_{50} (mg. kg^{-1})	Number of spores^{2,3}(10^5)	Spore (%)^{2,3}	Inhibition ED_{50} (mg. kg^{-1})
Nano CC-H	0	5.00^{a1}	–	1.78	29.25^{a}	–	13.03
	3	2.31^{b}	53.75^{c}		4.00^{b}	86.16^{d}	
	5	1.25^{c}	78.75^{b}		1.50^{cd}	94.81^{bc}	
	10	0.56^{d}	88.75^{a}		0.50^{cd}	97.49^{ab}	
	15	0.50^{d}	90.00^{a}		0.50^{cd}	98.38^{a}	
Nano CC-E	0	5.00^{a}	–	1.51	29.25^{a}	–	11.01
	3	2.25^{b}	55.00^{c}		2.00^{c}	93.08^{c}	
	5	1.22^{c}	75.50^{b}		1.50^{cd}	94.81^{bc}	
	10	0.50^{d}	90.00^{a}		0.50^{cd}	98.36^{a}	
	15	0.50^{d}	90.00^{a}		0.25^{d}	99.16^{a}	
Nano CC-M	0	5.00^{a}	–	1.19	29.25^{a}	–	16.48
	3	2.31^{b}	53.75^{c}		4.50^{b}	84.70^{d}	
	5	0.56^{d}	88.75^{a}		1.50^{cd}	94.83^{bc}	
	10	0.50^{d}	90.00^{a}		0.75^{cd}	98.21^{ab}	
	15	0.50^{d}	90.00^{a}		0.50^{cd}	98.33^{a}	
C.V. (%)		6.39	4.52		13.62	2.89	

[1]Average of four replications. Means followed by a common letter are not significantly different by DMRT at $P = 0.05$

[2]Average of four replications. Means followed by a common letter are not significantly different by DMRT at $P = 0.01$

[3]Inhibition (%) = R1 – R2/R1 × 100 where R1 is the colony diameter of the pathogen in the control and R2 the colony diameter of pathogen in treated plates

Table 11.7 Growth parameters of tangerine fruits treated with the nanoelicitor

Option	Fruit weight, g	Peel weight, %	Juice output, %	SDS[a], %
Control	41.3±7.72	20.0±3.1	52.9±2.8	10.3±0.5
Obstactin	39.6±7.01	19.0±0.6	45.9±1.0	10.4±0.4
Siliplant	45.9±8.35	20.9±2.1	49.2±3.2	9.9±0.5
Nanoelicitor	44.8±7.99	20.7±0.0	54.5±0.0	11.0±0.7
Zerebra agro	48.2±8.65	20.1±2.9	49.9±2.9	9.9±0.5
LSD ($p \leq 0.05$)	*1.15*	*0.59*	*1.21*	*0.97*

Source: Kunina et al. (2023)
[a]*SDS* soluble dry substances

15 µg/mL after three days, resulting in the detection of oryzalexin C at an Rf of 0.61 in rice leaves (Fig. 11.10). Oryzalexin C is a phytoalexin that is isolated from rice leaves treated with elicitors (Akatsuka et al. 1983). Nano-CBH, nano-CBE, and nano-CBM expressed antifungal activity to inhibit *D. oryzae* (brown spot of rice) (Fig. 11.11), with $ED_{50 \text{ values}}$ of 5.8, 4.9, and 2.8 µg/ml, respectively (Vareeket et al. 2023).

Natural product nanofibers constructed from *Trichoderma hamatum* were used to treat inoculated durian plants with *Phytophthora palmivora*, which can be used to detect phytoalexin in durian leaves, which was confirmed to be an elicitor that induces plant immunity, leading to the recovery of plants from rot disease (Phal et al. 2024). Phytoalexin accumulation in host plants after nanofiber application acts as a defense mechanism, leading to a decrease in disease prevalence. Moreover, nano-TK01M, which is expressed as a growth regulator, enhances physiological parameters in plants, including chlorophyll synthesis and carotenoids, to promote plant growth. The TK01M nanofibers inhibited sporangia and colonies of the tested pathogens at ED_{50} values of 3 and 11 µg/mL, respectively. *T. hamatum* K01 produces active metabolites such as pyrone 6-pentyl-2H-Pyran-2-one and sorbicillin. Pyrone and sorbicillin, which may react as natural products to decrease durian rot caused by *P. palmivora, were first reported* (Figs. 11.12 and 11.13) (Table 11.8).

Nanofibers constructed from *T. hamatum* K01, namely, nano-TK01E, nano-TK01M, and nano-TK01M, were measured by scanning electron microscopy and ranged from 57 to 242, 75 to 168, and 90 to 256 nm, respectively. The results indicated that nano-TK01M inhibited the growth of *P. palmivora sporangia* at an ED_{50} of 3.13 µg/mL. Sporangia did not grow at 15 µg/mL. Nano-TK01E suppressed the growth of sporangia at an ED_{50} of 5.48 µg/mL. The inoculated plants treated with nano-TK01M or metalaxyl did not present any symptoms. The inoculated durian plants treated with nano-TK01M appeared at an R_f of 0.87, which was confirmed to be associated with scopoletin phytoalexin production against *Phytophthora* rot in durian, as shown in Fig. 11.14 (Phal et al. 2024).

Table 11.8 Effect of nanofibers derived from *Trichoderma hamatum* K01 on *Phytophthora palmivora*

Nanofibres	Concentrations (μg/mL)	Colony growth (cm)	Growth inhibition (%)	ED_{50} (μg/mL)	Number of spores (10^5)	Inhibition (%)	ED_{50} (μg/mL)
TK01H	0	4.77[a]	–	19.96	27.00[a]	–	9.83
	3	3.70[b]	22.43[f]		26.00[ab]	3.70[e]	
	5	3.52[b]	26.21[f]		22.50[ab]	16.67[e]	
	10	3.02[c]	36.69[de]		10.00[cd]	62.96[bc]	
	15	2.67[cde]	44.03[bcd]		5.50[de]	79.63[ab]	
TK01E	0	4.77[a]	–	16.66	27.00[a]	–	5.48
	3	3.62[b]	24.11[f]		20.50[b]	24.07[d]	
	5	3.40[bc]	28.72[ef]		13.75[c]	49.07[cd]	
	10	3.02[cd]	36.69[de]		3.75[e]	86.11[ab]	
	15	2.40[e]	49.69[bc]		0.25[e]	99.07[a]	
TK01M	0	4.77[a]	–	11.06	27.00[a]	–	3.13
	3	2.85[cd]	40.25[cd]		14.50[c]	46.30[cd]	
	5	2.47[de]	48.22[bc]		4.75[de]	82.41[ab]	
	10	2.35[e]	50.73[b]		1.50[e]	94.44[a]	
	15	1.80[f]	62.26[a]		0.00[e]	100.00[a]	
CV%	-	7.89	17.72		29.28	32.87	

Source: Phal et al. (2024)
The average of four replications, the means followed by the same letters are not significantly different between treatments based on Duncan's multiple range test (DMRT) at $p < 0.05$

Phal et al. (2024) reported that the highest chlorophyll and carotenoid contents were detected in the nano-TK01M treatment, at 25 and 3 μg/g, respectively, which were higher than those in the metalaxyl application, which were 16.30 and 2.15 μg/g, respectively. Compared with those of the nontreated control plants, the chlorophyll and carotenoid contents of the treated plants were 10 and 1.65 μg/g, respectively, at 60 days after application (Table 11.9).

11.10 Nanofibers from *Trichoderma hamatum* Against Citrus Anthracnose

Phal et al. (2023) reported that TK01-MeOH extract from *Trichoderma hamatum* K01 (Fig. 11.15) actively inhibited *Colletotrichum gloeosporioides* C01 (Fig. 11.16), causing leaf anthracnose of citrus plants to inhibit conidia formation and colony growth at ED_{50} values of 355 and 273 ppm, respectively. *T. hamatum* produces

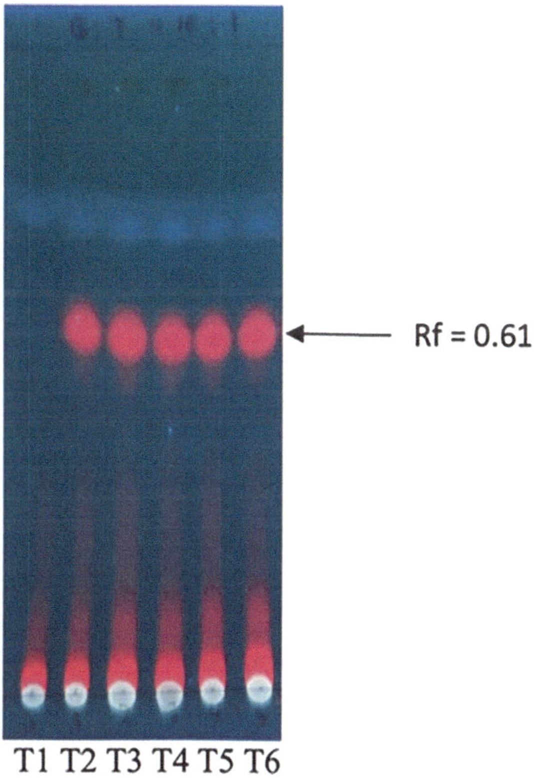

Fig. 11.10 Oryzalexin C was synthesized with benzene:ethyl acetate (10:1) under 365 nm UV light. The plants were inoculated with pathogen (T1), standard (T2), nano-CBH (T3), pathogen and nano-CBE (T4), and pathogen and nano-CBM (T5) propiconazole (T6). (Vareeket et al. 2023)

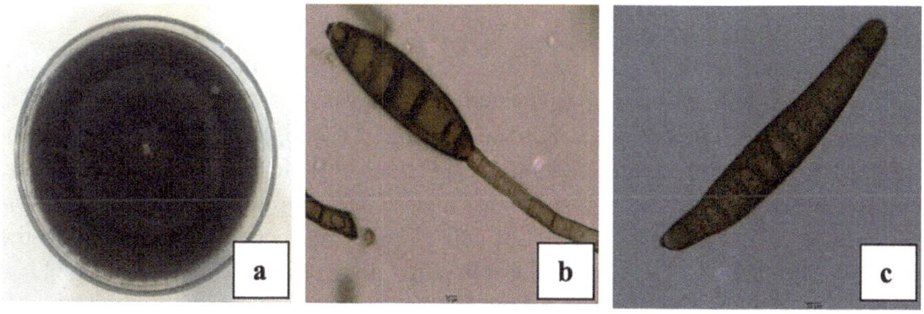

Fig. 11.11 *Drechslera oryzae*, (**a**) pure culture on PDA, (**b**) conidiophore and conidium, and (**c**) conidium. (Vareeket et al. 2023)

pyrone and sorbicillin, which were analyzed via GC–MS. Moreover, a nanofiber, TK01M, constructed from a crude metabolite of *T. hamatum,* inhibited colonies and conidia at ED_{50} values of 13 and 3 ppm, respectively. The performance of the crude

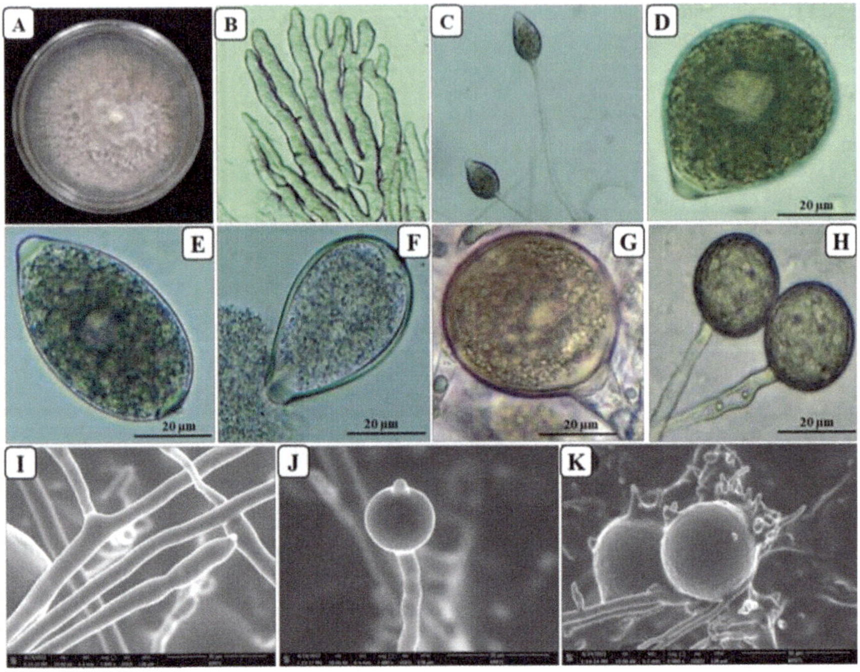

Fig. 11.12 Characterization of *Phytophthora palmivora*, (**a**) colony on PDA at 10 days, (**b**, **i**) branching mycelia, (**d**, **e**) sporangia, (**f**) zoospores, (**g**) one antheridium and oogonium, (**h**, **b**) chlamydospore, (**b**, **h**) under light microscope, and (**i**, **k**) scanning electron microscope. (Phal et al. 2024)

metabolites and nanofibers derived from *T. hamatum* K01, including diameter size and nanofibers, under a scanning electron microscope is shown in Fig. 11.17. The average diameter sizes of the nanofibers of nano-TK01E, nano-TK01H, and nano-TK01M were 187, 122, and 136 nm, respectively (Table 11.10). The nanofibers, namely, nano-TK01H, nano-TK01E, and nano-TK01M, inhibited *C. gloeosporioides* growth at ED50 values of 33, 28, and 13 ppm, respectively, and inhibited the growth of conidia at ED_{50} values of 15, 9, and 3 ppm, respectively (Table 11.11).

Natural product nanoelicitors have been developed from either pure compounds or active metabolites of *Ch. cupreum*, *Ch. brasiliense*, *Ch. cochlioides*, *Ch. elatum*, *Ch. globosum*, *Ch. lucknowense, Emericella nidulans, T. hamatum*, and *T. harzianum*, which are used by many growers to induce plant immunity to diseases, for example, chili and coffee anthracnose (*Colletotrichum capsici*, *Colletotrichum. gloeosporioides*), tomato wilt (*F. oxysporum*), leaf spot and blight of rice (*Curvularia lunata* and *Drechslera oryzae*), rice blast (*M. oryzae*), and root rot of durian and

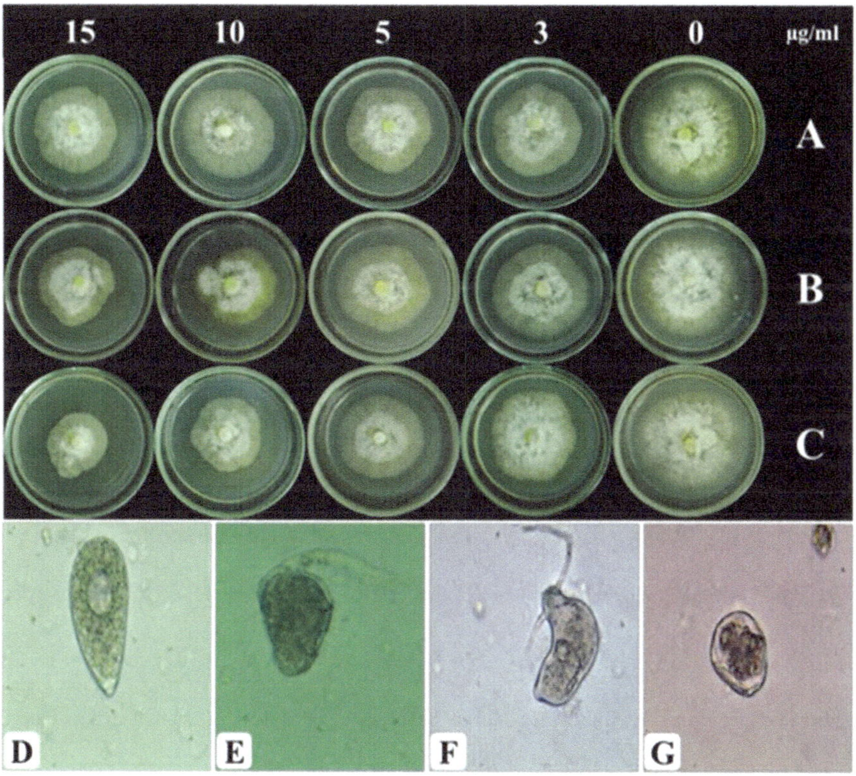

Fig. 11.13 Nanofibers constructed from *Trichoderma hamatum* K01 against *Phytophthora palmivora*. (**a–c**) nano-TK01H, nano-TK01E, and nano-TK01M with *P. palmivora*, (**d**) normal sporangia, (**e–g**) abnormal sporangia affected by nano-TK01H, nano-TK01E, and nano-TK01M. (Phal et al. 2024)

citrus (*Pythium* and *Phytophthora*). Each nanoelectrode expresses phytoalexin, for example, capsidiol, alpha tomatine, sakuranetin, oryzalexin B, scoparone, and scopoletin, in infected plants. Further research on natural product nanoelicitors for plant immunity to diseases, nanoelicitors for insect control and nanoelicitors for stimulating plant growth, is being conducted (Table 11.12).

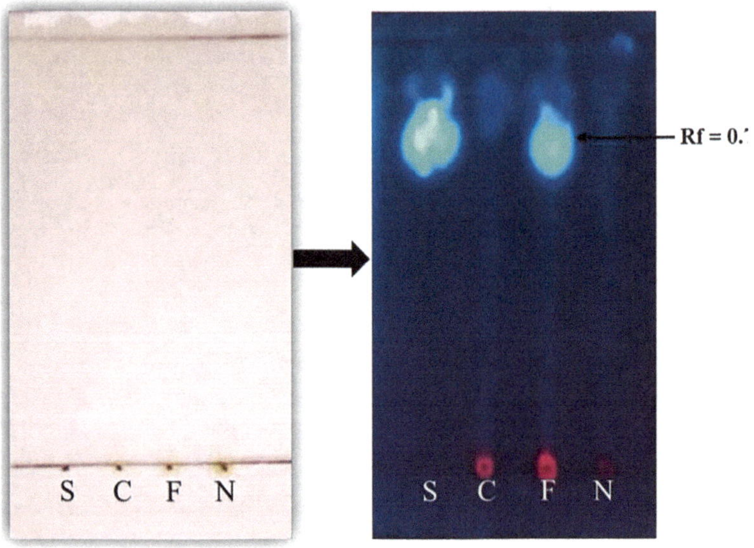

Fig. 11.14 Scopoletin phytoalexin at an R_f value of 0.87, S = scopoletin, C = control, F = nano-TK01M, and N = metalaxyl

Table 11.9 Comparison of the effects of nano-TK01 and metalaxyl on chlorophyll in durian leaves

	Chlorophyll (Chl) contents (μg/g)			
Treatments	Chl_a	Chl_b	Total Chl	Carotenoid
Control	7.49^{b}	2.57^{b}	10.07^{b}	1.65^{a}
Nano-TK01M	18.84^{a}	7.14^{a}	25.98^{a}	3.70^{a}
Metalaxyl	12.13^{ab}	4.16^{ab}	16.30^{ab}	2.15^{a}
CV%	27.56	31.43	28.13	30.14

Source: Phal et al. (2024)
Averages followed by the same latter are not significantly different

11.11 Conclusion Remarks

The methods for nanofiber construction must consider sustainable development for safe use of nanomaterials. Biologically based nanofibers are constructed from natural resources, and bioactive metabolites from microorganisms (bacteria, algae, fungi) and plants can be safely applied. Plant- or microbe-based nanofibers constructed from cellulose, chitin/chitosan, and lignin could be developed as biodegradable nanomaterials. These nanomaterials can help revitalize the environment through the accumulation of toxic agrochemicals and heavy metals in soil and water (Saraswat et al. 2023). The lowest toxicity and highest stability are clearly shown for biological and microbial substances (Bolade et al. 2020).

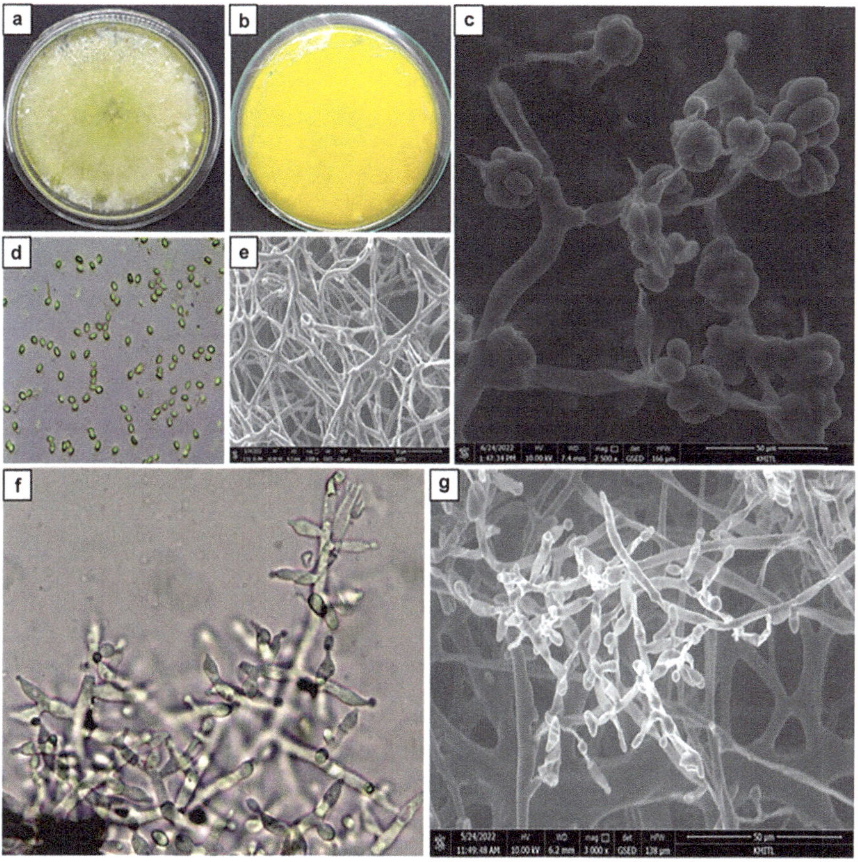

Fig. 11.15 Morphology of *T. hamadum*; (**a**) colony, (**b**) upside down of colony, (**c** and **d**) conidia, (**e**) hyphae, (**f, g**) conidiophore, (**d**, **f**) under light compound microscope, and (**c**, **e**, **g**) under SEM. (Phal et al. 2023)

Biological-based nanomaterials have been clearly demonstrated for sustainable development with less harmful effects on humans and the surrounding environment (Li et al. 2011a, b). These nanomaterials or nanofibers can be constructed from metabolites derived from different potent microorganisms (Mallikarjuna et al. 2011). Natural products derived from *Ch. brasiliense*, *Ch. cochliodes*, *Ch. cupreum*, *Ch. elatum*, *Ch. globosum*, *Ch. lucknowense*, and *T. harzianum*, *T. hamatum*, and *E. nidulans* are constructed as microbe-based nanomaterials that were developed to be nanoelicitors to induce plant immunity through phytoalexin production. Each nanoelicitor can induce phytoalexin in plants, for example, scoparone for *Phytophthora* or *Pythium* rot of citrus, scopoletin and anthrocyaidin for *Phytophthora*

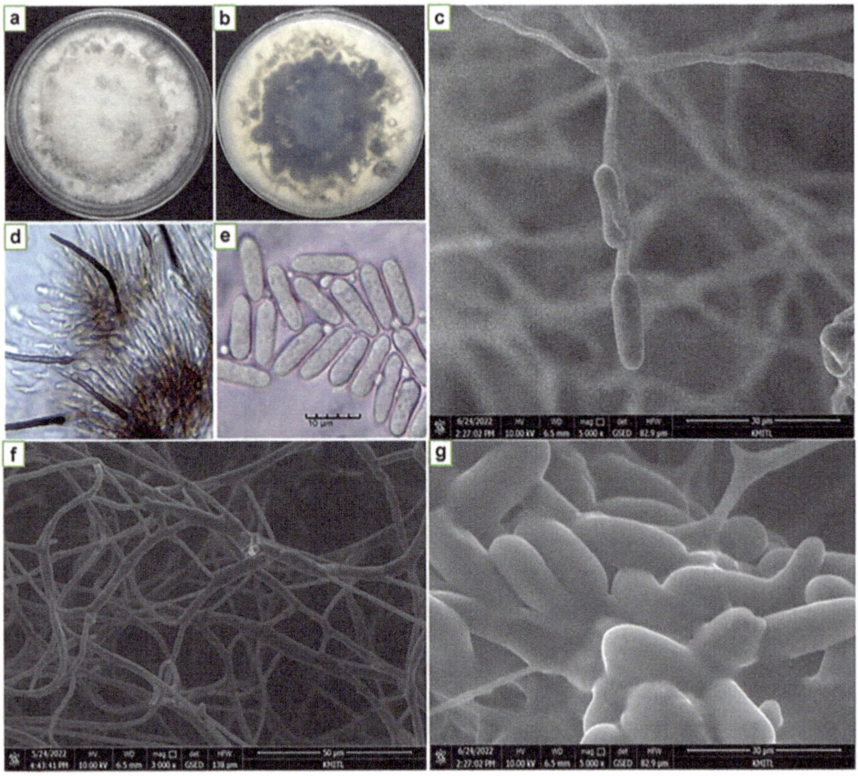

Fig. 11.16 Colony patterns of *C. gloeosporioides*. (**a**) Colony, (**b**) upside down of the colony, (**c**, **d**) setae and conidiophores, (**e**, **g**) hyphae, (**d**, **e**) under a light compound microscope (scale bar 10 μm, 40X magnification), and (**c**, **f**, **g**) under a scanning electron microscope. (Phal et al. 2023)

or *Pythium* rot of durian, alpha-tourmaline for *Fusarium* wilt of tomato, capsidiol for chili anthracnose, and sakuranetin for rice blast. These microbe-based nanomaterials for plant protection are constructed from active metabolites, which were developed to be biodegradable nanomaterials. Microbe-based nanomaterials are highly stable and safe and have no harmful effects on the surrounding biodiversity or environment; thus, they can be applied as agricultural inputs for organic agriculture.

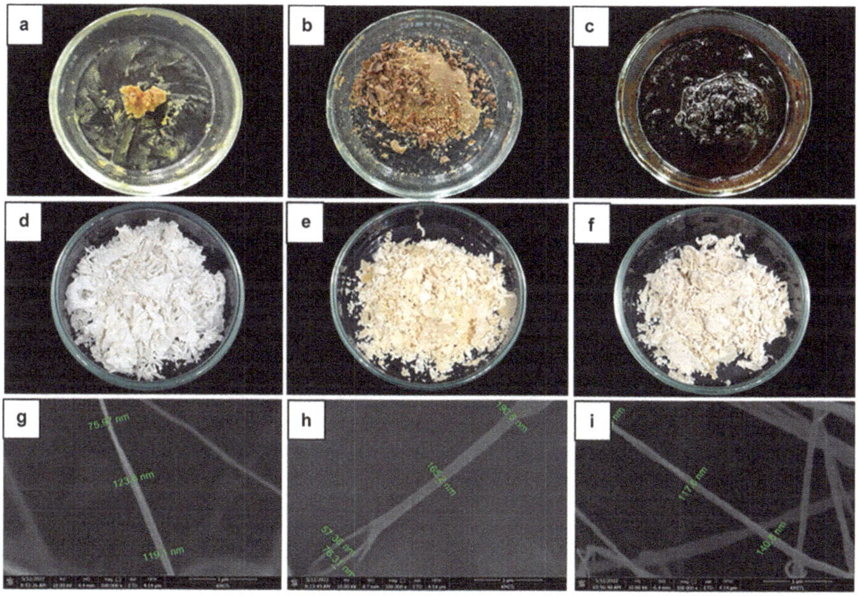

Fig. 11.17 Crude metabolites and nanofibers from *T. hamadum* (**a**) crude TK01-H, (**b**) crude TK01-E, (**c**) crude TK01-M, (**d**) nano-TK-01H, (**e**) nano-TK01E, (**f**) nano-TK01M, and (**g**, **h**, **i**) nanofibers under SEM. (Phal et al. 2023)

Table 11.10 The size of the nanofibers from *T. hamadum*

Nanoparticles	Mean (nm)	Max (nm)	Min (nm)
Nano-TK01H	122	168	75
Nano-TK01E	187	242	57
Nano-TK01M	136	256	90

Source: Phal et al. (2023)
Data were collected from 20 samples by measuring the size (diameter) of each nanofiber

Table 11.11 Effect of nanofibers from *T. hamatum* on the control *C. gloeosporioides*

Nanoparticles	Concentration (ppm)	Colony growth (cm)	Growth inhibition (%)	ED_{50} (ppm)	Conidia production (10^5)	Inhibition (%)	ED_{50} (ppm)
Nano-TK01H	0	4.90a	–	33	216.25a	–	15
	3	4.06b	17.04f		162.75ab	19.93d	
	5	3.82bc	21.89f		153.50ab	24.74de	
	10	3.70cd	24.49ef		141.00ab	31.35cd	
	15	3.38de	30.76de		84.24bc	58.59bc	
Nano-TK01E	0	4.90a	–	28	216.25a	–	9
	3	3.81bc	22.16f		154.00ab	23.91d	
	5	3.41de	30.29de		125bc	36.06cd	
	10	3.32ef	32.10cde		81.75bc	56.72bc	
	15	3.05fgh	37.72bcd		53.50cd	74.21bc	
Nano-TK01M	0	4.90a	–	13	216.25a	–	3
	3	3.28efg	32.81cd		137.25ab	37.20c	
	5	2.98gh	39.00bc		49.50cd	75.45ab	
	10	2.82h	42.33b		0.00d	100a	
	15	1.98i	59.37a		0.00d	100a	
CV%		4.46	14.49		31.91	40.39	

Source: Phal et al. (2023)
The average of four replications, the means followed by the common letters are not significantly different by DMRT at $p < 0.01$

Table 11.12 Applications of natural product nanoelicitors for plant disease immunity and growth stimulants

Species	Nanoelicitors	Antimicrobial (ED_{50}, ppm)	Disease immunity	Phytoalexin
Ch. globosum	Chaetoglobosin C	5–10	Chili anthracnose (*C. capsici*)	Capsidiol
Ch. lucknowense	Chaetomanone	5–10	Chili anthracnose (*C. capsici*)	Capsidiol
T. harzianum	Trichotoxin A50	5–10	Chili anthracnose (*C. capsici*)	Capsidiol
Ch. cupreum	Nano-rotiorinol	5–10	Chili anthracnose (*C. capsici*)	Capsidiol
Ch. globosum	Chaetoglobosin C	5–10	Wilt (*F. oxysporum*)	Alpha tomatine
Ch. lucknowense	Chaetomanone	5–10	Tomato wilt (*F. oxysporum*)	Alpha tomatine
T. harzianum	Trichotoxin A50	5–10	Tomato wilt (*F. oxysporum*)	Alpha tomatine
T. harzianum	Nano-trichotoxin	5–10	Coffee anthracnose (*C. gloeosporioides*)	–

(continued)

Table 11.12 (continued)

Species	Nanoelicitors	Antimicrobial (ED_{50}, ppm)	Disease immunity	Phytoalexin
Ch. globosum	Nano-CGH	1.21	Coffee anthracnose (*C. gloeosporioides*)	–
	Nano-CGE	1.19	Coffee anthracnose (*C. gloeosporioides*)	–
	Nano-CGM	1.93	Coffee anthracnose (*C. gloeosporioides*)	–
Ch. cochliodes	Nano-CCoH	33	Leaf spot (*Curvularia, Drechslera*)	–
	Nano-CCoE	9	Leaf spot (*Curvularia, Drechslera*)	–
	Nano-CCoM	16	Leaf spot (*Curvularia, Drechslera*)	–
Ch. elatum	Nano-CEH	7	Blast (*M. oryzae*)	–
	Nano-CEE	8	Blast (*M. oryzae*)	Sakuranetin and Oryzalexin B
	Nano-CEM	6	Blast (*M. oryzae*)	Sakuranetin and Oryzalexin B
Ch. brasiliense	Nano-CBH	6	Blast (*M. oryzae*)	Sakuranetin and Oryzalexin B
	Nano-CBE	9	Blast (*M. oryzae*)	Sakuranetin and oryzalexin B
	Nano-CBM	13	Blast (*M. oryzae*)	Sakuranetin and Oryzalexin B
Ch. brasiliense	Nano-CBH	3	Citrus rot (*Pythium, Phytophthora*)	Scoparone
	Nano-CBM	3.9	Citrus rot (*Pythium, Phytophthora*)	Scoparone
	Nano-CBH	2.6	Citrus rot	Scoparone
Ch. lucknowense	Nano-CLH	5	Blast (*M. oryzae*)	Sakuranetin and Oryzalexin B
	Nano-CLE	7	Blast (*M. oryzae*)	Sakuranetin and Oryzalexin B
	Nano-CLM	10	Blast (*M. oryzae*)	Sakuranetin and Oryzalexin B
Emericella nidulans	Nano-ENH	8.9	Blast (*M. oryzae*)	Sakuranetin and Oryzalexin B
	Nano-CLE	4.2	Blast (*M. oryzae*)	Sakuranetin and Oryzalexin B
	Nano-CLM	4.5	Blast (*M. oryzae*)	Sakuranetin and Oryzalexin B
T. hamatum	Nano-TK01H	9	Durian rot (*Phytophthora*)	Scopoletin
	Nano-TK01E	5	Durian rot (*Phytophthora*)	Scopoletin

(continued)

Table 11.12 (continued)

Species	Nanoelicitors	Antimicrobial (ED_{50}, ppm)	Disease immunity	Phytoalexin
	Nano-TK01M	3	Durian rot (*Phytophthora*)	Scopoletin
T. hamatum	Nano-TK01H	15	Citrus anthracnose (*C. gloeosporioides*)	–
	Nano-TK01E	9	Citrus anthracnose (*C. gloeosporioides*)	–
	Nano-TK01M	3	Citrus anthracnose (*C. gloeosporioides*)	–
Ch. brasiliense	Nano-CB	2–12	Citrus rot (*Pythium, Phytophthora*)	Scoparone
Ch. cochliodes	Nao-CCo	2–12	Citrus rot (*Pythium, Phytophthora*)	Scoparone
Ch. elatum	Nano-CE	2–12	Citrus rot (*Pythium, Phytophthora*)	Scoparone
Chaetomium spp.	Nanoelicitor	–	Plant growth stimulant	Active metabolites

References

Afek U, Sztejnberg A (1989) Effects of fosetyl-Al and phosphorous acid on scoparone, a phytoalexin associated with resistance of citrus to Phytophthora citrophthora. Phytopathology 79:736–739

Ahuja I, Kissen R, Bones AM (2012) Phytoalexins in defense against pathogens. Trends Plant Sci 17(2):73–90. https://doi.org/10.1016/j.tplants.2011.11.002

Akatsuka T, Kodama O, Kato H, Kono Y, Takeuchi S (1983) 3-Hydroxy-7-oxo sandaraco- pimaradiene (Oryzalexin a), a new Phytoalexin Isolated from Rice blast leaves. Agric Biol Chem Biosci Biotechnol Biochem 77:1141–1148

Angelova Z, Georgiev S, Roos W (2006) Elicitation of plants. Biotechnol Equip 20:72–83

Baby J, Raj JS, Biby ET, Sankarganesh P, Jeevitha MV, Ajisha SU, Rajan SS (2010) Toxic effect of heavy metals on aquatic environment. Int Biol Chem Sci 4:939–952

Bolade OP, Williams AB, Benson NU (2020) Green synthesis of iron-based nanomaterials for environmental remediation: a review. Environ Nanotechnol Monit Manag 13:100279

Brillas E, Martínez-Huitle CA (2015) Decontamination of wastewaters containing synthetic organic dyes by electrochemical methods, an updated review. Appl Catal B Environ 166(2015):603–643

Dar J, Soytong K (2014) Construction and characterization of copolymer nanomaterials loaded with bioactive compounds from *Chaetomium* species. J Agric Technol 10:823–831

Deverall BJ (1982) Introduction. In: Bailey JA, Mansfield JW (eds) Phytoalexins. Blackie, London, pp 1–20

Dillon VM, Overton J, Grayer RJ, Harbone J (1997) Differences in phytoalexin response among rice cultivars of different resistance to blast. Phytochemistry 44:559–603

Ditta A (2012) How helpful is nanotechnology in agriculture? Adv Nat Sci Nanosci Nanotechnol 3:033002

Elibol OH, Morisette DD, Denton JP, Bashir R (2003) Integrated nanoscale silicon sensors using top-down fabrication. Appl Phys Lett 83:4613–4615

Environmental Protection Agency (EPA) (2007) Nanotechnology white paper. EPA, Washington, DC

Glazebrook J, Ausubel FM (1994) Isolation of phytoalexin-deficient mutants of *Arabidopsis thaliana* and characterization of their interactions with bacterial pathogens. Proc Natl Acad Sci 91:8955–8959

Gnonlonfin GJB, Sammi A, Brimer L (2012) Review Scopoletin-A coumarin phytoalexin with medicinal properties. CRC Crit Rev Plant Sci 31:47–56

Hasegawa M, Mitsuhara I, Seo S, Okada K, Yamane H, Iwai T, Ohashi Y (2014) Analysis on blast fungus responsive characters of a flavonoid Phytoalexin Sakuranetin; accumulation in infected rice leaves, antifungal activity and detoxification by fungus. Molecules 19:11404–11418

Kodama O, Miyakawa J, Akatsuka T, Kiyosawa S (1992) Sakuranetin, a flavanone phytoalexin from ultraviolet-irradiated rice leaves. Phytochemistry 31:3807–3809

Kunina V, Soytong K, Song JJ, Belous O (2023) Biochemical composition of dwarf mandarin cv. 'Miagava-Vase' in the humid subtropics of Russia. Int J Agric Technol 19(1):121–128

Leewijit T, Pongnak W, Soytong K, Poeaim S (2016) Isolation of soil and endophytic fungi from rice (*Oryza sativa* L.). Int J Agric Technol 12:2191–2202

Lellis B, Favaro-Polonio CZ, Pamphile JA, Polonio JC (2019) Effects of textile dyes on health and the environment and bioremediation potential of living organisms. Biotechnol Res Inn 3(2019):275–290

Li M, Huang Q, Wu Y (2011a) A novel chitosan- poly (lactide) copolymer and its submicron particles as imidacloprid carriers. Pest Manag Sci 67:831–836

Li X, Xu H, Chen ZS, Chen G (2011b) Biosynthesis of nanoparticles by microorganisms and their applications. J Nanomater:1–16. https://doi.org/10.1155/2011/270974

Liu L, Punja ZK, Rahe JE (1995) Effect of *Pythium* spp. and glyphosate on phytoalexin production and exudation by bean (*Phaseolus vulgaris* L.) roots grown in different media. Physiol Mol Plant Pathol 47:391–405

Mallikarjuna K, Narasimha G, Dillip GR, Praveen B, Shreedhar B, Lakshmi CS, Reddy BVS, Raju BDP (2011) Green synthesis of silver nanoparticles using Ocimum leaf extract and their characterization. Dig J Nanomater Biostruct 6(2011):181–186

Maseko BOZ, Coutinho TA (2002) Pathogenicity of *Phytophthora* and *Pythium* species associated with citrus root rot in South Africa. S Afr J Bot 68:327–332

Mok SL (2009) Quantification of sakuranetin in paddy leaves and stem after elicitation with abiotic elicitors (UV, AgNO3, CuSO4). Mod Appl Sci 3:210–216

Moosophon P, Kanokmedhakul S, Kanokmedhakul K, Soytong K (2009) Prenylxanthones and a bicyclo[3.3.1] nona-2,6-diene derivative from the fungus *Emericella rugulosa*. J Nat Prod 72:1442–1446

Morales-Díaz AB, Ortega-Ortíz H, Juárez-Maldonado A, Cadenas-Pliego G, González-Morales S, Benavides-Mendoza A (2017) Application of nanoelements in plant nutrition and its impact in ecosystems. Adv Nat Sci-Nanosci Nanotechnol 8:013001 (13pp). https://doi.org/10.1088/2043-6254/8/1/013001

Mukhopadhyay SS (2014) Nanotechnology. Sci Appl 7:63

Nguyen HM, Hwang IC, Park JW, Park HJ (2012) Enhanced payload and photoprotection for pesticides using nanostructured lipid carriers with corn oil as liquid lipid. J Microencapsul 29:596–604

OECD (2014) Report of the questionnaire on regulatory. Regimes for Manufactured Nanomaterials 2010–2011 42:1–62

Pan C, Zhai J, Wang ZL (2019) Piezotronics and piezo-phototronics of third generation semiconductor nanowires. Chem Rev 119(15):9303–9359. https://doi.org/10.1021/acs.chemrev.8b00599

Perlatti B, de Souza Bergo PL, das Graças Fernandes da Silva MF, Fernandes JB, Forim MR (2013) Polymeric nanoparticle-based insecticides: a controlled release purpose for agrochemicals. In: Trdan S (ed) Insecticides – development of safer and more effective technologies. In Tech Open Access, Croatia. Ch. 20

Phal P, Soytong K, Poeaim S (2023) Natural product nanofibers derived from *Trichoderma hamatum* K01 to control citrus anthracnose caused by *Colletotrichum gloeosporioides*. Open Agriculture 8(1). https://doi.org/10.1515/opag-2022-0193

Phal P, Soytong K, Poeaim S (2024) A new nanofiber derived from *Trichoderma hamatum* K01 to control durian rot caused by *Phytophthora palmivora*. Asian J Agric Biol 2024(2):2023182. https://doi.org/10.35495/ajab.2023.182

Qin J-C, Zhang Y-M, Gao J-M, Bai M-S, Yang S-X, Laatsch H, Zhang A-L (2009) Bioactive metabolites produced by *Chaetomium globosum*, an endophytic fungus isolated from *Ginkgo biloba*. Bioorg Med Chem Lett 19(6):1572–1574

Rai M, Ingle A (2012) Role of nanotechnology in agriculture with special reference to management of insect-pest. Appl Micro- biol Biotechnol 94:287–293

Salata OV (2004) Application of nanoparticles in biology and medicine. J Nano- biotechnol 2:3

Saraswat P, Singh S, Prasad M, Misra R, Rajput DV, Ranjan R (2023) Applications of biobased nanomaterials in environment and agriculture: a review on recent progresses. Hybrid Adv 4:100097

Sibounnavong P, Soytong K (2011) Antifungal metabolites from antagonistic fungi used to control tomato wilt fungus *Fusarium oxysporum* f. sp. lycopersici. Afr J Biotechnol 10:19714–19722

Song JJ, Soytong K, Kanokmedhakul S, Kanokmedhakul K, Poeaim S (2020a) Natural product of nanoparticles constructed from *Chaetomium* spp to control rice blast disease caused by *Magnaporthe oryzae*. Int J Agric Biol 23:1013–1020

Song JJ, Soytong K, Kanokmedhakul S, Kanokmedhakul K, Poeaim S (2020b) Antifungal activity of microbial nanoparticles derived from *Chaetomium* spp against *Magnaporthe oryzae* causing rice blast. Plant Prot Sci 56(3):180–190

Song JJ, Soytong K, Thianrat L (2022a) Nanofibers derived from endophytic Chaetomium brasilense for growth stimulation of rice. Int J Agric Technol 18(2):843–856

Song JJ, Soytong K, Kanokmedhakul S (2022b) Control of rice blast disease caused by *Magnaporthe oryzae* by application of antifungal nanomaterials from *Emericella nidulans*. Plant Protect Sci 58:40–48

Soutter W (2012) Nanotechnology in agriculture. AZoNano.com Publishers. Available at http://www.azonano.com/article.aspx?ArticleID=3141#1%204. Accessed 25 May 2013

Soytong K, Kanokmedhakul S, Kukongviriyapa V, Isoba M (2001) Application of *Chaetomium* (Ketomium®) as a new broad spectrum biological fungicide for plant disease control: a review article. Fungal Divers 7:1–15

Soytong K, Kanokmedhakul S, Rattanacherdchai K, Chamaiporn C (2014) In: Sahayaraj K (ed) Microbial elicitors to induce immunity for plant disease control in chilli and tomato, pp 99–125 in basic and applied aspects of biopesticides. Springer. ISBN 978-81-322-1876-0. https://doi.org/10.1007/978-81-322-1877-7

Sulistyowati L, Keane PJ, Anderson JW (1990) Accumulation of the phytoalexin, 6,7-dimethoxycoumarin, in roots and stems of citrus seedlings following inoculation with *Phytophthora citrophthora*. Physiol Mol Plant Pathol 37(6):451–461. https://doi.org/10.1016/0885-5765(90)90036-WGetrightsandcontent

Tann H, Soytong K (2016) Effects of nanoparticles loaded with *Chaetomium globosum* KMITL-N0805 extracts against leaf spot of rice var. Sen Pidoa. Malaysian Appl Biol J 45:1–7

Tann H, Soytong K, Makhonpas C, Adthajadee A (2011) Comparison between organ- Ic, GAP and chemical methods for cultivation of rice varieties in Cambodia. J Agric Technol 7:2239–2245

Tongon R, Soytong K (2022) Nanoparticles derived from active metabolites of *Chaetomium cupreum* cc3003 against phytophthora rot of durian. Int J Agric Biol 27(1):19–27. https://doi.org/10.17957/IJAB/15.1894

Tongon R, Soytong K, Kanokmedhakul S, Kanokmedhakul K (2018) Nanoparticles from *Chaetomium brasiliense* to control *Phytophthora palmivora* caused root rot disease in durian var montong. Int J Agric Technol 14(7):2163–2170

Udompongsuk M, Soytong K, Kanokmedhakul S, Kanokmedhakul K (2018) In vitro efficacy of *Chaetomium brasiliense* against *Pythium* spp. causing root rot disease of tangerine. Int J Agric Technol 14(7):2181–2190

Uehara K (1958) On the production of phytoalexin by the host plant as a result of interaction between the rice plant and the blast fungus (*Pyricularia oryzae* Cav.). Ann Phytopathol Soc Jpn 23:127–130

VanEtten HD, Mansfield JW, Bailey JA, Farmer EE (1994) Two classes of plant antibiotics: phytoalexins versus Phytoanticipins. Plant Cell 6:1191

Vareeket R, Song JJ, Soytong K (2023) New discovery of natural product nanoparticles constructed from active metabolites from Chaetomium brasiliense for immunity to brown spot of rice var. RD 47. Int J Agric Technol 19(1):291–300

Vilavong S, Soytong K (2016) Application of bioformulation of *Chaetomium* for biocontrol of *Colletotrichum gloeosporioides* causing coffee anthracnose in Arabica variety in Laos. J Agric Technol 12:1239–1245

Wang GL, Valent B (2009) Advances in genetics, genomics and control of rice blast disease. Springer, New York

Wang W, Luo M, Fu Y, Wang S, Efferth T, Zu Y (2013) Glycyrrhizic acid nanoparticles inhibit LPS-induced inflammatory mediators in 264.7 mouse macrophage cells. Int J Nanomedicine 8:1377–1383

Yean HC, Atong M, Chong KP (2009) Lettucenin A and its role against *Xanthomonas campestris*. J Agri Sci 1:66–70

Chapter 12
Bacillus in Agriculture: Past, Present, and Future

Manigundan Kaari, Kishore Kumar Annamalai, Mekala Panneerselvam Karthik Prakash, Venugopal Gopikrishnan, and Radhakrishnan Manikkam

Abstract *Bacillus* in agriculture has developed dramatically, from early discoveries in pest control to their roles in plant growth promotion, soil health improvement, and sustainable farming techniques. This chapter focuses on the morphological differences, historical landmarks, and many processes by which *Bacillus* improves crop production. The emphasis is on its application in biofertilizers, biopesticides, and biostimulants, as well as its synergistic interactions with other beneficial microorganisms and incorporation with current technologies such as synthetic biology and precision agriculture. The chapter finishes with an overview of future opportunities, difficulties, and developments in the agriculture industry, highlighting *Bacillus* as a key microorganism of environmentally friendly and resilient agricultural systems.

Keywords Agriculture · Bacillus · PGP and crop production

12.1 Introduction

The application of microorganisms in agriculture has become a crucial strategy for increasing nutrient availability and promoting plant health. Among these beneficial microbes, certain bacterial species inhabiting the rhizosphere, commonly known as

M. Kaari (✉)
Centre for Drug Discovery and Development, Sathyabama Institute of Science and Technology, Chennai, Tamil Nadu, India

Department of Applied Bioscience, Dong-A University, Busan, South Korea
e-mail: manigundan.cddd@sathyabama.ac.in

K. K. Annamalai · M. P. Karthik Prakash · V. Gopikrishnan · R. Manikkam
Centre for Drug Discovery and Development, Sathyabama Institute of Science and Technology, Chennai, Tamil Nadu, India

V. Gopikrishnan et al. (eds.), *Harnessing Microbial Resources for Sustainable Agriculture*, Microorganisms for Sustainability 61,
https://doi.org/10.1007/978-981-95-4817-0_12

rhizobacteria, play pivotal roles in stimulating plant growth, increasing crop yields, and mitigating plant diseases (de Andrade et al. 2023; Alzate Zuluaga et al. 2024). These rhizobacteria, which are capable of efficiently colonizing plant roots, contribute significantly to nutrient cycling throughout a plant developmental stage. In particular, plant growth-promoting rhizobacteria (PGPRs) have emerged as essential tools in sustainable agriculture, offering eco-friendly alternatives to chemical inputs. Modern agricultural practices increasingly incorporate plant growth-promoting bacteria (PGPB) to improve crop productivity in an environmentally sustainable manner. Extensive research has validated the use of PGPB as biostimulants, demonstrating their ability to increase plant resilience against abiotic stresses while improving growth quality and yield. Notably, bacterial genera such as *Bacillus*, *Pseudomonas*, and *Streptomyces* have been identified as effective plant growth promoters. These bacteria are predominantly found in the rhizosphere but can also inhabit aquatic environments and exist within plants as endophytes (Feng et al. 2025; Thepbandit and Athinuwat 2024).

The genus *Bacillus* is particularly important because of its unique characteristics. As gram-positive, spore-forming bacteria, *Bacillus* species exhibit remarkable resilience, surviving extreme environmental conditions through endospore formation. Their metabolic diversity enables them to thrive in various soil ecosystems, producing a wide range of bioactive compounds, such as antibiotics, lytic enzymes, phytohormones, and siderophores. These compounds not only suppress phytopathogens but also increase nutrient uptake and improve the soil structure, reinforcing their importance in sustainable agriculture. Moreover, the formation of biofilms by *Bacillus* strains facilitates effective root colonization, offering lasting benefits to crops and strengthening interactions with other beneficial soil microorganisms (Mwamburi et al. 2024; Hashem et al. 2019; Rafanomezantsoa et al. 2025).

Several traits make *Bacillus* species particularly attractive as PGPR. Their ability to form heat- and desiccation-resistant endospores allows for the creation of stable, long-lasting dry formulations, increasing their commercial viability (Çam et al. 2023). As common members of vegetable crop microflora, *Bacillus* species efficiently outcompete phytopathogens in the rhizosphere. Importantly, they produce a variety of nonribosomal lipopeptides (LPs), which are key contributors to their biocontrol activities. For example, *Bacillus velezensis* produces lipopeptides such as fengycin and bacillomycin D, which synergistically inhibit the growth of fungal pathogens such as *Fusarium oxysporum* (Jang et al. 2023; Tsotetsi et al. 2022; Petkova et al. 2025).

The agricultural potential of *Bacillus* strains extends beyond pathogen control. Their ability to solubilize phosphorus, fix atmospheric nitrogen, and mobilize potassium significantly increases soil fertility (Mahapatra et al. 2022). Additionally, the production of volatile organic compounds (VOCs) and phytohormones such as auxins, gibberellins, and cytokinins promotes root development and overall plant vigor. *Bacillus* species also increase plant resilience to abiotic stresses, including drought, salinity, and heavy metal contamination, while fostering synergistic relationships within the soil microbiome (Saxena et al. 2020; Bargaz et al. 2021). Owing to these multifunctional traits, *Bacillus*-based products such as biofertilizers, biopesticides, and biostimulants are increasingly adopted in sustainable agriculture, offering an eco-friendly alternative to conventional chemical products and contributing to improved soil health and agricultural productivity.

12.2 Uniqueness of *Bacillus*

The distinctive morphology of *Bacillus* species plays a critical role in promoting soil health and enhancing plant growth, offering significant advantages for sustainable agriculture. These bacteria are rod shaped and known for their ability to form endospores, granting them remarkable resilience against harsh environmental conditions such as drought, heat, ultraviolet radiation, and chemical exposure (Sales and Rigobelo 2024). The ability to produce durable endospores enables *Bacillus* strains to persist in soil environments for extended periods, maintaining consistent plant-beneficial activities even under adverse conditions. Their rod-shaped cellular structure facilitates efficient colonization of the rhizosphere, promoting close interaction with plant roots. Furthermore, *Bacillus* species are capable of forming biofilms, which serve as protective barriers against soilborne pathogens while enhancing nutrient acquisition by plants. Motility, supported by flagella, allows these bacteria to move toward root exudates, strengthening their association with plant hosts (Jagadeesan et al. 2024; Vasantha-Srinivasan et al. 2025).

In addition to their physical resilience, *Bacillus* strains contribute significantly to soil fertility by decomposing organic matter, solubilizing key nutrients such as phosphorus and potassium, and fixing atmospheric nitrogen (Wang et al. 2024). Their structural flexibility supports beneficial plant–microbe interactions, including the induction of systemic resistance against pathogens and the promotion of plant growth through the synthesis of phytohormones. The genus *Bacillus* encompasses more than 60 species capable of producing endospores resistant to extreme physical factors such as high temperatures, radiation, and dehydration. Moreover, several species generate chemical compounds that further support their application as biofertilizers. *Bacillus*-based products, including biofertilizers, biopesticides, and biostimulants, are widely utilized to improve soil fertility, increase plant resilience, and promote eco-friendly, sustainable agricultural practices (Blanco Crivelli et al. 2024; Samantaray et al. 2024).

12.3 Historical Perspective

12.3.1 Discovery and Early Research

The study of *Bacillus* strains dates back to the nineteenth century, with *B. subtilis* among the first species to be thoroughly investigated. A major milestone in agricultural science was the discovery of *B. thuringiensis* (Bt), renowned for its production of insecticidal crystal proteins that transformed pest management practices (Schnepf et al. 1998). Initial agricultural applications of *Bacillus* focused on the use of these bacteria as soil inoculants to increase plant nutrition. The ability of these bacteria to fix atmospheric nitrogen, solubilize phosphate, and promote a balanced soil microbiome makes them ideal candidates for the development of biofertilizers. As research has progressed, the unique potential of *Bacillus* species has become increasingly evident. The commercialization of *Bacillus*-based biopesticides, especially products derived from Bt, has played a pivotal role in the advancement of integrated pest management (IPM) systems. In addition to pest control, early agricultural scientists recognized the broader impact of *Bacillus* strains on soil fertility and plant health. By the mid-twentieth century, the isolation of diverse *Bacillus* strains with both plant growth-promoting and biocontrol capabilities laid the groundwork for the development of microbe-based fertilizers and protective agricultural formulations. Currently, *Bacillus*-based bioformulations have become vital components of organic and sustainable farming systems, supporting eco-friendly agricultural practices worldwide. Initial studies focused on the role of *Bacillus* species in nitrogen fixation and phosphate solubilization, key processes that increase nutrient availability in the soil and promote robust plant growth (Wu et al. 2025; Timofeeva et al. 2023; Bakki et al. 2024).

A major breakthrough occurred in 1901 with the discovery of *B. thuringiensis* (Bt), a species renowned for producing insecticidal crystal proteins (cry toxins) that selectively target insect larvae. This finding introduced a highly effective biological alternative to chemical pesticides, revolutionizing pest control practices in agriculture (Schnepf et al. 1998). Additional *Bacillus* species, such as *B. subtilis* and *B. amyloliquefaciens,* were identified for their strong biocontrol properties, producing antimicrobial compounds capable of suppressing soilborne pathogens, including fungi and bacteria. Early agricultural applications of *Bacillus* also included seed inoculation, where the introduction of beneficial strains promoted improved germination rates, enhanced root development, and increased crop yields. As the emphasis on sustainable agriculture has increased, *Bacillus*-based biofertilizers and soil amendments have been developed to reduce reliance on chemical fertilizers while simultaneously enhancing soil structure and microbial biodiversity. These pioneering efforts laid the groundwork for the development of modern *Bacillus*-based bioproducts. Today, such products are integral to integrated pest management (IPM) systems, organic farming, and regenerative agriculture, contributing significantly to crop health and environmental sustainability (Miljaković et al. 2020; Dhanabalan et al. 2024; Sarwar et al. 2018).

12.3.2 Present Applications in Agriculture

In contemporary agriculture, *Bacillus* strains have become essential for promoting plant growth, enhancing soil health, and providing effective biocontrol solutions. Their ability to form resilient endospores enables them to survive harsh environmental conditions, making them particularly suitable for development into biofertilizers and biopesticides. Key species such as *B. subtilis*, *B. amyloliquefaciens*, and *B. velezensis* are widely recognized for their ability to suppress plant pathogens through the production of potent antimicrobial compounds (Sales and Rigobelo 2024; Vasantha-Srinivasan et al. 2025). Moreover, *B. thuringiensis* (Bt) remains a cornerstone of biological pest control, offering targeted protection against insect pests through its production of insecticidal crystal proteins. In addition to pest management, *Bacillus* species significantly increase nutrient availability in soils by solubilizing phosphorus, fixing atmospheric nitrogen, and synthesizing phytohormones that stimulate root development and increase crop yields. These bacteria also contribute to soil fertility by decomposing organic matter, detoxifying harmful pollutants, and maintaining a balanced microbial ecosystem (Golnari et al. 2024; Ongena and Jacques 2008).

Moreover, *Bacillus* strains play crucial roles in helping crops withstand abiotic stresses such as drought and salinity, making them valuable in the development of climate-resilient agricultural systems. Advances in synthetic biology, nanotechnology, and precision agriculture have further improved the effectiveness of *Bacillus*-based bioformulations, offering innovative solutions that reduce reliance on chemical inputs while increasing sustainability and ensuring global food security.

12.3.2.1 Biocontrol and Disease Suppression

Biological control, which utilizes beneficial microorganisms, offers a sustainable approach for managing plant diseases and improving crop productivity. Considerable research has focused on identifying microbial biocontrol agents capable of suppressing phytopathogens, particularly those responsible for soil-borne diseases (Ayaz et al. 2023; Galli et al. 2024). Among these agents, many *Bacillus* strains have demonstrated strong biocontrol potential, making them valuable tools for sustainable agriculture. *Bacillus* species, including *B. subtilis*, *B. amyloliquefaciens*, and *B. velezensis*, employ multiple mechanisms to inhibit pathogenic microbes. These mechanisms include the production of diverse antimicrobial compounds such as lipopeptides, polyketides, and bacteriocins, the ability to form resilient endospores, the development of biofilms on plant root surfaces, and the induction of systemic resistance in host plants. These strategies not only suppress pathogens directly but also strengthen a plant's own defenses, enhancing resilience against infections (Miljaković et al. 2020).

Several antimicrobial compounds produced by *Bacillus* strains play critical roles in disease suppression. Cyclic lipopeptides such as surfactin, bacillomycin, and fengycin exhibit strong antifungal and antibacterial activities. Surfactin, in particular, is an amphiphilic molecule capable of disrupting the cell membranes of target pathogens by integrating into lipid layers, a mechanism well documented for its biocontrol efficacy. Studies have shown that surfactin also participates in signaling activities and surface tension reduction, contributing to biofilm formation and root colonization. Interestingly, the biosynthesis of certain lipopeptides is interdependent. For example, mutants of *B. subtilis* deficient in surfactin production were found to be unable to synthesize bacillomycin, whereas supplementation with external surfactin restored bacillomycin production. Furthermore, combined deficiencies in both surfactin and bacillomycin significantly reduced biocontrol effectiveness against fungi such as *Rhizoctonia solani*, highlighting the synergistic role of these compounds (Ongena and Jacques 2008).

In addition to lipopeptides, *Bacillus* strains also secrete exoenzymes such as chitinases, proteases, and glucanases, which degrade fungal and bacterial cell walls. The chitinase-producing *B. subtilis* strain significantly reduced disease incidence in tomato seedlings during both greenhouse and field trials. Additionally, *Bacillus* species produce volatile organic compounds (VOCs) that act as chemical signals, enhancing plant immunity and inhibiting pathogen growth. Similarly, VOCs produced by *B. subtilis* could suppress spore germination and hyphal growth of *Botrytis cinerea* in vitro. Although the role of VOCs in *in-plant* biocontrol remains to be fully confirmed, their contribution to the overall biocontrol repertoire of *Bacillus* is significant. Moreover, exogenous application of acetoin, a VOC produced by *Bacillus*, triggered induced systemic resistance (ISR) in maize plants against *Pseudomonas syringae*. Intriguingly, the presence of maize root exudates increased the expression of acetolactate synthase, a key enzyme in acetoin biosynthesis, suggesting that plant root secretions can stimulate beneficial metabolite production in *Bacillus* (Gomaa 2012).

While significant progress has been made in understanding the biocontrol mechanisms of *Bacillus* species, the complex interactions among beneficial microbes, host plants, and pathogens require further investigation. Nevertheless, the diverse secondary metabolites and versatile biocontrol strategies of *Bacillus* continue to make them indispensable for environmentally friendly plant disease management.

12.3.2.2 Plant Growth Promotion and Biofertilizers

Bacillus-based biofertilizers play a vital role in enhancing plant growth by improving nutrient availability and supporting plant health under various environmental conditions. These bacteria contribute to nutrient cycling by facilitating nitrogen fixation, solubilizing phosphorus, and mobilizing potassium, increasing the accessibility of essential elements to plants. In addition, *Bacillus* strains produce phytohormones such as auxins, gibberellins, and cytokinins, which stimulate root development, increase plant vigor, and promote overall crop productivity. The secretion of organic acids, including citric acid, acetic acid, and oxalic acid, enables *Bacillus* species to break down complex soil minerals, improving nutrient uptake efficiency. Enzymes such as phosphatases further assist in converting insoluble forms of phosphorus, such as tricalcium phosphate, aluminum phosphate, and iron phosphate, into soluble forms available for plant absorption. Factors such as the phosphate source, temperature, and presence of organic carriers influence the effectiveness of phosphorus solubilization. The use of phosphate-solubilizing *Bacillus* in microbial fertilizers not only increases crop yields but also reduces reliance on chemical phosphorus fertilizers, thereby lowering production costs and minimizing environmental pollution (Kumar et al. 2021a, b; Samantaray et al. 2024; Sales and Rigobelo 2024).

Current research focuses on unraveling the internal molecular mechanisms governing phosphate solubilization in *Bacillus*, screening and genetically modifying high-efficiency strains, and understanding their interactions with plants. Future directions emphasize multiomics integration, improving environmental adaptability, and exploring synergistic interactions within microbial communities to enhance the agricultural application of phosphate-solubilizing *Bacillus* (Zulfiqar et al. 2025; Pan and Cai 2023). In addition to nutrient mobilization, *Bacillus* strains are instrumental in mitigating abiotic stresses such as drought, salinity, and heavy metal toxicity. They achieve this goal by modulating stress-related gene expression and increasing antioxidant enzyme activity in plants, thus improving stress tolerance and ensuring better yields under suboptimal conditions. The production of plant hormones is another key trait of plant growth-promoting *Bacillus*. *B. subtilis* has been shown to promote soybean growth through auxin production, whereas *B. cereus*, *B. macrolides*, and *B. pumilus* contribute to red pepper growth by synthesizing gibberellins (Kimotho and Maina 2024; Akinsemolu et al. 2024). Despite extensive research, gaps remain in the precise identification of microbial plant hormone analogs owing to limitations in analytical techniques such as mass spectrometry and incomplete metabolite databases.

Another important plant growth-promoting mechanism involves the production of 1-aminocyclopropane-1-carboxylate (ACC) deaminase by *Bacillus* strains. This enzyme, a hormone that can inhibit growth under stress conditions, reduces ethylene levels in plants. For example, *B. licheniformis* HSW-16 improves salt stress tolerance in wheat, whereas *B. amyloliquefaciens* MMR04 and *B. subtilis* MMR18 protect *Pennisetum glaucum* under drought stress by degrading stress-induced ethylene (Shahid et al. 2023; Naing et al. 2021). Moreover, strains such as *B. subtilis* PM32Y, *B. cereus* WZ3S1, *Bacillus* SM73, and *Bacillus* WZ3S3 have been successfully used alongside alfalfa to remediate petroleum- and hydrocarbon-contaminated soils through phytoremediation (Gamalero and Glick 2024). Research is likely to focus on developing more efficient and stable *Bacillus* strains with enhanced ACC deaminase activity, as well as exploring their synergistic effects when combined with other beneficial microorganisms to further support sustainable agricultural practices.

12.3.2.3 Soil Health and Nutrient Cycling

Bacillus strains play crucial roles in enhancing soil health and promoting nutrient cycling, making them indispensable for sustainable agricultural practices. These bacteria contribute to soil fertility by decomposing organic matter, solubilizing essential nutrients, and fostering a balanced microbial ecosystem. Through the secretion of exoenzymes and organic acids, *Bacillus* species breakdown complex organic compounds into bioavailable forms, facilitating the release of key nutrients such as phosphorus, potassium, and nitrogen necessary for plant growth. Notably, species such as *B. megaterium* and *B. mucilaginosus* are particularly effective at solubilizing phosphate, transforming insoluble phosphorus into forms readily absorbed by plants (Samantaray et al. 2024; Saxena et al. 2020; Vasantha-Srinivasan et al. 2025).

Additionally, certain *Bacillus* strains participate in biological nitrogen fixation, reducing the reliance on synthetic nitrogen fertilizers and promoting a more sustainable nutrient supply. In addition to nutrient mobilization, *Bacillus* strains significantly improve soil structure. By producing biofilms and exopolysaccharides, they increase soil aggregation, water retention, and aeration, increasing the resistance of the soil to erosion and degradation. This structural improvement not only supports healthier plant growth but also enhances the overall resilience of agricultural systems. Furthermore, *Bacillus* species contribute to soil detoxification by breaking down harmful substances such as heavy metals and agrochemical residues, thereby maintaining a cleaner and more balanced soil environment. By improving nutrient cycling, increasing microbial diversity, and enhancing soil physical properties, *Bacillus* strains ensure the long-term sustainability of agricultural soils, ultimately leading to increased crop productivity and improved environmental stewardship (Mahapatra et al. 2022; Sales et al. 2024).

12.3.2.4 Direct and Indirect Mechanisms of *Bacillus* in Agriculture

Bacillus strains increase agricultural productivity through direct and indirect mechanisms that improve plant growth, soil health, and disease resistance. Directly, they solubilize essential nutrients such as phosphorus, potassium, and zinc, making them bioavailable to plants, while certain species fix atmospheric nitrogen, reducing the need for synthetic fertilizers. They also produce phytohormones such as auxins, gibberellins, and cytokinins, stimulating root growth and seed germination. Additionally, *Bacillus* strains improve the soil structure by producing exopolysaccharides (EPSs) and forming biofilms that increase water retention and aeration. Indirectly, they act as biocontrol agents, producing antimicrobial compounds that suppress fungal and bacterial pathogens while triggering induced systemic resistance (ISR) in plants, enhancing their natural defenses. *Bacillus* strains also release volatile organic compounds (VOCs) that promote plant growth and detoxify soil contaminants by breaking down agrochemical residues and heavy metals. These combined mechanisms make *Bacillus* an essential bioagent for sustainable agriculture, reducing chemical dependency and improving soil fertility while ensuring crop resilience against environmental stresses (Tsotetsi et al. 2022; Rafanomezantsoa et al. 2025; Meena et al. 2016; Liu et al. 2023).

12.3.2.5 Interaction Between *Bacillus* Strains in the Soil and the Plant Environment

Bacillus strains interact dynamically within the soil and plant environment, forming complex relationships that increase plant growth, nutrient availability, and disease resistance. In the rhizosphere, *Bacillus* strains establish mutualistic associations with plant roots, colonizing root surfaces and forming biofilms that protect plants from pathogens while promoting beneficial microbial communities. These bacteria communicate through quorum sensing, coordinating their activities, such as biofilm formation, enzyme production, and antibiotic secretion. *Bacillus* strains also exhibit synergistic interactions, where different species complement each other's functions, improving nutrient cycling and plant growth promotion. For example, *Bacillus subtilis* and *Bacillus megaterium* work together to solubilize phosphorus and mobilize potassium, making these essential nutrients more accessible to plants (Nishisaka et al. 2024). Additionally, *Bacillus* strains compete with soil-borne pathogens through antagonistic interactions, producing antimicrobial compounds that suppress harmful fungi and bacteria and reduce the incidence of diseases such as *Fusarium* wilt and bacterial blight. These bacteria also interact with other beneficial microbes, such as nitrogen-fixing bacteria and mycorrhizal fungi, enhancing soil fertility and promoting plant resilience to abiotic stresses such as drought and salinity. Through these interactions, *Bacillus* strains contribute to a balanced soil microbiome, fostering a healthier plant environment and ensuring sustainable agricultural productivity (Xie et al. 2024; Boubsi et al. 2023; Ali et al. 2024).

12.3.2.6 Interaction Between *Bacillus* and Other Beneficial Microbes for Plant Health

Bacillus strains interact synergistically with other beneficial microbes in the rhizosphere, enhancing plant health, soil fertility, and resilience to environmental stress. They establish mutualistic relationships with mycorrhizal fungi, such as *arbuscular mycorrhizal fungi* (AMF), by improving root colonization and nutrient exchange (Khaliq et al. 2022). *Bacillus* produces exoenzymes and organic acids that help mycorrhizal fungi access phosphorus and other nutrients, leading to improved plant growth. Additionally, *Bacillus* enhances the efficiency of nitrogen-fixing bacteria, such as *Rhizobium* and *Azospirillum*, by creating a more favorable soil environment through biofilm formation and exopolysaccharide production, ensuring better nitrogen availability for plants. *Bacillus* also interacts with plant growth-promoting rhizobacteria (PGPRs), such as *Pseudomonas* and *Actinobacteria*, which work together to suppress pathogens through antibiotic production, competition for nutrients, and quorum sensing regulation. This microbial cooperation strengthens induced systemic resistance (ISR) in plants, increasing their resistance to fungal and bacterial infections. Moreover, *Bacillus* strains play a role in soil microbiome stabilization, fostering a balance between microbial communities that contribute to organic matter decomposition and soil structure improvement. The interactions of these bacteria with phosphate-solubilizing bacteria (PSB) ensure efficient nutrient mobilization, whereas coinoculation with biocontrol agents further enhances plant protection against soil-borne diseases (Sagar et al. 2021; Laishram et al. 2024). By forming these beneficial microbial alliances, *Bacillus* strains contribute to a more resilient, nutrient-rich, and disease-free plant environment, supporting sustainable and eco-friendly agricultural practices.

12.4 *Bacillus*-Based Synthetic Community for Increased Productivity

The *Bacillus*-based synthetic community is an engineered consortium of multiple *Bacillus* strains that are carefully selected to work together *to increase* agricultural productivity by increasing plant growth, soil fertility, and disease resistance. These communities are designed to perform multiple functions *synergistically*, such as nutrient solubilization, biocontrol, stress tolerance, and phytohormone production, offering a more robust and sustainable alternative to single-strain applications (Tariq et al. 2025). By combining different functional *Bacillus* strains, such as *B. subtilis* for root colonization and pathogen suppression, *B. megaterium* for phosphorus solubilization, and *B. amyloliquefaciens* for biotic stress resistance, synthetic communities can maximize crop yield and soil health. These strains communicate via quorum sensing, regulating their activity on the basis of plant needs and environmental conditions. Additionally, *Bacillus* synthetic communities increase microbial diversity in the soil, ensuring a balanced microbiome that promotes sustainable plant growth. They improve abiotic stress tolerance, such as resistance to drought and salinity, by producing osmoprotectants and exopolysaccharides that help plants retain moisture and nutrients. Furthermore, these communities reduce dependency on chemical fertilizers and pesticides, making agriculture more eco-friendly and cost-effective (Ali et al. 2025; Zhang et al. 2023). With advances in synthetic biology and precision microbiome engineering, *Bacillus* synthetic communities are being optimized for crop-specific applications, ensuring long-term soil fertility and increased agricultural productivity while promoting sustainable farming practices.

12.4.1 Commercial Bacillus-*Based Products*

Bacillus-based commercial products have gained widespread use in sustainable agriculture because of their role in biocontrol, biofertilization, plant growth promotion, and stress tolerance. These products are widely available in the form of biopesticides, biofertilizers, and biostimulants, reducing the need for chemical inputs while improving soil and crop health. *Bacillus* strains are extensively used in biopesticides because of their ability to suppress plant pathogens through the production of antimicrobial compounds, enzymes, and biofilms (Wu et al. 2025). Some of the most widely used *Bacillus*-based biopesticides include *B. thuringiensis,* which is among the most well-known microbial insecticides and produces Cry and Vip proteins, which act as toxins against insect larvae, especially Lepidoptera (caterpillars), Diptera (mosquitoes), and Coleoptera (beetles) (Palma et al. 2014). Bt-based pesticides are used in crops such as cotton, corn, soybean, and vegetables to combat pests such as bollworms, armyworms, and rootworms. Commercial Bt products include Dipel (Bt var. *kurstaki*), which controls caterpillars in fruits and vegetables. Xentari targets caterpillar pests in forestry and horticulture. Another product, thuricide, is a widely used garden pesticide for caterpillar control (Roh et al. 2007; Razzaq et al. 2023). Similarly, *B. subtilis* and *B. amyloliquefaciens* biopesticides are known for their ability to suppress fungal and bacterial diseases through the production of lipopeptides, antibiotics, and volatile organic compounds (VOCs). They are used against diseases such as powdery mildew, *Fusarium* wilt, *Pythium* root rot, and bacterial blight (Jiao et al. 2021; Ajuna et al. 2024). Popular commercial products, including Serenade® (*B. subtilis* QST 713), are widely used biofungicides that are effective against fungal and bacterial plant pathogens (Lahlali et al. 2011). Double nickel® (*B. amyloliquefaciens* D747) is another biopesticide used for managing soil-borne and foliar diseases (Pethybridge et al. 2019). Similarly, Companion® is a *Bacillus*-based biofungicide that helps suppress root diseases (Nihorimbere et al. 2024). Additionally, Sonata® was developed by *B. pumilus* QST 2808), which controls fungal diseases such as powdery mildew, and Rhizovital® was developed by *B. velezensis* FZB42, which is also a biopesticide that promotes root growth (Radhakrishnan et al. 2017).

Bacillus-based biofertilizers improve soil fertility by solubilizing phosphorus, fixing nitrogen, and *increasing* nutrient uptake. They act as eco-friendly alternatives to chemical fertilizers (Nosheen et al. 2021; Daniel et al. 2022). *Bacillus* spp. such as *B. megaterium* and *B. mucilaginosus* solubilize inorganic phosphorus, making it available to plants. Common products include Biophos®, a *Bacillus*-based phosphorus-solubilizing biofertilizer used for improving plant nutrition, and PhosphoMax, another product that increases phosphorus availability in acidic soils (Prakash and Arora 2019). Some *Bacillus* species, such as *B. polymyxa*, contribute to biological nitrogen fixation, reducing the need for synthetic nitrogen fertilizers. For example, NitroBac is a *Bacillus*-based nitrogen-fixing biofertilizer for legumes and cereals (Li et al. 2022). Moreover, *B. circulans* and *B. mucilaginosus* promote potassium mobilization, increasing plant access to potassium. PotashGrow is a

Bacillus-based potassium solubilizer that improves crop nutrient absorption (Pramanik et al. 2019; Yasin et al. 2016).

Biostimulants, which include *Bacillus* species, play crucial roles in increasing plant stress tolerance, root development, and overall crop productivity. These products help improve plant resilience to environmental stressors such as drought, salinity, and temperature fluctuations. *Bacillus* strains produce substances such as exopolysaccharides (EPS), osmoprotectants, and stress-related proteins, which help plants cope with abiotic stress (Lahlali et al. 2022; Ali et al. 2024; Etesami et al. 2023). A notable example is Rhizovital®, derived from *B. velezensis* FZB42, which promotes root growth and enhances drought tolerance (Chowdhury et al. 2013; Fan et al. 2018). Another product, ProAct®, which was developed from *B. subtilis*, enhances plant defense mechanisms. *Bacillus* strains also stimulate root elongation through the production of auxins and other phytohormones, and they are commonly used in seed treatments to facilitate germination and protect seedlings from soil pathogens. TagTeam®, a *Bacillus*-based seed inoculant, enhances seedling growth by increasing nutrient uptake, whereas Optimize® promotes early root development in cereals and legumes through seed coating. Despite these benefits, the adoption of *Bacillus*-based formulations has been slow, primarily owing to a limited understanding of the chemical and biochemical mechanisms underlying their actions, as well as challenges in defining their effectiveness under field conditions. The performance of these rhizobacteria often varies between controlled environments and real-world field settings. There is a pressing need for more fundamental research to explore the metabolism and biochemistry of *Bacillus* strains, as well as to elucidate the biochemical and molecular interactions between *Bacillus* and plants at different levels, encompassing both bacterial and host plant systems.

12.4.2 **Bacillus** *in Precision and Smart Agriculture*

With advances in biotechnology, *Bacillus*-based products are now integrated with precision agriculture, nanotechnology, and bioinformatics to increase efficiency. Nanoencapsulated *Bacillus* products improve the shelf life, stability, and controlled release of *Bacillus* strains. Smart microbial formulations are tailored *Bacillus* consortia designed for specific crops and environmental conditions to optimize plant growth. *Bacillus*-based commercial products play a vital role in modern sustainable agriculture, offering effective alternatives to chemical pesticides and fertilizers. From biopesticides such as Bt formulations to biofertilizers that improve nutrient availability and biostimulants that *increase* plant resilience, these products are key to eco-friendly farming practices. As technology advances, precision agriculture and synthetic biology will further optimize *Bacillus*-based formulations, making agriculture more efficient, productive, and environmentally sustainable (Kishorkumar et al. 2024; Gómez-Lama Cabanás and Mercado-Blanco 2025; Kumar et al. 2021a, b).

12.5 Future Prospects: Innovations and Challenges

12.5.1 Advances in Bacillus*-Based Bioformulations*

The future prospects of *Bacillus*-based bioformulations are promising, driven by advancements in synthetic biology, precision agriculture, and biotechnological innovations. Researchers are developing genetically engineered *Bacillus* strains with enhanced traits for improved nutrient solubilization, biocontrol efficacy, and abiotic stress tolerance. CRISPR-based gene editing has been explored to fine-tune *Bacillus* metabolic pathways, optimizing their ability to produce phytohormones, antimicrobial compounds, and stress-mitigating enzymes (Guo et al. 2025). Additionally, *nanoencapsulation* technologies enhance the stability and efficiency of *Bacillus* bioformulations, ensuring controlled release and prolonged field activity. The integration of smart microbial consortia, where multiple *Bacillus* strains are combined for complementary functions, is also gaining traction, improving soil health, plant resilience, and crop productivity. However, challenges remain, including strain stability under varying environmental conditions, large-scale production costs, and regulatory approvals for commercial applications. Addressing these challenges through advanced formulation techniques, precision microbiome engineering, and sustainable scaling strategies will enable *Bacillus*-based bioformulations to become a key component of future eco-friendly and high-yield agricultural systems.

12.5.2 Bacillus *Synthetic Community for Future Agricultural Productivity*

The *Bacillus* synthetic community is an emerging approach in agriculture aimed at enhancing crop productivity, soil health, and plant resilience by leveraging the synergistic interactions of multiple *Bacillus* strains. Unlike single-strain applications, synthetic communities are designed to combine diverse *Bacillus* species with complementary functions, such as nutrient solubilization, biocontrol, stress resistance, and plant growth promotion (Tariq et al. 2025; Vasantha-Srinivasan et al. 2025). For example, *B. subtilis* improves root colonization and disease suppression, whereas *B. megaterium* enhances phosphorus solubilization, and *B. amyloliquefaciens* produces antifungal compounds. These communities work collectively, improving soil microbiome stability, reducing pathogen pressure, and increasing plant tolerance to abiotic stresses such as drought and salinity (Etesami et al. 2023; Singh and Shyu 2024; Maslennikova et al. 2023). Advances in synthetic biology and precision microbiome engineering are further optimizing these communities for crop-specific applications, ensuring targeted nutrient delivery and sustainable plant growth. However, challenges such as strain compatibility, formulation stability, and large-scale application feasibility need to be addressed for widespread adoption. As agricultural practices shift toward *biobased* solutions, *Bacillus* synthetic communities have the potential to revolutionize sustainable farming, reducing dependency on chemical inputs while increasing productivity and environmental resilience.

12.5.3 Integration with Modern Technologies

The integration of *Bacillus*-based solutions with modern agricultural technologies is transforming sustainable farming by increasing efficiency, precision, and adaptability. Advances in synthetic biology, nanotechnology, and precision agriculture are optimizing *Bacillus* bioformulations for improved nutrient delivery, biocontrol, and stress resistance. CRISPR gene editing is being explored to enhance *Bacillus* strains, tailoring their metabolic pathways for *increased* enzyme production, improved nitrogen fixation, and *increased* pathogen resistance. *Nanoencapsulation* techniques improve the stability and controlled release of *Bacillus*-based biofertilizers and biopesticides, ensuring a longer shelf life and more effective field performance (Sales et al. 2024; Ragasruthi et al. 2024). In precision agriculture, *Bacillus* inoculants are integrated with sensor-based soil monitoring systems, allowing farmers to apply microbial bioformulations only where and when needed, reducing waste and maximizing benefits. The use of AI and big data analytics further refines *Bacillus* applications by predicting soil health trends and optimizing microbial consortia for specific crops and environmental conditions. Despite these advancements, scaling production, maintaining strain viability *under* diverse conditions, and *providing* regulatory *approval* remain challenges. However, as biotechnology continues to evolve, *Bacillus*-based formulations will play a critical role in next-generation sustainable agriculture, reducing dependency on chemical inputs while ensuring higher crop yields and soil fertility.

12.5.4 Challenges and Limitations

Despite the wide-ranging benefits of *Bacillus* spp. in agriculture, several challenges and limitations hinder their full-scale adoption in agricultural research and commercial applications. One of the primary challenges is strain stability and survival under diverse environmental conditions, as *Bacillus* performance can be affected by soil pH, moisture levels, temperature fluctuations, and microbial competition. Another key limitation is inconsistency in field efficacy, where results from controlled laboratory studies do not always translate to real-world agricultural settings, leading to variable outcomes. Additionally, the formulation and large-scale production of *Bacillus*-based biofertilizers and biopesticides require advanced stabilization techniques, such as nanoencapsulation, to ensure longer shelf life and viability. Regulatory barriers and approval processes also slow the commercialization of *Bacillus*-based products, as stringent guidelines are needed to assess their safety, environmental impact, and efficiency before market entry. Furthermore, *interactions* with native soil microbiota pose another challenge, as introduced *Bacillus* strains may not always integrate well with existing microbial communities, affecting their effectiveness. Finally, farmer awareness and adoption remain concerns, as many growers are still reliant on conventional chemical inputs owing to a lack of knowledge and accessibility to *Bacillus*-based alternatives. Addressing these challenges through advanced formulation technologies, regulatory simplifications, and field trials will be crucial to unlocking the full potential of *Bacillus* in sustainable agriculture.

12.6 Conclusion

Bacillus strains have played a transformative role in agriculture, evolving from early discoveries in biocontrol and soil enrichment to their current widespread use in biofertilizers, biopesticides, and biostimulants. Historically, *Bacillus* species such as *B. thuringiensis* have revolutionized biological pest control, whereas other strains have contributed to soil fertility and *increased* plant growth. Currently, advancements in *Bacillus*-based formulations have made them a critical component of sustainable agriculture, reducing the dependence on synthetic fertilizers and pesticides while improving soil health, plant resilience, and crop productivity. With increasing research, *Bacillus*-based products are now integrated into precision agriculture, synthetic biology, and nanotechnology, offering more efficient and targeted agricultural solutions. However, challenges such as strain stability, large-scale production, regulatory hurdles, and field performance consistency remain barriers to widespread adoption. In the future, genetic engineering, AI-driven microbiome optimization, and next-generation bioformulations will further increase the effectiveness of *Bacillus* applications, driving eco-friendly and high-yield agricultural systems. As agriculture transitions toward biological solutions, *Bacillus* will continue to be a key player in promoting global food security, sustainable farming, and environmental resilience.

References

Ajuna HB, Lim HI, Moon JH, Won SJ, Choub V, Choi SI et al (2024) The prospect of antimicrobial peptides from *Bacillus* species with biological control potential against insect pests and diseases of economic importance in agriculture, forestry and fruit tree production. Biotechnol Biotechnol Equip 38(1). https://doi.org/10.1080/13102818.2024.2312115

Akinsemolu AA, Onyeaka H, Odion S, Adebanjo I (2024) Exploring *Bacillus subtilis*: ecology, biotechnological applications, and future prospects. J Basic Microbiol 64(6):e2300614. https://doi.org/10.1002/jobm.202300614

Ali S, Akhtar MS, Siraj M, Zaman W (2024) Molecular communication of microbial plant biostimulants in the rhizosphere under abiotic stress conditions. Int J Mol Sci 25(22):12424. https://doi.org/10.3390/ijms252212424

Ali Q, Khan AR, Yujie W, Chenjie Y, Zhao Q, Ayaz M, Raza W, Gu Q, Wu H, Mundra S, Gao X (2025) Antimicrobial metabolites of *Bacillus* velezensis FZB42 reshape rice rhizosphere microbial community composition and induce host resistance against *Rhizoctonia solani*. Curr Plant Biol *41*:100440. https://doi.org/10.1016/j.cpb.2025.100440

Alzate Zuluaga MY, Fattorini R, Cesco S, Pii Y (2024) Plant–microbe interactions in the rhizosphere for smarter and more sustainable crop fertilization: the case of PGPR-based biofertilizers. Front Microbiol 15:1440978. https://doi.org/10.3389/fmicb.2024.1440978

Ayaz M, Li C-H, Ali Q, Zhao W, Chi Y-K, Shafiq M, Ali F, Yu X-Y, Yu Q, Zhao J-T et al (2023) Bacterial and fungal biocontrol agents for plant disease protection: journey from lab to field, current status, challenges, and global perspectives. Molecules 28(18):6735. https://doi.org/10.3390/molecules28186735

Bakki M, Banane B, Marhane O, Esmaeel Q, Hatimi A, Barka EA, Azim K, Bouizgarne B (2024) Phosphate solubilizing *Pseudomonas* and *Bacillus* combined with rock phosphates promoting tomato growth and reducing bacterial canker disease. Front Microbiol 15:1289466. https://doi.org/10.3389/fmicb.2024.1289466

Bargaz A, Elhaissoufi W, Khourchi S, Benmrid B, Borden KA, Rchiad Z (2021) Benefits of phosphate solubilizing bacteria on belowground crop performance for improved crop acquisition of phosphorus. Microbiol Res 252:126842. https://doi.org/10.1016/j.micres.2021.126842

Blanco Crivelli X, Cundon C, Bonino MP, Sanin MS, Bentancor A (2024) The complex and changing genus *Bacillus*: a diverse bacterial powerhouse for many applications. Bacteria 3(3):256–270. https://doi.org/10.3390/bacteria3030017

Boubsi F, Hoff G, Arguelles Arias A, Steels S, Andrić S, Anckaert A, Roulard R, Rigolet A, van Wuytswinkel O, Ongena M (2023) Pectic homogalacturonan sensed by *Bacillus* acts as host associated cue to promote establishment and persistence in the rhizosphere. iScience 26(10):107925. https://doi.org/10.1016/j.isci.2023.107925

Çam S, Küçük Ç, Almaca A (2023) *Bacillus* strains exhibit various plant growth promoting traits and their biofilm-forming capability correlates to their salt stress alleviation effect on maize seedlings. J Biotechnol 369:35–42. https://doi.org/10.1016/j.jbiotec.2023.05.004

Chowdhury SP, Dietel K, Rändler M, Schmid M, Junge H, Borriss R et al (2013) Effects of *Bacillus amyloliquefaciens* FZB42 on lettuce growth and health under pathogen pressure and its impact on the rhizosphere bacterial community. PloS One 8(7):e68818. https://doi.org/10.1371/journal.pone.0068818

Daniel AI, Fadaka AO, Gokul A, Bakare OO, Aina O, Fisher S, Burt AF, Mavumengwana V, Keyster M, Klein A (2022) Biofertilizer: the future of food security and food safety. Microorganisms 10(6):1220. https://doi.org/10.3390/microorganisms10061220

de Andrade LA, Santos CHB, Frezarin ET, Sales LR, Rigobelo EC (2023) Plant growth-promoting rhizobacteria for sustainable agricultural production. Microorganisms 11(4):1088. https://doi.org/10.3390/microorganisms11041088

Dhanabalan S, Muthusamy K, Iruthayasamy J, Kumaresan PV, Ravikumar C, Kandasamy R, Natesan S, Periyannan S (2024) Unleashing *Bacillus* species as versatile antagonists: harnessing the biocontrol potentials of the plant growth-promoting rhizobacteria to combat Macrophomina phaseolina infection in Gloriosa superba. Microbiol Res 283:127678. https://doi.org/10.1016/j.micres.2024.127678

Etesami H, Jeong BR, Glick BR (2023) Potential use of *Bacillus* spp. as an effective biostimulant against abiotic stresses in crops-a review. Curr Res Biotechnol 5:100128. https://doi.org/10.1016/j.crbiot.2023.100128

Fan B, Wang C, Song X, Ding X, Wu L, Wu H, Gao X, Borriss R (2018) *Bacillus velezensis* FZB42 in 2018: the gram-positive model strain for plant growth promotion and biocontrol. Front Microbiol 9:2491. https://doi.org/10.3389/fmicb.2018.02491

Feng Q, Luo Y, Liang M, Cao Y, Wang L, Liu C, Zhang X, Ren L, Wang Y, Wang D, Zhu Y, Zhang Y, Xiao B, Li N (2025) Rhizobacteria protective hydrogel to promote plant growth and adaptation to acidic soil. Nat Commun 16(1):1684. https://doi.org/10.1038/s41467-025-56988-3

Galli M, Feldmann F, Vogler UK et al (2024) Can biocontrol be the game-changer in integrated pest management? A review of definitions, methods and strategies. J Plant Dis Prot 131:265–291. https://doi.org/10.1007/s41348-024-00878-1

Gamalero E, Glick BR (2024) Use of plant growth-promoting bacteria to facilitate phytoremediation. AIMS Microbiol 10(2):415–448. https://doi.org/10.3934/microbiol.2024021

Golnari M, Bahrami N, Milanian Z, Rabbani Khorasgani M, Asadollahi MA, Shafiei R, Fatemi SS (2024) Isolation and characterization of novel *Bacillus* strains with superior probiotic potential: comparative analysis and safety evaluation. Sci Rep 14(1):1457. https://doi.org/10.1038/s41598-024-51823-z

Gomaa EZ (2012) Chitinase production by *Bacillus thuringiensis* and *Bacillus licheniformis*: their potential in antifungal biocontrol. J Microbiol 50:103–111. https://doi.org/10.1007/s12275-012-1343-y

Gómez-Lama Cabanás C, Mercado-Blanco J (2025) Groundbreaking technologies and the biocontrol of fungal vascular plant pathogens. J Fungi 11(1):77. https://doi.org/10.3390/jof11010077

Guo N, Wang S, Whitfield CT, Batchelor WD, Wang Y, Blersch D, Higgins BT, Feng Y, Liles MR, de-Bashan LE, Wang Y, Ma Y (2025) High-efficiency CRISPR-Cas9 genome editing unveils biofilm insights and enhances antimicrobial activity in *Bacillus* velezensis FZB42. Biotechnol Bioeng 122(4):983–994. https://doi.org/10.1002/bit.28933

Hashem A, Tabassum B, Fathi Abd Allah E (2019) *Bacillus subtilis*: a plant-growth promoting rhizobacterium that also impacts biotic stress. Saudi J Biol Sci 26(6):1291–1297. https://doi.org/10.1016/j.sjbs.2019.05.004

Jagadeesan Y, Meenakshisundaram S, Pichaimuthu S, Balaiah A (2024) A scientific version of understanding "why did the chickens cross the road"? – a guided journey through *Bacillus* spp. toward sustainable agriculture, circular economy and biofortification. Environ Res 244:117907. https://doi.org/10.1016/j.envres.2023.117907

Jang S, Choi SK, Zhang H, Zhang S, Ryu CM, Kloepper JW (2023) History of a model plant growth-promoting rhizobacterium, *Bacillus velezensis* GB03: from isolation to commercialization. Front Plant Sci 14:1279896. https://doi.org/10.3389/fpls.2023.1279896

Jiao R, Cai Y, He P, Munir S, Li X, Wu Y, Wang J, Xia M, He P, Wang G, Yang H, Karunarathna SC, Xie Y, He Y (2021) *Bacillus amyloliquefaciens* YN201732 produces Lipopeptides with promising biocontrol activity against fungal pathogen *Erysiphe cichoracearum*. Front Cell Infect Microbiol 11:598999. https://doi.org/10.3389/fcimb.2021.598999

Khaliq A, Perveen S, Alamer KH, Zia Ul Haq M, Rafique Z, Alsudays IM, Althobaiti AT, Saleh MA, Hussain S, Attia H (2022) Arbuscular mycorrhizal fungi symbiosis to enhance plant–soil interaction. Sustainability 14(13):7840. https://doi.org/10.3390/su14137840

Kimotho RN, Maina S (2024) Unraveling plant–microbe interactions: can integrated omics approaches offer concrete answers? J Exp Bot 75(5):1289–1313. https://doi.org/10.1093/jxb/erad448

Kishorkumar C, Harish S, Karthikeyan G, Sharmila DJS, Nivedha M (2024) Harnessing Nanoencapsulated *Bacillus* spp. consortia to combat groundnut bud necrosis Orthotospovirus in tomato. ACS Appl Mater Interfaces 16(9):11185–11193. https://doi.org/10.1021/acsami.3c16145

Kumar S, Diksha, Sindhu SS, Kumar R (2021a) Biofertilizers: an ecofriendly technology for nutrient recycling and environmental sustainability. Curr Res Microb Sci 3:100094. https://doi.org/10.1016/j.crmicr.2021.100094

Kumar P, Pandhi S, Mahato DK et al (2021b) *Bacillus*-based nanobioformulations for phytopathogens and insect–pest management. Egypt J Biol Pest Control 31:128. https://doi.org/10.1186/s41938-021-00475-6

Lahlali R, Peng G, McGregor L, Gossen BD, Hwang SF, McDonald M (2011) Mechanisms of the biofungicide Serenade (*Bacillus subtilis* QST713) in suppressing clubroot. Biocontrol Sci Tech 21(11):1351–1362. https://doi.org/10.1080/09583157.2011.618263

Lahlali R et al (2022) *Bacillus* spp.-mediated drought stress tolerance in plants: current and future prospects. In: Islam MT, Rahman M, Pandey P (eds) Bacilli in agrobiotechnology. Bacilli in climate resilient agriculture and bioprospecting. Springer, Cham. https://doi.org/10.1007/978-3-030-85465-2_21

Laishram B, Devi OR, Dutta R, Senthilkumar T, Goyal G, Paliwal DK, Panotra N, Rasool A (2024) Plant–microbe interactions: PGPM as microbial inoculants/biofertilizers for sustaining crop productivity and soil fertility. Curr Res Microb Sci 8:100333. https://doi.org/10.1016/j.crmicr.2024.100333

Li Q, Liu S, Li Y et al (2022) Nitrogen fixation by *PaeniBacillus polymyxa* WLY78 is responsible for cucumber growth promotion. Plant Soil 473:507–516. https://doi.org/10.1007/s11104-022-05307-6

Liu Y, Yue Z, Sun Z, Li C (2023) Harnessing native *Bacillus* spp. for sustainable wheat production. Appl Environ Microbiol 89(2):e0124722. https://doi.org/10.1128/aem.01247-22

Mahapatra S, Yadav R, Ramakrishna W (2022) *Bacillus subtilis* impact on plant growth, soil health and environment: Dr. Jekyll and Mr. Hyde. J Appl Microbiol 132(5):3543–3562. https://doi.org/10.1111/jam.15480

Maslennikova VS, Tsvetkova VP, Shelikhova EV, Selyuk MP, Alikina TY, Kabilov MR, Dubovskiy IM (2023) *Bacillus subtilis* and *Bacillus amyloliquefaciens* mix suppresses Rhizoctonia disease and improves rhizosphere microbiome, growth and yield of potato (*Solanum tuberosum* L.). J Fungi (Basel, Switzerland) 9(12):1142. https://doi.org/10.3390/jof9121142

Meena VS et al (2016) Can *Bacillus* species enhance nutrient availability in agricultural soils? In: Islam M, Rahman M, Pandey P, Jha C, Aeron A (eds) Bacilli and agrobiotechnology. Springer, Cham. https://doi.org/10.1007/978-3-319-44409-3_16

Miljaković D, Marinković J, Balešević-Tubić S (2020) The significance of *Bacillus* spp. in disease suppression and growth promotion of field and vegetable crops. Microorganisms 8(7):1037. https://doi.org/10.3390/microorganisms8071037

Mwamburi SM, Islam SI, Dinh-Hung N, Dangsawat O, Sowanpreecha R, Khang LTP, Montha N, Therdtatha P, Dwinanti SH, Permpoonpattana P, Linh NV (2024) Genomic characterization of *Bacillus* sp. THPS1: a hot spring-derived species with functional features and biotechnological potential. Microorganisms 12(12):2476. https://doi.org/10.3390/microorganisms12122476

Naing AH, Maung TT, Kim CK (2021) The ACC deaminase-producing plant growth-promoting bacteria: influences of bacterial strains and ACC deaminase activities in plant tolerance to abiotic stress. Physiol Plant 173(4):1992–2012. https://doi.org/10.1111/ppl.13545

Nihorimbere G, Korangi Alleluya V, Nimbeshaho F, Nihorimbere V, Legrève A, Ongena M (2024) *Bacillus*-based biocontrol beyond chemical control in Central Africa: the challenge of turning myth into reality. Front Plant Sci 15:1349357. https://doi.org/10.3389/fpls.2024.1349357

Nishisaka CS, Ventura JP, Bais HP, Mendes R (2024) Role of *Bacillus subtilis* exopolymeric genes in modulating rhizosphere microbiome assembly. Environ Microbiome 19(1):33. https://doi.org/10.1186/s40793-024-00567-4

Nosheen S, Ajmal I, Song Y (2021) Microbes as biofertilizers, a potential approach for sustainable crop production. Sustainability 13(4):1868. https://doi.org/10.3390/su13041868

Ongena M, Jacques P (2008) *Bacillus* lipopeptides: versatile weapons for plant disease biocontrol. Trends Microbiol 16(3):115–125. https://doi.org/10.1016/j.tim.2007.12.009

Palma L, Muñoz D, Berry C, Murillo J, Caballero P (2014) *Bacillus thuringiensis* toxins: an overview of their biocidal activity. Toxins 6(12):3296–3325. https://doi.org/10.3390/toxins6123296

Pan L, Cai B (2023) Phosphate-solubilizing bacteria: advances in their physiology, molecular mechanisms and microbial community effects. Microorganisms 11(12):2904. https://doi.org/10.3390/microorganisms11122904

Pethybridge SJ, Gugino BK, Kikkert JR (2019) Efficacy of double Nickel LC (*Bacillus* amyloliquefaciens D747 strain) for management of white mold in snap and dry bean. Plant Health Prog 20:61–66

Petkova M, Marcheva M, Petrova A-L, Slavova V, Shilev S (2025) Plant growth-promoting and biocontrol characteristics of four *Bacillus* strains and evaluation of their effects on wheat (*Tr. Aestivum* L.). Int J Plant Biol 16(1):1. https://doi.org/10.3390/ijpb16010001

Prakash J, Arora NK (2019) Phosphate-solubilizing *Bacillus* sp. enhances growth, phosphorus uptake and oil yield of *Mentha arvensis* L. 3 Biotech 9(4):126. https://doi.org/10.1007/s13205-019-1660-5

Pramanik P, Goswami AJ, Ghosh S, Kalita C (2019) An indigenous strain of potassium-solubilizing bacteria *Bacillus* pseudomycoides enhanced potassium uptake in tea plants by increasing potassium availability in the mica waste-treated soil of North–East India. J Appl Microbiol 126(1):215–222. https://doi.org/10.1111/jam.14130

Radhakrishnan R, Hashem A, Abd Allah EF (2017) *Bacillus*: a biological tool for crop improvement through bio-molecular changes in adverse environments. Front Physiol 8:667. https://doi.org/10.3389/fphys.2017.00667

Rafanomezantsoa P, El-Hasan A, Voegele RT (2025) Potential of *Bacillus halotolerans* in mitigating biotic and abiotic stresses: a comprehensive review. Stress (Amsterdam Neth) 5(2):24. https://doi.org/10.3390/stresses5020024

Ragasruthi M, Balakrishnan N, Murugan M, Swarnakumari N, Harish S, Sharmila DJS (2024) *Bacillus thuringiensis* (Bt)-based biopesticide: navigating success, challenges, and future horizons in sustainable pest control. Sci Total Environ 954:176594. https://doi.org/10.1016/j.scitotenv.2024.176594

Razzaq A, Zafar MM, Ali A, Li P, Qadir F, Zahra LT, Shaukat F, Laghari AH, Yuan Y, Gong W (2023) Biotechnology and solutions: insect-pest-resistance management for improvement and development of Bt cotton (*Gossypium hirsutum* L.). Plants (Basel) 12(23):4071. https://doi.org/10.3390/plants12234071

Roh JY, Choi JY, Li MS, Jin BR, Je YH (2007) *Bacillus thuringiensis* as a specific, safe, and effective tool for insect pest control. J Microbiol Biotechnol 17(4):547–559

Sagar A, Rathore P, Ramteke PW, Ramakrishna W, Reddy MS, Pecoraro L (2021) Plant growth promoting Rhizobacteria, Arbuscular mycorrhizal fungi and their synergistic interactions to counteract the negative effects of saline soil on agriculture: key macromolecules and mechanisms. Microorganisms 9(7):1491. https://doi.org/10.3390/microorganisms9071491

Sales LR, Rigobelo EC (2024) The role of *Bacillus* sp. in reducing chemical inputs for sustainable crop production. Agronomy 14(11):2723. https://doi.org/10.3390/agronomy14112723

Samantaray A, Chattaraj S, Mitra D, Ganguly A, Kumar R, Gaur A, Mohapatra PKD, Santos-Villalobos SL, Rani A, Thatoi H (2024) Advances in microbial based bioinoculum for amelioration of soil health and sustainable crop production. Curr Res Microb Sci 7:100251. https://doi.org/10.1016/j.crmicr.2024.100251

Sarwar A, Hassan MN, Imran M, Iqbal M, Majeed S, Brader G, Sessitsch A, Hafeez FY (2018) Biocontrol activity of surfactin a purified from *Bacillus* NH-100 and NH-217 against rice bakanae disease. Microbiol Res 209:1–13. https://doi.org/10.1016/j.micres.2018.01.006

Saxena AK, Kumar M, Chakdar H, Anuroopa N, Bagyaraj DJ (2020) *Bacillus* species in soil as a natural resource for plant health and nutrition. J Appl Microbiol 128(6):1583–1594. https://doi.org/10.1111/jam.14506

Schnepf E, Crickmore N, Van Rie J, Lereclus D, Baum J, Feitelson J, Zeigler DR, Dean DH (1998) *Bacillus thuringiensis* and its pesticidal crystal proteins. Microbiol Mol Biol Rev 62(3):775–806. https://doi.org/10.1128/MMBR.62.3.775-806.1998

Shahid M, Singh UB, Khan MS, Singh P, Kumar R, Singh RN, Kumar A, Singh HV (2023) Bacterial ACC deaminase: insights into enzymology, biochemistry, genetics, and potential role in amelioration of environmental stress in crop plants. Front Microbiol 14:1132770. https://doi.org/10.3389/fmicb.2023.1132770

Singh S, Shyu DJH (2024) Perspective on utilization of *Bacillus* species as plant probiotics for different crops in adverse conditions. AIMS Microbiol 10(1):220–238. https://doi.org/10.3934/microbiol.2024011

Tariq A, Guo S, Farhat F, Shen X (2025) Engineering synthetic microbial communities: diversity and applications in soil for plant resilience. Agronomy 15(3):513. https://doi.org/10.3390/agronomy15030513

Thepbandit W, Athinuwat D (2024) Rhizosphere microorganisms supply availability of soil nutrients and induce plant defense. Microorganisms 12(3):558. https://doi.org/10.3390/microorganisms12030558

Timofeeva AM, Galyamova MR, Sedykh SE (2023) Plant growth-promoting soil bacteria: nitrogen fixation, phosphate solubilization, siderophore production, and other biological activities. Plants 12(24):4074. https://doi.org/10.3390/plants12244074

Tsotetsi T, Nephali L, Malebe M, Tugizimana F (2022) *Bacillus* for plant growth promotion and stress resilience: what have we learned? Plants 11(19):2482. https://doi.org/10.3390/plants11192482

Vasantha-Srinivasan P, Park KB, Kim KY, Jung WJ, Han YS (2025) The role of *Bacillus* species in the management of plant-parasitic nematodes. Front Microbiol 15:1510036. https://doi.org/10.3389/fmicb.2024.1510036

Wang X, Chi Y, Song S (2024) Important soil microbiota's effects on plants and soils: a comprehensive 30-year systematic literature review. Front Microbiol 15:1347745. https://doi.org/10.3389/fmicb.2024.1347745

Wu X, Liu Y, Jia B, Tao L, Li H, Wang J, Yuan Z, Sun X, Yao Y (2025) Four decades of *Bacillus* biofertilizers: advances and future prospects in agriculture. Microorganisms 13(1):187. https://doi.org/10.3390/microorganisms13010187

Xie J, Sun X, Xia Y, Tao L, Tan T, Zhang N, Xun W, Zhang R, Kovács ÁT, Xu Z, Shen Q (2024) Bridging the gap: biofilm-mediated establishment of *Bacillus velezensis* on *Trichoderma guizhouense* mycelia. Biofilm 8:100239. https://doi.org/10.1016/j.bioflm.2024.100239

Yasin M, Munir I, Faisal M (2016) Can *Bacillus* spp. enhance K^+ uptake in crop species. In: Meena V, Maurya B, Verma J, Meena R (eds) Potassium solubilizing microorganisms for sustainable agriculture. Springer, New Delhi. https://doi.org/10.1007/978-81-322-2776-2_12

Zhang N, Wang Z, Shao J, Xu Z, Liu Y, Xun W, Miao Y, Shen Q, Zhang R (2023) Biocontrol mechanisms of *Bacillus*: improving the efficiency of green agriculture. Microb Biotechnol 16(12):2250–2263. https://doi.org/10.1111/1751-7915.14348

Zulfiqar S, Gu R, Liu Y, Zhang Y (2025) From genes to traits: maximizing phosphorus utilization efficiency in crop plants. Front Plant Sci 16:1527547. https://doi.org/10.3389/fpls.2025.1527547

Chapter 13
Microbial Biosurfactants for Agriculture Applications

Kishore Kumar Annamalai, Eshwarnath Vallipuram Sudhakar, Manigundan Kaari, Venugopal Gopikrishnan, Narasimha Golla, and Radhakrishnan Manikkam

Abstract Sustainable agriculture is increasingly critical in addressing global challenges such as food security, environmental degradation, and climate change. The overreliance on chemical pesticides and agrochemicals has led to soil and water pollution, reduced soil fertility, and the emergence of pesticide-resistant pathogens. These issues underscore the need for ecofriendly alternatives to enhance agricultural productivity while minimizing environmental harm. Microbial biosurfactants, produced by bacteria, fungi, and yeasts, have emerged as promising tools in sustainable agriculture due to their biodegradability, low toxicity, and multifunctional properties. These surface-active compounds play a vital role in soil remediation, pest and disease control, plant growth promotion, and stress tolerance in crops. Biosurfactants improve soil quality by enhancing microbial activity, nutrient availability, and organic matter decomposition, while also mitigating the effects of salinity, drought, and heavy metal contamination. Additionally, they exhibit antimicrobial properties, making them effective biocontrol agents against plant pathogens. Despite their potential, challenges such as high production costs, regulatory hurdles, and limited field-scale applications hinder their widespread adoption. However, advancements in bioprocess engineering, genetic modification, and the use of low-cost substrates offer promising avenues for scaling up biosurfactant production. The global market for biosurfactants is projected to grow significantly, driven by increasing demand for sustainable agricultural practices and environmentally friendly products. This review highlights the role of microbial biosurfactants in sustainable agriculture, explores their applications, and discusses the challenges and future prospects for their integration into agricultural systems. By leveraging biosurfactants,

K. K. Annamalai · E. V. Sudhakar · V. Gopikrishnan · R. Manikkam (✉)
Centre for Drug Discovery and Development, Sathyabama Institute of Science and Technology, Chennai, Tamil Nadu, India

M. Kaari
Department of Applied Bioscience, Dong-A University, Pusan, South Korea

N. Golla
Department of Virology, Sri Venkateswara University, Tirupati, India

V. Gopikrishnan et al. (eds.), *Harnessing Microbial Resources for Sustainable Agriculture*, Microorganisms for Sustainability 61,
https://doi.org/10.1007/978-981-95-4817-0_13

agriculture can move toward a more sustainable and resilient future, ensuring food security and environmental stewardship for generations to come.

Keywords Biosurfactant · Agriculture · Biocontrol · Sustainable · Drought · Biodegradability

13.1 Introduction

13.1.1 Overview of Sustainable Agriculture

The agriculture, food, and food processing sectors face significant challenges in meeting the increasing food demands of a rapidly growing population. Consequently, it is crucial to adopt various effective strategies to enhance crop provisions and yields. Commonly employed methods encompass the application of detrimental or severe agrochemicals, fertilizers, and rapidly cultivated crop varieties. A variety of agrochemicals are commonly employed due to their effectiveness as bactericidal, fungicidal, insecticidal, herbicidal, and nematocidal agents. Significant agricultural products including wheat, rice, maize, and sorghum (essential food sources), along with sugarcane (a cash crop), soybean, groundnuts, cotton, fruits, and vegetables face threats from various pathogens and pests. Chemical origin organophosphate (profenofos, chlorpyrifos, and glyphosate), organochlorine (lindane, mirex, and chlordane), and carbamate-based pesticides are routinely applied in agriculture to manage pest growth (Lamilla et al. 2021; Mdeni et al. 2022). Worldwide, approximately two million tons of pesticides are utilized annually, with 50% comprising herbicides, 30% insecticides, 18% fungicides, and around 2% consisting of rodenticides and nematicides (Sharma et al. 2019; Sarkar et al. 2021). Organophosphorus and organochlorine-based pesticides lack specificity for their intended targets and persist in the ecosystem, adhering to both soil and water bodies. The widespread application of chemical pesticides poses detrimental effects not only on soil health, fertility, microbial diversity, and the environment but also on the health of humans and animals. Conditions like cancer, pulmonary dysfunction, and immune system deficiencies are often associated with these practices (Gupta et al. 2004; Sabarwal et al. 2018; Rani et al. 2021). Significant levels of pesticide residues have been detected in milk, meat, and various food products (Gill et al. 2020). Boedeker et al. (2020) indicated that approximately 44% of the farming population experiences poisoning each year due to these synthetic chemicals. The detrimental effects of agrochemicals and pesticides, including their cancerous, teratogenic, and immunosuppressive properties, pose significant risks to human health (Choudhary et al. 2018).

To address these challenges, there has been a notable increase in the use of microbes and their byproducts to improve agricultural productivity and crop yield, as well as to reduce the usage of toxic and hazardous agrochemicals (Popp et al. 2013).

Biosurfactants represent a recently investigated category of microbial biomolecules, commonly applied across diverse sectors such as agriculture, waste management, and pharmaceuticals. They serve as essential raw materials for lubrication, wetting, foaming, emulsion formulations, and stabilizing dispersions. Biosurfactants are defined as surface-acting agents that enhance surface-surface interactions by forming micelles derived from natural sources, including plants, microbes, and animals. Furthermore, biosurfactants have been employed in applications to lower the interfacial surface tension between the solution and the surface, as well as at air/water or oil/water interfaces. The incorporation of surfactants into an oil/water or water/air system leads to a decrease in surface tension until a threshold is reached, at which surfactants begin to organize into structures like micelles, vesicles, and bilayers. This threshold is typically referred to as the critical micelle concentration (C.M.C.) (Muthusamy et al. 2008; Singh et al. 2021; Rahman and Gakpe 2008). Microbial surfactants are natural, biodegradable, and nontoxic, leading to a consistent rise in market demand. Microorganisms that produce biosurfactants can exhibit both high molecular weights, including proteins, lipopolysaccharides, lipoproteins, and polysaccharides, as well as low molecular weights, such as glycolipids and lipopeptides. Biosurfactants with high molecular weight demonstrate enhanced effectiveness in stabilizing oil-water emulsions, whereas those with low molecular weight excel at reducing surface tension (Eras-Muñoz et al. 2022).

13.2 Microbial Biosurfactants

13.2.1 Properties and Types

Biosurfactants are mainly categorized according to their structural composition and origin. The hydrophilic top typically consists of amino acids, peptides, or mono-, di-, or polysaccharides. The hydrophobic end typically consists of saturated, unsaturated, linear, branched, or hydroxylated fatty acids. Classification of biosurfactants follows according to their chemical composition and the microorganisms from which they are derived (Desai and Banat 1997; Christofi and Ivshina 2002; Muthusamy et al. 2008). Microbial biosurfactants can be classified based on their chemical nature, ionic charge, and the microorganism that produces them. Typically, microbial biosurfactants are organized into four main categories based on their polymeric composition: (i) glycolipids, (ii) lipopeptides and lipoproteins, (iii) glycopeptides and glycolipopeptides, and (iv) bioemulsifiers with biosurfactant capacity, including phospholipids, polymeric biosurfactants, and particulate biosurfactants

(Shekhar et al. 2015) Table 13.1 represents the Bacteria which produce biosurfactants and their key properties.

13.2.1.1 Glycolipids

This type of biosurfactant is widely recognized for its accessibility and is composed of carbohydrates combined with long-chain aliphatic acids or hydroxyaliphatic acids. The linkage occurs through either an ether or an ester group. Rhamnolipids, trihalolipids, and sophorolipids represent the most recognized biosurfactants within the category of glycolipids. The most extensively investigated glycolipids include rhamnolipids produced by *Pseudomonas* sp. (Jarvis and Johnson 1949), mannosylerythritol lipids generated by *Pseudozyma antarctica* (Kitamoto et al. 1990), and trehalose lipids synthesized by *Rhodococcus* sp., *Nocardia* sp., *Arthrobacter* sp., and *Mycobacterium* sp. (Lang and Wagner, 1996). Cellobiolipids are produced by *Ustilago maydis* (Teichmann et al. 2007) and sophorolipids produced by *Candida* sp. (Cooper and Paddock, 1984).

13.2.1.2 Lipopeptides and Lipoprotein

A significant variety of cyclic lipopeptides, such as decapeptide antibiotics and lipopeptide antibiotics, are generated. These consist of a lipid associated with a polypeptide chain. Surfactin, a lipopeptide, is synthesized by *Bacillus* species and consists of seven amino acids linked to a carboxyl and hydroxyl group of a 14-carbon fatty acid. A notable feature of surfactin is its capacity to lyse mammalian erythrocytes and to generate spheroplasts (Bernheimer and Avigad 1970). This class of biosurfactants typically comprises molecules that consist of cyclic peptides connected to a fatty acid. Numerous bacteria have been identified as producers of antibiotic-like molecules, including *Bacillus subtilis*, which synthesizes surfactin and subtilisin. Peptide-lipid and lichenysin are produced by *Bacillus licheniformis* and *Bacillus subtilis* (Yakimov et al. 1997; Begley et al. 2009). Viscosin is synthesized by *Pseudomonas fluorescens* (Banat et al. 2010), while Serrawettin is produced by *Serratia marcescens* (Lai et al. 2013). Fengycin is synthesized by *Bacillus* sp. (Vanittanakom et al. 1986), and Arthrofactine is produced by *Arthrobacter* sp. (Morikawa et al. 1993). Gramicidins are synthesized by *Bacillus brevis* and *Brevibacterium brevis* (Krauss and Chan 1983). Polymyxins are produced by *Bacillus polymyxa* and *Brevibacterium polymyxa*. Ornithine lipids are synthesized by *Myroides* sp., *Pseudomonas* sp., *Thiobacillus* sp., *Agrobacterium* sp., and *Gluconobacter* sp. (Desai and Banat 1997).

Table 13.1 Types of biosurfactant from bacteria and their key properties

S. no	Type of biosurfactant	Producer organism(s)	Chemical structure	Surface/emulsifying activity	Other key Properties	References
1.	Glycolipids	*Pseudomonas aeruginosa*, *Burkholderia* spp.	Carbohydrate + fatty acid	Strong reduction of surface tension (to ~25–30 mN/m); excellent emulsification	High biodegradability, low toxicity; rhamnolipids and sophorolipids are most studied	Lang and Wullbrandt (1999), Cameotra and Makkar (2004)
2.	Lipopeptides and lipoproteins	*Bacillus subtilis*, *B. licheniformis*	Peptide ring + lipid tail	Good surface activity; stable under high salinity and pH	Potent antimicrobial activity; surfactin is a well-known example	Seydlová and Svobodová (2008), Mnif and Ghribi (2015)
3.	Phospholipids	*Acinetobacter* spp., *Pseudomonas* spp.	Phosphate-containing lipids	Moderate surface activity	Found naturally in some microbial cell membranes; contribute to emulsification	Kosaric (1992), Desai and Banat (1997)
4.	Polymeric biosurfactants	*Acinetobacter calcoaceticus*, *Alcaligenes faecalis*	Polysaccharides, proteins, or a mix	Low surface activity; high emulsifying capacity	Viscosity-enhancing; suitable for soil and heavy metal remediation applications	Mulligan et al. (2001), Ron and Rosenberg (2001)
5.	Particulate biosurfactants (whole cells)	*Mycobacterium*, *Corynebacterium*	Cell-bound surfactants	Variable surface activity	Effective in hydrocarbon uptake and biodegradation	Rosenberg and Ron (1999)

13.2.1.3 Fatty Acids, Phospholipids, and Neutral Lipids

A significant number of bacteria and yeast produce substantial quantities of fatty acids and phospholipid surfactants while growing on n-alkanes (Cirigliano and Carman 1985). The hydrophilic and lipophilic balance (HLB) correlates positively with the length of the hydrocarbon chain present in their structures. A variety of fungi, yeasts, and bacteria such as *Nocardia erythropolis, Thiobacillus thiooxidans, Candida lepus, Acinetobacter* sp., *Pseudomonas* sp., *Micrococcus* sp., *Micrococcus* sp., *Candida* sp., *Penicillium* sp., and *Aspergillus* sp. (Käppeli and Finnerty 1979) capable of thriving on hydrophobic substrates like alkanes produce significant quantities of phospholipids, fatty acids, or neutral lipids to enhance the absorption of the carbon source. In *Acinetobacter* sp., vesicles rich in phosphatidylethanolamine are synthesized (Käppeli and Finnerty 1979), leading to the formation of optically clear microemulsions of alkanes in water. Bile salt produced by *Myroides* sp. (Desai and Banat 1997) and phosphatidylethanolamine generated by *Rhodococcus erythropolis* cultivated on n-alkane reduce the interfacial tension between hexadecane and water to below 1 mN/m, with a critical micelle concentration (CMC) of 30 mg/l (Kretschmer et al. 1982).

13.2.1.4 Polymeric Biosurfactants

The most extensively investigated polymeric biosurfactants include alasan, liposan, lipomanan, emulsan, and various other polysaccharide-protein complexes. Acinetobacter calcoaceticus RAG-1 produces a highly effective extracellular polyanionic amphipathic heteropolysaccharide bioemulsifier (Rosenberg et al. 1979). Emulsan serves as a potent emulsifying agent for hydrocarbons in aqueous solutions (Zosim et al. 1982), even at concentrations as minimal as 0.001% to 0.01%. Liposan is an extracellular water-soluble emulsifier synthesized by *Candida lipolytica*, comprising 83% carbohydrate and 17% protein, a carbohydrate-protein complex produced by the yeast *Yarrowia lipolytica*. An additional instance is the Mannan lipid protein produced by Candida tropicalis, along with Protein PA generated by *Pseudomonas aeruginosa* (Cirigliano and Carman 1984). Polymeric biosurfactants consist of multiple components. Emulsan and biodispersan, synthesized by *Acinetobacter calcoaceticus*, represent the most extensively studied compounds. They consist of a heteropolysaccharide backbone to which fatty acids are covalently linked (Rosenberg and Ron 1999).

13.2.1.5 Particulate Biosurfactants

Extracellular membrane vesicles partition hydrocarbons to form a microemulsion, which is crucial for alkane uptake by microbial cells. Vesicles of *Acinetobacter* sp. exhibit a diameter ranging from 20 to 50 nm and possess a buoyant density of 1.158

cubic g/cm. These vesicles are composed of proteins, phospholipids, and lipopolysaccharides (Käppeli and Finnerty 1979).

13.3 Microorganisms Producing Biosurfactants

A wide array of microorganisms—including bacteria, fungi, and yeasts—are recognized for producing biosurfactants, with the yield and nature of these compounds significantly influenced by the source of isolation. Polluted soils, industrial effluents, and wastewater discharge points serve as common reservoirs, providing conditions that support biosurfactant-producing strains capable of metabolizing toxic substrates.

The production and quality of microbial biosurfactants are governed by several factors such as carbon source composition and concentrations of nitrogen, phosphorus, iron, magnesium, and manganese, as well as environmental conditions like pH, temperature, agitation, and dilution rates. For applications like microbial-enhanced oil recovery (MEOR), parameters such as temperature, pH, pressure, and salinity are critical (Khire 1994).

13.3.1 Key Bacterial Genera

Bacillus species are widely known for producing lipopeptides like surfactin—one of the most studied biosurfactants. These molecules consist of a lipid linked to a peptide, giving them potent surface-active properties. Recent research has explored the molecular genetics regulating their biosynthesis (Li et al. 2010). *Pseudomonas* spp. is second only to *Bacillus* in biosurfactant research. They produce glycolipids like rhamnolipids and lipopeptides such as arthrofactin, viscosin, putisolvin, and amphisin, with each compound linked to specific strains (Jyoti et al. 2021). Acinetobacter species synthesize high molecular weight biosurfactants like emulsan and alasan. Emulsan from *Acinetobacter calcoaceticus* contains a complex of proteins and lipoheteropolysaccharides, while alasan from *A. radioresistens* features anionic heteropolysaccharides rich in alanine. *Serratia* spp. produce cyclic depsipeptides called serrawettins, with distinct strains responsible for each variant. Novel lipids such as rubiwettin R1 and RG1 have also been reported from *Serratia rubidaea*. *Rhodococcus* spp., especially *R. erythropolis,* produce glycolipids and trehalose dimycolates with excellent emulsification properties, particularly in oil degradation applications (Peng et al. 2007). *Halomonas* spp., isolated from extreme environments like Antarctic Sea ice, are known for generating emulsifying exopolysaccharides and glycoproteins with promising surface-active properties (Pepi et al. 2005). *Myroides* spp., isolated from oil-contaminated waters, produce extracellular L-ornithine lipids that exhibit strong oil displacement and emulsification abilities, maintaining stability under varied temperatures but sensitive to high salt and alkaline conditions (Maneerat and Phetrong 2007).

13.3.2 Fungal and Yeast Producers

Among fungi, *Candida bombicola* is a well-documented producer of sophorolipids. *Yarrowia lipolytica* synthesizes lipid- and protein-based bioemulsifiers during the stationary phase, contributing to cell hydrophobicity and emulsification. Yeast species like *Pseudozyma* and *Torulopsis petrophilum* are prominent producers of glycolipid biosurfactants. *Kurtzmanomyces, Ustilago maydis*, and *Schizonella melanogramma* have also been shown to produce mannosylerythritol lipids (MELs) with potential industrial applications (Cooper 1985; Kakugawa et al. 2002).

13.4 Microbial Intervention in Sustainable Agriculture

13.4.1 Role of Microbial Biosurfactants

They are microbial surfactants extensively utilized in the petroleum, medicinal, pharmaceutical, agricultural, and environmental industries and also as main ingredient in cosmetics, food additives, and detergents (Bhardwaj et al. 2013). The harmful effluents generated during the production of synthetic surfactants adversely affect the environment which affect its commercial appeal. The global market for chemical surfactants is anticipated to achieve a compound annual growth rate (CAGR) of 5.3% from 2020 to 2027 (Dixit et al. 2019), whereas biosurfactants are projected to grow at a CAGR exceeding 5.5% between 2020 and 2026, particularly rhamnolipids, which may surpass USD 145 million. The biological functions and applications of different biosurfactants are closely linked to the chemical alterations in their molecular structures (Morita et al. 2015). Certain studies indicate that fungi and yeast are additionally engaged in the synthesis of biosurfactants. Table 13.2 indicates the biosurfactant multifaceted application.

13.5 Applications in Agriculture

13.5.1 Soil Improvement

Biosurfactants represent a promising tool for improving soil quality, especially in the context of sustainable agriculture and environmental remediation. Due to their unique chemical and physiological properties, both natural and synthetic surfactants

Table 13.2 Biosurfactants and their multifaceted application

S. no	Function	Description	Example microbes	Agricultural impact	References
1.	Soil conditioning	Enhance soil structure and permeability via emulsification and aggregate formation	*Bacillus subtilis*, *Pseudomonas* spp.	Improved aeration, root penetration, and water retention	Mulligan (2005), Geys et al. (2014)
2.	Nutrient mobilization	Solubilize hydrophobic nutrients (e.g., phosphorus, iron)	*Pseudomonas fluorescens*, *Rhizobium* spp.	Increased nutrient availability for plant uptake	Ghosh et al. (2021), Kour et al. (2021)
3.	Biocontrol agent	Disrupt pathogen biofilms and membranes; induce systemic resistance	*Bacillus amyloliquefaciens*, *Streptomyces* spp.	Reduced plant disease and pathogen suppression	Raaijmakers et al. (2010) Sharma et al. (2022)
4.	Plant growth promotion	Biosurfactants help in phytohormone production and better rhizosphere colonization	*Pseudomonas putida*, *Azospirillum brasilense*	Enhanced seed germination, growth, and crop yield	Gurav et al. (2025), Singh and Cameotra (2004)
5.	Stress mitigation	Reduce metal toxicity and osmotic stress (drought/salinity)	*Burkholderia cepacia*, *Serratia marcescens*	Promotes tolerance under abiotic stress	Pathak and Keharia (2014), Pacwa-Płociniczak et al. (2011)
6.	Ecofriendly alternatives	Replace chemical surfactants, fertilizers, and pesticides	Various indigenous soil microbes	Biodegradable, sustainable, and low-toxicity inputs	Banat et al. 2014a, b), Marchant and Banat (2012)

can enhance soil structure and facilitate the bioremediation of pollutants (Sachdev and Cameotra, 2013). The in situ production of biosurfactants is influenced by several factors, including soil composition and carbon and nitrogen availability, as well as environmental conditions such as temperature, pH, and oxidation-reduction potential. For example, Vecino Bello et al. (2012) demonstrated that biosurfactant activity from *Lactobacillus pentosus* varied with operation time, pH, salinity, and temperature. Within a pH range of 3.0–5.5, lower salinity and temperature synergistically reduced surface tension, whereas at pH 8.0, the combined increase in these parameters produced a similar effect. The biosurfactant's emulsifying ability also depended on time, temperature, and salt concentration (Vecino et al. 2014). Further optimization was achieved through salt-free aqueous extraction using phosphate buffer. Notably, biosurfactant efficiency is closely linked to the organic matter present in soil (Benoit et al. 2008). These findings support the broader potential of biosurfactants in soil improvement and their future role in managing degraded or contaminated soils.

Beyond soil structure enhancement, biosurfactants play a vital role in stimulating beneficial microbial communities, particularly in the rhizosphere. Kosaric (2001) emphasized their biotechnological importance in promoting microbial growth. Ali and Sandouka (2012) investigated the impact of a biosurfactant derived from sunflower seed meal on microbial populations in soils deliberately contaminated with crude petroleum. In the absence of biosurfactant treatment, total heterotrophic bacteria (THB) levels were $15.0 \pm 1.40 \times 10^7$ cfu/g in rhizosphere soil and $5.43 \pm 0.20 \times 10^7$ cfu/g in nonrhizosphere soil. After biosurfactant application, these levels rose significantly to $43.2 \pm 1.4 \times 10^7$ cfu/g and $155.7 \pm 18 \times 10^7$ cfu/g, respectively. However, certain biosurfactants have also shown inhibitory effects on specific bacterial strains (Sotirova et al. 2008), highlighting the need for context-specific assessments.

In addition to promoting microbial diversity, biosurfactants contribute to the breakdown of organic matter and enhancement of enzymatic activity in soil. Organic decomposition tends to slow down in soils rich in cellulose or lignin. Surfactants such as Tween 80 have been shown to enhance the production of cellulase (Pardo et al. 1996), amylase (Goes and Sheppard 1999), and phytase (Mandviwala and Khire 2000) and improve cellulose hydrolysis (Eriksson et al. 2002). Jin et al. (2023) explored the role of rhamnolipids in composting and observed increased populations of bacteria, actinomycetes, and fungi in biosurfactant-treated soils. Rhamnolipids also stabilized the hydrolysis of carboxymethyl cellulose and enhanced xylanase activity, accelerating the degradation of cellulose and hemicellulose. This led to a rise in water-soluble carbon compounds such as glucose and xylose, which supported more robust microbial communities throughout the composting process. Biosurfactants produced by plants and microorganisms have also been employed in remediating soils contaminated with pesticides, heavy metals, and hydrocarbons (Beal and Betts 2000; Kumar et al. 2006; Mulligan 2005; Ochoa-Loza et al. 2007). These effects are largely achieved by reducing the hydrophobicity of soil contaminants. For instance, biosurfactants from *L. pentosus* reduced water repellence in various plant substrates and improved octane biodegradation (Moldes et al. 2013; Vecino et al. 2013). Collectively, biosurfactant-producing microbes offer a valuable strategy for restoring soils affected by salinity and hydrophobic pollutants.

13.5.2 Biocontrol of Pests and Diseases

In agricultural practices, plants are vulnerable to attacks from phytopathogens—mainly fungi and bacteria—throughout their lifecycle, from planting to storage. These infections often lead to considerable economic losses. The introduction of synthetic pesticides provided an effective solution, gaining widespread acceptance due to their immediate results and ease of use (Oliveira et al. 2015; Mishra and Arora 2018). Many of these pesticides contain chemical surfactants that act as emulsifiers, dispersants, and wetting agents, thereby enhancing their overall efficacy (Rostas and Blassmann 2009). However, surfactants such as alcohol ethoxylates (AEOs), alkylamine ethoxylates (ANEOs), polyethoxylated tallowamines (POEAs), organosilicons, and trisiloxanes are known to accumulate in soil and are classified as organic pollutants. Their persistence in the environment has raised serious concerns regarding soil health, ecosystem balance, and human safety. As a result, researchers are exploring ecofriendly alternatives.

Biosurfactants have attracted attention as potential substitutes due to their biodegradable nature and low toxicity. They can serve as both adjuvants and active compounds in plant protection (Kosaric and Vardar-Sukan 2014). Sophorolipids, a type of biosurfactant, exhibit antimicrobial properties due to their amphiphilic structure, which enables interactions between the sophorose and fatty acid components. This interaction reduces surface and interfacial tension, destabilizing pathogen membranes and increasing their permeability (Seung-Hak et al. 2003; Lahkar et al. 2015; Valotteau et al. 2017). The effectiveness of sophorolipids depends on their structural form, with lactonic variants displaying stronger antimicrobial action than acidic forms. Their activity also varies based on the producing microorganism and the target pathogen (Sharma et al. 2016; Zhang et al. 2017), showing greater efficacy against gram-positive bacteria compared to gram-negative bacteria and fungi (Ribeiro et al. 2013). Several in vitro studies support their effectiveness. Lang et al. (1989) reported that sophorolipids inhibited the germination of *Glomerella cingulate conidia*—a pathogen affecting apples, grapes, and other fruits. Yoo et al. (2005) highlighted their action against *Botrytis cinerea*, the cause of gray mold in vegetables like strawberries, grapes, and tomatoes. Similarly, Sen et al. (2017) showed that sophorolipids from *Rhodotorula babjevae* YS3 inhibited multiple plant pathogens, including *Colletotrichum gloeosporioides, Fusarium* spp., *Corynespora cassiicola,* and *Trichophyton rubrum.*

Beyond sophorolipids, numerous biosurfactants produced by microorganisms, particularly rhizobacteria, exhibit antimicrobial properties and hold promise as biocontrol agents in sustainable agriculture (Nihorimbere et al. 2011a). These biosurfactants enhance the efficacy of plant growth-promoting microbes by supporting mechanisms such as parasitism, antibiosis, competition, induced systemic resistance, and hypovirulence (Singh et al. 2007). Both synthetic and biological surfactants are used extensively in agriculture to strengthen microbial antagonism and support the application of microbial products (Jazzar and Hammad 2003; Kim et al. 2004). Studies have shown their contribution to improving the insecticidal efficiency of various systems in both lab and field conditions (Jazzar and Hammad 2003; Gronwald et al. 2002; Krishnayya and Grewal 2002). Surfactants are also used alongside *Myrothecium verrucaria* fungi to control weed species that reduce land productivity and threaten biodiversity (Boyette et al. 2002). Furthermore, they have been shown to inhibit aflatoxin production by *Aspergillus* species, which contaminate crops like peanuts, cottonseed, and maize during storage and other stages of crop development (Rodriguez and Mahoney 1994). Overall, surfactants—whether synthetic or biological—play diverse roles in managing plant pathogens and contribute to a range of agricultural practices aimed at improving crop protection and productivity.

13.5.3 Plant Growth Enhancement

Microbial biosurfactants and plant growth-promoting rhizobacteria (PGPR) serve as effective bioinoculants in agriculture. They aid in nutrient solubilization, nitrogen fixation, and hormone production (Chopra et al. 2020a, b; Lal et al. 2018). Microorganisms such as *Pseudomonas, Bacillus,* and *Burkholderia* are commonly associated with the rhizospheres of crops like tea, tomato, and wheat (Villa-Rodriguez et al. 2019; Cochard et al. 2022). Their biosurfactants show promise as biocontrol agents against phytopathogens, contributing to sustainable farming practices (Arifiyanto et al. 2020).

For plant-beneficial rhizobacteria to be effective, they must interact closely with plant root surfaces (Nihorimbere et al. 2011a, b). Traits such as motility, biofilm formation, and production of quorum sensing molecules like acyl homoserine lactone (AHL) are key to this interaction. These molecules, more concentrated in the rhizosphere than bulk soil, are essential for antifungal compound production and biofilm regulation (Newton and Fray 2004; Loh et al. 2002). Rhamnolipids from *Pseudomonas* spp. has been shown to influence quorum sensing, microbial motility, and biofilm formation (Dusane et al. 2010; Ron and Rosenberg 2011). These biosurfactants not only facilitate root colonization but also enhance nutrient availability and improve soil wettability, which contributes to better plant growth.

Recent studies have highlighted the potential of biosurfactant-producing PGPR strains such as *P. aeruginosa* RTE4 and *P. guariconensis* LE3. The RTE4 strain, isolated from tea rhizosphere, showed RL-based biocontrol activity against multiple pathogens including *Corticium invisum* and *Fusarium solani* (Chopra et al. 2020b). Similarly, the LE3 strain exhibited strong antifungal action against *Macrophomina phaseolina*, along with beneficial traits like phosphate solubilization and IAA, siderophore, and ACC deaminase production (Khare and Arora 2021). Bioformulations using LE3 and its biosurfactants significantly improved sunflower yield and biocontrol efficacy, enhancing root colonization and plant immunity.

These studies emphasize the value of biosurfactants in crop protection. Beyond antimicrobial activity, rhamnolipids and lipopeptides also prime plant immune responses, offering broad protection against pathogens (Delaunois et al. 2014; Bardin et al. 2015; Keswani et al. 2019). Khare and Arora (2021) proposed the development of crop-specific formulations with robust rhizosphere performance, environmental safety, and wide-spectrum antimicrobial activity—an approach that aligns with the goals of sustainable agriculture.

13.6 Stress Tolerance in Crops

13.6.1 Salinity and Drought Stress

Plants, being sessile organisms, are highly vulnerable to both biotic and abiotic stressors throughout their lifecycle. Biotic stresses—caused by bacteria, fungi, viruses, insects, and nematodes—are a major reason for reduced global crop yields (Murali et al. 2021; Gowtham et al. 2022; Hamidian et al. 2023). Abiotic stresses such as drought, salinity, extreme temperatures, flooding, poor light conditions, heavy metal toxicity, and radiation also significantly affect plant growth and development (Sarker et al. 2021; Ahmad et al. 2022).

These environmental pressures disrupt key physiological, molecular, and biochemical functions in plants, often causing oxidative stress, hormonal imbalances, ionic toxicity, and impaired nutrient uptake (Ma et al. 2020; Munir et al. 2022). Stress often alters gene expression tied to metabolic pathways, ultimately reducing crop productivity (Ullah et al. 2021). Alarmingly, around 90% of arable lands are susceptible to such stressors, which may decrease yields by up to 70% (Waqas et al. 2019).

Drought and salinity are among the most critical stressors, especially in arid and semiarid regions. Global warming and improper irrigation practices are intensifying soil salinity, reducing water availability and leading to salt accumulation (Khan 2022). High salinity affects nearly 20% of irrigated agricultural lands worldwide, while drought stress continues to expand due to changing climate patterns (Ullah et al. 2021).

Both drought and salinity disrupt plant water relations and ion balance, particularly due to excess sodium (Na^+) which competes with essential potassium (K^+) for cellular uptake (Desoky et al. 2020; Hasanuzzaman et al. 2022). These stresses also lead to reactive oxygen species (ROS) production, damaging cell membranes and affecting photosynthesis, thereby lowering yield and crop quality (Khan 2022). To cope, plants employ strategies such as sodium exclusion or vacuolar sequestration, membrane modifications, hormonal regulation, and activation of antioxidant systems (Ullah et al. 2021; Khan 2022). However, genetic and transgenic solutions are often time-consuming and costly.

In this context, plant growth-promoting rhizobacteria (PGPR) are gaining attention for their ability to enhance stress resilience in a sustainable manner. PGPR help plants tolerate drought and salinity by modulating phytohormones, enhancing antioxidant enzyme activity, improving root architecture, and regulating ion uptake (Kim et al. 2014; Singh et al. 2017; Khan et al. 2021; Zhao et al. 2022).

While earlier studies have explored PGPR functions, a consolidated understanding of recent developments and mechanistic insights remains limited. Addressing this gap is critical for future-ready agriculture, especially as food production needs to increase by 70% to meet the demands of an estimated 2.3 billion more people by 2050.

Thus, exploring PGPR as an ecofriendly alternative to agrochemicals may offer promising strategies to improve crop productivity and soil health under challenging environmental conditions (Singh et al. 2019a; Hoseini et al. 2022; Mawar et al. 2023).

13.6.2 Heavy Metal Mitigation

Bioremediation involves the use of microorganisms to eliminate toxic pollutants from contaminated environments. Microbes such as bacteria and fungi are naturally equipped with stress-adaptive mechanisms like biosorption, bioaccumulation, and oxidation that allow them to tolerate and transform heavy metals (Javanbakht et al. 2014; Rajendran et al. 2003). One of the most promising microbial products in this context is biosurfactants—amphiphilic molecules that reduce surface and interfacial tension and facilitate the removal of metal contaminants from soils.

Biosurfactants, particularly those produced by bacteria, are increasingly being recognized for their role in remediating heavy metal pollution. These compounds stand out due to their low toxicity, high biodegradability, and resistance to environmental stresses like extreme pH, temperature, and salinity (Usman et al. 2016). For example, rhamnolipids—a type of biosurfactant—have shown substantial efficiency in removing metals like Cu, Cd, Pb, and Cr from sediments, with removal efficiencies up to 87% depending on factors such as biosurfactant concentration, pH, and duration of exposure (Chen et al. 2017).

Innovative techniques like electrokinetic treatment combined with rhamnolipids and complexing agents have further enhanced metal removal from sewage sludge, achieving removal rates above 80% for several metals (Tang et al. 2017). Similarly, biosurfactants from *Burkholderia* species, paired with flocculation agents, have been shown to effectively extract metals like Zn, Mn, and Cd from contaminated soils (Yang et al. 2018). These methods offer ecofriendly alternatives to chemical surfactants and conventional remediation techniques.

Mechanistically, biosurfactants operate via adsorption and complexation with free metal ions, often forming stable metal-micelle structures that can be separated through filtration techniques like MEUF (Wu et al. 2017; Tortora et al. 2016). For instance, sophorolipids from *Starmerella bombicola* have achieved over 90% removal of Cd and more than 50% for Pb in soil washing processes (Qi et al. 2018).

The effectiveness of biosurfactants depends on multiple factors including their molecular structure, charge, size, and the characteristics of the contaminated soil—such as pH, cation exchange capacity, and pore size (Wan et al. 2017; Xue et al. 2018; Pourfadakari et al. 2019). Given their versatility and environmental compatibility, biosurfactants represent a sustainable and cost-effective option for tackling heavy metal pollution in soils.

13.7 Benefits and Challenges

13.7.1 Environmental and Economic Advantages

Biosurfactants (BioSs) are ecofriendly alternatives to chemical surfactants, offering numerous advantages in the remediation of oil-contaminated sites. Unlike synthetic surfactants, they do not produce harmful secondary pollutants. Their key features—biodegradability, low toxicity, high surface activity, and effectiveness across varied environmental conditions—make them ideal for such applications (Nikolopoulou et al. 2008; Singh et al. 2019a, b). In fact, several studies have shown that biosurfactants can achieve up to 80% hydrocarbon removal within a week of application, outperforming chemical dispersants in aquatic environments (Pi et al. 2017).

However, despite their advantages, biosurfactants remain underutilized in commercial sectors, mainly due to high production costs—estimated to be around 12 times higher than synthetic surfactants (Mohanty et al. 2021). Most conventional surfactants used globally still rely on petroleum-based sources, making cost reduction in biosurfactant production a critical focus. Strategies to tackle this include using low-cost feedstocks like agroindustrial waste, oily residues, dairy waste, and municipal byproducts (Effendi et al. 2018; de Oliveira Schmidt et al. 2021).

The global biosurfactant market has seen steady growth, with revenues expected to rise from USD 1.8 billion in 2016 to USD 2.6 billion by 2023 (Mohanty et al. 2021). Production volumes are projected to increase from approximately 344,000 tons in 2013 to nearly 462,000 tons by 2020, reflecting growing interest in replacing chemical surfactants (Sidkey et al. 2016).

Nevertheless, major challenges persist, particularly in large-scale production, downstream processing, and raw material availability (Henkel et al. 2012; Banat et al. 2014a). Feedstock cost alone can account for up to 50% of the total production cost (Ruggeri et al. 2009). Hence, sourcing affordable and regionally available substrates is vital. For example, Oman uses date molasses—a cheap and abundant byproduct—as a carbon source for lipopeptide biosurfactant production (AlBahry et al. 2013).

Reducing costs further depends on identifying efficient microbial strains, optimizing culture conditions, genetic modification for yield improvement, and streamlining downstream processes (Sekhon et al. 2012; Partovi et al. 2013). Continued research is essential to unlock the full potential of biosurfactants in petroleum remediation and broader industrial and environmental applications (de França et al. 2015; Ferreira et al. 2017; Dalili et al. 2015).

13.8 Current Limitations

Biosurfactants, produced by microorganisms, present sustainable solutions in agriculture for soil remediation, plant protection, and biofertilization. However, their widespread use is limited by several key challenges. One of the primary issues is their high production cost compared to synthetic surfactants. Fermentation-based production requires optimized growth conditions, specific nutrients, and expensive downstream processing (Satpute et al. 2010; Gudiña et al. 2016). The typically low yield of biosurfactants further adds to the cost, making large-scale agricultural use difficult.

Efforts to reduce costs by using industrial byproducts like waste oils and molasses have shown potential, yet maintaining consistent quality and yield remains problematic (Banat et al. 2014a, b). Additionally, the bulk availability of such substrates varies by region, limiting scalability.

Regulatory approval is another major hurdle. Despite their microbial origin, biosurfactants often face lengthy and costly safety assessments due to the absence of standardized agricultural regulations (Kiran et al. 2016). Some compounds may also interact with native soil microbes, necessitating detailed ecological risk studies (Nitschke and Silva 2018). Environmental factors further affect their reliability. Biosurfactants can degrade under extreme temperatures, pH variations, or UV exposure, reducing their effectiveness in the field (Pacwa-Płociniczak et al. 2011). While biodegradability supports sustainability, it also shortens their functional lifespan in soils, requiring repeated applications (Marchut-Mikołajczyk et al. 2021).

Most current studies are confined to lab or greenhouse trials, and few have explored biosurfactants under real field conditions. Challenges such as diverse soil types, microbial interactions, and inconsistent results have slowed field-scale implementation (Jimoh and Lin 2019). The limited availability of biosurfactant-based agricultural products also restricts farmer adoption. Moreover, biosurfactants can influence soil microbial communities—positively or negatively—depending on the context. While some enhance beneficial microbes, others may suppress key microbial groups essential for nutrient cycling and plant growth (Mukherjee et al. 2006). Their interaction with root exudates and nutrient uptake pathways also remains under-researched. Synthetic surfactants continue to dominate due to their low cost, chemical stability, and established efficacy in applications like pesticide formulation and soil wetting (Mulligan 2005). Until biosurfactants can match this affordability and consistency, they will struggle to replace synthetic alternatives.

To overcome these barriers, future work should focus on improving biosurfactant yields through microbial strain engineering, using agroindustrial wastes as feedstock, and developing formulations that enhance stability and field performance.

Acknowledgment Authors thank the authorities of the Sathyabama Institute of Science and Technology, Chennai, India, for their constant support and encouragement.

References

Ahmad A, Rasul S, Imran M, Shah T, Akhtar M, Maqsood MH (2022) Abiotic stress and plant responses: a physiological and molecular overview. Curr Plant Biol 30:100248

AlBahry SN, Al-Wahaibi YM, Al-Bemani AS, Joshi SJ, Elshafie AE, Al-Bahry HN, Al-Sulaimani H (2013) Biosurfactant production by Bacillus subtilis B30 and its application in enhancing oil recovery. Colloids Surf B: Biointerfaces 101:243–250

Arifiyanto A, Surtiningsih T, Agustina D, Alami NH (2020) Antimicrobial activity of biosurfactants produced by actinomycetes isolated from rhizosphere of Sidoarjo mud region. J Appl Microbiol 129(2):456–468. https://doi.org/10.1111/jam.14602

Banat IM, Franzetti A, Gandolfi I, Bestetti G, Martinotti MG, Fracchia L et al (2010) Microbial biosurfactants production, applications and future potential. Appl Microbiol Biotechnol 87:427–444

Banat IM, Satpute SK, Cameotra SS, Patil R, Nyayanit NV (2014a) Cost effective technologies and renewable substrates for biosurfactants' production. Front Microbiol 5:697

Banat IM, Franzetti A, Gandolfi I, Bestetti G, Martinotti MG, Fracchia L, Smyth TJP, Marchant R (2014b) Microbial biosurfactants production, applications and future potential. Appl Microbiol Biotechnol 98(24):9915–9926. https://doi.org/10.1007/s00253-014-6169-6

Bardin M, Ajouz S, Comby M, Lopez-Ferber M, Graillot B, Siegwart M, Nicot PC (2015) Is the efficacy of biological control against plant diseases likely to be more durable than that of chemical pesticides? Front Plant Sci 6:566

Beal R, Betts WB (2000) Role of rhamnolipid biosurfactants in the uptake and mineralization of hexadecane in Pseudomonas aeruginosa. J Appl Microbiol 89(1):158–168

Begley M, Cotter PD, Hill C, Ross RP (2009) Identification of a novel two-peptide lantibiotic, lichenicidin, following rational genome mining for LanM proteins. Appl Environ Microbiol 75(17):5451–5460

Benoit M, Desnues B, Mege JL (2008) Macrophage polarization in bacterial infections. J Immunol 181(6):3733–3739

Bernheimer AW, Avigad LS (1970) Nature and properties of a cytolytic agent produced by Bacillus subtilis. Microbiology 61(3):361–369

Bhardwaj H, Gupta R, Tiwari A (2013) Communities of microbial enzymes associated with biodegradation of plastics. J Polym Environ 21:575–579

Boedeker W, Watts M, Clausing P, Marquez E (2020) The global distribution of acute unintentional pesticide poisoning: estimations based on a systematic review. BMC Public Health 20:1–19

Boyette CD, Hoagland RE, Weaver MA (2002) Interaction of Myrothecium verrucaria, a bioherbicide for kudzu, with a glyphosate formulation. Biocontrol Sci Tech 12(4):409–417

Cameotra SS, Makkar RS (2004) Recent applications of biosurfactants as biological and immunological molecules. Curr Opin Microbiol 7(3):262–266

Chen M, Xu P, Zeng G, Yang C, Huang D, Zhang J (2017) Biosurfactant-producing bacteria for remediation of contaminated soils: a review. Sci Total Environ 577:132–144

Chopra S, Arora NK, Khare E (2020a) Microbial biosurfactants as ecofriendly agrochemicals for plant health and productivity. Rhizosphere 14:100202

Chopra S, Arora NK, Khare E (2020b) Pseudomonas aeruginosa RTE4: a tea rhizobacterium with potential for plant growth promotion and biocontrol of fungal pathogens. Front Bioeng Biotechnol 8:861

Choudhary M, Jat HS, Datta A, Yadav AK, Sapkota TB, Mondal S et al (2018) Sustainable intensification influences soil quality, biota, and productivity in cereal-based agroecosystems. Appl Soil Ecol 126:189–198

Christofi N, Ivshina IB (2002) Microbial surfactants and their use in field studies of soil remediation. J Appl Microbiol 93(6):915–929

Cirigliano MC, Carman GM (1984) Isolation of a bioemulsifier from Candida lipolytica. Appl Environ Microbiol 48(4):747–750

Cirigliano MC, Carman GM (1985) Purification and characterization of liposan, a bioemulsifier from Candida lipolytica. Appl Environ Microbiol 50(4):846–850
Cochard B, Gagné-Bourque F, Antoun H (2022) The application of plant growth-promoting rhizobacteria in Solanum lycopersicum: a review. Microorganisms 10(4):765. https://doi.org/10.3390/microorganisms10040765
Cooper DG, & Paddock D (1984) Production of a biosurfactant from Torulopsis bombicola. Appl Environ Microbiol 47(1):173–176
Cooper BH (1985) Phialophora verrucosa and other chromoblastomycotic fungi. In: Fungal dimorphism: with emphasis on fungi pathogenic for humans. Springer US, Boston, MA, pp 263–280
Dalili D, Rasekh B, Ebrahimipour G, Ghasemi Y (2015) Characterization of biosurfactant produced by Bacillus sp. isolated from oil contaminated soil. Iranian J Chem Chem Eng (IJCCE) 34(4):95–104
de França IWL, Lima AS, Melo VMM (2015) Production of a biosurfactant by Bacillus aryabhattai using cashew apple juice and its application in the remediation of oil-contaminated soil. Front Microbiol 6:1020
de Oliveira Schmidt F, dos Santos CG, da Silva LG, Meleiro LP (2021) Biosurfactant production from agro-industrial residues: a sustainable alternative for bioremediation and bioprocesses. Environ Technol Rev 10(1):91–110
Delaunois B, Farace G, Jeandet P, Clément C, Baillieul F, Dorey S (2014) Elicitors as alternative strategy to pesticides in grapevine? Current knowledge on their mode of action from controlled conditions to vineyard. Environ Sci Pollut Res 21(7):4837–4846
Desai JD, Banat IM (1997) Microbial production of surfactants and their commercial potential. Microbiol Mol Biol Rev 61(1):47–64
Desoky ESM, Saad AM, El-Saadony MT, Merwad ARM, Rady MM, AbuQamar SF (2020) Impact of ozone on maize productivity and physiological traits under water deficit. Plants 9(2):237
Diab A, & Sandouka M (2012) Effect of phytogenic biosurfactant on the microbial community and on the bioremediation of highly oil-polluted desert soil. J Am Sci 8(1):544–550
Dixit R, Pandey M, Pandey P, Shukla VK (2019) Chemotherapeutic drugs and gallbladder cancer: market potential in India. In Bioeconomy Sustain Dev, pp 379–388
Dusane DH, Damare SR, Zinjarde SS, Venugopalan VP (2010) Quorum sensing: implications on rhamnolipid biosurfactant production. Biotechnol Genet Eng Rev 27(1):159–184
Effendi AJ, Abdullah AZ, Bakar MHA (2018) Utilization of palm oil mill effluent (POME) for biosurfactant production: progress and prospects. J Clean Prod 199:468–480
Eras-Muñoz E, Farré A, Sánchez A, Font X, Gea T (2022) Microbial biosurfactants: a review of recent environmental applications. Bioengineered 13(5):12365–12391
Eriksson T, Karlsson J, Tjerneld F (2002) A model explaining declining rate in hydrolysis of lignocellulose substrates with cellobiohydrolase I (Cel7A) and endoglucanase I (Cel7B) of Trichoderma reesei. Appl Biochem Biotechnol 101:41–60
Ferreira INS, Silveira E, Rufino RD (2017) Biosurfactant production by Candida tropicalis on industrial residues. Biotechnol Prog 33(2):339–346
Geys R, Soetaert W, Van Bogaert I (2014) Biotechnological opportunities in biosurfactant production. Curr Opin Biotechnol 30:66–72
Gill D, Whitehead C, Wondimagegn D (2020) Challenges to medical education at a time of physical distancing. Lancet (London, England) 396(10244):77
Goes AP, Sheppard JD (1999) Effect of surfactants on α-amylase production in a solid substrate fermentation process. J Chem Technol Biotechnol Int Res Process Environ Clean Technol 74(7):709–712
Gowtham HG, Dhananjaya BL, Kannan S, Prasad RD (2022) Biotic stress management in crop plants: current perspectives and future directions. Physiol Mol Biol Plants 28(5):773–790
Ghosh RK, Gupta S, Singh V, & Ward PS (2021) Demand for crop insurance in developing countries: New evidence from India. J Agric Econ 72(1):293–320
Gronwald JW, Eberlein CV, Guttieri MJ (2002) Use of adjuvants to enhance herbicide activity. Weed Technol 16(1):1–10

Gudiña EJ, Rangarajan V, Sen R, Rodrigues LR (2016) Potential therapeutic applications of biosurfactants. Trends Pharmacol Sci 37(3):233–241

Gupta A, Nelson JM, Barrett TJ, Tauxe RV, Rossiter SP, Friedman CR et al (2004) Antimicrobial resistance among campylobacter strains, United States, 1997–2001. Emerg Infect Dis 10(6):1102

Gurav R, Hwang S, Bhatia SK, Kshirsagar AD, Mandal S, Yang YH (2025) Exopolysaccharide production using pinewood hydrolysate as a substrate for psychrotrophic bacterium isolated from Svalbard glacier soil. Biomass Convers Biorefinery 15(2):2263–2273

Hamidian H, Safaie N, Shams-Bakhsh M (2023) A comprehensive review on biotic stress tolerance in crop plants: approaches for improvement. Biol Res 56(1):4. https://doi.org/10.1186/s40659-023-00388-3

Hasanuzzaman M, Bhuiyan TF, Mahmud JA, Parvin K, Fujita M (2022) Drought and salinity stress in plants: an overview on physiological effects and adaptive mechanisms. Environ Exp Bot 203:105025. https://doi.org/10.1016/j.envexpbot.2022.105025

Henkel M, Müller MM, Kügler JH, Lovaglio RB, Contiero J, Syldatk C, Hausmann R (2012) Rhamnolipids as biosurfactants from renewable resources: concepts for next-generation rhamnolipid production. Process Biochem 47(8):1207–1219

Hoseini SM, Maleki R, Maleki M (2022) Plant growth-promoting rhizobacteria: a sustainable approach to improve crop productivity under abiotic stress. Environ Technol Innov 28:102671. https://doi.org/10.1016/j.eti.2022.102671

Jarvis FG, Johnson MJ (1949) A glyco-lipide produced by Pseudomonas aeruginosa. J Am Chem Soc 71(12):4124–4126

Javanbakht V, Alavi SA, Zilouei H (2014) Mechanisms of heavy metal biosorption by brown algae: a review. J Chem Eng Data 59(8):2741–2750

Jazzar C, Hammad EA (2003) Surfactants and insecticides interaction and effects on pest control. Commun Agric Appl Biol Sci 68(4):317–325

Jimoh AA, Lin J (2019) Biosurfactants: multifunctional biomolecules of the 21st century. Int J Multidiscip Res Dev 6(1):32–40

Jin Y, Chen J, Zhang Q, Farooq U, Lu T, Wang B et al (2023) Biosurfactant-affected mobility of oxytetracycline and its variations with surface chemical heterogeneity in saturated porous media. Water Res 244:120509

Jyoti K, Pattnaik P, Singh T (2021) Green synthesis of silver nanoparticles using sustainable resources and their use as antibacterial agents: a review. Curr Mater Sci Formerly: Recent Patents on Materials Science 14(1):40–52

Kakugawa K, Tamai M, Imamura K, Miyamoto K, Miyoshi S, Morinaga Y et al (2002) Isolation of yeast Kurtzmanomyces sp. I-11, novel producer of mannosylerythritol lipid. Biosci Biotechnol Biochem 66(1):188–191

Käppeli O, Finnerty WR (1979) Partition of alkane by an extracellular vesicle derived from hexadecane-grown Acinetobacter. J Bacteriol 140(2):707–712

Keswani C, Singh SP, Cueto L, García-Estrada C, Meza A, Glare TR (2019) Auxins of microbial origin and their use in agriculture. Appl Microbiol Biotechnol 103(3):1179–1195

Khan A (2022) Salinity stress in plants: a comprehensive review on physiological and molecular responses. Plant Stress 4:100068. https://doi.org/10.1016/j.stress.2022.100068

Khan N, Bano A, Rahman MA, Guo J, Kang Z, Babar MA (2021) Comparative physiological and metabolic analysis reveals a complex mechanism involved in drought tolerance in chickpea (Cicer arietinum L.) induced by PGPR and PGRs. Sci Rep 11(1):20949

Khare E, Arora NK (2021) Biosurfactant based formulation of Pseudomonas guariconensis LE3 with multifarious plant growth promoting traits controls charcoal rot disease in Helianthus annuus. Environ Sustain 4(2):213–222. https://doi.org/10.1007/s42398-021-00159-4

Khire JM (1994) Production of moderately halophilic amylase by newly isolated Micrococcus sp. 4 from a salt-pan. Lett Appl Microbiol 19(4):210–212

Kim DS, Park BS, Kim SG (2004) Antifungal activity of biosurfactants produced by Bacillus subtilis against rice blast fungus (Magnaporthe grisea). Plant Pathol J 20(1):36–39

Kim K, Jang YJ, Kim YC, Cho MC, Cho YS, Kim IS (2014) Enhancement of drought tolerance in plants by plant growth-promoting rhizobacteria. J Microbiol Biotechnol 24(9):1183–1190
Kiran GS, Hema TA, Gandhimathi R, Rani ME, Selvin J, Thomas TA, Lipton AP (2016) Optimization and production of a biosurfactant from the sponge-associated marine fungus Aspergillus ustus and its potential application in controlling Pseudomonas aeruginosa biofilm. Colloids Surf B: Biointerfaces 154:63–72
Kitamoto D, Akiba S, Hioki C, Tabuchi T (1990) Extracellular accumulation of mannosylerythritol lipids by a strain of Candida Antarctica. Agric Biol Chem 54(1):31–36
Kosaric N (1992) Biosurfactants in industry. Pure Appl Chem 64(11):1731–1737
Kosaric N (2001) Biosurfactants and their application for soil bioremediation. Food Technol Biotechnol 39(4):295–304
Kosaric N, Vardar-Sukan F (2014) Biosurfactants: production and utilization—processes, technologies, and economics. CRC Press
Kour D, Kaur T, Devi R, Yadav A, Singh M, Joshi D et al (2021) Beneficial microbiomes for bioremediation of diverse contaminated environments for environmental sustainability: present status and future challenges. Environ Sci Pollut Res 28:24917–24939
Krauss EM, Chan SI (1983) Complexation and phase transfer on nucleotides by gramicidin S. Biochemistry 22(18):4280–4291
Kretschmer A, Bock H, Wagner F (1982) Chemical and physical characterization of interfacial-active lipids from Rhodococcus erythropolis grown on n-alkanes. Appl Environ Microbiol 44(4):864–870
Krishnayya NSR, Grewal PS (2002) Synergistic and additive interactions among entomopathogenic nematodes and chemical insecticides against white grubs (Coleoptera: Scarabaeidae). Biol Control 25(2):125–133
Kumar A, Roberts D, Wood KE, Light B, Parrillo JE, Sharma S et al (2006) Duration of hypotension before initiation of effective antimicrobial therapy is the critical determinant of survival in human septic shock. Crit Care Med 34(6):1589–1596
Lahkar J, Saikia RR, Bordoloi S, Deka S, Kalita MC (2015) Sophorolipid biosurfactant against dermatophytes: a new approach in antifungal therapy. Biocatal Agric Biotechnol 4(3):289–295
Lang A, & Wagner GA (1996) Infrared stimulated luminescence dating of archaeosediments. Archaeom 38(1):129–141
Lai Q, Wang J, Gu L, Zheng T, Shao Z (2013) Alcanivorax marinus sp. nov., isolated from deep-sea water. Int J Syst Evol Microbiol 63(Pt_12):4428–4432
Lal R, Pandey AK, Sharma S (2018) Insight into the role of PGPR in sustainable agriculture and environment. Front Sustain Food Syst 2:1–7
Lamilla C, Schalchli H, Briceño G, Leiva B, Donoso-Piñol P, Barrientos L et al (2021) A pesticide biopurification system: a source of biosurfactant-producing bacteria with environmental biotechnology applications. Agronomy 11(4):624
Lang S, Wullbrandt D (1999) Rhamnose lipids–biosynthesis, microbial production and application potential. Appl Microbiol Biotechnol 51:22–32
Lang S, Wullbrandt D, Wagner F (1989) Sophorose lipid biosurfactants. In: Kosaric N (ed) Biosurfactants. Marcel Dekker, New York, pp 70–93
Li Y, Han D, Hu G, Sommerfeld M, Hu Q (2010) Inhibition of starch synthesis results in overproduction of lipids in Chlamydomonas reinhardtii. Biotechnol Bioeng 107(2):258–268
Loh J, Pierson EA, Pierson LS, Stacey G, Chatterjee A (2002) Quorum sensing in plant-associated bacteria. Curr Opin Plant Biol 5(4):285–290
Ma Y, Oliveira RS, Freitas H, Zhang C (2020) Biochemical and molecular mechanisms of plant-microbe-metal interactions: relevance for phytoremediation. Front Plant Sci 11:1360
Mandviwala TN, Khire JM (2000) Production of high activity thermostable phytase from thermotolerant Aspergillus niger in solid state fermentation. J Ind Microbiol Biotechnol 24(4):237–243
Maneerat S, Phetrong K (2007) Isolation of biosurfactant-producing marine bacteria and characteristics of selected biosurfactant. Songklanakarin J Sci Technol 29(3):781–791

Marchant R, Banat IM (2012) Biosurfactants: a sustainable replacement for chemical surfactants? Biotechnol Lett 34:1597–1605

Marchut-Mikołajczyk O, Drożdżyńska A, Różalska S (2021) Biosurfactants produced by microorganisms: potential applications in agriculture. Int J Mol Sci 22(19):10863

Mawar R, Gopal M, Vyas B (2023) Current advances and future prospects of plant growth-promoting rhizobacteria (PGPR) for enhancing crop productivity under stress conditions. Arch Microbiol 205(3):108

Mdeni NL, Adeniji AO, Okoh AI, Okoh OO (2022) Analytical evaluation of carbamate and organophosphate pesticides in human and environmental matrices: a review. Molecules 27(3):618

Mishra S, Arora NK (2018) Secondary metabolites of fluorescent pseudomonads in biocontrol of phytopathogens for sustainable agriculture. Appl Soil Ecol 125:35–45

Mnif I, Ghribi D (2015) Review lipopeptides biosurfactants: mean classes and new insights for industrial, biomedical, and environmental applications. Pept Sci 104(3):129–147

Mohanty SS, Jena S, Nayak S, Das S (2021) Biosurfactants: a green and sustainable technology for environmental remediation and industrial applications. J Clean Prod 284:124703

Moldes AB, Paradelo R, Vecino X, Cruz JM, Gudiña E, Rodrigues L et al (2013) Partial characterization of biosurfactant from Lactobacillus pentosus and comparison with sodium dodecyl sulphate for the bioremediation of hydrocarbon contaminated soil. Biomed Res Int 2013(1):961842

Morita T, Fukuoka T, Imura T, Kitamoto D (2015) Mannosylerythritol lipids: production and applications. J Oleo Sci 64(2):133–141

Morikawa M, Daido H, Takao T, Murata S, Shimonishi Y, & Imanaka T (1993) A new lipopeptide biosurfactant produced by Arthrobacter sp. strain MIS38. J Bacteriol 175(20):6459–6466

Mukherjee S, Das P, Sen R (2006) Towards commercial production of microbial surfactants. Trends Biotechnol 24(11):509–515

Mulligan CN (2005) Environmental applications for biosurfactants. Environ Pollut 133(2):183–198

Mulligan CN, Yong RN, Gibbs BF (2001) Surfactant-enhanced remediation of contaminated soil: a review. Eng Geol 60(1–4):371–380

Munir I, Rehman S, Ahmad F, Khan ZU (2022) Response of plants to abiotic stress: role of reactive oxygen species and antioxidants. Plant Physiol Rep 27(3):353–368

Murali M, Kumar A, Senthil-Kumar M (2021) Plant–pathogen interactions: exploring biocontrol mechanisms of plant growth-promoting rhizobacteria. Microorganisms 9(3):682

Muthusamy K, Gopalakrishnan S, Ravi TK, Sivachidambaram P (2008) Biosurfactants: properties, commercial production and application. Curr Sci 94:736–747

Newton JA, Fray RG (2004) Integration of environmental and host-derived signals with quorum sensing during plant–microbe interactions. Cell Microbiol 6(3):213–224

Nihorimbere V, Cawoy H, Seyer A, Brunelle A, Thonart P, Ongena M (2011a) Impact of rhizosphere factors on cyclic lipopeptide signature from the plant-beneficial strain Bacillus amyloliquefaciens S499. FEMS Microbiol Ecol 79(1):176–191

Nihorimbere V, Cawoy H, Seyer A, Brunelle A, Thonart P, Ongena M (2011b) Impact of rhizosphere factors on the antibacterial activity of Bacillus amyloliquefaciens S499 against plant pathogenic bacteria. FEMS Microbiol Ecol 76(3):395–410

Nikolopoulou M, Pasadakis N, Kalogerakis N, Venieri D (2008) Enhanced ex situ bioremediation of crude oil contaminated beach sand by supplementation with nutrients and rhamnolipids. Mar Pollut Bull 56(11):1855–1862

Nitschke M, Silva SS (2018) Recent food applications of microbial surfactants. Crit Rev Food Sci Nutr 58(4):631–638

Ochoa-Loza FJ, Noordman WH, Jannsen DB, Brusseau ML, Maier RM (2007) Effect of clays, metal oxides, and organic matter on rhamnolipid biosurfactant sorption by soil. Chemosphere 66(9):1634–1642

Oliveira A, Domingues CM, Silva AM (2015) Application of pesticides and its impact on human health: a review. Environ Sci Pollut Res 22(19):13804–13819

Pacwa-Płociniczak M, Płaza GA, Piotrowska-Seget Z, Cameotra SS (2011) Environmental applications of biosurfactants: recent advances. Int J Mol Sci 12(1):633–654
Pathak KV, & Keharia H (2014) Application of extracellular lipopeptide biosurfactant produced by endophytic Bacillus subtilis K1 isolated from aerial roots of banyan (Ficus benghalensis) in microbially enhanced oil recovery (MEOR). 3 Biotech 4(1):41–48
Pardo A, Selman M, Ridge K, Barrios R, Sznajder JI (1996) Increased expression of gelatinases and collagenase in rat lungs exposed to 100% oxygen. Am J Respir Crit Care Med 154(4):1067–1075
Partovi M, Shokrollahzadeh S, Shojaosadati SA (2013) Biosurfactant production from Pseudomonas aeruginosa PTCC 1074 using whey as substrate. J Environ Health Sci Eng 11(1):1–7
Peng P, Zhai MG, Guo JH, Kusky T, Zhao TP (2007) Nature of mantle source contributions and crystal differentiation in the petrogenesis of the 1.78 Ga mafic dykes in the central North China craton. Gondwana Res 12(1–2):29–46
Pepi M, Cesàro A, Liut G, Baldi F (2005) An antarctic psychrotrophic bacterium Halomonas sp. ANT-3b, growing on n-hexadecane, produces a new emulsyfying glycolipid. FEMS Microbiol Ecol 53(1):157–166
Pi Y, Zhang D, Liu Y, Chen J, Wang J (2017) Applications of biosurfactants in environmental pollution control: a review. Crit Rev Biotechnol 37(4):543–554
Popp J, Pető K, Nagy J (2013) Pesticide productivity and food security. A review. Agron Sustain Dev 33:243–255
Pourfadakari S, Ghassemi M, Moussavi G (2019) Biosurfactant-enhanced electrokinetic remediation of heavy metals-contaminated soil using rhamnolipid: optimization and modeling. Sep Purif Technol 210:935–942
Qi Y, Wei W, Dai J, Wang Y, Liu C (2018) Application of sophorolipids from Starmerella bombicola for heavy metal removal in soil washing process. J Environ Manag 217:544–551
Raaijmakers JM, De Bruijn I, Nybroe O, Ongena M (2010) Natural functions of lipopeptides from Bacillus and Pseudomonas: more than surfactants and antibiotics. FEMS Microbiol Rev 34(6):1037–1062
Rahman PK, Gakpe E (2008) Production, characterisation and applications of biosurfactants-review. Biotechnol 7:360
Rajendran P, Muthukrishnan J, Gunasekaran P (2003) Microbial remediation of heavy metal contaminated ground water. Environ Monit Assess 87(3):293–300
Rani L, Thapa K, Kanojia N, Sharma N, Singh S, Grewal AS et al (2021) An extensive review on the consequences of chemical pesticides on human health and environment. J Clean Prod 283:124657
Ribeiro IAC, Faustino MAF, Neves MGPMS, Tome AC, Fernandes R, Almeida A (2013) Photodynamic inactivation of bacteria using modified porphyrins and sophorolipids. Eur J Med Chem 65:292–297
Rodriguez SB, Mahoney NE (1994) Inhibition of aflatoxin production by surfactants. Appl Environ Microbiol 60(3):1063–1070
Ron EZ, Rosenberg E (2001) Natural roles of biosurfactants: Minireview. Environ Microbiol 3(4):229–236
Ron EZ, Rosenberg E (2011) Natural roles of biosurfactants. Environ Microbiol 13(2):229–234
Rosenberg E, Ron EZ (1999) High-and low-molecular-mass microbial surfactants. Appl Microbiol Biotechnol 52:154–162
Rosenberg E, Zuckerberg A, Rubinovitz C, Gutnick D (1979) Emulsifier of Arthrobacter RAG-1: isolation and emulsifying properties. Appl Environ Microbiol 37(3):402–408
Rostas M, Blassmann K (2009) Insects had it first: surfing the interface between plants and surfactants. Plant Signal Behav 4(5):445–446
Ruggeri C, Veglio F, Pagnanelli F (2009) Process optimization for biosurfactant production by Pseudomonas aeruginosa using agro-industrial wastes. Chem Eng Trans 17:1063–1068
Sabarwal A, Kumar K, Singh RP (2018) Hazardous effects of chemical pesticides on human health–cancer and other associated disorders. Environ Toxicol Pharmacol 63:103–114

Sachdev DP, & Cameotra SS (2013) Biosurfactants in agriculture. Appl Microbiol Biotechnol 97(3):1005–1016

Sarkar DJ, Sarkar SD, Das BK, Sahoo BK, Das A, Nag SK et al (2021) Occurrence, fate and removal of microplastics as heavy metal vector in natural wastewater treatment wetland system. Water Res 192:116853

Sarker U, Oba S, Daramy MA, Nugraha R (2021) Plant responses to abiotic stress and amelioration strategies: a review. Agronomy 11(8):1545

Satpute SK, Banat IM, Dhakephalkar PK, Banpurkar AG, Chopade BA (2010) Biosurfactants, bioemulsifiers and exopolysaccharides from marine microorganisms. Biotechnol Adv 28(4):436–450

Sekhon KK, Khanna S, Cameotra SS (2012) Biosurfactant production using molasses and whey under thermophilic conditions. Int J Dairy Technol 65(4):602–607

Sen R, Tripathi A, Kaur T (2017) Sophorolipid production from Rhodotorula babjevae YS3 and its application in controlling phytopathogens. J Basic Microbiol 57(12):1051–1062

Seung-Hak Y, Song JJ, Jeon EY, Song JM (2003) Antimicrobial activity of sophorolipid. J Microbiol Biotechnol 13(4):585–589

Seydlová G, Svobodová J (2008) Review of surfactin chemical properties and the potential biomedical applications. Cent Eur J Med 3:123–133

Sharma D, Saharan BS, Chauhan N, Procházka E, Saini HS (2016) Production and structural characterization of biosurfactant produced by Bacillus subtilis MTCC 9763. Int J Sci Technol Res 5(5):50–55

Sharma N, Manhas RK, Ohri P (2022) Streptomyces hydrogenans strain DH-16 alleviates negative impact of Meloidogyne incognita stress by modifying physio-biochemical attributes in Solanum lycopersicum plants. Sci Rep 12(1):15214

Shekhar S, Sundaramanickam A, Balasubramanian T (2015) Biosurfactant producing microbes and their potential applications: a review. Crit Rev Environ Sci Technol 45(14):1522–1554

Sharma S, Verma R, & Pandey LM (2019) Crude oil degradation and biosurfactant production abilities of isolated Agrobacterium fabrum SLAJ731. Biocatal Agric Biotechnol 21:101322

Sidkey NM, Abo-Shadi MA, Badr HA, Shabeb AA (2016) Biosurfactant production by endophytic bacteria isolated from medicinal plant Ocimum basilicum L. J Genet Eng Biotechnol 14(2):251–256

Singh RK, Singh P, Sharma L (2017) Plant growth-promoting rhizobacteria: an eco-friendly approach for sustainable agriculture. Biocatal Agric Biotechnol 11:90–100. https://doi.org/10.1016/j.bcab.2017.06.010

Singh P, & Cameotra SS (2004) Potential applications of microbial surfactants in biomedical sciences. TRENDS Biotechnol 22(3):142–146

Singh A, Van Hamme JD, & Ward OP (2007) Surfactants in microbiology and biotechnology: Part 2. Application aspects. Biotechnol Adv 25(1):99–121

Singh S, Singh V, Upadhyay RS (2019a) Rhizospheric microbial diversity and PGPR functions in sustainable agriculture. World J Microbiol Biotechnol 35(12):187

Singh R, Kumar M, Mittal A, Mehta PK (2019b) Microbial surfactants: a promising bioresource for remediation of heavy metals and xenobiotics. Environ Chem Lett 17:597–613

Sotirova AV, Spasova DI, Galabova DN, Karpenko E, Shulga A (2008) Rhamnolipid–biosurfactant permeabilizing effects on gram-positive and gram-negative bacterial strains. Curr Microbiol 56:639–644

Tang X, Li J, Li K, Lin Y, Lin H (2017) Electrokinetic removal of heavy metals from sewage sludge with the aid of complexing agents. J Hazard Mater 322:400–407

Teichmann B, Linne U, Hewald S, Marahiel MA, Bölker M (2007) A biosynthetic gene cluster for a secreted cellobiose lipid with antifungal activity from Ustilago maydis. Mol Microbiol 66(2):525–533

Tortora M, Pansini M, Pezzella C, Marzocchella A, d'Antonio G (2016) Biosurfactant micelle-enhanced ultrafiltration for wastewater treatment. J Membr Sci 510:1–8

Ullah A, Sun H, Yang X, Zhang X (2021) Drought coping strategies in cotton: increased crop per drop. Plant Biotechnol J 19(4):773–788
Usman M, Ibrahim M, Khan M (2016) Biosurfactants: an eco-friendly alternative for remediation of heavy metals. Environ Sci Pollut Res 23(13):13451–13467
Valotteau C, Baccile N, Roelants SLKW, Uyttersprot K, Vanbroekhoven K, Soetaert W (2017) Self-assembly properties of acidic, lactonic, and acetylated sophorolipids: a comparative physico-chemical and functional study. Langmuir 33(35):8754–8764
Vanittanakom N, Loeffler W, Koch U, Jung G (1986) Fengycin-a novel antifungal lipopeptide antibiotic produced by Bacillus subtilis F-29-3. J Antibiot 39(7):888–901
Vecino Bello X, Devesa-Rey R, Cruz JM, Moldes AB (2012) Study of the synergistic effects of salinity, pH, and temperature on the surface-active properties of biosurfactants produced by Lactobacillus pentosus. J Agric Food Chem 60(5):1258–1265
Vecino X, Devesa-Rey R, Cruz JM, Moldes AB (2013) Evaluation of biosurfactant obtained from Lactobacillus pentosus as foaming agent in froth flotation. J Environ Manag 128:655–660
Vecino X, Barbosa-Pereira L, Devesa-Rey R, Cruz JM, Moldes AB (2014) Study of the surfactant properties of aqueous stream from the corn milling industry. J Agric Food Chem 62(24):5451–5457
Villa-Rodriguez E, Mena-Violante HG, Orozco-Mosqueda MC, Santoyo G (2019) Plant probiotic bacteria: mechanisms and potential applications in the rhizosphere. Microbiol Res 229:126331
Wan Y, Wang Y, Fan T, Jin X (2017) Influence of soil properties on the removal of heavy metals from contaminated soil by biosurfactants. Environ Sci Pollut Res 24(6):5650–5661
Waqas M, Wang W, Zafar M, Khan MA, Khan AL (2019) Abiotic stress tolerance and sustainable agriculture. In: Hasanuzzaman M et al (eds) Plant abiotic stress tolerance. Springer, pp 1–23
Wu J, Li Y, Zhang D, Chen S, Liu J (2017) Mechanism and application of micelle-enhanced ultrafiltration in removal of heavy metals: a review. Chemosphere 168:1467–1476
Xue Z, He H, Gao H (2018) Biosurfactants for bioremediation of heavy metals: recent advances and future perspectives. Environ Pollut 240:338–348
Yakimov MM, Amro MM, Bock M, Boseker K, Fredrickson HL, Kessel DG, Timmis KN (1997) The potential of Bacillus licheniformis strains for in situ enhanced oil recovery. J Pet Sci Eng 18(1–2):147–160
Yang Z, Zhou F, Jin M, Zhao Y, Wang Q (2018) Application of biosurfactant produced by Burkholderia sp. for heavy metals removal from contaminated soil. Environ Sci Pollut Res 25(23):22726–22735
Yoo MS, Choi JN, Park JW (2005) Antifungal activity of sophorolipid from Candida bombicola against plant pathogenic fungi. J Microbiol Biotechnol 15(1):68–71
Zhang QQ, Yan YQ, Dai XQ, Liu YQ, Zhang YZ (2017) Antimicrobial activity of sophorolipid biosurfactants produced by Wickerhamiella domercqiae. AMB Express 7(1):120
Zhao Y, Wang H, Chen W, Liu J, Liu Y (2022) Role of PGPR in improving crop tolerance to drought and salinity stress. Agronomy 12(5):1054
Zosim Z, Gutnick D, Rosenberg E (1982) Properties of hydrocarbon-in-water emulsions stabilized by Acinetobacter RAG-1 emulsan. Biotechnol Bioeng 24(2):281–292

Chapter 14
Endophytes as Biological Control Agents of Plant Diseases

Ibrahim Elsaid Ibrahim Elshahawy and Wafaa Mohamed Haggag

Abstract The term "endophyte" refers to nonpathogenic microorganisms found within plants. These microorganisms are numerous, comprising several orders of both fungi and bacteria, and can operate as commensalistic or mutualistic symbionts with their hosts. Endophytes are distinguished by their ability to infiltrate plant tissues, persist, and interact with the host in a variety of ways. They enter plant tissues from within and can act as control agents or elicitors, causing resistance and alleviating the negative consequences of abiotic stress. They can improve plant fitness by lowering biotic and abiotic stress, encouraging the growth and development of plants, and decreasing agricultural output losses.

This review examined the probable mechanisms of endophyte–host interactions. Endophytes create phytohormones that increase plant health and provide extra benefits such as nutrient intake, nitrogen fixation, and preservation under stress from abiotic as well as biotic settings. Some endophytes can actively protect their host plant from infections by secreting substances that kill or limit the growth of pathogenic microbes or by activating the plant immune response through the activation of the host plant's natural signaling pathways. Endophytes inhibit infections via a variety of mechanisms, including mycoparasitism, competition, antibiosis, and plant defense induction, demonstrating their ability to control a wide spectrum of phytopathogenic organisms. They can also produce a wide range of bioactive compounds that are beneficial to plants. In this study, we detail the isolation techniques as well as various in vitro and in vivo studies on endophytes as biological control agents against pre- and postharvest illnesses. Furthermore, registration and regulatory strategies for endophytic biocontrol agents are discussed. Thus, isolating and identifying endophytic microorganisms from unstudied plants can lead to the discovery of novel species capable of producing chemicals of interest for application in disease control, induced resistance elicitors, and abiotic stress tolerance induction.

Keywords Endophytes · Plant diseases · Biocontrol agents · Abiotic stress

I. E. I. Elshahawy (✉) · W. M. Haggag
Department of Plant Pathology, National Research Centre, Giza, Egypt

V. Gopikrishnan et al. (eds.), *Harnessing Microbial Resources for Sustainable Agriculture*, Microorganisms for Sustainability 61,
https://doi.org/10.1007/978-981-95-4817-0_14

14.1 Introduction

Food security is one of the most important concerns of the world's growing population. The world's population is predicted to reach 9.7 billion by 2050, necessitating more than a 50% increase in agricultural production above current levels. Furthermore, the requirement to ensure food supply has increased as a result of limited resources such as agricultural land, deforestation, shifting climatic patterns, and agricultural productivity losses caused by pre- or postharvest pathogen attacks or plant illnesses (Peng et al. 2021). Plants are constantly subjected to a variety of biotic and abiotic stresses that have a negative impact on their growth and survival (Haggag 2024). Plant diseases or pathogen attacks are predicted to account for 15–30% of total production from agricultural losses (Ristaino et al. 2021).

Nonetheless, a significant population relies on synthetic pesticides to combat plant diseases; nonetheless, the frequent and reckless use of commercial insecticides has a severe influence on the quality of food, soil texture, native microbiota, and human health (Aktar et al. 2009). In addition, contamination of soil and water ecosystems, as well as the evolution of insect resistance, has emerged as a significant impediment to chemical pesticide use. Microbial strains have evolved as long-term, nontoxic, and ecologically benign solutions to biotic stress problems (Haggag et al. 2017; Haggag and Ali 2019; Morales-Cedeno et al. 2021).

Since they have shown potency against a variety of plant pathogens at low concentrations, endophytic strains have been considered a nontoxic and genetically stable approach for antagonistic or biocontrol agents (BCAs) (Strobel 2018). Endophytic strains also encourage plant growth, which further benefits plants. Many microbial species, including bacteria, fungi, and actinomycetes, have been discovered and employed as BCAs in different parts of the world during the past 20 years to reduce agricultural crop losses before or after harvest (Kumari et al. 2022). Endophytic microorganisms are among the most suitable methods since they are more capable of colonizing and acclimatizing than epiphytic microorganisms, which are found above or below the plant surface (Priyashantha et al. 2023).

Endophytes use a variety of direct and indirect ways to regulate plant diseases and limit phytopathogen proliferation. Induced systemic resistance, siderophore production, nutrition and space competition, and antimicrobial chemical and volatile release are a few important strategies. Furthermore, the use of endophytes alters the local microbial community, hence considerably affecting agricultural productivity (Anand et al. 2023). The biology of endophytes, their penetration into plant tissues, their colonization, and the processes behind endophytic biocontrol agents are examined in this article. Additionally, screening techniques, isolation procedures, and the potential use of endophytic biocontrol agents in the control of plant diseases are included.

14.2 Definition of Endophytes

The word "endon" means "within," whereas "phyton" means "plant." These two Greek words are where the word "endophyte" originated. De Bary called "any microorganism living among plants tissues" an "endophyte" in 1866. "Organisms that cause asymptomatic infections within plant tissues" is how Carroll described "endophytes" in 1986 (Carroll 1988). The term "endophyte" was defined by Petrini at the close of the twentieth century to encompass "all microorganisms inhabiting plant organs that at some time in their life may invade internal plant tissues without causing apparent harm to their host" (Potshangbam et al. 2017).

14.3 Types of Endophytes

Research indicates that endophytes are present in every plant species at some point during its life cycle (Verma et al. 2021). The genotypes, organs, seasons, and environmental conditions of plants affect the composition of the endophytic microbial population. On the other hand, the most prevalent bacterial phylum was Proteobacteria, followed by Actinobacteria, Firmicutes, and Pseudomonas. In plants or their organs, the three most prevalent endophytic bacterial genera found are *Enterobacter*, *Acinetobacter*, and *Pantoea*. The main endophytic fungal taxa found in different host plants are *Glomeromycota*, *Ascomycota*, *Basidiomycota*, and arbuscular mycorrhizal fungi (AMFs) (Kumar et al. 2020; Verma et al. 2021; Kashyap et al. 2023).

14.4 Entry of Endophytes into Plant Tissues

These bacteria are vital to the health and productivity of plants because they are part of the complex microbial community that exists above, outside, and inside their organs. When certain bacteria that reside on the surface of plants invade plant tissues, they transform into endophytes (Santoyo et al. 2016). Numerous proteins and signaling molecules aid in the complex process of endophyte entry into plant tissues (Fig. 14.1). In wounds and naturally occurring apertures, such as stomata, cotyledon hydathodes, and aerial locations in root zones, microorganisms most frequently invade host tissues (Kumar et al. 2020). Although secretory products or natural apertures can allow microbial strains to enter, the latter is referred to as active, and the former is referred to as passive (Kumar et al. 2020).

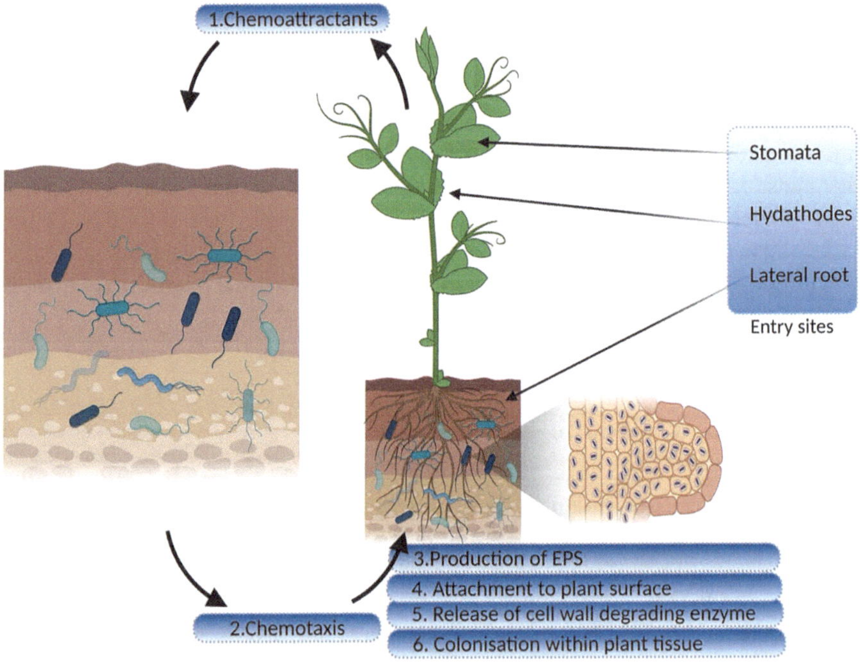

Fig. 14.1 The way that endophytes penetrate different plant tissues. They first penetrate the roots, which are home to nodules, lesions, fractures, and both primary and auxiliary root hair cells. Second, these bacteria adhere to the exopolysaccharides (EPSs) found in leaf stomata, as modified by Rani et al. (2022)

14.5 Colonization and Transmission of Endophytes

There are several ways to successfully colonize microbial strains (Kumar et al. 2020). For example, the rhizospheres of plants release copious amounts of exudates containing organic acids as well as proteins, lipids, amino acids, and flavonoids. The chemotactic response of organisms, which guides their movement toward the plant surface and eventual colonization, is caused by these exudates. Importantly, every microbial strain is specific to a certain exudate component (Oku et al. 2012). Root exudates demonstrate a two-way relationship between the endophyte and its host plant by inducing the expression of genes encoding proteins involved in adhesion, chemotaxis, and motility (Jha et al. 2018). Chemotaxis is an essential mechanism that takes place inside roots and in the rhizosphere for mobility and invasion (Kawasaki et al. 2016).

Owing to their critical role in exopolysaccharide (EPS) synthesis and biofilm formation, mutant strains of *Azorhizobium caulinodans* that lack a chemotaxis gene cluster (che) have been demonstrated to exhibit defective invasion (Liu et al. 2018).

Furthermore, a number of factors, including plant genotype and the type of microorganisms present in the surrounding environment—which may be pathogenic or symbiotic—determine how well endophytes colonize (Kumar et al. 2020). Certain endophytes have the capacity to proliferate throughout the plant, eventually making their way to the seeds, fruits, and flowers. Some endophytes travel via the xylem of their host roots to reach their meristems, where they are assisted by the movement of flagella and the plant transpiration current (Compant et al. 2011). To take advantage of nutrient-rich intracellular areas, other endophytes must produce enzymes that breakdown cell walls (Truyens et al. 2014).

Endophyte transmission within the host can occur via a variety of paths, including vertical, horizontal, and mixed pathways (Fig. 14.2). The vertical mode, which is usually followed by endophyte species, describes the transfer of endophytes via seeds, pollen, and parents. On the other hand, the horizontal mode illustrates how endophytic bacteria or fungi proliferate through the soil, air, and surroundings. According to Frank et al. (2017), some microbial strains are categorized in mixed mode because they use both modes of transmission.

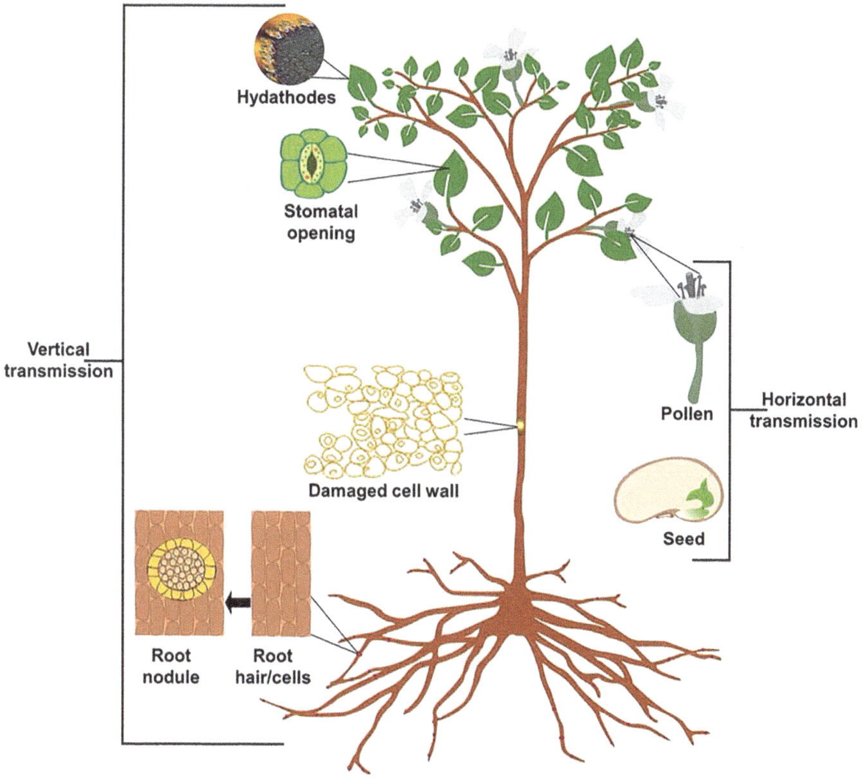

Fig. 14.2 Transfer of endophytes to the host plant. While the horizontal mode, as used by Compant et al. (2011), represents the transfer of either bacterial or fungal endophytic germs by air, soil, and the environment, the vertical mode illustrates endophyte transmission by seeds, pollen, and parent plants

14.6 The Possibility of Pathogenic Versus Endophytism: A Fine Distinction Between Two Different Lifestyles

Despite the positive effects of this plant–endophyte relationship, plants may also receive signals from microorganisms and respond appropriately by activating their immune system (Zipfel and Oldroyd 2017). Microbial strains must initially enter plant tissues through intracellular or intercellular gaps to become endophytes, but during this process, bacteria are exposed to the oxidative environment of the plant. To lessen the oxidative burst, they produce antioxidant machinery enzymes such as glutathione S-transferases, peroxidase, alkyl-hydroperoxide reductase C, catalase, and superoxide dismutase (Zeidler et al. 2004). The function of antioxidant enzymes generated by *Enterobacter* species in the successful colonization of poplar plants was examined by Balsanelli et al. (2016).

Studies have shown that while harmful and symbiotic endophytes exhibit somewhat similar colonization patterns, their defensive mechanisms are significantly different. Similarly, the number of enzymes that degrade cell walls differs between symbiotics and pathogens. For example, viruses release large quantities of enzymes that breakdown cell walls, which initiates the host's immunological response, whereas symbiotic endophytes release small amounts of enzymes that do not cause an immune response (Kumar et al. 2023).

To recognize microbial signaling molecules, plants rely on innate immunity. This leads to two different defense mechanisms: the first activates MAMP-triggered immunity (MTI) by recognizing microbe-associated molecular patterns (MAMPs) via cell surface-localized pattern recognition receptors (PRRs), and the second activates effector-induced immunity (ETI) by utilizing intracellular receptors to identify molecules produced by microorganisms (also known as effectors) (Mendoza-Mendoza et al. 2018). According to Zamioudis and Pieterse (2012), beneficial bacteria can avoid ETI by modifying or eliminating known effectors, permitting effective colonization of host roots, or avoiding host mechanisms of defense while in the effector-triggered sensitivity stage.

Plants react similarly to pathogenic fungal invasion and first-endophyte fungal infection in terms of the MTI defense mechanism. In this system, endophytic fungi from the community of microbes may modify their interactions with the host plant to "selfishly" increase their own fitness (Vandenkoornhuyse et al. 2015). Two examples are structural modifications and changes in reaction targets. β-glucan, which is produced by fungal cell walls, can activate plant immunity. A lectin secreted by the root endophyte *Piriformospora indica* only binds to β-glucan. This lectin can change the fungal cell wall and reduce β-glucan-induced immunity in plant hosts, according to Wawra et al. (2016). The current hypotheses and mechanisms underlying the relationships between endophytic fungi and plants are illustrated in Fig. 14.3.

The initial hypothesis of "balanced antagonism" was put forth by Schulz et al. (1999). The balanced antagonistic theory states that plant defensive mechanisms and the harmful effects of endophytes on plants are in balance. By doing so, endophytes are able to survive inside their hosts without showing any symptoms of

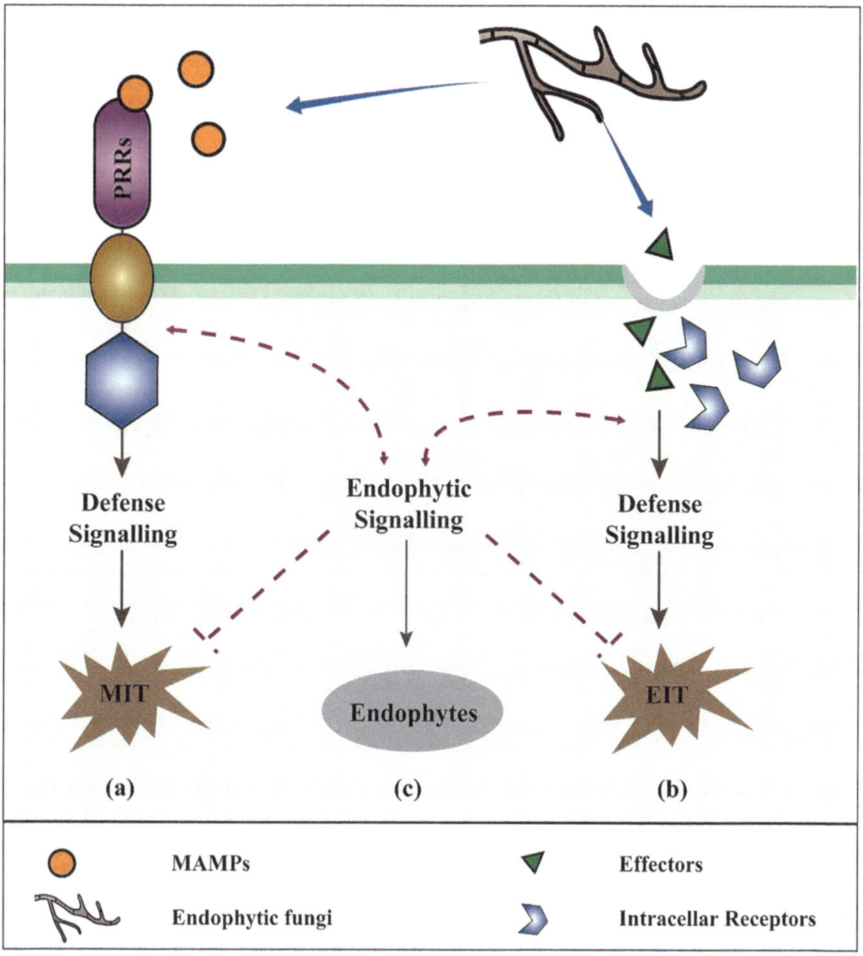

Fig. 14.3 Two distinct defense responses are produced by plant immune systems in response to their recognition of microbial signaling molecules. (**a**) MAMP-triggered immunity (MTI) is the consequence of plants using cellular surface-localized pattern recognition receptors (PRRs) to identify microbe-associated molecular patterns (MAMPs) in the host. (**b**) Plant intracellular receptors are able to identify substances called effectors that are produced by microorganisms, leading to effector-triggered immunity. (**c**) Beneficial bacteria can successfully colonize roots by surviving or evading plant defenses by remaining in an effector-triggered susceptible state (as modified by Yan et al. 2019). The interaction of endophytes with a living plant partner requires the regulation and management of plant immunity, which developed to shield plants from undesirable and damaging invaders (Ruhe et al. 2016)

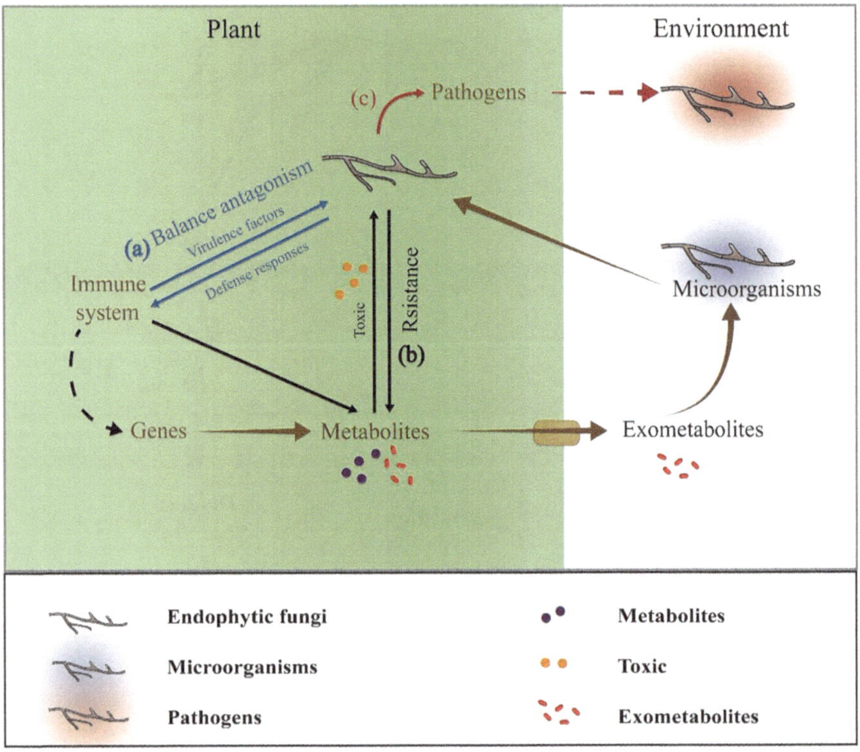

Fig. 14.4 Suggested mechanisms for endophyte–host–plant interactions. Endophytes (brown arrows) are microorganisms that migrate inside plants and release chemicals that control microbial populations. (**a**) Endophytes can multiply despite the host's defense response because they have a variety of harmful characteristics (blue arrows). (**b**) Endophytes actively inhibit the stimulation of the plant immune response via several complex strategies (black arrows). In response to specific environmental or intrinsic conditions, endophytes can develop into latent pathogens (red arrows), as Yan et al. (2019) have adapted

illness, avoiding being rendered incompetent by the toxic substances in their hosts (Fig. 14.4a). Partnerships between plants and endophytes remain asymptomatic as long as fungal virulence and the host defenses are balanced.

The plant initially developed a temporary immune response in response to endophytes and diseases, which it perceived as potential invaders. To address the host's defense response, these potential invaders possess a range of pathogenic traits. Yan et al. (2018) reported that if fungal virulence and plant defense are in balance, colonization is asymptomatic. Endophytes either infect plants or die otherwise.

14.7 Positive Impact of Endophyte Colonization on Plants

14.7.1 Promotion of Plant Growth (PGP)

Endophytic microorganisms can increase plant growth by supplying nutrients, directly generating phytohormones and growth regulators, or controlling phytohormone concentrations.

14.7.1.1 Fixation of Nitrogen

One of the most important nutrients for plant growth is nitrogen (N). It is an essential part of every living thing's protoplasm. According to Kim and Rees (1994), nitrogen-fixing bacteria transform atmospheric N2 into ammonia, which plants can use through a process called biological nitrogen fixation (BNF) and a complex enzyme system called nitrogenase. In addition to nonleguminous plants (such as Francia) and leguminous plants (such as Rhizobia), nitrogen-fixing microbes called symbiotic bacteria produce BNF at moderate temperatures (Kim and Rees 1994). The focus of this symbiotic group was leguminous plants. The sugars on the bacterial cell wall and the lectins from the leguminous plant are responsible for this specialization (Kim and Rees 1994).

The elongation of root hairs is caused by bacteria that are present in large quantities close to the root and convert tryptophan from secreted root exudates to indole acetic acid (IAA). The bacteria then create an infection thread of bacterial cells, cellulose, and hemicellulose molecules by penetrating the root hair at its weakest location (Kim and Rees 1994). Nodules start to form when bacteria reach the cortex (Fig. 14.5). In the fixation process, the nitrogenase enzyme takes part in the following ways: 2HN3 is produced by the nitrogenase enzyme (N2 + 3H2). The structure of the two components of this enzyme was established by Dean and Jacobson

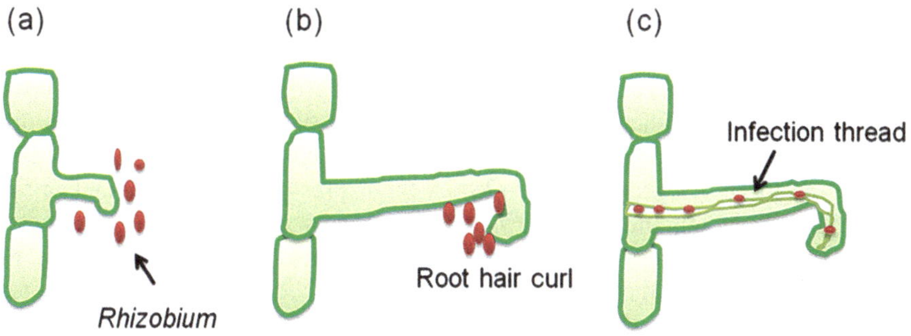

Fig. 14.5 (**a**) Rhizobia adhere to root cells through a process called nodulation. (**b**) Rhizobia generate nod factors that encourage the curling of root hairs. (**c**) According to Kim and Rees (1994), a bacterial infection thread penetrates the hair on the root and initiates bacteroid conditions, resulting in nodules

(1992). Strong reducing power is provided to electrons by the first component, dinitrogenase reductase. These electrons are used by the second enzyme, dinitrogenase, to convert N2 into NH3. Nitrogen fixation genes, or nif genes, are present in both symbiotic and free-living PGPR (Kim and Rees 1994).

14.7.1.2 Solubility of Phosphate

The next limiting element for plants is phosphorus. They affect signaling and cell metabolism (Vance et al. 2003). The plants cannot use more than 90% of the phosphorus in the soil.

Phosphorus is absorbed by plants as monobasic H2 PO4− and diabasic HPO42−; however, it is unavailable due to its bonding to organic or inorganic substances (Vance et al. 2003). Ecology is negatively impacted by the overuse and inappropriate application of phosphorus, even when it is utilized as a chemical fertilizer. Endophytes are microbes that solubilize phosphorus and make it accessible to plants. In part, the production of organic acids such as gluconic acid releases phosphorus from a bound state (Rodriguez et al. 2006). Furthermore, enzymes, including phosphonates, phytases, and C-P lyases, help transform insoluble phosphorus into accessible phosphorus.

Arbuscular mycorrhizal fungi, also known as roots, have been present for more than 400 million years, making this symbiotic relationship one of the oldest and most common (Kobae 2019). Corn, wheat, and soybeans are just a few of the commercially important crops that exhibit this relationship. According to Smith et al. (2011), plants are able to absorb phosphate further from the root surface when arbuscular mycorrhizal fungi hyphae form outside of the roots. Numerous arbuscular mycorrhizal fungal propagules, including spores, hyphae, and root remnants, are found in soil, and crop roots are frequently mycorrhizal (Sanders et al. 1996). AMF penetration into root cortical cells and the development of an arbuscule, a densely branched hyphal structure, are characteristics of arbuscular mycorrhizal fungal symbiosis (Sanders et al. 1996).

Even though AMF live inside root cells, the periarbuscular membranes around them that attach to plant cell plasma membranes show that they are outside of cells and cannot be fully absorbed as plant organelles (Harrison and Ivanov 2017). An increase in host phosphate transporter gene expression during arbuscule development facilitates phosphate absorption from arbuscules (Pumplin et al. 2012). According to Harrison and Ivanov (2017), the expressed protein is found in the periarbuscular membrane (Fig. 14.6). A carbon supply is necessary for arbuscular mycorrhizal production in host plants to promote growth (Sawers et al. 2017). To offset the expense of other types of metabolism, AMF development requires lipids and carbohydrates from the host plants (Sawers et al. 2017). As a result, the arbuscules nearest the tip of the growing hyphae are the newest, and those nearest the penetration site are the oldest. Arbuscules collapse in decreasing order of age because of their brief existence (1–2 days). A plant membrane known as the

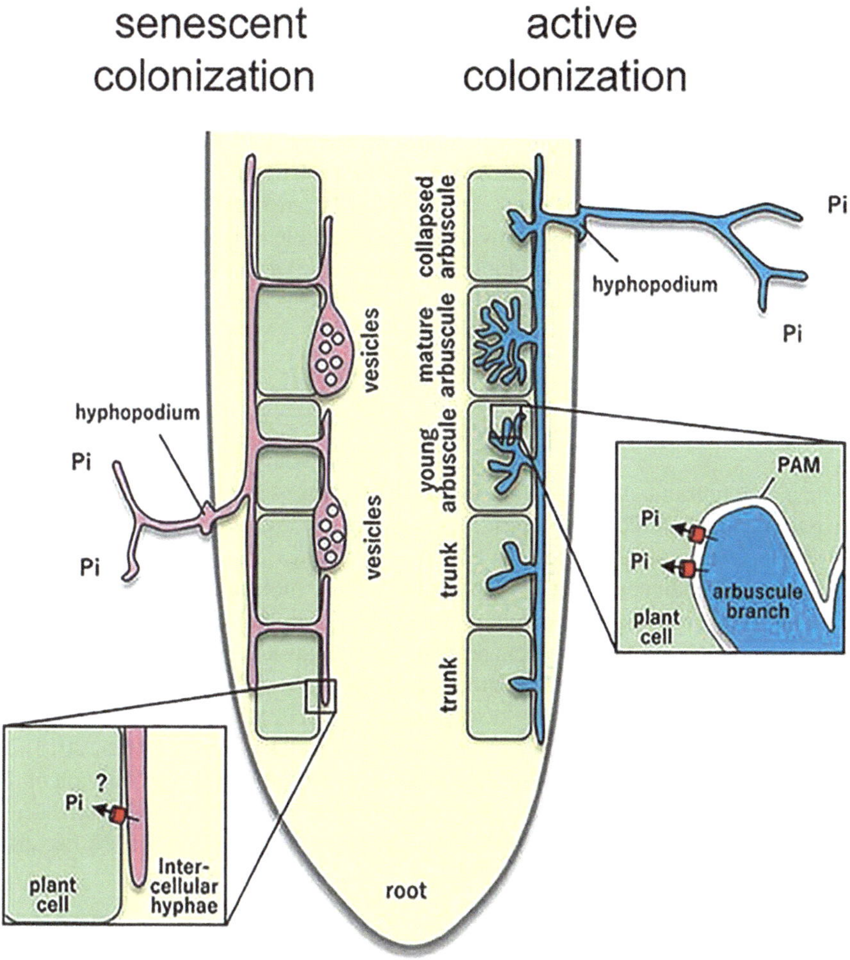

Fig. 14.6 Life cycle of intraradical colonization and expression of phosphate transporters in arbuscular mycorrhizal symbiosis. A region colonized by a single AMF hypha is referred to as an "infection unit." Upon entering the root epidermis, hyphae proliferate either intracellularly (Paris type, not depicted) or intercellularly (Arum type) within cortical cells. As modified by Kobae (2019), arbuscules that participate in nutrient exchange develop in cortical cells according to their initial penetration position (hyphopodium) in these processes

periarbuscular membrane envelops arbuscules. The proteins called plant phosphate transporter proteins are found in this membrane and are in charge of phosphate absorption from the arbuscules. Vesicles and intercellular hyphae are frequently observed during the senescent colonization stages of an infected unit.

14.7.1.3 Potassium Solubilization

According to Sindhu et al. (2016), approximately 90% of the potassium (K) in soil is found in insoluble rocks and silicate minerals. Insufficient potassium causes plants to produce few seeds, have underdeveloped root systems, and yield little (Sindhu et al. 2016). *Acidithiobacillus ferrooxidans*, *Bacillus edaphicus*, *Bacillus mucilaginosus*, *Burkholderia*, and *Paenibacillus* sp. are plant bacterial endophytes that can dissolve potassium rock by producing and releasing natural acids (Sindhu et al. 2016). These potassium-soluble bacteria emit a variety of organic acids, exopolysaccharides, and metal complexing ligands that facilitate the release of potassium from minerals.

14.7.1.4 Siderophore Formation

Iron, a vital nutrient, is found in aerobic environments as Fe3+ and produces oxyhydroxides and insoluble hydroxides. Plants and microbes are unable to absorb these insoluble forms. Small molecules called siderophores are typically produced by endophytes. Vendan et al. (2010) reported that these molecules sequester Fe3+ and mobilize existing ions. Johnson et al. (2013) reported that certain endophytes produce hydroxamate, whereas others produce catecholate siderophores. Water-soluble siderophores can be classified as either intracellular or extracellular. For intracellular iron storage, the former are located inside the cell, whereas the latter are discharged as iron-free siderophores for cellular iron uptake (Johnson et al. 2013). In situations where iron is scarce, specific proteins contribute to the transmission of the iron siderophore complex. Siderophore development enables PGP and plant disease suppression (Johnson et al. 2013). Although many plant species absorb bacterial siderophores (Fe3+), their role in PGP is still unclear.

14.7.1.5 Growth Regulators

Other hormones that impact plant growth and development include auxins, cytokinins, gibberellins, ethylene, and abscisic acid, which are produced by endophytic microbes. Auxins: As per Grossmann (2010), auxin is an essential hormone that controls the growth and development of plants. Indole-3-acetic acid (IAA) is an auxin produced by bacteria, fungi, and plants. According to McSteen (2010), IAA promotes adventitious root development in monocots and lateral root production in dicots. Vascular development and cambial expansion are combined in IAA. According to Grossmann (2010), auxins thicken the secondary wall and increase the number of xylem cells. Following their passage through the phloem in concentration gradients, they gather in various tissues (Tromas and Perrot-Rechenmann 2010).

Cytokinins Zeatin is a cytokinin family member that participates in callus development and helps plant cells divide when auxin is present (Salome et al. 2001). The differentiation of roots and shoots is aided by auxin and cytokinins, respectively.

Gibberellins: According to Fulchieri et al. (1993), gibberellins increase the density of hairs on the roots in root zones that are involved in the uptake of water and nutrients, hence promoting plant growth. Gibberellins and other terpenoid families are involved primarily in the division of cells, internode elongation, and cell elongation. Among other physiological functions in plants, abscisic acid is a stress hormone that influences stromal closure, seed germination, and resistance to abiotic stress. Additionally, it acts as an abiotic element elicitor, causing plants to produce bioactive compounds (Sun et al. 2012).

ACC deaminase: 1-aminocyclopropane-1-carboxylate (ACC) synthase serves two purposes in the synthesis of ethylene: it inhibits primary root elongation and lateral root development while encouraging the growth of root hairs. A stressed plant releases more ethylene (Glick 2005). Another name is the "stress hormone." Numerous endophytes produce an enzyme known as 1-aminocyclopropane-1-carboxylate (ACC) deaminase, which converts ACC to α-ketobutyrate and ammonia (Glick et al. 2007). Lowering ethylene levels, which is accomplished by lowering ACC levels, reduces plant stress.

14.7.2 The Role of Endophytes in the Development of Abiotic Stress Tolerance

Abiotic stresses has a detrimental effect on the morphology and physiology of plants by interfering with genetic regulation (Egamberdieva et al. 2017). This leads to changes in hormones, membrane malfunction, and the accumulation of reactive oxygen species (ROS). In numerous studies, endophytes have been shown to increase resilience to abiotic stresses (Ali et al. 2018). Endophytes can aid in the growth and abiotic adaptation of their host plants by producing plant hormones and absorbing nutrients (Fig. 14.7).

14.7.2.1 Resistance to Heat Stress

According to Ali et al. (2018), new agricultural methods are needed to increase crop output at relatively high temperatures as the consequences of climate change worsen. Research has shown that endophytes confer heat tolerance to plants such as *Cucumis sativus* (Ali et al. 2018), *Helianthus annuus* and *Glycine max* (Ismail et al. 2020), and *Adiantum capillus-veneris* (Ismail et al. 2020).

Compared with untreated cucumber plants under heat stress, the plants treated with the heat-tolerant thermophilic endophytic fungus *Thermomyces* sp. presented significant increases in total sugars, flavonoids, saponins, soluble proteins, and

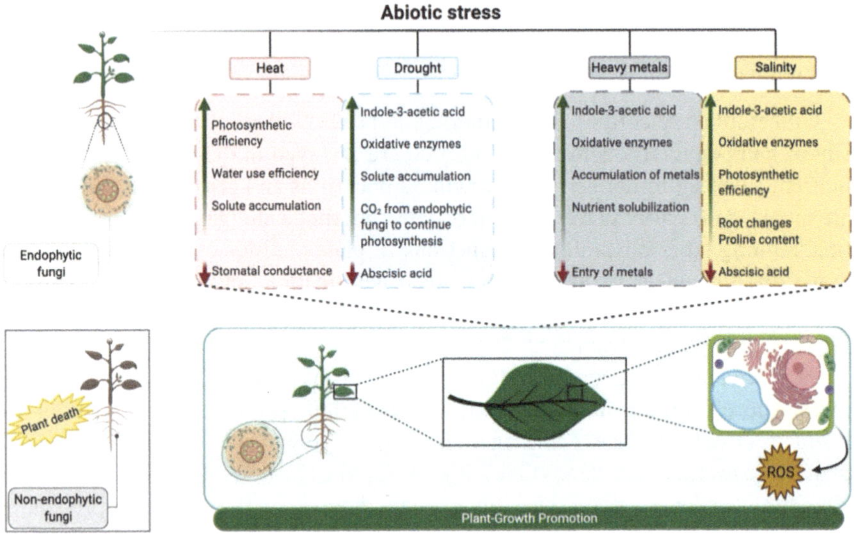

Fig. 14.7 Possible reactions of plants to abiotic stresses when they interact with endophytic fungi. The increase in the specified attributes is represented by the green arrow, and their loss is indicated by the red arrow. (Adapted from Fontana et al. (2021))

antioxidant enzyme activity; conversely, the plants treated with Thermomyces sp. presented longer roots than did the untreated cucumber plants. Under low-water conditions, this phenological response enables the plant to more efficiently penetrate and absorb nutrients and moisture from the soil, making it an essential adaptation trait in arid environments (Ali et al. 2018).

However, compared with untreated cucumber plants, those treated with *Thermomyces* sp. presented longer roots. This phenological response is crucial for adaptation in arid situations because it enables plants to more effectively penetrate and absorb nutrients and moisture from the soil under low-water conditions (Ali et al. 2018). When a plant experiences temperature stress, its stomatal conductance quickly increases, which encourages a high rate of transpiration.

Even under these circumstances, these plants exhibit poor transpiration rates and delayed stomatal opening when treated with endophytes (Ali et al. 2018). Under heat stress, the endophytic *Thermomyces* sp. stabilized the leaf water content and increased the water usage efficiency. Additionally, stomatal closure is a physiologically adaptive strategy utilized by thermophilic fungi to store water before additional damage from high-temperature stressors occurs (Ali et al. 2018). Ali et al. (2018) claim that this fungus encourages the accumulation of both primary and secondary metabolites. Numerous interactions between plants and microorganisms cause sugars and flavonoids to accumulate in plant tissues. These compounds function as signaling molecules and scavengers of reactive oxygen species (ROS), which aid plants in thriving and overcoming biotic and abiotic stresses (Abdelrahman et al. 2015).

14.7.2.2 Resistance to Drought Stress

Drought significantly affects productivity in agriculture worldwide, and its frequency is predicted to rise annually due to climate change. This invariably results in the accumulation of ROS, which is detrimental to plant health. Endophytes can greatly increase a plant's tolerance to drought. The generation of phytohormones by endophytes is a fundamental process (Khan et al. 2013). Abscisic acid (ABA) has been shown to protect plants against drought-induced oxidative damage when it is administered exogenously by activating stress-responsive genes and strengthening the antioxidant system of plants (Khan et al. 2013; Bano et al. 2012).

Additionally, the introduction of the endophyte *Bacillus pumilus* into *Glycyrrhiza uralensis* resulted in a significant increase in total flavonoids, total polysaccharides, glycyrrhizic acid, oxygen species, and antioxidants, as well as a significant increase in the expression of key enzymes involved in the production of glycyrrhizic acid. These secondary metabolites produced by endophyte-infected plants also help to manage drought stress (Xie et al. 2019).

Endophyte-mediated mechanisms are thought to help plants adapt to drought tolerance through the production of volatile compounds, phytohormones, reactive oxygen species (ROS), exopolysaccharide, and 1-aminocyclopropane-1-carboxylate deaminase, as well as by altering the morphology of roots, producing antistress metabolites, and positively regulating the stress-responsive genes of host plants (Eid et al. 2019). Moreover, grasses that contain endophytes have been shown to accumulate solutes when exposed to water stress (Malinowski and Belesky 2000). One way that endophytes may help host plants tolerate water stress is by using the CO_2 they produce to sustain photosynthesis, which reduces the amount of CO_2 produced by stressed plants whose stomas have closed (Bloemen et al. 2013); *Populus deltoides* plants receive and utilize 2.7% of the CO_2 released by endophytes throughout the photosynthetic process.

14.7.2.3 Resistance to Heavy Metal Stress

Endophytes, such as *Glycine max* (Bilal et al. 2020), *Lycopersicon esculentum* (Zhu et al. 2018), and *Triticum aestivum* (Ikram et al. 2018), have been demonstrated to help plants withstand heavy metal stress. By secreting IAA and preventing the transport of heavy metals from the soil to plants, the endophytic fungus *Penicillium roqueforti*, for example, enhanced resistance in *T. aestivum* plants growing in soil contaminated with heavy metals. Additionally, following the injection of the endophytic fungus, wheat plants irrigated with residual water presented increased growth, reduced levels of heavy metals in the roots and shoots, and improved nutrient absorption.

Wheat plants that were not inoculated with *P. roqueforti* and subjected to heavy metal stress presented symptoms of chlorosis and decreased growth. By forming a symbiotic connection with host plants, *P. roqueforti* inoculation can help regulate heavy metal levels in highly contaminated soil and promote plant growth (Malik et al. 2020). *Exophiala pisciphila*, an endophytic fungus associated with *Zea mays* roots, is another example. It is highly resistant to cadmium (Cd) and has been shown to reduce Cd phytotoxicity while promoting maize development (Wang et al. 2016).

Furthermore, through enzymatic activity, endophytes promote cell detoxification and improve the host plant's capacity to absorb heavy metals either directly or indirectly. Endophytes may indirectly benefit the host plant by encouraging plant development and stress tolerance, or they may directly benefit the plant by increasing heavy metal mobilization and reducing metal toxicity in plants (Babu and Reddy 2011). Endophytes increase the amount of nutrients that the host plant may receive from contaminated soil. Furthermore, according to Khan and Doty (2011), these endophytes have the ability to decompose pollutants in polluted soil and transform them into nontoxic forms. Endophyte phytohormones play a major role in enhancing plant resistance to heavy metal stress. Endophytes may emit exogenous phytohormones that alter the host plant's physiology in a positive way and make it more resilient to stress. IAA increased root volume, surface area, and diameter, reducing the detrimental effects of lead on sunflower growth, according to Fässler et al. (2010). According to studies by Ahmad et al. (2011), SA signaling may also result in a hypersensitive reaction that follows the development of "systemic acquired resistance" that reduces heavy metal stress. In addition to phytohormones, the capacity of endophytic fungi for biofertilization might increase the amount of nutrients available to the host plant in soil contaminated with heavy metals through solubilization (Khan et al. 2010).

14.7.2.4 Resistance to Salinity Stress

Endophyte microorganisms, including *Z. mays* (Jan et al. 2019), *S. lycopersicum* (Abdelaziz et al. 2019), *O. sativa* (Jogawat et al. 2013), *T. aestivum* (Zhang et al. 2016), Cucumis sativus (Waqas et al. 2012), and G. max (Radhakrishnan et al. 2015), among others, have been found to contribute to salt stress tolerance in plant cultures. For example, under salt stress conditions, the endophytic fungus *P. indica* increased the development and production of *S. lycopersicum*, resulting in several morphological and biochemical alterations that mitigated the consequences of salt stress.

This endophyte increased plant fresh and dry mass, fruit yield, root branching, the expression of enzymes such as catalase and superoxide dismutase, and the levels of chlorophyll and IAA by 65% when subjected to salt stress. Additionally, tomato plants colonized with endophytes presented lower levels of proline and abscisic acid (ABA) than noncolonized plants did (Abdelaziz et al. 2019).

Under salt stress, many plants produce high amounts of proline; however, when plants are inoculated with endophytic fungi, these levels can be reduced (Abdelaziz et al. 2019). According to Cao et al. (2014), abscisic acid (ABA) regulates proline synthesis to reduce cytoplasmic osmotic stress caused by high salt levels in the root zone. Thus, under saline stress conditions, ABA levels and proline contents decrease by approximately 30% (Khalid et al. 2018). Oxidative stress, drought, flooding, salinity, and heat stress are examples of abiotic stressors that are interrelated and produce reactive oxygen species (ROS), which damage cells and eventually cause death when they are exposed to them for prolonged periods of time (Waqas et al. 2012).

According to Mittler (2002), elevated ROS levels in plant cells result in oxidative stress, lipid peroxidation, and the oxidative degradation of DNA and RNA. The ROS signal directly affects the transcription, translation, and redox balance of regulatory proteins, according to Choudhury et al. (2017). Plants respond to this by reducing the negative impacts of stress and controlling the body's levels of reactive oxygen species (ROS).

According to one theory, endophytes initially emit comparatively few ROS, including hydrogen peroxide, which activates the host's antioxidant enzymes (White and Torres 2010). Small amounts of ROS released continuously reduce cell hypersensitivity to ROS, enhance other endosymbiotic relationships with the host, and facilitate the uptake of plant nutrients (phosphorus, potassium, magnesium, and calcium) (Overmyer et al. 2003). The buildup of proline, methionine, flavonoids, and other phenolic compounds is one of the main ways in which plant tissues react to the presence of ROS generated by endophytes to increase resistance (Overmyer et al. 2003). Hypersensitive responses and host-acquired systemic resistance are believed to be caused by endophyte–host–plant crosstalk as well as the production of ROS and antioxidants (Tanaka et al. 2006). According to Rodriguez and Redman (2008), certain fungal endophytes produce ROS to draw resources from host cells and maintain mutualistic relationships with plants, whereas others lower ROS levels to lessen the negative effects of abiotic stresses on their hosts. As previously mentioned, Fig. 14.7 offers a comprehensive summary of research on endophytic fungi that shows how they affect abiotic stresses.

14.7.3 Endophytes as Biocontrol Agents of Plant Diseases

The use of biological processes such as endophytes, which suppress or oppose phytopathogens, can help minimize the need for agrochemicals in the prevention of plant diseases. Endophytic biocontrol mechanisms can be one of the following:

14.7.3.1 Space Competition with Pathogens

Some endophytes rapidly invade plant tissues, competing with pathogens for nutrition and space (Fig. 14.8). For example, two diseases that can cause root rot in older plants and damping off of seedlings are *Rhizoctonia solani* and *Pythium myriotylum*. The fungus *Beauveria bassiana* can endophytically colonize tomato and cotton seeds, which can lessen disease. This suggests that space competition is a likely biological control technique for root rot pathogens and other diseases (Ownley et al. 2008). AMF usually occupy the same ecological niche and invasion site as

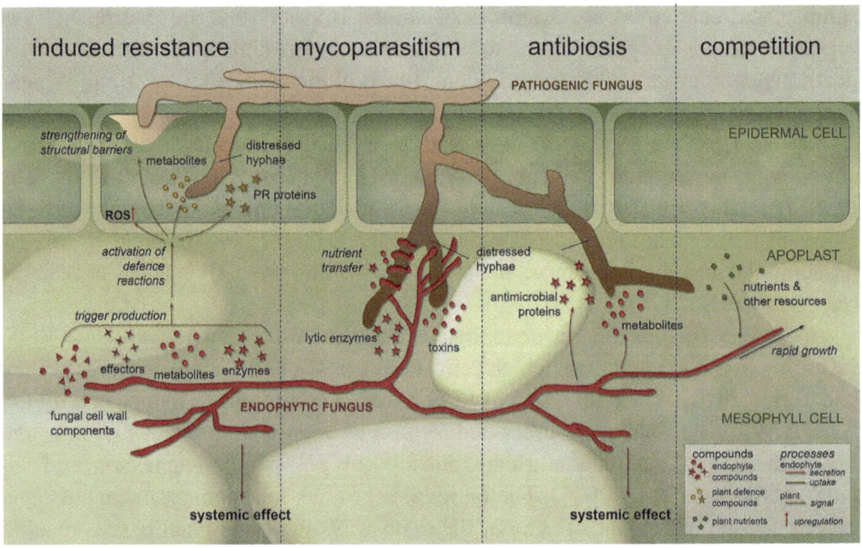

Fig. 14.8 Plant diseases are biologically controlled by fungal endophytes. Both direct and indirect endophyte suppressions of the pathogen through competition, antibiosis, or mycoparasitism are suggested mechanisms involved in the tripartite interaction between the endophyte, host plant, and pathogen. Created opposition. During colonization, the endophyte may release enzymes, metabolites, and effectors from the cell wall, including chitin and β-glucans, which may function as inducing agents and cause the plant to produce a signal. The transmission and (systemic) transfer of a signal cause the defense mechanisms of the pathogen assault site to become more activated

soil-borne illnesses. As a result, AMF and pathogens need to be spatially competitive, and their main biocontrol role is to prevent early plant root infection by wilt and root rot pathogens (Weng et al. 2022).

14.7.3.2 Antimicrobial Metabolite Production

The growth of plant pathogenic microorganisms is stopped by secondary metabolites produced by endophytes, some of which have antifungal and antibacterial properties (Mousa et al. 2015). Endophytes contain a wide range of antimicrobial metabolites, including peptides, flavonoids, quinones, phenols, alkaloids, steroids, polyketides, and terpenes (Munir et al. 2020; Haggag and Eid 2022).

The production and release of metabolites with a variety of uses, including disease prevention, are encouraged by the host–endophyte relationship (Yan et al. 2018). *Aspergillus*, *Penicillium*, *Fusarium*, and *Phoma* are four endophytic fungal isolates from finger millet plants that can produce anti-*Fusarium* chemicals, for example (Mousa et al. 2015). The antifungal activity of endophytic fungi produced from the roots of *Sophora tonkinensis* is approximately the same as that of carbendazim wettable powders, and these fungi reduce the incidence of three fungal infections in *Panax notoginseng* plants by more than 50% in vitro (Yao et al. 2017).

Beneficial endophytic fungi have the ability to produce or release antimicrobial compounds in response to elicitors, such as chemicals derived from pathogens or other plant inhabitants (Soliman and Raizada 2013; Soliman et al. 2015). Paraconiothyrium SSM001 is a well-known example of this method; it functions as a fungicide against host infections by producing Taxol, which stops other fungi from mitosizing. It is feasible to increase Taxol synthesis in vitro by coculturing *Paraconiothyrium* SSM001 with other fungi from its host plant, such as *Alternaria* and *Phomopsis*, either directly or indirectly through their metabolites (Soliman and Raizada 2013). Diffusible substances such as chitin and chloromethane, as well as pathogens, can cause the expression of genes that produce Taxol, which is subsequently released by SSM001 (Soliman et al. 2015).

14.7.3.3 Predation and Parasitism

Fungal cells can be broken down and used as food by exoenzymes generated by endophyte-control agents (Fig. 14.8). These enzymes contribute to the disintegration of the cellular walls of phytopathogenic bacteria and are among the most common biological control techniques, especially with respect to fungal infections (Villarreal-Delgado et al. 2018). Glycoproteins, polysaccharides, and other substances that vary depending on the fungal species make up the cellular walls of all fungi, including those that infect plants (Bowman and Free 2006). Eighty percent of the cell wall is composed of polysaccharides (Latge 2007). These polysaccharides use a vast network of glycosidic linkages to increase the structural rigidity of the cell wall.

The capacity of lytic enzymes to disrupt these bonds weakens the cell wall of phytopathogenic fungi, causing lysis and cell death (Jadhav et al. 2017). The structure of cell walls is altered or broken down by lytic enzymes such as chitinases, cellulases, proteases, and β-1,3-glucanases, according to Moto et al. (2017). Enzymes of this type are produced by biological control agents.

14.7.3.4 Mycoparasitism

Endophytes also employ direct attack of the detected pathogen or its propagules as a crucial tactic to safeguard host ecology (Huang et al. 2020). Through cell lysis and hyphal penetration, endophytic fungi degrade pathogen cell walls (Fig. 14.8). This mechanism has four stages. The first and second steps involve the identification of fungal pathogens and the chemicalotropic growth of endophytic fungal mycelia toward pathogenic fungi. The third and fourth steps of fungal cell penetration by endophytic fungi into pathogenic fungi cause fungal cell walls to disintegrate (Huang et al. 2020). The hyphae of the pathogenic fungus *Rhizoctonia solani* can be penetrated and destroyed by some *Trichoderma* species used in biocontrol (Grosch et al. 2006).

Similarly, endophytic fungi isolated from common reeds hinder the formation of soil-borne fungal infections by coiling around their hyphae and killing the hyphal cytoplasm once it has entered the cells. Several enzymes that break down fungal hyphae are secreted by endophytic fungi (Cao et al. 2009). A variety of fungal antagonists, including *Gliocladium virens*, *Ampelomyces quisqualis*, *Acrodontium crateriforme*, and *Acremonium alternatum*, can parasitize powdery mildew infestations (Kiss 2003).

14.7.3.5 Production of Siderophores

Endophyte-producing siderophores can suppress certain phytopathogens by depriving them of iron (Wang et al. 1993). Endophytes employ this tactic to stop the growth of fungal diseases by depleting the rhizosphere of iron. Plant growth is not significantly impacted by these siderophores since plants have a limited amount of iron. The plant can also utilize the chelated iron that endophytes have (Wang et al. 1993).

14.7.3.6 Hydrogen Cyanide (HCN)

HCN is a volatile component of endophytes that can suppress the majority of phytopathogenic fungi (Siddiqui et al. 2006). Certain endophytes produce HCN from glycine. *Pseudomonas* sp. is more successful at producing cyanide. Other *Bacillus* and *Rhizobium* species also contribute to cyanide HCN production. In vitro, Schippers et al. (1990) reported that 50% of *Pseudomonas* isolates from potato and wheat may produce HCN. Siddiqui et al. (2006) linked the disease-protective effect of black rot in tobacco roots caused by *Thielaviopsis basicola* to HCN.

14.7.3.7 Acting as a Component of Plant Defense

Endophytes may aid in plant defense in addition to the previously outlined methods (Liu et al. 2017). When Taxol compounds are released from young tree plant tissues that are diseased, the endophytic fungus *Paraconiothyrium* SSM001 can reach the site of pathogen invasion. Typically, exocytosis traps Taxol in intracellular hydrophobic bodies. Together with the *Paraconiothyrium* SSM001 mycelium, these Taxol hydrophobic bodies create an extracellular barrier that stops infections from growing inside the tissues of yew trees (Soliman et al. 2015). A similar method was demonstrated with the bacterial endophyte *Enterobacter* sp., which was isolated from finger millet plants. This endophyte swarms toward the *Fusarium graminearum* entry site, bending the root hairs to form a multilayer root-hair endophyte stack (RHESt). Prior to eliminating *F. graminearum*, RHESt develops. This endophyte swarms toward the *Fusarium graminearum* entry site, bending the root hairs to form a multilayer root-hair endophyte stack (RHESt). RHESt creates a physical barrier that prevents *F. graminearum* from developing inside roots prior to being killed (Mousa et al. 2016).

Fungi that infiltrate plant roots enter through the cell wall. AMF symbiosis can promote the growth, thickening, and branching of the host plant's root system while also thickening the root tip epidermis and accelerating the lignification of the cell wall. Additionally, it can alter the physical structure of roots and add new cell layers, both of which effectively slow the pathogen infection process (Boutaj et al. 2019). In response to *Verticillium dahliae* stress, the xylem structure of *Gossypium hirsutum* plants, which live symbiotically with the AMF *G. mosseae* and *G. etunicatum*, increased.

Additionally, the vessels and palisade tissue were distorted; gelatinous materials developed at the vessels; the cells shrank and distorted; the color became more intense; and the cell walls noticeably thickened. The loss of mitochondrial inner folds, lignification, a notable decrease in vacuoles, and various structural alterations to the root system are all advantageous for increasing host plant resistance to *Verticillium dahliae* (Boutaj et al. 2019). In the root epidermis and endodermis of tomato plants, AMF can produce a mycelial network, callose, and a papillary structure stacked by nonesterified pectin. In this way, pathogens are prevented from infecting root cell tissues. The kinetics of *Phytophthora* infection are altered by modifications in the tomato root structure (Pozo et al. 2002).

Changes in the root systems of mycorrhizated plants are reflected in the promotion of the formation of hydroxyproline-rich glycoproteins (HRGPs) in cell walls (Balestrini and Bonfante 2014). HRGP is a linear sugar protein found in plant cell walls. It can fortify the host plant's cell wall as a structural element, increasing resistance to the breakdown of the cell wall caused by enzymes such as proteases, cellulases, and hemicellulases that are generated during pathogen infection. On the other hand, HRGP acts as a lectin during pathogenic invasion of plants by binding to the cell wall and blocking pathogens from entering plant cells (Balestrini and Bonfante 2014).

14.7.3.8 Induction of Resistance

It is thought that endophytes increase a plant's defenses against illness and insect infestation. Plants can operate more effectively and strengthen their defense responses against subsequent pathogen invasions via this activation mechanism, which is also referred to as defense priming or induced systemic resistance (Poveda et al. 2020). Pathogenicity-related (PR) protein overexpression is not involved in the regulation of induced systemic resistance (ISR), which is frequently regulated by ethylene or jasmonate acid (Ramamoorthy et al. 2001). Salicylic acid is the mediator of systemic acquired resistance (SAR), which is frequently linked to pathogen infection and causes the accumulation of the PR protein (Tripathi et al. 2019).

These PR proteins aid in the fight against infection and cell death because they contain enzymes such as chitinase and beta 1,3-glucanase, which degrade invasive pathogen cells and strengthen the cell wall barrier (Gao et al. 2010). However, endophyte-induced systemic resistance might also be associated with the amplification of genes associated with illness (Mishra et al. 2018).

Fusarium solani, which is isolated from tomato plants, has been demonstrated to induce ISR against foliar disease caused by *Septoria lycopersici* by activating the PR, PR7, and PR5 genes in the roots (Kavroulakis et al. 2007). In a different study, nonpathogenic mutants of *Colletotrichum magna* strains developed resistance in *Cucumis sativus* and *Citrullus lanatus* plants by producing large amounts of lignin deposition, peroxidase, and the enzyme phenylalanine ammonia lyase. This prevents the plants from being infected by *F. oxysporum* and *C. orbiculare* (Redman et al. 2022). Additionally, by increasing superoxide dismutase and peroxidase activity in the host plant, the interaction of the endophyte *Neotyphodium lolii* with several pathogens decreases the symptoms of necrotic patches and leaf lesions (Tian et al. 2008). Endophytes can act in a variety of ways, by creating different metabolites that compete with pathogens and cause plant resistance; however, further study is needed to fully comprehend this mechanism (Arnold and Herre 2003).

The reactive oxygen species (ROS) production, the fortification of structural barriers that prevent pathogen penetration, and the release of antimicrobial metabolites and proteins that inhibit pathogen growth are examples of defense responses. Endophytes may directly display pathogenicity. Mycoparasitism. Endophytes can mycoparasitize pathogens directly (by coiling or penetrating prey) or indirectly (without making physical contact). In both cases, the endophyte can extract resources from the pathogen by generating lytic enzymes and toxins. Once a plant becomes aware of these compounds, resistance may develop an antibiotic. The pathogen is inhibited by endophyte-produced compounds and antimicrobial proteins. The transfer of these compounds may provide systemic protection for the plant.

Competition. According to Latz et al. (2018), the endophyte competes with invasive pathogens for this niche, as it develops inside the apoplast by consuming nutrients or other plant resources.

14.8 Screening of Endophytes as Biological Control Agents

14.8.1 In Vitro Screening

Plants are both endophytes and epiphytes, supporting a diverse range of microbial communities. By preventing pathogen growth, some of these communities even shield fruits and plants against disease invasion. Therefore, choosing microbial strains is the first step in biocontrol screening (Kashyap et al. 2023). Additionally, strains are categorized according to a number of factors, including the production of siderophores, antibiotics, enzymes, bioactive chemicals, volatile emissions, etc. (Cao et al. 2018). The epiphytic microbial populations present in plant samples must be eliminated to detect endophytic bacteria. Sample tissue needs to be routinely treated or sanitized with distilled water, alcohol, or other disinfectants to eliminate epiphytic microorganisms.

Commonly utilized substances include sodium hypochlorite (NaOCl), hydrogen peroxide (H_2O_2), and 70% ethanol. $HgCl_2$ and other heavy metal salts are also employed as disinfectants to eliminate epiphytic microorganisms (Nurul et al. 2012).

Dual Culture Assays The plant pathogen and biocontrol agent were cocultured in vitro in this experiment, either in liquid cultures or on solid agar. When a biocontrol agent is present, the reduction in disease symptoms or the inhibition of pathogen development are evaluated (Raymaekers et al. 2020) (Fig. 14.9).

Assays for antagonism: This assay evaluates the ability of a biocontrol agent to directly obstruct the growth or development of plant diseases via a range of methods, such as growth inhibition assessments, confrontation assays, and inhibition zone tests (Sales et al. 2017).

Enzymatic Assays The biocontrol activity of any endophytic microbial strain can be ascertained by searching for secreted enzymes such as chitinase, glucanase, protease, or cellulase. The effectiveness of biocontrol agents can be evaluated since these enzymes directly aid in the degradation of toxic materials or the inhibition of pathogen growth (Solanki et al. 2012).

Assessment of volatile substances: Assessing the volatiles generated by microbial antagonists is another common method of biocontrol screening. Microbial antagonists and pathogens are grown on separate agar plates that are parafilm-sealed to avoid volatile diffusion during analysis (Ezra et al. 2004).

14.9 In Vivo Method

One of the most important methods for biocontrol screening is this in vivo technique. Microbial antagonists and pathogens are added to the fruit and plant components during in vivo investigations. Furthermore, the severity of the disease was examined by measuring the lesion diameter to estimate the efficiency of the endophytes. However, during inoculation, either pathogens or endophytes are administered at predetermined intervals (Raymaekers et al. 2020).

Seedling Growth Assays By evaluating the germination rate, length of roots, length of shoots, and overall seedling vigor, Chavéz-Díaz et al. (2022) evaluated the biocontrol efficacy of endophytic strains on seedling germination and growth.

Foliar Pathogen Inoculation Assays This test assesses the effectiveness of an endophyte agent following its application to plant leaves and subsequent plant disease inoculation. The development of disease markers, including lesion formation or disease severity, is tracked to evaluate the efficacy of endophytes (Pandit et al. 2022).

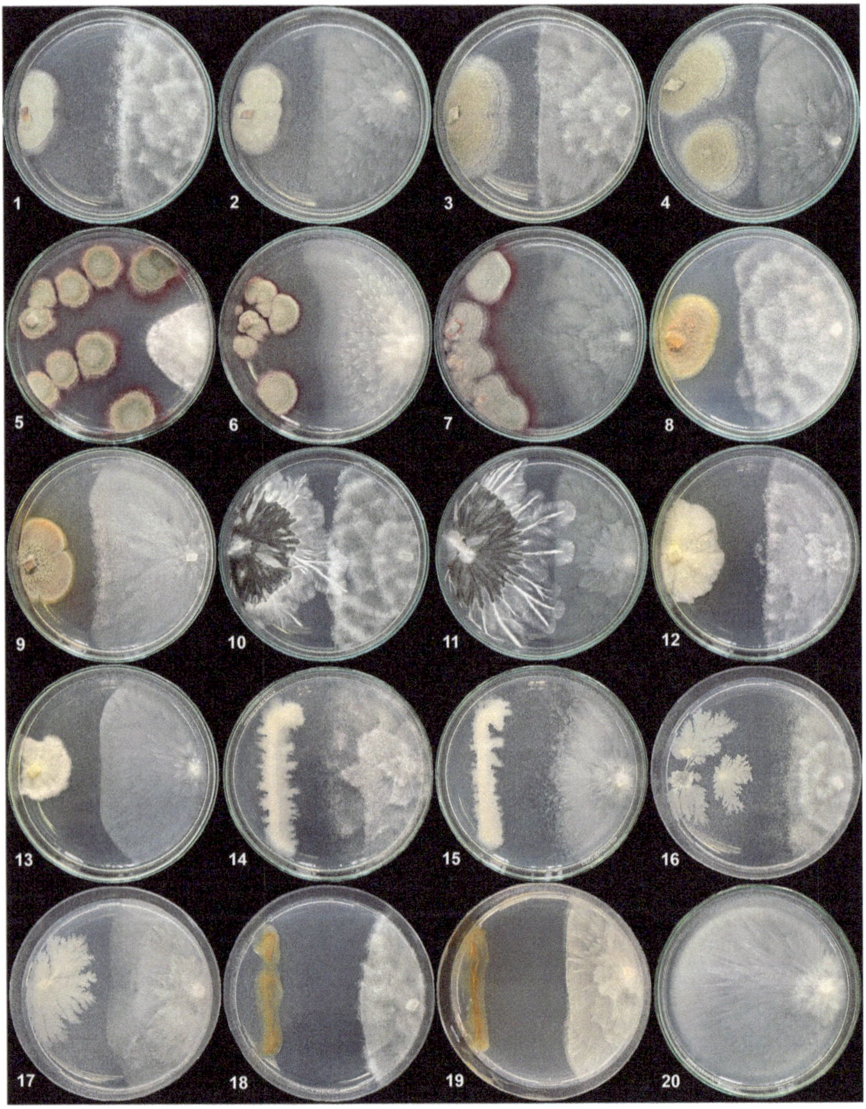

Fig. 14.9 Suppression of bacterial and fungal endophyte strains on medium after 10 days at 25 °C. *Ph. cinnamomi-Penicillium citrinum, Ph. citrinum-Ph. multibullata, P. fusisporum-Penicillium citrinum,* and . *Fifth, Ph. cinnamomi; sixth, Ph. multibullata; seventh, Ph. vanyenensis; Ph. vanyenensis and P. austrosinicum; P. austrosinicum and Ph. cinnamomi; the species Xylaria curta, which is also known as Ph. cinnamomi, and Ph. multibullata are Clonostachys rosea—Ph. cinnamomi; Clonostachys rosea, Ph. vanyenensis; Achromobacter* sp.—*Ph. cinnamomi; Achromobacter* sp.—*Ph. vanyenensis; Bacillus* sp.*; Ph. cinnamomi;. Bacillus* sp.*; Ph. Multibullata; Burkholderia gladioli, Ph. cinnamomi; Burkholderia gladioli, Ph. Vanyenensis;* and control—Ph. *vanyenensis.* (Adapted by Dang et al. (2023))

Root Drench or Soil Application Assays This process involves applying an endophyte substance directly to plant roots or to the soil. Moreover, the pathogens are introduced by root colonization or soil inoculation. Dynamics are utilized to evaluate the success of endophytes after analyzing disease incidence, severity, or pathogen population (Dhar Purkayastha et al. 2018).

Greenhouse and Field Trials Research conducted in greenhouses and outdoors assesses endophyte biocontrol agents in authentic environments. After being treated with endophytes, plants are monitored for growth, disease incidence, yield, and other pertinent characteristics in a controlled or field environment over a predetermined length of time (Dhar Purkayastha et al. 2018). In vivo experiments have been conducted to determine effective BCAs and their effects on plants, taking into consideration a number of additional factors, including the amount of water, stress from oxidative damage, and other physiological circumstances of plants and fruits (Lecomte et al. 2016).

Figure 14.10 shows the detailed procedure of isolating endophytes from fruit samples, sterilization, BCA screening, and application in pre- and postharvest management.

14.9.1 Apprehending the Endophytes

The host–parasite relationship cannot be accurately depicted via conventional methods for detection, isolation, and identification. Some bacteria are not included in cultivation-based research because the ideal circumstances needed for the majority of microbial development cannot be replicated or maintained (Abbas et al. 2024). However, more sophisticated methods, such as enhanced microscopic techniques, have replaced conventional isolation and cultivation methodologies in the study of endophytes.

The only direct method of examining endophytes that helps determine the type of infection, concentration unique to a certain tissue, degree of colonization, and response of the plant is microscopy. We can examine the ultrastructures of endophytes via electron microscopy (Pimentel Esposito-Polesi et al. 2017). According to Ramanujam et al. (2016), scanning electron microscopy (SEM) and electron microscopy with transmission (TEM) can both yield vital details about a sample's surface and internal makeup, respectively. According to Ramanujam et al. (2016), electron microscopy has proven to be highly effective in recognizing endophytes, assessing the degree of colonization, interacting with hosts, and developing themselves in the plant environment. Our knowledge of plant–endophyte interactions has significantly increased owing to advancements in genetics, bioinformatics, and next-generation sequencing. Several molecular methods are used to characterize endophytic bacteria, such as random amplified polymorphic DNA (RAPD), repetitive extragenic palindromic sequence, denaturing gradient gel electrophoresis

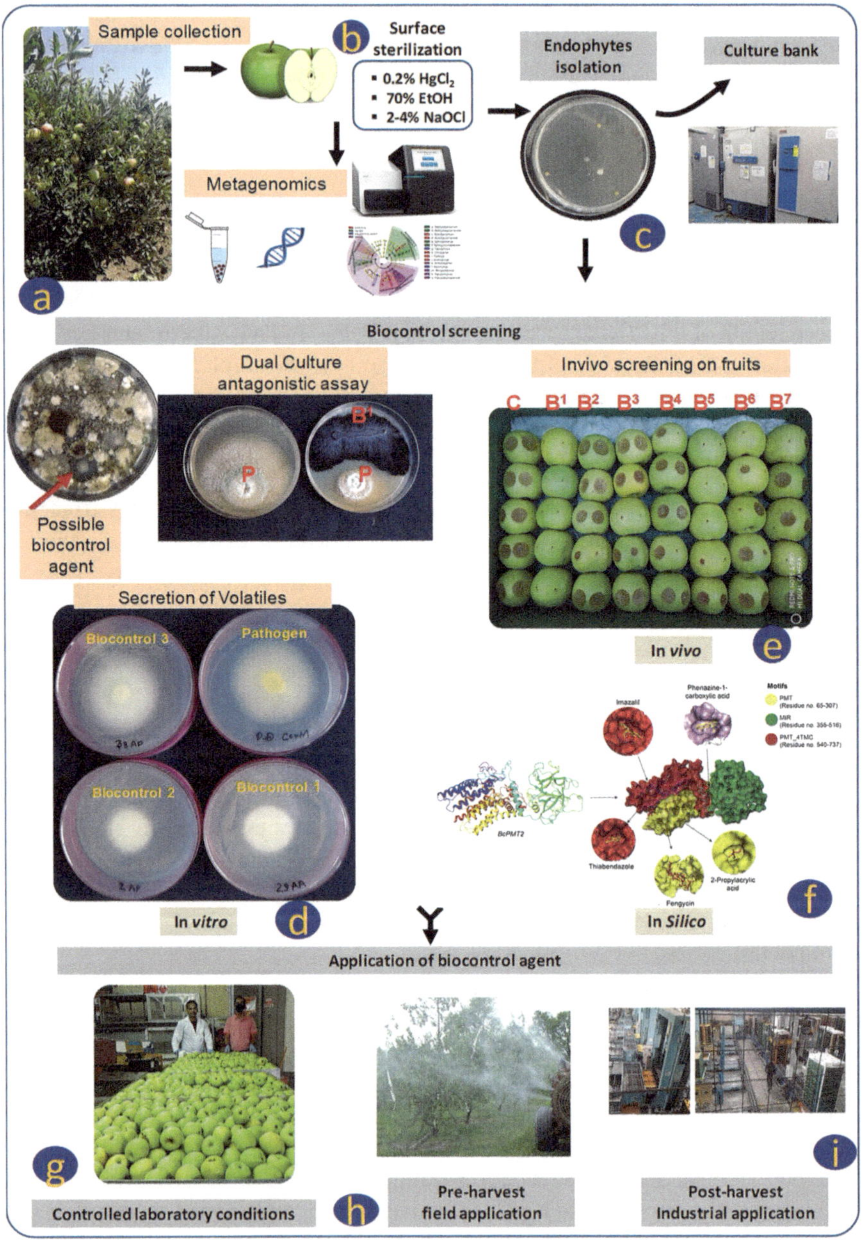

Fig. 14.10 Biocontrol screening and potential applications; (**a**) endophyte isolation via sample collection; (**b**) surface sterilization to eliminate epiphytic microflora; (**c**) endophyte isolates; (**d**) biocontrol screening; in vitro, potential biocontrol agent: the clear zone surrounding the bacterial colony can be considered for future biocontrol agents, dual culture antagonistic assay; strain B1 inhibits the growth of the pathogen and the secretion of volatiles: biocontrol agents 1 and 2 reduce the growth of the pathogen via volatile synthesis; (**e**) in vivo, strain B1 (Bacteria1) can be considered biocontrol agents because the strain completely inhibits the growth of the pathogen in contrast with (C) control (fruits containing the pathogen only); (**f**) in silico approach to biocontrol screening; (**g**) application of biocontrol agent under controlled laboratory conditions; and (**h**, **i**) the use of biocontrol agents both before and after harvest. (According to Kashyap et al. (2023), B2–B7 represent Bacteria 1–7, and P represents the pathogen)

(PCR-DGGE), amplified ribosomal DNA restriction analysis, repetitive extragenic palindromic sequence, BOX-A1R-based repetitive extragenic palindromic PCR, and enterobacterial repetitive intergeneric consensus (Rani et al. 2022).

Compared with other imaging techniques, such as TEM and fluorescence microscopy, ROS staining is more cost-effective and useful. Nitroblue tetrazolium (NBT) and 3,30-diaminobenzidine tetrachloride (DAB) are used to detect radicals such as superoxide (O_{2-}) and hydrogen peroxide (H_2O_2), which are connected to the microbial invasion of eukaryotic cells, much like ROS. DAB results in brown pigmentation, which suggests the presence of endophytes in tissues (White et al. 2014). A dark blue, water-insoluble formazan is created when superoxide radicals break down NBT (Shinogi et al. 2003). In contrast to DAB, which is unable to penetrate shoot tissues and thus cannot detect endophytes, NBT is able to recognize endophytes in a variety of different plant tissues.

14.9.2 *Endophytes as Biocontrol Agents: Registration and Regulation*

Typically, microbial endophytes need to be registered or authorized before being used or sold. The European Food Safety Authority in the European Union and the Environmental Protection Agency (EPA) in the USA are two regulatory agencies that monitor and assist with the process of registration. There are several steps involved in this process, including labeling and documentation. The identity of the product, usage guidelines, safety precautions, and storage requirements must all be accurately stated on labels for microbial endophytes. Additionally, regulatory submissions often require quality, safety, and efficacy data.

After approval and distribution into the market, postmarket surveillance is usually conducted to guarantee the ongoing safety and efficacy of microbial biocontrol agents. Legal requirements and safety measures, however, could differ by nation and region. To identify more details and specific necessities regarding the approval and safety of microbial endophytes, regulatory bodies are needed (Sundh and Goettel 2013; Sundh and Eilenberg 2021).

14.10 Endophyte Difficulties in Agricultural Activities

The use of specific endophytic strains as biocontrol agents may have drawbacks, which should be mentioned. The root fungus endophyte *Cryptosporiopsis* sp. has been shown in a recent study to significantly reduce Norway spruce seedling growth (Terhonen et al. 2016). Fecundity and the ability of natural enemies to survive as adults have other negative effects. "This is because endophyte produced poisons, also called loline alkaloids, are passed from invaded plants to parasitoids or natural enemies at different levels of the food chain," Lopez and Sword (2015) explain. Furthermore, in some intrinsic or natural environments, endophytes can act as latent pathogens (Eaton et al. 2011). Endophytic fungi have been shown to induce varying degrees of sickness in spice corms when intentionally introduced, both in vitro and in vivo (Wani et al. 2016).

14.11 Examples of Studied Endophytes

Table 14.1 lists recently studied endophytes that promote plant development and have biocontrol effects on plant hosts.

14.12 Summary

A significant annual loss of agricultural productivity due to diseases of plants is one of the largest barriers to meeting food supply targets for the world's growing population. Agrochemicals are essential for managing plant diseases to preserve a sustainable and profitable agricultural system. However, excessive chemical use has adverse effects on ecological functioning, human health, and agricultural sustainability. Numerous agricultural applications are improved by the interactions between microorganisms and plants. Plants are home to the vast majority of microorganisms in the environment.

Plant–endophyte interactions, which are generally viewed as beneficial since they promote growth and development and increase the host plant's resilience to a variety of biotic hazards, have important effects on the physiology and overall performance of the host plant.

Plant–endophyte interactions, which are generally viewed as beneficial since they promote growth and development and increase the tolerance of the host plant to a variety of biotic and abiotic stressors, significantly affect the physiology and overall performance of the host plant. These methods substantially contribute to agricultural sustainability by providing environmentally friendly inputs that improve crop quality and production while reducing the use of dangerous synthetic fertilizers.

Endophytes can be mixed with target cultures to produce tolerance to abiotic stresses. This has positive benefits and allows plants to grow in places where they would normally struggle to develop without the endophytic agent.

Endophytes, on the other hand, are an appealing alternative to chemical pesticides since they may provide plant disease management strategies that support sustainable agriculture. Despite their numerous benefits in plant disease management, endophytes have limited applications in the field. Endophytes have few practical uses in plant disease management, despite their many advantages. Stress-tolerant endophytes should be screened more thoroughly, as should their transfer from the

Table 14.1 Plant growth-promoting PGPs and the biocontrol properties of some endophytes

Endophytes	Host	PGP/biocontrol	References
Penicillium sp.	Wheat	P solubilization	Wakelin et al. (2004)
F. oxysporum	Banana	ISR against *Radopholus similis*	Vu et al. (2006)
Streptomyces olivochromogenes, *Microbispora rosea* subsp. *Rosea*	Chinese cabbage	Biocontrol of *Plasmodiophora brassicae*	Lee et al. (2008)
S. griseus	Wheat	IAA	Hamdali et al. (2008)
Piriformospora indica	Barley	Ethylene/phytohormone production	Schafer et al. (2009)
Streptomyces sp. MBCu-56	Cucumber	*Colletotrichum orbiculare*	Shimizu et al. (2009)
Cladosporium sp.	Cucumber	Gibberellins	Hamayun et al. (2010)
Acinetobacter johnsonii	Sugar beet	IAA, P solubilization	Yingwu et al. (2011)
B. subtilis NA-108, *B. subtilis* NA-120, *Enterobacter* sp. EMB-79	Strawberry	IAA, siderophore, N fixation	De Melo Pereira et al. (2012)
Paecilomyces formosus	Cucumber	Gibberellin	Khan et al. (2012)
Streptomyces albosporus R13	Rice	Siderophore	Gangwar et al. (2012)
Penicillium sp., *Phoma glomerata*	Cucumber	IAA, gibberellins, jasmonic acid	Waqas et al. (2012)
Scolecobasidium humicola	Tomato	N fixation	Mahmoud and Narisawa (2013)
Pestalotiopsis sp.	Tomato	IAA	Hoffman et al. (2013)
Streptomyces sp.	Maize	Biocontrol of *Pythium aphanidermatum*	Costa et al. (2013)
Streptomyces sp., *Nocardia* sp., *Nocardiopsis* sp., *Spirillospora* sp., *Microbispora* sp., *Micromonospora* sp.	Mandarin	IAA	Shutsrirung et al. (2013)
Aspergillus flavipes CanS-34A, *Chaetomium globosum* CanS-73, *Clonostachys rosea* CanS-43, *Leptosphaeria biglobosa* CanS-51	Oilseed rape	Biocontrol of *Sclerotinia sclerotiorum*, *Botrytis cinerea*	Zhang et al. (2014)
Enterobacter sp. strain FD17	Maize	IAA and ACC, P solubilization, siderophore	Naveed et al. (2014)
Trichoderma gamsii	Lentil	P solubilization, chitinase, ammonia, salicylic acid	Rinu et al. (2014)

(continued)

Table 14.1 (continued)

Endophytes	Host	PGP/biocontrol	References
Streptomyces sp.	Wheat	P solubilization, IAA, phytase, chitinase, siderophore	Jog et al. (2014)
Burkholderia, Klebsiella, Novosphingobium, Sphingomonas	Rice	IAA, P solubilization, siderophore	Rangjaroen et al. (2014)
Bacillus subtilis var. *amyloliquefaciens* 72β24	Rice	Biocontrol of *Rhizoctonia solani*	Nagendran et al. (2014)
Bacillus sp., *Enterobacter* sp.	Corn	Nitrogen fixation, IAA, siderophore	Szilagyi-Zecchin et al. (2014)
Bacillus, Paenibacillus, Klebsiella, Acinetobacter	Wheat	IAA, P solubilization, siderophore, biocontrol of *Gaeumannomyces graminis*	Duran et al. (2014)
Burkholderia, Azospirillum, Ideonella, Pseudacidovorax, Bradyrhizobium	Potatoes	N fixation, phytohormone production, BCAs of *Fusarium, Koribacter, Pectobacterium*	Pageni et al. (2014)
Gluconacetobacter diazotrophicus	Sugarcane	Systemic defense	Idogawa et al. (2014)
Bacillus megaterium	Rice	IAA, ACC deaminase, N fixation	Cheng et al. (2020)
Chaetomium globosum	Maize	Biocontrol of *Cephalosporium maydis*	Elshahawy and Khattab (2022)
Bacillus velezensis, Bacillus megaterium, and *Herbaspirillum huttiense*	Tomato	Biocontrol of Rhizoctonia root-rot disease	Abbas et al. (2024)

laboratory to the field. In vitro characterization of putative endophytic strains is a crucial first stage in the growth and development of bioagents, as is the combination of field testing and biomass production following biocontrol agent commercialization, which is ultimately necessary to guarantee agricultural security worldwide.

References

Abbas MM, Ismael WH, Mahfouz AY, Daigham GE, Attia MS (2024) Efficacy of endophytic bacteria as promising inducers for enhancing immune responses in tomato plants and managing *Rhizoctonia* root-rot disease. Sci Rep 14:1331

Abdelaziz ME, Abdelsattar M, Abdeldaym EA, Atia MAM, Mahmoud AWM, Saad MM, Hirt H (2019) *Piriformospora indica* alters Na^+/K^+ homeostasis, antioxidant enzymes and LeNHX1 expression of greenhouse tomato grown under salt stress. Sci Hortic 256:108532

Abdelrahman M, Sawada Y, Nakabayashi R, Sato S, Hirakawa H, El-Sayed M, Hirai MY, Saito K, Yamauchi N, Shigyo M (2015) Integrating transcriptome and target metabolome variability in doubled haploids of *Allium cepa* for abiotic stress protection. Mol Breed 35:195

Ahmad P, Nabi G, Ashraf M (2011) Cadmium-induced oxidative damage in mustard [*Brassica juncea* (L.) Czern. & Coss.] Salicylic acid can alleviate these effects in plants. S Afr J Bot 77(1):36–44
Aktar MW, Sengupta D, Chowdhury A (2009) Impact of pesticides use in agriculture: their benefits and hazards. Interdiscip Toxicol 2(1):1–12
Ali AH, Abdelrahman M, Radwan U, El-Zayat S, El-Sayed MA (2018) Effects of a *Thermomyces* fungal endophyte isolated from extremely hot desert-adapted plants on the heat stress tolerance of cucumber. Appl Soil Ecol 124:155–162
Anand U, Pal T, Yadav N, Singh VK, Tripathi V, Choudhary KK, Shukla AK, Sunita K, Kumar A, Bontempi E et al (2023) Current scenario and future prospects of endophytic microbes: promising candidates for abiotic and biotic stress management for agricultural and environmental sustainability. Microb Ecol 86:1455–1486
Arnold AE, Herre EA (2003) Canopy cover and leaf age affect colonization by tropical fungal endophytes: ecological pattern and process in *Theobroma cacao* (*Malvaceae*). Mycologia 95:388–398
Babu AG, Reddy MS (2011) Dual inoculation of arbuscular mycorrhizal and phosphate solubilizing fungi contributes in sustainable maintenance of plant health in fly ash ponds. Water Air Soil Pollut 219:3–10
Balestrini R, Bonfante P (2014) Cell wall remodeling in mycorrhizal symbiosis: a way toward biotrophism. Front Plant Sci 5:237
Balsanelli E, Tadra-Sfeir MZ, Faoro H, Pankievicz VC, de Baura VA, Pedrosa FO, de Souza EM, Dixon R, Monteiro RA (2016) Molecular adaptations of *Herbaspirillum seropedicae* during colonization of the maize rhizosphere. Environ Microbiol 18:2343–2356
Bano A, Ullah F, Nosheen A (2012) Role of abscisic acid and drought stress on the activities of antioxidant enzymes in wheat. Plant Soil Environ 58(4):181–185
Bilal S, Shahzad R, Imran M, Jan R, Kim KM, Lee IJ (2020) Synergistic association of endophytic fungi enhances *Glycine max* L. Resilience to combined abiotic stresses: heavy metals, high temperature and drought stress. Ind Crop Prod 143:111931
Bloemen J, McGuire MA, Aubrey DP, Teskey RO, Steppe K (2013) Transport of root-respired CO_2 via the transpiration stream affects aboveground carbon assimilation and CO_2 efflux in trees. New Phytol 197:555–565
Boutaj H, Meddich A, Wahbi S, Moukhli A, El Alaoui-Talibi Z, Douira A, Filali-Maltouf A, El Modafar C (2019) Effect of arbuscular mycorrhizal fungi on Verticillium wilt development of olive trees caused by *Verticillium dahliae*. Res J Biotechnol 14:8
Bowman SM, Free SJ (2006) The structure and synthesis of the fungal cell wall. Bioessays 28:799–808
Cao H, Fu TM, Zhang L, Henze DK, Miller CC, Lerot C, Abad GG, De Smedt I, Zhang Q, van Roozendael M et al (2018) Adjoint inversion of Chinese nonmethane volatile organic com pound emissions using space-based observations of formaldehyde and glyoxal. Atmos Chem Phys 18:15017–15046
Cao MJ, Wang Z, Zhao Q, Mao JL, Speiser A, Wirtz M, Hell R, Zhu JK, Xiang CB (2014) Sulfate availability affects ABA levels and germination response to ABA and salt stress in *Arabidopsis thaliana*. Plant J 77:604–615
Cao R, Liu X, Gao K, Mendgen K, Kang Z, Gao J, Dai Y, Wang X (2009) Mycoparasitism of endophytic fungi isolated from reed on soilborne phytopathogenic fungi and production of cell wall-degrading enzymes in vitro. Curr Microbiol 59:584–592
Carroll G (1988) Fungal endophytes in stems and leaves: from latent pathogen to mutualistic symbiont. Ecology 69:2–9
Chavéz-Díaz IF, Cruz-Cárdenas CI, Sandoval-Cancino G, Calvillo-Aguilar FF, Ruíz-Ramírez S, Blanco-Camarillo M, Rojas-Anaya E, Ramírez-Vega H, Arteaga-Garibay RI, Zelaya-Molina LX (2022) Seedling growth promotion and potential biocontrol against phytopathogenic *Fusarium* by native rhizospheric *pseudomonas* spp. strains from *Amarillo Zamorano* maize landrace. Rhizosphere 24:100601

Cheng T, Yao X, Wu C, Zhang W, He W, Dai C (2020) Endophytic *bacillus megaterium* triggers salicylic acid-dependent resistance and improves the rhizosphere bacterial community to mitigate rice spikelet rot disease. Appl Soil Ecol 156:103710

Choudhury FK, Rivero RM, Blumwald E, Mittler R (2017) Reactive oxygen species, abiotic stress and stress combination. Plant J 90:856–867

Compant S, Mitter B, Coli-Mull JG, Gangl H, Sessitsch A (2011) Endophytes of grapevine flowers, berries, and seeds: identification of cultivable bacteria, comparison with other plant parts, and visualization of niches of colonization. Microbial Ecol 62:188–197

Costa FG, Zucchi TD, de Melo IS (2013) Biological control of phytopathogenic fungi by endophytic actinomycetes isolated from maize (*Zea mays* L.). Braz Arch Biol Technol 56(6):948–955

Dang QN, Burgess TI, McComb J, Pham TQ, Le BV, Tran TV, Nguyen LT, Hardy GEJ (2023) Fungal and bacterial endophytes antagonistic to *Phytophthora* species causing root rot in *Cinnamomum cassia*. Mycol Prog 22:28

De Melo Pereira GV, Magalhaes KT, Lorenzetii ER, Souza TP, Schwan RF (2012) A multiphasic approach for the identification of endophytic bacterial in strawberry fruit and their potential for plant growth promotion. Microb Ecol 63(2):405–417

Dean DR, Jacobson MR (1992) Biochemical genetics of nitrogenase. In: Stacey G, Burris RH, Evans HJ (eds) Biological nitrogen fixation. Chapman and Hall, New York, pp 763–834

Dhar Purkayastha G, Mangar P, Saha A, Saha D (2018) Evaluation of the biocontrol efficacy of a *Serratia marcescens* strain indigenous to tea rhizosphere for the management of root rot disease in tea. PLoS One 13:e0191761

Duran P, Jacquelinne J, Acuña-Milko A, Azcón JR, Paredes C, Rengel Z, Luz Mora M (2014) Endophytic bacteria from selenium-supplemented wheat plants could be useful for plant-growth promotion, biofortification and *Gaeumannomyces graminis* biocontrol in wheat production. Biol Fertil Soils 50(6):983–990

Eaton CJ, Cox MP, Scott B (2011) What triggers grass endophytes to switch from mutualism to pathogenism? Plant Sci 180(2):190–195

Egamberdieva D, Wirth S, Alqarawi A, Abd Allah EF, Hashem A (2017) Phytohormones and beneficial microbes: essential components for plants to balance stress and fitness. Front Microbiol 8:2104

Eid AM, Salim SS, Hassan SED, Ismail MA, Fouda A (2019) Role of endophytes in plant health and abiotic stress management. In: Kumar V, Prasad R, Kumar M, Choudhary DK (eds) Microbiome in plant health and disease. Springer, pp 119–144

Elshahawy IE, Khattab AA (2022) Endophyte *Chaetomium globosum* improves the growth of maize plants and induces their resistance to late wilt disease. J Plant Dis Prot 129:1125–1144

Ezra D, Hess WM, Strobel GA (2004) New endophytic isolates of *Muscodor albus*, a volatile-antibiotic-producing fungus. Microbiology 150:4023–4031

Fässler E, Evangelou MW, Robinson BH, Schulin R (2010) Effects of indole-3-acetic acid (IAA) on sunflower growth and heavy metal uptake in combination with ethylene diamine disuccinic acid (EDDS). Chemosphere 80(8):901–907

Fontana DC, de Paula S, Torres AG, de Souza VHM, Pascholati SF, Schmidt D, Dourado Neto D (2021) Endophytic fungi: biological control and induced resistance to phytopathogens and abiotic stresses. Pathogens 10:570

Frank AC, Saldierna Guzmán JP, Shay JE (2017) Transmission of bacterial endophytes. Microorganisms 5:70

Fulchieri M, Lucangeli C, Bottini R (1993) Inoculation with *Azospirillum lipoferum* affects growth and gibberellin status of corn seedling roots. Plant Cell Physiol 34:1305–1309

Gangwar M, Rani S, Sharma N (2012) Diversity of endophytic actinomycetes from wheat and its potential as plant growth promoting and biocontrol agents. J Adv Lab Res Biol 3(1):18–23

Gao FK, Dai CC, Liu XZ (2010) Mechanisms of fungal endophytes in plant protection against pathogens. Afr J Microbiol Res 4:1346–1351

Glick BR (2005) Modulation of plant ethylene levels by the bacterial enzyme ACC deaminase. FEMS Microbiol Lett 251:1–7

Glick BR, Cheng Z, Czarny J, Duan J (2007) Promotion of plant growth by ACC deaminase-producing soil bacteria. Eur J Plant Pathol 119:329–339

Grosch R, Scherwinski K, Lottmann J, Berg G (2006) Fungal antagonists of the plant pathogen *Rhizoctonia solani*: selection, control efficacy and influence on the indigenous microbial community. Mycol Res 110:1464–1474

Grossmann K (2010) Auxin herbicides: current status of mechanism and mode of action. Pest Manag Sci 66:113–120

Haggag WM (2024) Plant biotic and abiotic stresses combinations and their management: review article. Int J Agricu Technol 20(1):105–122

Haggag WM, Tawfik MM, Abouziena HF, Abd El Wahed MSA, Ali RR (2017) Enhancing wheat production under arid climate stresses using bioelicitors. Gesunde Pflanz 69(3):149–158

Haggag WM, Eid MM (2022) Antifungal and antioxidant activities of Ag@FeONPs@Chitosan preparation by endophyte *Streptomyces aureofaciens*. Int J Agric Technol 18(2):535–548

Haggag WM, Ali RR (2019) Microorganisms for wheat improvement under biotic stress and dry climate. Agric Eng Int CIGR J 21(2):118–126

Hamayun M, Afzal Khan S, Latif Khan A, Rehman G, Kim Y, Iqbal I, Hussain J, Sohn E, Lee I (2010) Gibberellin production and plant growth promotion from pure cultures of *Cladosporium* sp. MH-6 isolated from cucumber (*Cucumis sativus* L.). Mycologia 102(5):989–995

Hamdali H, Hafi di M, Virolle MJ, Ouhdouch Y (2008) Rock phosphate-solubilizing *Actinomycetes*: screening for plant growth-promoting activities. World J Microbiol Biotechnol 24(11):2565–2575

Harrison MJ, Ivanov S (2017) Exocytosis for endosymbiosis: membrane trafficking pathways for development of symbiotic membrane compartments. Curr Opin Plant Biol 38:101–108

Hoffman MT, Gunatilaka MK, Wijeratne K, Gunatilaka L, Arnold AE (2013) Endohyphal bacterium enhances production of indole-3-acetic acid by a foliar fungal endophyte. PLoS One 8(9):e73132

Huang LQ, Niu YC, Su L, Deng H, Lyu H (2020) The potential of endophytic fungi isolated from cucurbit plants for biocontrol of soil borne fungal diseases of cucumber. Microbiol Res 231:126369

Idogawa N, Amamoto R, Murata K, Kawai S (2014) Phosphate enhances levan production in the endophytic bacterium *Gluconacetobacter diazotrophicus* Pal5. Bioengineered 5(3):173–179

Ikram M, Ali N, Jan G, Jan FG, Rahman IU, Iqbal A, Hamayun M (2018) IAA producing fungal endophyte *Penicillium roqueforti* Thom., enhances stress tolerance and nutrients uptake in wheat plants grown on heavy metal contaminated soils. PLoS One 13:e0208150

Ismail I, Hussain A, Mehmood A, Qadir M, Husna H, Iqbal A, Hamayun M, Khan N (2020) Thermal stress alleviating potential of endophytic fungus *Rhizopus oryzae* inoculated to sunflower (*Helianthus annuus* L.) and soybean (*Glycine max* L.). Pak J Bot 52(5):1857–1865

Jadhav HP, Shaikh SS, Sayyed RZ (2017) Role of hydrolytic enzymes of rhizoflora in biocontrol of fungal phytopathogens: an overview. In: Rhizotrophs: plant growth promotion to bioremediation. Springer, pp 183–203

Jan GF, Hamayun M, Hussain A, Jan G, Iqbal A, Khan A, Lee IJ (2019) An endophytic isolate of the fungus *Yarrowia lipolytica* produces metabolites that ameliorate the negative impact of salt stress on the physiology of maize. BMC Microbiol 19:3

Jha P, Panwar J, Jha PN (2018) Mechanistic insights on plant root colonization by bacterial endophytes: a symbiotic relationship for sustainable agriculture. Environ Sustain 1:25–38

Jog R, Pandya M, Nareshkumar G, Rajkumar S (2014) Mechanism of phosphate solubilization and antifungal activity of *Streptomyces* spp. isolated from wheat roots and rhizosphere and their application in improving plant growth. Microbiology 160:778–788

Jogawat A, Saha S, Bakshi M, Dayaman V, Kumar M, Dua M, Varma A, Oelmüller R, Tuteja N, Johri AK (2013) *Piriformospora indica* rescues growth diminution of rice seedlings during high salt stress. Plant Signal Behav 8:e26891

Johnson LJ, Koulman A, Christensen M, Lane GA, Fraser K, Forester N, Johnson RD, Bryan GT, Rasmussen S (2013) An extracellular siderophore is required to maintain the mutualistic interaction of *Epichloë festucae* with *Lolium perenne*. PLoS Pathog 9(5):e1003332

Kashyap N, Singh SK, Yadav N, Singh VK, Kumari M, Kumar D, Shukla L, Kaushalendra L, Bhardwaj N, Kumar A (2023) Biocontrol screening of endophytes: applications and limitations. Plants 12:2480

Kavroulakis N, Ntougias S, Zervakis GI, Ehaliotis C, Haralampidis K, Papadopoulou KK (2007) Role of ethylene in the protection of tomato plants against soil-borne fungal pathogens conferred by an endophytic *Fusarium solani* strain. J Exp Botany 58:3853–3864

Kawasaki A, Donn S, Ryan PR, Mathesius U, Devilla R, Jones A et al (2016) Microbiome and exudates of the root and rhizosphere of *Brachypodium distachyon*, a model for wheat. PLoS One 11:e0164533

Khalid M, Hassani D, Liao J, Xiong X, Bilal M, Huang D (2018) An endosymbiont *Piriformospora indica* reduces adverse effects of salinity by regulating cation transporter genes, phytohormones, and antioxidants in *Brassica campestris* ssp. Chinensis Environ Exp Bot 153:89–99

Khan AL, Hamayun M, Kang S, Kim Y, Jung H, Lee J, Lee I (2012) Endophytic fungal association via gibberellins and can improve plant growth under abiotic stress: an example of *Paecilomyces formosus* LHL10. BMC Microbiol 12:3

Khan AL, Hussain J, Al-Harrasi A, Al-Rawahi A, Lee IJ (2013) Endophytic fungi: resource for gibberellins and crop abiotic stress resistance. Crit Rev Biotechnol 35(1):62–74

Khan MS, Zaidi A, Ahemad M, Oves M, Wani PA (2010) Plant growth promotion by phosphate solubilizing fungi-current perspective. Arch Agron Soil Sci 56:73–98

Khan Z, Doty S (2011) Endophyte-assisted phytoremediation. Curr Top Plant Biol 12:97–105

Kim J, Rees DC (1994) Nitrogenase and biological nitrogen fixation. Biochemistry 33:389–397

Kiss L (2003) A review of fungal antagonists of powdery mildews and their potential as biocontrol agents. Pest Manag Sci Former Pestic Sci 59:475–483

Kobae Y (2019) Dynamic phosphate uptake in arbuscular mycorrhizal roots under field conditions. Front Environ Sci 6:159

Kumar A, Droby S, Singh VK, Singh SK, White JF (2020) Entry, colonization, and distribution of endophytic microorganisms in plants. In: Microbial endophytes. Elsevier, Cambridge, MA, pp 1–33

Kumar A, Santoyo G, White JF, Mishra VK (2023) Special issue "Microbial endophytes: functional biology and applications": editorial. Microorganisms 11:918

Kumari M, Qureshi KA, Jaremko M, White JF, Singh SK, Sharma VK, Singh KK, Santoyo G, Puopolo G, Kumar A (2022) Deciphering the role of endophytic microbiome in postharvest diseases management of fruits: opportunity areas in commercial upscale production. Front Plant Sci 13:1026575

Latge JP (2007) The cell wall: a carbohydrate armour for the fungal cell. Mol Microbiol 66:279–290

Latz MAC, Jensen B, Collinge DB, Jørgensen HJL (2018) Endophytic fungi as biocontrol agents: elucidating mechanisms in disease suppression. Plant Ecol Divers 11(5–6):555–567

Lecomte C, Alabouvette C, Edel-Hermann V, Robert F, Steinberg C (2016) Biological control of ornamental plant diseases caused by *Fusarium oxysporum*: a review. Biol Control 101:17–30

Lee SO, Choi GJ, Choi YH, Jang KS, Park DJ, Kim CJ, Kim JC (2008) Isolation and characterization of endophytic actinomycetes from Chinese cabbage roots as antagonists to *Plasmodiophora brassicae*. J Microbiol Biotechnol 18:1741–1746

Liu HW, Carvalhais LC, Crawford M, Singh E, Dennis PG, Pieterse CMJ, Schenk PM (2017) Inner plant values: diversity, colonization and benefits from endophytic bacteria. Front Microbiol 8:2552

Liu X, Liu W, Sun Y, Xia C, Elmerich C, Xie Z (2018) A cheZ-like gene in *Azorhizobium caulinodans* is a key gene in the control of chemotaxis and colonization of the host plant. Appl Environ Microbiol 84:1827–1844

Lopez DC, Sword GA (2015) The endophytic fungal *entomopathogens Beauveria bassiana* and *Purpureocillium lilacinum* enhance the growth of cultivated cotton (*Gossypium hirsutum*) and negatively affect survival of the cotton bollworm *(Helicoverpa zea)*. Biol Control 89:53–60

Mahmoud RS, Narisawa K (2013) A new fungal endophyte, *Scolecobasidium humicola*, promotes tomato growth under organic nitrogen conditions. PLoS One 8(1):78746

Malik A, Butt TA, Naqvi STA, Yousaf S, Qureshi MK, Zafar MI, Farooq G, Nawaz I, Iqbal M (2020) Lead tolerant endophyte *Trametes hirsuta* improved the growth and lead accumulation in the vegetative parts of *Triticum aestivum* L. Heliyon 6:e04188

Malinowski DP, Belesky DP (2000) Adaptations of endophyte-infected cool-season grasses to environmental stresses: mechanisms of drought and mineral stress tolerance. Crop Sci 40:923–940

McSteen P (2010) Auxin and monocot development. Cold Spring Harb Perspect Biol 2:1–2

Mendoza-Mendoza A, Zaid R, Lawry R, Hermosa R, Monte E, Horwitz BA, Mukherjee PK (2018) Molecular dialogs between *Trichoderma* and roots: role of the fungal secretome. Fungal Biol Rev 32(2):62–85

Mishra A, Singh SP, Mahfooz S, Singh SP, Bhattacharya A, Mishra N, Nautiyal C (2018) Endophyte-mediated modulation of defense-related genes and systemic resistance in *Withania somnifera* (L.) Dunal under *Alternaria alternata* stress. Appl Environ Microbiol 84(8):e02845–e02817

Mittler R (2002) Oxidative stress, antioxidants and stress tolerance. Trends Plant Sci 7:405–410

Morales-Cedeno LR, Orozco-Mosqueda MC, Loeza-Lara PD, Parra-Cota FI, Santos-Villalobos S, Santoyo G (2021) Plant growth-promoting bacterial endophytes as biocontrol agents of pre and postharvest diseases: fundamentals, methods of application and future perspectives. Microbiol Res 242:126612

Moto MS, Gomes CB, Souza Júnior IT, Moura AB (2017) Bacterial selection for biological control of plant disease: criterion determination and validation. Braz J Microbiol 48:62–70

Mousa WK, Adrian S, Jeffrey D, Philip S, Liu H, Zhou T, France-Isabelle A, Raizada MN (2015) An endophytic fungus isolated from finger millet (*Eleusine coracana*) produces anti-fungal natural products. Front Microbiol 6:1157

Mousa WK, Shearer C, Limayrios V, Ettinger CL, Eisen JA, Raizada MN (2016) Root-hair endophyte stacking in finger millet creates a physicochemical barrier to trap the fungal pathogen. Nat Microbiol 1:16167

Munir S, Li Y, He P, He P, Ahmed A, Wu Y, He Y (2020) Unraveling the metabolite signature of citrus showing defense response toward *Candidatus liberibacter* asiaticus after application of endophyte *Bacillus subtilis* L1-21. Microbiol Res 234:126425

Nagendran K, Karthikeyan G, Faisal PM, Kalaiselvi P, Raveendran M, Prabakar K, Raguchander T (2014) Exploiting endophytic bacteria for the management of sheath blight disease in rice. Biol Agric Hortic 30(1):8–23

Naveed M, Mitter B, Yousaf S, Pastar M, Afzal M, Sessitsch A (2014) The endophyte *Enterobacter* sp. FD17: a maize growth enhancer selected based on rigorous testing of plant beneficial traits and colonization characteristics. Biol Fertil Soils 50(2):249–262

Nurul HD, Shashita J, Rozi M (2012) An improved surface sterilization technique for introducing leaf, nodal and seed explants of *Aquilaria malaccensis* from field sources into tissue culture. Asia Pac J Mol Biol Biotechnol 20:55–58

Oku S, Komatsu A, Tajima T, Nakashimada Y, Kato J (2012) Identification of chemotaxis sensory proteins for amino acids in *Pseudomonas fluorescens* Pf0-1 and their involvement in chemotaxis to tomato root exudate and root colonization. Microbes Environ 27:462–469

Overmyer K, Brosché M, Kangasjärvi J (2003) Reactive oxygen species and hormonal control of cell death. Trends Plant Sci 8:335–342

Ownley BH, Griffin MR, Klingeman WE, Gwinn KD, Molton JK, Pereira RM (2008) *Beauveria bassiana*: endophytic colonization and plant disease control. J Invertebr Pathol 98(3):267–270

Pageni BB, Lupwayi NZ, Akter Z, Larney FJ, Kawchuk LM, Gan YT (2014) Plant growth-promoting and phytopathogen- antagonistic properties of bacterial endophytes from potato (*Solanum tuberosum* L.) cropping systems. Can J Plant Sci 94(5):835–844

Pandit MA, Kumar J, Gulati S, Bhandari N, Mehta P, Katyal R, Rawat CD, Mishra V, Kaur J (2022) Major biological control strategies for plant pathogens. Pathogens 11:273

Peng Y, Li SJ, Yan J, Tang Y, Cheng JP, Gao AJ, Yao X, Ruan JJ, Xu BL (2021) Research progress on phytopathogenic fungi and their role as biocontrol agents. Front Microbiol 12:670135

Pimentel Esposito-Polesi N, Fiori De Abreu-Tarazi M, Vieira De Almeida C, Tsai SM, De Almeida M (2017) Investigation of endophytic bacterial community in supposedly axenic cultures of pineapple and orchids with evidence on abundant intracellular bacteria. Curr Microbiol 74:103–113

Potshangbam M, Devi SI, Sahoo D, Strobel GA (2017) Functional characterization of endophytic fungal community associated with *Oryza sativa* L. and *Zea mays* L. Front Microbiol 8:325e330

Poveda J, Abril-Urias P, Escobar C (2020) Biological control of plant-parasitic nematodes by filamentous fungi inducers of resistance: *Trichoderma*, mycorrhizal and endophytic fungi. Front Microbiol 11:992

Pozo MJ, Cordier C, Dumas-Gaudot E, Gianinazzi S, Barea JM, Azcón-Aguilar C (2002) Localized versus systemic effect of arbuscular mycorrhizal fungi on defense responses to Phytophthora infection in tomato plants. J Exp Bot 53:525–534

Priyashantha AKH, Karunarathna SC, Lu L, Tibpromma S (2023) Fungal endophytes: an alternative biocontrol agent against phytopathogenic fungi. Encyclopedia 3:759–780

Pumplin N, Zhang X, Noar RD, Harrison MJ (2012) Polar localization of a symbiosis-specific phosphate transporter is mediated by a transient reorientation of secretion. Proc Natl Acad Sci USA 109:E665–E672

Radhakrishnan R, Khan AL, Kang SM, Lee IJ (2015) A comparative study of phosphate solubilization and the host plant growth promotion ability of *Fusarium verticillioides* RK01 and *Humicola* sp. KNU01 under salt stress. Ann Microbiol 65:585–593

Ramamoorthy V, Viswanathan R, Raguchander T, Prakasam V, Samiyappan R (2001) Induction of systemic resistance by plant growth promoting rhizobacteria in crop plants against pests and diseases. Crop Prot 20:1–11

Ramanujam B, Renuka S, Shylesha AN (2016) Electron microscopic studies for confirmation of endophytic colonization of *Beauveria bassiana* in maize. J Pure Appl Microbiol 10:3017–3021

Rangjaroen C, Rerkasem B, Teaumroong N, Noisangiam R, Lumyong S (2014) Promoting plant growth in a commercial rice cultivar by endophytic diazotrophic bacteria isolated from rice landraces. Ann Microbiol 65:253–266

Rani S, Kumar P, Dahiya P, Maheshwari R, Dang AS, Suneja P (2022) Endophytism: a multidimensional approach to plant–prokaryotic microbe interaction. Front Microbiol 13:861235

Raymaekers K, Ponet L, Holtappels D, Berckmans B, Cammue BP (2020) Screening for novel biocontrol agents applicable in plant disease management–a review. Biol Control 144:104240

Redman RS, Anderson JA, Biaggi TM, Malmberg KE, Rienstra MN, Weaver JL, Rodriguez RJ (2022) Symbiotic modulation as a driver of niche expansion of coastal plants in the San Juan Archipelago of Washington state. Front Microbiol 13:868081

Rinu K, Sati P, Pandey A (2014) *Trichoderma gamsii* (NFCCI 2177): a newly isolated endophytic, psychrotolerant, plant growth promoting, and antagonistic fungal strain. J Basic Microbiol 54:408–417

Ristaino JB, Anderson PK, Bebber DP, Brauman KA, Cunniffe NJ, Fedoroff NV, Finegold C, Garrett KA, Gilligan CA, Jones CM et al (2021) The persistent threat of emerging plant disease pandemics to global food security. Proc Natl Acad Sci USA 118:e2022239118

Rodriguez H, Fraga R, Gonzalez T, Bashan Y (2006) Genetics of phosphate solubilization and its potential applications for improving plant growth-promoting bacteria. Plant Soil 287:15–21

Rodriguez R, Redman R (2008) More than 400 million years of evolution and some plants still can't make it on their own: plant stress tolerance via fungal symbiosis. J Exp Botany 59:1109–1114

Ruhe J, Agler MT, Placzek A, Kramer K, Finkemeier I, Kemen EM (2016) Obligate biotroph pathogens of the genus *Albugo* are better adapted to active host defense compared to niche competitors. Front Plant Sci 7:820

Sales VM, Ferguson-Smith AC, Patti ME (2017) Epigenetic mechanisms of transmission of metabolic disease across generations. Cell Metabol 25:559–571

Salome GIE, Hynes RK, Nelson LM (2001) Cytokinin production by plant growth promoting rhizobacteria and selected mutants. Can J Microbiol 47:404–411

Sanders IR, Clapp JP, Wiemken A (1996) The genetic diversity of arbuscular mycorrhizal fungi in natural ecosystems-a key to understanding the ecology and functioning of the mycorrhizal symbiosis. New Phytol 133:123–134

Santoyo G, Moreno-Hagelsieb G, del Carmen Orozco-Mosqueda M, Glick BR (2016) Plant growth-promoting bacterial endophytes. Microbiol Res 183:92–99

Sawers RJ, Svane SF, Quan C, Grønlund M, Wozniak B, Gebreselassie MN et al (2017) Phosphorus acquisition efficiency in arbuscular mycorrhizal maize is correlated with the abundance of root-external hyphae and the accumulation of transcripts encoding PHT1 phosphate transporters. New Phytol 214:632–643

Schafer P, Pfiffi S, Voll LM, Zajic D, Chandler PM, Waller F, Scholz U, Pons-Kuhnemann J, Sonnewald S, Sonnewald U, Kogel KH (2009) Manipulation of plant innate immunity and gibberellin as factor of compatibility in the mutualistic association of barley roots with *Piriformospora indica*. Plant J 59:461–474

Schippers B, Bakker AW, Bakker PAHM, Van Peer R (1990) Beneficial and deleterious effects of HCN-producing *Pseudomonads* on rhizosphere interactions. Plant Soil 129:75–83

Schulz B, Römmert AK, Dammann U, Aust HJ, Strack D (1999) The endophyte–host interaction: a balanced antagonism. Myco Res 103:1275–1283

Shimizu M, Yazawa S, Ushijima Y (2009) A promising strain of endophytic *Streptomyces* sp. for biological control of cucumber anthracnose. J Gen Plant Pathol 75:27–36

Shinogi T, Suzuki T, Kurihara T, Narusaka Y, Park P, Shinogi T et al (2003) Microscopic detection of reactive oxygen species generation in the compatible and incompatible interactions of *Alternaria alternata* Japanese pear pathotype and host plants. J Gen Plant Pathol 69:7–16

Shutsrirung A, Chromkaewa Y, Pathom-Areeb W, Choonluchanona S, Boonkerdc N (2013) Diversity of endophytic actinomycetes in mandarin grown in northern Thailand, their phytohormone production potential and plant growth promoting activity. Soil Sci Plant Nutr 59(3):322–330

Siddiqui S, Siddiqui ZA, Ahmad I (2006) Evaluation of fluorescent *Pseudomonads* and *Bacillus* isolates for the biocontrol of a wilt disease complex of pigeonpea. World J Microbiol Biotechnol 21(5):729–732

Sindhu SS, Parmar P, Phour M, Sehrawat A (2016) Potassium-solubilizing microorganisms (KSMs) and its effect on plant growth improvement. In: Potassium solubilizing microorganisms for sustainable agriculture. Springer, New Delhi, pp 171–185

Smith SE, Jakobsen I, Grønlund M, Smith FA (2011) Roles of arbuscular mycorrhizas in plant phosphorus nutrition: interactions between pathways of phosphorus uptake in arbuscular mycorrhizal roots have important implications for understanding and manipulating plant phosphorus acquisition. Plant Physiol 156:1050–1057

Solanki MK, Robert AS, Singh RK, Kumar S, Pandey AK, Srivastava AK, Arora DK (2012) Characterization of mycolytic enzymes of *Bacillus* strains and their bioprotection role against *Rhizoctonia solani* in tomato. Curr Microbiol 65:330–336

Soliman SSM, Greenwood JS, Bombarely A, Mueller LA, Tsao R, Mosser DD, Raizada MN (2015) An endophyte constructs fungicide-containing extracellular barriers for its host plant. Curr Biol 25(19):2570–2576

Soliman SSM, Raizada MN (2013) Interactions between cohabitating fungi elicit synthesis of Taxol from an endophytic fungus in host taxus plants. Front Microbiol 4(3):1–14

Strobel G (2018) The emergence of endophytic microbes and their biological promise. J Fungi 4:57

Sun LN, Su H, Zhu Y, Xu MJ (2012) Involvement of abscisic acid in ozone-induced puerarin production of *Pueraria thomsonii* Benth suspension cell cultures. Plant Cell Rep 31:179–185

Sundh I, Eilenberg J (2021) Why has the authorization of microbial biological control agents been slower in the EU than in comparable jurisdictions? Pest Manag Sci 77:2170–2178

Sundh I, Goettel MS (2013) Regulating biocontrol agents: a historical perspective and a critical examination comparing microbial and macrobial agents. BioControl 58:575–593

Szilagyi-Zecchin VJ, Ikeda AC, Hungria M, Adamoski D, Kava-Cordeiro V, Glienke C, Galli-Terasawa LV (2014) Identification and characterization of endophytic bacteria from corn (*Zea mays* L.) roots with biotechnological potential in agriculture. AMB Express 4:26

Tanaka A, Christensen MJ, Takemoto D, Park P, Scott B (2006) Reactive oxygen species play a role in regulating a fungus perennial ryegrass mutualistic interaction. Plant Cell 18:1052–1066

Terhonen E, Sipari N, Asiegbu FO (2016) Inhibition of phytopathogens by fungal root endophytes of Norway spruce. Biol Control 99:53–63

Tian P, Nan Z, Li C, Spangenberg G (2008) Effect of the endophyte *Neotyphodium lolii* on susceptibility and host physiological response of perennial ryegrass to fungal pathogens. Eur J Plant Pathol 122:593–602

Tripathi D, Raikhy G, Kumar D (2019) Chemical elicitors of systemic acquired resistance-Salicylic acid and its functional analogs. Curr Plant Biol 17:48–59

Tromas A, Perrot-Rechenmann C (2010) Recent progress in auxin biology. C R Biol 333:297–306

Truyens S, Weyens N, Cuypers A, Vangronsveld J (2014) Bacterial seed endophytes: genera, vertical transmission and interaction with plants. Environ Microbiol Rep 7:40–50

Vance CP, Ehde-Stone C, Allan DL (2003) Phosphorous acquisition and use: critical adaptations by plants for screening a renewable resource. New Phytol 157:423–447

Vandenkoornhuyse P, Quaiser A, Duhamel M, Van AL, Dufresne A (2015) The importance of the microbiome of the plant holobiont. New Phytol 206(4):1196–1206

Vendan RT, Yu YJ, Lee SH, Rhee YH (2010) Diversity of endophytic bacteria in ginseng and their potential for plant growth promotion. J Microbiol 48(5):559–565

Verma H, Kumar D, Kumar V, Kumari M, Singh SK, Sharma VK, Droby S, Santoyo G, White JF, Kumar A (2021) The potential application of endophytes in management of stress from drought and salinity in crop plants. Microorganisms 9:1729

Villarreal-Delgado MF, Villa-Rodríguez ED, Cira-Chávez LA, Estrada-Alvarado MI, Parra-Cota FI, de los Santos-Villalobos S (2018) The genus *Bacillus* as a biological control agent and its implications in the agricultural biosecurity. Mex J Phytopathol 36:95–130

Vu T, Hauschild R, Sikora RA (2006) *Fusarium oxysporum* endophytes induced systemic resistance against *Radopholus similis* on banana. Nematology 8:847–852

Wakelin SA, Warren RA, Harvey PR, Ryder MH (2004) Phosphate solubilization by *Penicillium* spp. closely associated with wheat roots. Biol Fertil Soils 40:36–43

Wang JL, Li T, Liu GY, Smith JM, Zhao ZW (2016) Unraveling the role of dark septate endophyte (DSE) colonizing maize (*Zea mays*) under cadmium stress: physiological, cytological and genic aspects. Sci Rep 6:22028

Wang Y, Brown HN, Crowley DE, Szaniszlo PJ (1993) Evidence for direct utilization of a siderophore, ferrioxamine B, in axenically grown cucumber. Plant Cell Environ 16:579–585

Wani ZA, Mirza DN, Arora P, Riyaz-Ul-Hassan S (2016) Molecular phylogeny, diversity, community structure, and plant growth promoting properties of fungal endophytes associated with the corms of saffron plant: an insight into the microbiome of *Crocus sativus* Linn. Fungal Biol (UK) 120(12):1509–1524

Waqas M, Khan AL, Kamran M, Hamayun M, Kang SM, Kim YH, Lee IJ (2012) Endophytic fungi produce gibberellins and indoleacetic acid and promotes host-plant growth during stress. Molecules 17:10754–10773

Wawra S, Fesel P, Widmer H, Timm M, Seibel J, Leson L, Kesseler L, Nostadt R, Hilbert M, Langen G, Zuccaro A (2016) The fungal specific beta-glucan-binding lectin FGB1 alters cell-wall composition and suppresses glucan-triggered immunity in plants. Nat Commun 7:13188

Weng W, Yan J, Zhou M, Yao X, Gao A, Ma C, Cheng J, Ruan J (2022) Roles of arbuscular mycorrhizal fungi as a biocontrol agent in the control of plant diseases. Microorganisms 10:1266

White JF, Torres MS, Somu MP, Johnson H, Irizarry I, Chen Q et al (2014) Hydrogen peroxide staining to visualize intracellular bacterial infections of seedling root cells. Microsc Res Tech 77:566–573

White JF, Torres MS (2010) Is plant endophyte-mediated defensive mutualism the result of oxidative stress protection? Physiol Plant 138:440–446

Xie ZC, K CY, Zhang WJ, Lang DY, Zhang XH (2019) *Bacillus pumilus* alleviates drought stress and increases metabolite accumulation in *Glycyrrhiza uralensis* Fisch. Environ Exp Bot 158:99–106

Yan L, Zhao H, Zhao X, Xu X, Di Y, Jiang C, Shi J, Shao D, Huang Q, Yang H (2018) Production of bioproducts by endophytic fungi: chemical ecology, biotechnological applications, bottlenecks, and solutions. Appl Microbiol Biotechnol 102:6279–6298

Yan L, Zhu J, Zhao X, Shi J, Jiang C, Shao D (2019) Beneficial effects of endophytic fungal colonization on plants. Appl Microbiol Biotechnol 103:3327–3340

Yao YQ, Lan F, Qiao YM, Wei JG, Huang RS, Li LB (2017) Endophytic fungi harbored in the root of *Sophora tonkinensis* Gapnep: diversity and biocontrol potential against phytopathogens. Microbiology 6(3):e00437

Yingwu S, Kai L, Chun L (2011) Growth promotion effects of the endophyte *Acinetobacter johnsonii* strain 3-1 on sugar beet. Symbiosis 54:159–166

Zamioudis C, Pieterse CM (2012) Modulation of host immunity by beneficial microbes. Mol Plant Microbe Interact 25(2):139–150

Zeidler D, Zähringer U, Gerber I, Dubery I, Hartung T, Bors W et al (2004) Innate immunity in *Arabidopsis thaliana*: lipopolysaccharides activate nitric oxide synthase (NOS) and induce defense genes. Proc Natl Acad Sci USA 101:15811–15816

Zhang Q, Zhang J, Yang L, Zhang L, Jiang D, Chen W, Li G (2014) Diversity and biocontrol potential of endophytic fungi in *Brassica napus*. Biol Control 72:98–108

Zhang S, Gan Y, Xu B (2016) The application of the plant growth-promoting fungus *Trichoderma longibrachiatum* T6 enhances the tolerance of wheat to salt stress through improvements in the antioxidative defense system and gene expression. Front Plant Sci 7:1405

Zhu L, Li T, Wang C, Zhang X, Xu L, Xu R, Zhao Z (2018) Effects of dark septate endophyte (DSE) inoculation on tomato seedlings under Zn and Cd stress. Environ Sci Pollut Res 8(25):35232–35241

Zipfel C, Oldroyd GE (2017) Plant signaling in symbiosis and immunity. Nature 543(7645):328–336

Chapter 15
The Role of Microorganisms in Activating Plant Immunity Against Viral Pathogens

Wafaa Mohamed Haggag, Ibrahim Elsaid Ibrahim Elshahawy, and Mohamed G. Farhat

Abstract Plant viruses threaten sustainable agriculture and cause large financial losses. The generally utilized procedures cannot completely eradicate viral infection in affected plants. Many novel methods are currently being used to prevent plant viral infection; however, none of these treatments have proven to be effective. The generally utilized procedures cannot completely eradicate the viral load in affected plants. This review focuses on the role of plant growth-promoting rhizobacteria/endophytic bacteria, actinomycetes, and fungal species in preventing plant viral infection. Cross-protection mechanisms, ISR, and the accumulation of defensive enzymes, phenolic compounds, lipopeptides, proteases, and RNases in response to plant virus infection have been addressed.

Keywords Microorganisms · Plant viral diseases · Immunity · Management

15.1 Introduction

Viral infections are a substantial threat to sustainable and profitable agriculture worldwide, resulting in annual losses of several billion dollars (Mumford et al. 2016). Viruses cause a variety of pathogenic changes that affect all aspects of plant life. Most viral infections induce systemic harm, which means that the virus spreads from the initial site of infection to other parts of the plant (Wang et al. 2022). Infected plants typically carry the virus until they die, and through vegetative replication, the virus is transferred to offspring. Plants are the first manifestation of viral infection. The physiology and biochemistry of host cells and tissues are altered by viruses (Hull 2014). The disease diagnostic signs can vary greatly depending on the host plant's characteristics, the virulence and aggression of the virus strain, the

W. M. Haggag (✉) · I. E. I. Elshahawy · M. G. Farhat
Department of Plant Pathology, National Research Centre, Giza, Egypt

V. Gopikrishnan et al. (eds.), *Harnessing Microbial Resources for Sustainable Agriculture*, Microorganisms for Sustainability 61,
https://doi.org/10.1007/978-981-95-4817-0_15

duration of the infection, and environmental factors (Jones and Naidu 2019). These elements also affect how long the incubation period lasts. Although it can last up to a year in herbaceous plants, the incubation period usually lasts a few days or weeks. The greatest impact is caused by emerging diseases, which sharply increase in occurrence, geographic distribution, and/or toxicity. The following are the primary causes of virus emergence: (i) monocrop-based agricultural systems that are more susceptible to pathogens and pests owing to their low genetic diversity and high plant density; (ii) the global trade in plant material (germplasm and live plants), which transports viruses, hosts, and vectors to new areas and environments; (iii) the impact of climate change on the distribution area of hosts and vectors; and (iv) the rapid evolution and adaptation of viruses (Elena et al. 2014). Unlike bacteria or fungi, which may be treated with antibacterial or antifungal medications, plants cannot be cured after contracting a virus. Therefore, the goal of disease management is to prevent viruses from infecting plants or to prevent plants from becoming resistant to viral infection. This calls for the creation of a variety of strategies that are specific to each virus, host, and environment (pathosystem). The rapid development and parasitic nature of viruses make it difficult to combat viral infection. When viruses enter cell nuclei, they use the plant's own resources to replicate (Jones and Naidu 2019).

Breeding programs to produce virus-resistant forms are based on the marker-assisted selection method to assess a variety of genetic defenses, and the creation of transgenic varieties with high resistance to viral diseases is a promising strategy against plant viruses (Akhter et al. 2021). Gene silencing problems and acquired resistance, which are frequently specific and transient, are drawbacks of genetically engineered methods (Li and Wang 2022). Raising genetically modified plants that are immune to certain viruses has the drawback of possibly redistributing to the virus species structure, which could lead to the spread of other dangerous viral infections. One of the most widely used methods for recovering virus-infected plants is the use of apical meristems. By using the apical virus-free zone, this technique creates a healthy initial plant that will act as the starting point for the production of primary potato seeds. For a range of plant crops, the effectiveness of this approach has been continuously demonstrated (Zhang et al. 2019). The challenge of obtaining the original virus-free material is one drawback of this process. The complexity of regenerating meristems as small as 0.1 mm sets the apical meristem method apart. As meristem size increases, their engraftment and regeneration capacities also increase, but the risk of viral infection also increases (Moses et al. 2017). Microplants and potato tubers were treated via chemotherapeutic and thermotherapeutic methods. Exposure modes might vary on the basis of the traits of the variety. According to Oves and Gaytova (2016), the effectiveness of rehabilitation for potato microplants during thermal treatment is 2.4 times greater than that for tubers. Thermotherapy and chemotherapy are added to the apical meristem method in vitro. These techniques essentially represent a single, highly complex biotechnological approach to plant virus recovery. This method has been used extensively worldwide to revitalize a variety of plant species (Wang et al. 2021). The possibility of self-clonal mutations is a serious drawback of tissue culture-based rehabilitation.

These plants may lose their varietal identity as a result of morphological and characteristic differences from the original variety. Given that the valuable economic features of cultivated varieties are still being evaluated and that they might be given mutants that they do not have, this presents a serious risk to primary seed production (Kesiraju and Sreevathsa 2017). The growing emphasis on the ecologization of agricultural crop production has made the search for biosafe phytopathogen management techniques relevant. Identifying ways to improve the plant immune system and create a plant defense system based on pesticide application and the reserve capacity of the organism is crucial because it creates new prospects for the development of biotechnology (Maksimov et al. 2019).

There is an urgent need for new methods to control viral infections (Akhter et al. 2021; Anikina et al. 2023). One tactic is plant immunization, which involves increasing the body's defense mechanisms. In-depth studies of the molecular and genetic underpinnings of plant resistance to viruses, as well as methods for inducing defense reactions in plants, have increased the potential for increasing plant resistance by enhancing the immune system. A variety of substances referred to as elicitors are essential for the production of an immunological response. Attenuated cultures of pathogens, nonpathogens, or metabolites can be used to biologically immunize plants (Kothari and Patel 2004; Dyakov et al. 2007). The use of microbial inoculants to create systemic resistance to viral infections is intriguing. *Pseudomonas fluorescens* and *Bacillus* sp. isolated from banana roots are among the microbial inoculants that researchers have shown to have significant bioprotective effects on banana virus sickness caused by the banana bunchy top virus (BBTV). In addition to the endophytic bacterial strains EPB5 and EPB22, the *P. fluorescens* Pf1 and CHA0 strains were used to inoculate banana plants. Compared with untreated plants, these plants presented a 60% lower rate of BBTV infection when grown in soil (Nowak and Shulaev 2003). According to Maksimov et al. (2019), isolates of *Pseudomonas* species, *Bacillus* species, *Pseudomonas fluorescens*, *Pseudomonas putida*, *Pseudomonas aeruginosa*, and *Pseudomonas taiwanensis* help shield papaya plants from the ring spot virus (PRSV). Maksimov et al. (2019) reported that the production of phytosporin-M, which is based on *Bacillus subtilis* 26D, was highly effective in protecting plants from viral diseases. The activity of these microorganisms has been verified against a variety of diseases, including bacterial, fungal, and viral diseases. By preventing the growth of the melon aphid *Aphis gossypii*, which is the disease vector, an isolate of *B. subtilis* BS3A25 and its culture filtrate prevented the development of cucumber mosaic virus in tomato plants. The activation of plant defense mechanisms against insects and/or viruses that are influenced by bacteria (Maksimov et al. 2019) and the oxidation of surfactants produced by bacteria (Rodrıguez et al. 2018) may be connected to this. The titer of iris yellow spot virus (IYSV) particles in plants was significantly decreased by the endophytic fungus *Hypocrea lixii* (F3ST1), which colonizes interior onion tissues. This led to less damage by the virus's primary transmitter, *T. tabaci* Lind (Muvea et al. 2018). Identifying bioagents that are environmentally safe, have beneficial activity, and combine a variety of biocidal strains to combat major pests and pathogens while triggering immune responses in plants through the activation of particular signaling cascades, induced systemic resistance (ISR), and secondary metabolites in virus-infected plants is an intriguing task. In contrast to chemical pesticides and transgenic approaches, beneficial and heterogeneous groups such as actinomycetes, endophytic microbes, plant growth-promoting bacteria, and fungal agents can suppress viral activity and inhibit the egg-laying capacity of insect vectors, although numerous researchers have highlighted the importance of beneficial microorganisms in protecting plants from pests and pathogens, including plant viral diseases (LeCocq et al. 2017).

15.2 Microorganisms Used in the Management of Plant Virus Diseases

15.2.1 Actinomycete-Based Control of Plant Viruses

In addition to their potential as plant probiotics and biocontrol agents, actinomycetes are intriguing candidates for the manufacture of secondary metabolite chemicals, commonly known as antibiotics, owing to their capacity to colonize plants and synthesize antibiotics in situ (Pandey et al. 2021). Actinomycetes, including *Streptomyces ahygroscopicus* (Salwan and Sharma 2018), *Streptomyces* sp. ZX01 (Zhang et al. 2015), and *Streptomyces noursei* var. *xichangensis* (Han et al. 2014), have been shown by a number of researchers to have antiviral efficacy against TMV infection in tobacco. Actinomycetes suppress viruses in a number of ways, such as by triggering defensive mechanisms in plants and generating signaling chemicals. Marine actinomycetes have been identified as possible sources of numerous bioactive compounds, and marine animals differ from terrestrial dwellers in terms of metabolism and physiology (Lam 2006). Latake and Borkar (2017) evaluated the antiviral effectiveness of metabolites from 28 marine actinomycete isolates against cucumber mosaic virus (CMV). When used separately as a seed and spray treatment, *Streptomyces olivaceus* metabolites were found to be highly effective in reducing CMV infection in both open field and in vitro settings. When *Streptomyces* culture filtrates and virus sap combination were applied to *Nicotiana glutinosa* and *Chenopodium quinoa*, respectively, a decreasing tendency in different necrotic lesions of tobacco mosaic virus (TMV) and potato virus Y (PVY) was noted (Mohamed and Galal 2005). Similarly, by adding various components, such as fermented actinomycete broth, with TMV sap at different times, Xing et al. (2012) reported TMV inactivation in *Nicotiana glutinosa* and *N. tabacum*. *Streptomyces ahygroscopicus* produces ε-poly-L-lysine (ε-PL), an antiviral molecule with a molecular weight of 3454–4352 Da that has anti-TMV properties (Chen et al. 2019).

15.2.2 Bacteria-Based Control of Plant Viruses

Under simulated conditions, numerous bacterial strains, including *Pseudomonas aeruginosa*, *Burkholderia* sp., and *Bacillus* sp., were tested against cotton leaf curl virus (CLCuV) alone and in diverse combinations. According to estimates, the viral load in challenged plants was 74% greater than that in control plants, and the greatest suppression of CLCuV was shown in plants treated with a combination of bacterial strains of only 0.4%. The characteristics, viral load, and disease incidence were found to be extremely strongly correlated, according to the results of principal component biplot analysis (PCA) (Ramzan et al. 2016). *Pseudomonas fluorescens* has been reported to have field efficacy for CMV and tomato mottle virus (ToMoV) in *Solanum lycopersicum*, as well as biocontrol action against *Tospovirus* (Kandan et al. 2003) and lesion-inducing tobacco necrosis virus (TNV) disease in *Nicotiana tabacum* (Murphy et al. 2000). These findings led to a decrease in the number of TNV-induced lesions in plants treated with *P. fluorescens* (Kandan et al. 2003). Bacterial microbial consortia are also thought to suppress a variety of viruses. A reduction in CMV infection in *S. lycopersicum* was observed with bacterial consortia, specifically *B. subtilis* GB03 + *B. pumilus* SE34/+ *B. amyloliquefaciens* IN937a/+ *B. subtilis* IN937b/*B. pumilus* INR7/*B. pumilus* T4 (Jetiyanon et al. 2003). *Cucumis sativus* CMV infection was decreased by *Serratia marcescens* 90–166 + *P. putida* 89B-61/*B. pumilus* T4 (Murphy et al. 2003). *Helianthus annuus* infection with tobacco streak virus (TSV), which causes sunflower necrosis disease (SND), is decreased by *B. licheniformis* MML2501 + *Bacillus* spp. MML2551 + *Pseudomonas aeruginosa* MML2212 + *Streptomyces fradiae* MML1042 (Srinivasan and Mathivanan 2009). *B. amyloliquefaciens* IN937a + *B. pumilus* SE34 + *B. pumilus* T4 showed antiviral activities against papaya ring spot virus (PRSV-W) and tomato chlorotic spot virus (TCSV) in *Carica papaya/S. lycopersicum* (Abdalla et al. 2017). *Pseudomonas fluorescens* Pf1. + *Bacillus* spp. EPB22 showed antiviral activity against banana bunchy top virus (BBTV) in *Musa* spp. (Harish et al. 2008). Combinations of chemical immune regulators and individual bacterial isolates or consortia, such as *P. fluorescens* CHAO + chitin, decreased BBTV infection in *Musa* spp. (Kavino et al. 2008); *Pseudomonas* sp. 206(4) + B-15+JK-16+chitosan inhibited ToLCV infection in *S. lycopersicum* (Mishra et al. 2014); *B. polymyxa* + *P. fluorescens* + chitin were used to control squash mosaic virus (SqMV) in *Cucumis sativus* (Firmansyah and Widodo Hidayat 2017); and *B. pumilus* INR7 + benzothiadiazole were used to control CMV in *Capsicum annuum* (Yi et al. 2013).

15.2.3 Fungi-Based Control of Plant Viruses

Plant gene expression is regulated by endophytes and fungal bioagents, which can identify physiological changes in stress-induced plants as defense mechanisms (Liu et al. 2017). Muvea et al. (2018) reported that inoculation with endophytic fungi (*Hypocrea lixii* F3ST1) decreased the number of thrips vectors, which led to a higher mortality rate for the vector population. Additionally, because of the agents' antixenotic repellent activity, there was a decrease in oviposition and feeding behavior. Additionally, endophyte-colonized onions may have less vector feeding activity, which could slow the transmission of viruses. Iris yellow spot virus (IYSV) (genus: *Tospovirus*) is spread by *T. tabaci*, and its endophytic contact makes viruliferous thrips less effective at spreading it. This has an adverse effect on IYSV reproduction in affected plants. This could be because endophytes, like fungi, can activate gene expression in defense cascade pathways, which in turn promotes the buildup of defensive chemicals for resistance development (Pieterse et al. 2014). The mechanism of fungal–viral interactions may involve the influence of fungal secondary metabolites produced via endophytic associations. The alkaloids generated by the application of fungal endophytes possess antiviral qualities (Selim et al. 2012). Onion seedling endophytic colonization may prevent viruliferous thrips from causing damage to food. Additionally, endophytic colonization has a detrimental effect on IYSV replication in plants and increases the futility of viruliferous thrips that transmit IYSV. Endophytic fungi may therefore constitute a crucial part of research on tripartite (plant–endophyte–insect) relationships. When fungal endophytes (*Neotyphodium uncinatum*)-infected meadow ryegrass (*Lolium pratensis*) were compared with endophyte-free plants, comparable outcomes were observed in aphids, particularly *Rhopalosiphum padi*, which transmits barley yellow dwarf virus (BYDY). These findings suggest that endophyte production of alkaloids, such as lolines, may help protect meadow ryegrass from BYDV infection (Lehtonen et al. 2005). One potential explanation for the low virus load observed among the plants is that coat protein-encoding gene expression is inhibited or downregulated, which hinders virus replication in the early stages of infection. For example, peanut bud necrosis virus can also be controlled by the application of bioagents to stop the systemic spread of the virus in wild *Arachis* (Reddy et al. 2000). The ISR may result from plant defense signals interfering with virus reproduction and systemic transmission via cell-to-cell migration (Kandan et al. 2005). Among the several biocontrol techniques, ISR is the most successful at reducing viral illness in infected plants. Many plant species have PGPR colonizing their rhizosphere, which benefits host plants by increasing their vigor and resilience to diseases caused by bacteria, viruses, nematodes, fungi, and viroids (Kloepper et al. 2004).

15.2.4 Virus-Based Control of Plant Viruses

Secondary volatiles and other chemicals that are very conducive to vector attraction may be released by the viral load in infected plants. In addition to influencing insect vectors to feed on healthy plants, the virus can create an antifeeding material when the vector tends to feed on the plant, which will cause the vector to flee to a new plant (Mauck et al. 2012). Insects can also suffer greatly from viruses. TSWV-infected plant-feeding whiteflies have limited fecundity and delayed growth (Pan et al. 2013). Minor and moderate strains of plant viruses, as well as the pepino mosaic virus (PepMV) of *S. lycopersicum*, which depends on mild variants of PepMV for induction, have been shown to act as elicitors in plants (Schenk et al. 2010). One or both mild viral strains are present in commercially accessible products on the market.

15.3 Microbe-Induced Antiviral Mechanisms Against Plant Virus Infection

15.3.1 Cross-Protection

One type of induced resistance that plants acquire to protect themselves from plant viruses is called cross-protection. Despite reports that phylogenetically diverse viruses protect against one another, it works best among closely related strains of the same virus. Weak strain cross-protection (MSCP), which occurs when host plants are immunized with a weak strain to shield them from being challenged by aggressive or severe strains of the same plant virus, was first identified by Dutch Thung and Englishman Salaman approximately 70 years ago. This prevents future-proofing of production systems (Pechinger et al. 2019). Cross-protection has been successful in combating *Citrus tristeza virus* (CTV), a member of the *Closterovirus* genus carried by aphid vectors (Folimonova 2013). Although the exact mechanisms of cross-protection are still unknown, a number of tenable theories have been proposed, including the formation of antibodies, specific adsorption by novel cell compounds, depletion of essential metabolites, competition between main and severe viruses for host components, the point of replication sites within the cells, interference with disassembly, secondary virus translation or replication (Lee et al. 2005), and aggravated RNA silencing by the protector virus that leads to sequence-specific deprivation of the test virus RNA (Ratcliff et al. 1999). RNA silencing is one of the frequently discussed cross-protection mechanisms. The mechanism involves the recognition of double-stranded RNA (dsRNA) produced during viral replication by dicer-like (DCL) enzymes, which then cut the dsRNA into tiny pieces ranging in length from 21 to 26 nucleotides (Pooggin 2018). Short nucleotide fragments known as "small-interfering RNAs" (siRNAs) can increase RNA silencing in plants by generating an RNA-induced silencing complex (RISC) composed of Argonaute proteins. When siRNAs are dispersed throughout a plant, they cause the virus to be silenced in areas where there is no previous viral interaction, leading to complete silencing. All plant sections contain RNAs that are comparable in sequence to the RNA that initiates silencing during widespread silencing. More research is needed on the role of *Argonaute* proteins in the translational suppression of viral RNAs, which is initiated by a combination of host resistance factors and virus-eliciting agents (Sanfaçon 2015). Additionally, ZYMV in squash, melon, and watermelon (Hughes and Ollennu 1994), cocoa swollen shoot virus in cocoa (Hughes and Ollennu 1994), tomato mosaic virus (ToMV) in *S. lycopersicum* and pepper, and papaya ringspot virus in papaya (Yeh et al. 1988) are among the viruses for which such claims have been found to be effective.

15.3.2 Antiviral/Antibiotic Compounds

The primary antiviral substances found in fungi include proteins, polysaccharides, and polysaccharide peptides. Polysaccharides and their peptides with antiviral activities have recently been explored in fungal organisms such as *Coriolus versicolor*, *Coprinus comatus*, *Lentinus edodes* (Berk.) sing, *Pleurotus ostreatus*, and *Flammulina velutipes* (Zhao et al. 2016). Lower doses of a polysaccharide peptide derived from *Coriolus versicolor* demonstrated anti-TMV activities of 85.4% and 64.8% (Zhao et al. 2015a). At a concentration of 10 g mL^{-1}, the lentinan component demonstrated anti-TMV efficacy, with a restorative percentage of 58.7% (Wang et al. 2013). By promoting the expression of peroxidase, phenylalanine ammonia-lyase, and pathogenesis-related (PR) proteins in host plants, lentinan and polysaccharide peptide components may enhance the innate resistance of plants to a variety of diseases. Other fungi with antiviral-active compounds include *Agrocybe aegerita, Flammulina velutipes, Lentinus edodes, Pleurotus citrinopileatus*, and *P. eryngii*. Additionally, methanolic extracts of *Penicillium oxalicum* and *Neosartorya fischeri* have shown repressive efficacy against TMV infections (Shen et al. 2009). The antimicrobial compounds of *Trichoderma pseudokoningii* SMF2, including peptaibols and trichokonins, are ISR-determining factors that confer resistance against TMV (Shen et al. 2009). Alkaloids have also been shown to have antiviral properties (Wang et al. 2004). Similarly, through ISR, the expression of TMV symptoms in *S. lycopersicum* is decreased in *Trichoderma harzianum* (Kolase and Sawant 2007). For many plant viruses, the application of antiviral substances that trigger systemic resistance in plants has been documented. To manage plant viral infections, microbe-derived antiviral medicines provide effective, efficient, and biodegradable alternatives to conventional chemical agents. Their methods for combating plant viruses can either directly affect the proteins or nucleic acids of the virus or, ultimately, control the host's response to the virus (Calil and Fontes 2016).

The 80-genus actinomycete group produces a wide range of secondary metabolites that are very suggestive of crop protection. Actinomycetes are the source of more than half of known antibiotics (Magarvey et al. 2004). Ningnanmycin (NNM), an antiviral substance that is most efficient at preventing TMV infections, which are frequent in crop production, is derived from *Actinomycetes–Streptomyces noursei var. xichangensis*, a new cytosine nucleoside peptide antibiotic (Zhao et al. 2015a, b). According to established research, ningnanmycin may be responsible for heightened resistance to TMV-infected plants by activating a number of plant defense pathways; inducing the activity of peroxidase (POD), phenylalanine ammonia-lyase (PAL), and superoxide dismutase (SOD); and activating a group of acidic PR proteins. Additionally, the expression of NPR1 and Jaz3 has been shown to prevent TMV CP [22]. In China, the cytosine peptide, which was isolated from *Streptomyces ahygroscopicus*, has been marketed, registered, and promoted as an antiplant-viral drug because of its effective virucidal action (Han et al. 2014). The novel glycoprotein GP-1, which was isolated from *Streptomyces* sp. ZX01, was also reported by Zhang et al. (2015) to have an 80% anti-TMV rate at a dose of 1 mg mL^{-1}. As an authoritative defensive enzyme, peroxidase is involved in several physiological reactions to biotic stressors in plants (Abd El-Rahim et al. 2017). *Streptomyces* strains cause systemic acquired resistance (SAR) to viral infections, according to Galal (2006), whereas *P. aeruginosa* increases the resistance of tobacco plants to TMV (Meyer et al. 1999) and an antiviral substance from *S. noursei* var. *xichangensis*, which is known to cause ISR (Han et al. 2014). Blasticidin S, dextromycin and mitomycin C (Hirai 1962), oxytetracycline and streptomycin (Mace 1965), puromycin (Milchenko 1969), laurisin or formycin (Huang and Misato 1970), cycloheximide (Raychaudhuri 1970), chloramphenicol (Kataoka et al. 1969), and other medications have been used to inhibit TMV. Actinomycin D and blasticidin S have been used to control bromegrass mosaic virus (Kümmert and Semal 1971). For the use of ohyamycin and blasticidin S against PVX, root dipping and foliar sprays have been suggested. DPB can be used to treat tomato leaf curl (Manjunatha et al. 2010), and it also increases *S. lycopersicum* yield. Through the temporary production of MAMP genes, the bacterial bioagent *Bacillus amyloliquefaciens* (VB7) effectively suppressed the expression of GBNV illness symptoms by up to 84%, induced an immune response, and decreased the virus titer (Vanthana et al. 2019).

The algae group includes a wide variety of plants, ranging from seaweeds that stretch over 30 m to diatoms that are minuscule and unicellular. Microalgae against HIV were the first natural chemicals found (Schaeffer and Krylov 2000). Seaweed polysaccharides, namely, sulfated polysaccharides derived from brown algae, exhibit potent virucidal effects (Pardee et al. 2004). At a concentration of 10 mg mL^{-1}, alginate, a plentiful polysaccharide found in brown algae, demonstrated 95% inhibition of PVX (Pardee et al. 2004). At a dosage of 10 mg mL^{-1}, only 6 of the 30 species of marine algae presented above 80% inhibition rates against PVX. Nagorskaia et al. (2008) reported that the red marine alga *Tichocarpus crinitus* and the sea alga *Ulva pertusa* both release kappa/carrageenan through their lectins, which can reduce TMV infection in Xanthi-NC tobacco leaves.

15.3.3 Lipopeptides

Many bacteria use nonribosomal synthesis to create lipopeptides, which are antimicrobial peptides. Peptaibols, cyclopeptides, and pseudopeptides are examples of secondary metabolites (Stein 2005). Numerous researchers have identified bioactive chemicals that suppress plant viruses by screening bacterial agents for virucidal action. According to Zhou et al. (2008), the bacterial strain ZH14 generates bioactive proteins with stable and long-lasting resistance to TMV. The enhanced control efficacy of *P. fluorescens* CZ against TMV was defined by Shen et al. (2014). The *Serratia marcescens* culture filtrate possesses persistent antiviral ability against CMV, according to Thapa et al. (2009). *Pseudomonas chlororaphis* O6-synthesized cyclic peptide in cell-free supernatant showed potent antiviral action against TMV (Ju et al. 2012).

Bacillus and *Pseudomonas* species are the main producers of lipopeptides, which are relatively low in antimicrobial, 1000–2000 Da cyclic compounds (Cai et al. 2013). In actuality, the most crucial elements that support their efficacy as biocontrol agents in PGPMs are lipopeptides. Lipopeptides are produced via a multienzyme manufacturing process using precursors of specific gene clusters from nonribosomal peptide synthetases (NRPs) (Stein 2005). Furthermore, a variety of lipopeptides, including surfactin, iturin, sophoro lipids, trehalose lipids, rhamnolipid lipids, and mannosylerythritol lipids, have shown promise as efficient biocontrol agent alternatives for a wide range of agricultural production applications (Cortes-Sanchez et al. 2013). *Bacillus* species create three primary families of surfactin, iturin, and fengycin, which are composed of hydrophilic amino acids (7–10 amino acids) connected by a hydrophobic fatty acid tail. These substances act as effective antagonists, blocking plant infections in a number of ways by activating and strengthening the plant defense mechanism in defense-related systems known as ISR (Shekhar et al. 2015). Recently, certain cyclic lipopeptides, such as the ISR, have been shown to trigger defense reactions in plants. It has been demonstrated that the volatile chemicals emitted by *Bacillus* species, including 2,3-butanediol (Ryu et al. 2004), work as elicitors to activate ISR in plants. Owing to their remarkable ion networks and aperture, lipopeptide surfactants have the potential to disrupt the usual integrity and absorptivity of the lipid layer of the plasma membrane. Lipopeptides provide their main antibacterial activity against a variety of microorganisms, including bacteria, fungi, viruses, and mycoplasma, due to their capacity to disrupt the physical integrity of the living membrane (Inès and Dhouha 2015). By reducing infection by the virus vector *P. betae*, *B. amyloliquefaciens* S499 effectively prevents rhizomania illness by releasing cyclic lipopeptides in sugar beetroot leaves and roots (Desoignies et al. 2012). The beet necrotic yellow vein virus (BNYVV), which causes rhizomania, is spread by the obligatory root-infecting parasite *Polymyxa betae* (McGrann et al. 2009). *Bacillus* cyclic lipopeptides (CLPs) significantly reduce *P. betae* infection in sugar beet through ISR. Sugar beets with lower virus concentrations exhibit insufficient resistance to *P. betae*, as demonstrated by McGrann et al. (2009) and Barr et al. (1995). CLPs seem to offer a novel strategy for managing rhizomania disease over the long run.

15.3.4 Ribonucleases Against Plant RNA Viruses

In various crucial transcriptional stages, ribonucleic acid (RNA) can be dimerized and broken down into smaller pieces by a class of hydrolytic enzymes called ribonucleases, which can also prevent viruses from reproducing. The traditional immune system components are ribonucleases (Ilinskaya and Mahmud 2014). Bacteria can produce a variety of ribonucleases, including binases, baRNases, and baliphases, which are enzymes that can leak into the surrounding environment. High quantities of ribonucleases have antiviral qualities and break down viral RNA, whereas low concentrations aid in plant growth. The expression of the pac1 ribonuclease gene from *Schizosaccharomyces pombe* has improved resistance to CMV, ToMV, PVY, and TSWV in chrysanthemum, tobacco, and impatiens plants, although the induction of resistance to viruses via the expression of a double strand-specific RNase gene has changed the resistance levels from a delay in the appearance of disease symptoms to complete protection against viruses (Ogawa et al. 2005). Tobacco plants exhibit resistance to certain viruses through the production of the bacterial double strand-specific RNase gene (Langenberg et al. 1997).

Endophytic strains that can produce RNases prevent viruses from spreading and influence how viral symptoms manifest. To make biocontrol agents, bacterial strains such as *Bacillus* sp. TS2 and *B. subtilis* 26D, which have high endophytic rates and RNase activity, are used. Sorokan et al. (2020) reported that the culture media of *B. thuringiensis* B-6066, *Bacillus* sp. TS2, *Bacillus* sp. STL-7, and *B. subtilis* 26D, as well as strains that have an exceptional capacity to colonize internal plant tissues, presented increased RNase activity. These strains also presented a decreased incidence and severity of viral disease caused by the potato viruses M, S, and Y. According to Pakniat-Jahromy et al. (2010), approximately one-third of the transgenic tobacco clones that expressed the baRNase gene from *B. amyloliquefaciens* presented complete resistance to ToLCV infection. Thus, *Bacillus* and other bacteria can protect plants from viral infection by interfering with vectors of viral particles, such as insects and plant diseases, according to this study.

15.3.5 Plant Growth-Promoting Rhizobacterium-Induced Resistance in Plants

PGPRs are endophytic or rhizospheric bacteria that protect plants from invasive diseases, promote growth, and improve yield sustainability (Sahran and Nehra 2011). As part of synergistic action, PGPRs aid other microorganisms in suppressing pathogens or enhancing their ability to promote plant growth. The substances generated by plant growth-promoting microorganisms (PGPMs) have the potential to interact with the host immune system and cause crop plants to develop systemic resistance to phytopathogens (Maksimov et al. 2015). In general, receptors composed of leucine-rich repeats (LRRs) (Sorokina 2012) identify PAMPs (pathogen-associated molecular patterns) such as flagellin and lipopeptides of endophytic bacteria (Boris and Kochieva 2013) or virus CPs (Boris and Kochieva 2013). When plants are driven by rhizobacteria and their metabolic products, as well as in response to infections and fungal viruses, they express genes that generate pathogenesis-related (PR) proteins, such as the antiviral PR-4 and PR-10 (Bai et al. 2013). As a result, PGPM might significantly affect how plants react to viral infection.

When bioagents or endophytes are present in plants, they can recognize the proteins of plant viruses, especially viral RNA and the coat protein (CP) of a virus. Furthermore, plant cell receptors are able to identify pathogen-associated molecular patterns (PAMPs), which are small conserved molecular motifs present in microbes. These motifs cause the development of plant defense reactions (Niehl et al. 2016), which include the rapid generation of reactive oxygen species (ROS), modifications to phytohormone composition, and the synthesis of a variety of metabolites, including the regulation of defensive genes during local and systemic expression (Zvereva et al. 2016). Under greenhouse conditions, Li et al. (2016) reported that *Enterobacter asburiae* triggered TYLCV resistance and reduced illness rates, which approached 42% even at 45 dpi. In *S. lycopersicum* plants, the various strains of *B. amyloliquefaciens* regulate disease resistance against ToMoV (Murphy et al. 2000). ISR is the term for the phenomenon whereby certain rhizobacteria that interact with plant roots develop resistance to bacteria, fungi, and viruses, resulting in plant diseases. Additionally, ethylene and jasmonate signaling are components of the ISR that strengthen the host plant's defenses against a range of plant diseases (Glick 2012). PGPRs induce systemic resistance via the ethylene and jasmonic acid ISR route or the salicylic acid-dependent SAR pathway (Sahran and Nehra 2011). The hypersensitive reaction (HR) in plants is characterized by localized necrosis and disruption of the ability of the virus to spread throughout the body. HRs shield uninfected tissues by trapping the phytopathogen inside necrotic lesions or ringspots that develop at the infection site on leaves, stems, and fruits. HR is promoted by changes in the plant cell wall, including increased amounts of calcium ions, nitric oxide and superoxide radicals, and SA, JA, and H_2O_2. In the *Solanum tuberosum* crop, the potato Nytbr and Nctbr genes control HRs against the PVYC and PVYO strains of PVY, respectively (Makarova et al. 2017). By releasing ribonucleases and phytohormones, the two endophytic bacterial strains, *B. subtilis* 26D and *B. subtilis* Ttl2, inhibited PVX and PVY in *S. lycopersicum* plants. Antimicrobial action against phytopathogens is significantly influenced by a variety of PR proteins (Van Loon et al. 1994). Treatment of pepper plants containing *B. amyloliquefaciens* 5B6 decreased the prevalence of CMV by inducing the expression of transcription-encoding genes, including the PR-4, PR-5, and PR-10 proteins (Lee and Ryu 2016). In addition to phenolic compounds, the endophytic *Bacillus* spp. EPB22 bacterial combination and rhizospheric *P. fluorescens* Pf1-treated banana plants induced the expression of defense enzymes such as PAL, PO, and PPO, which decreased the prevalence of BBTV by up to 80% (Harish et al. 2008). Banana plants infected with BBTV (Tollenaere et al. 2017), *S. lycopersicum* plants infected with TSWV (Tollenaere et al. 2017), and black gram plants infected with urdbean leaf crinkle virus (ULCV) (Karthikeyan et al. 2009) also exhibited similar reactions. According to Montebianco et al. (2020), the fungal phytopathogen *Cladosporium herbarum*'s cell wall glycoprotein peptidogalactomannan (pGM) impairs viral infection by promoting the expression of defense-related genes involved in the development of SAR and ROS and their accumulation in BY2 tobacco cells. Higher levels of expression are shown for PR-1a (unknown function), PR-2 (-1-3 endoglucanase), PR-3 (chitinase), PR-5 (thaumatin-like protein), PAL (phenylpropanoid pathway gene), and genes associated with innate resistance and plant stress responses, including LOX1 (lipoxygenase) and NtPrxN1 (peroxidase). *P. fluorescens* strains are excellent at producing large amounts of protective enzymes in bananas. The synthesis of phenolics and other secondary metabolic chemicals, which are believed to be important in ISR against BBTV illness and increase yield, is linked to the activities of these enzymes (Firmansyah et al. 2017). Chirkov et al. (2001) determined the connection between callose, β-1,3 glucanase, and ribonuclease induction as a defense mechanism against PVX after chitosan (1 mg/L) inoculation.

It is unclear how ROS function in plant–virus interactions. During TMV infection, ROS are expected to function as defensive molecules (Abo-Zaid et al. 2020), and H_2O_2 functions as a systemic antiviral signaling unit. Increased ROS production is also thought to be a biological marker of SAR induction. *T. harzianum*, *T. polysporum*, *T. atroviride*, and their consortia prevent whiteflies from being infected with pepper leaf curl virus (PepLCV). With a significant increase in phenolic content (183% higher), they endophytically inhabit pepper plants and trigger innate host immunity through activated phenylpropanoid production. The production of salicylic acid by *Trichoderma* bioagents seems to be a major factor in the suppression of PepLCV (up to 50%) and the accumulation of ROS at the infection site, which restricts the ability of the virus to spread (Rochal et al. 2021). By supplying nutrients and plant hormones and secreting allelochemicals, ePGPB protects crops from a variety of pathogens, including viruses. By promoting ISR in the host plant, which in turn triggers defense-related genes via the metabolism of salicylic acid, ethylene, and jasmonic acid, it also participates in indirect biocontrol. Biological control with PGPR microbes can be recommended to protect against viral pathogens because PGPRs play both direct and indirect roles in the sustainable management of crop plants through improved seed germination and emergence, plant growth promotion, biological nitrogen fixation, phosphate solubilization, increased yield, yield components, and nutrient uptake, as well as by inducing ISR and other defensive compounds and enzymes that are essential for disease resistance activity.

15.4 Conclusions

There are now several results concerning the involvement of bacteria, fungi, actinomycetes, and other organisms in defensive mechanisms against plant diseases, in addition to viruses. In essence, the transmission and multiplication of viruses are hindered by defense mechanisms that bacteria and their metabolites generate against viral infection. Predictions of the use of bacteria and plant RNase to stop viral infection in plants are difficult and uncharted research topics. Therefore, a useful strategy for plant defense is the discovery of biological agents with antiviral qualities that are harmless to the environment and can protect plants from viral infections. PGPRs directly combat viruses by generating RNases or SR inducers that live on surfaces or within plant tissues. These microorganisms also indirectly reduce the viral load in agroecological systems by controlling vectors with "RNA biocides" that are specific to crop pests. To effectively control plant viruses, it is necessary to study antiviral-inducing microbial populations, including the ideal number of microorganisms, the period needed to reduce the virus inoculum, and the impact of the virus vector population on antiviral characteristics. The influence of many ecological characteristics, including temperature, light hours, rainfall, and relative humidity (RH), on interactions with antagonistic bacteria has not yet been well studied. This poses a serious obstacle to the development of an antiviral resistance mechanism. Identification of the plant microbial components that produce resistance to antiviral defense requires extensive upgrades in the realms of proteomics and metabolomics owing to the lack of a molecular basis for antiviral resistance generated by microbial agents.

References

Abd El-Rahim WM, Moawad H, Azeiz AZA, Sadowsky MJ (2017) Optimization of conditions for decolorization of azo-based textile dyes by multiple fungal species. J Biotechnol 260:11–17. [CrossRef] [PubMed]

Abdalla OA, Bibi S, Zhang S (2017) Application of plant growth-promoting rhizobacteria to control Papaya ring spot virus and Tomato chlorotic spot virus. Arch Phytopathol Plant Protect 50:584–597. [CrossRef]

Abo-Zaid GA, Matar SM, Abdelkhalek A (2020) Induction of plant resistance against tobacco mosaic virus using the biocontrol agent *Streptomyces cellulosae* isolate Actino 48. Agronomy 10:1620. [CrossRef]

Akhter M, Nakahara KS, Masuta C (2021) Resistance induction based on the understanding of molecular interactions between plant viruses and host plants. Virol J 18(1):1–12. https://doi.org/10.1186/s12985-021-01647-4

Anikina I, Kamarova A, Issayeva K, Issakhanova S, Mustafayeva N, Insebayeva M, Mukhamedzhanova A, Khan SM, Ahmad Z, Lho LH, Han H, Raposo A (2023) Plant protection from virus: a review of different approaches. Front Plant Sci 14:1163270. https://doi.org/10.3389/fpls.2023.1163270

Bai S, Dong C, Li B, Dai H (2013) A PR-4 gene identified from *Malus domestica* is involved in the defense responses against Botryosphaeria dothidea. Plant Physiol Biochem 62:23–32. [CrossRef] [PubMed]

Barr KJ, Asher MJ, Lewis BG (1995) Resistance to Polymyxa betae in wild Beta species. Plant Pathol 44:301–307. [CrossRef]

Boris KV, Kochieva EZ (2013) NBS-LRR resistance genes to potato virus X. Biol Bull Rev 133:124–132. [CrossRef]

Cai XC, Li H, Xue YR, Liu CH (2013) Study of endophytic Bacillus amyloliquefaciens CC09 and its antifungal CLPs. J Appl Biol Biotechnol 1:1–5

Calil IP, Fontes EPB (2016) Plant immunity against viruses: antiviral immune receptors in focus. Ann Bot 119:711–723. [CrossRef]

Chen J, Liu H, Xia Z, Zhao X, Wu Y, An M (2019) Purification and structural analysis of the effective anti-TMV compound ε-Poly-L-lysine produced by Streptomyces ahygroscopicus. Molecules 24:1156. [CrossRef] [PubMed]

Chirkov SN, Iiina AV, Surgucheva NA, Letunova EV, Varitsev YA, Tatarinova NY, Varlamov VP (2001) Effect of chitosan on systemic viral infection and some defense responses in potato plants. Rus J Plant Physiol 48:774–779. [CrossRef]

Cortes-Sanchez AJ, Hernandez-Sanchez H, Jaramillo-Flores ME (2013) Biological activity of glycolipids produced by microorganisms: new trends and possible therapeutic alternatives. Microbiol Res 168:22–32. [CrossRef]

Desoignies N, Schramme F, Ongena M, Legrève A (2012) Systemic resistance induced by Bacillus lipopeptides in Beta vulgaris reduces infection by the rhizomania disease vector Polymyxa betae. Mol Plant Pathol 14:416–421. [CrossRef]

Dyakov YT, Dzhavakhiya V, Korpela T (2007) Molecular basis of plant immunization. In: Comprehensive and molecular phytopathology. Elsevier, pp 423–438. https://doi.org/10.1016/B978-044452132-3/50020-1

Elena SF, Fraile A, Garcıá –Arenal, F. (2014) Evolution and emergence of plant viruses. Adv Virus Res 88:161–191. https://doi.org/10.1016/B978-0-12-800098-4.00003-9

Firmansyah D, Widodo Hidayat SH (2017) Chitosan and plant growth promoting rhizobacteria application to control squash mosaic virus on cucumber plants. Asian J Plant Pathol 11:148–155. [CrossRef]

Folimonova SY (2013) Developing an understanding of cross-protection by Citrus tristeza virus. Front Microbiol 4:76. [CrossRef]

Galal AM (2006) Induction of systemic acquired resistance in cucumber plant against cucumber mosaic cucumovirus by local Streptomyces strains. Plant Pathol J 5:343–349. [CrossRef]

Glick BR (2012) Plant growth-promoting bacteria: mechanisms and applications. Hindawi Publishing Corporation, Scientifica, London, UK

Han Y, Luo Y, Qin S, Xi L, Wan B, Du L (2014) Induction of systemic resistance against tobacco mosaic virus by ningnanmycin in tobacco. Pestic Biochem Phys 111:14–18. https://doi.org/10.1016/j.pestbp.2014.04.008

Harish S, Kavino M, Kumar N, Saravanakumar D, Soorianatha sundaram K, Samiyappan R (2008) Biohardening with plant growth promoting rhizosphere and endophytic bacteria induces systemic resistance against Banana bunchy top virus. Appl Soil Ecol 39:187–200. [CrossRef]

Hirai T (1962) Studies on the chemotherapy for the plant virus diseases. Ann Phytopathol Soc Jpn 27:115–121. [CrossRef]

Huang KT, Misato T (1970) Agricultural antibiotics. Rev Pl Prot Res 3:12–23

Hughes JdA, Ollennu LAA (1994) Mild strain protection of cocoa in Ghana against cocoa swollen shoot virus—a review. Plant Pathol 43:442–457. [CrossRef]

Hull R (2014) Plant virology, 5th edn. Academic Press, London, p 854. Available at: https://www.elsevier.com/books/plant-virology/hull/978-0-12-384871-0

Ilinskaya ON, Mahmud RS (2014) Ribonucleases as antiviral agents. Mol Biol 48:615–623. [CrossRef]

Inès M, Dhouha G (2015) Lipopeptide surfactants: production, recovery and pore forming capacity. Peptides 71:100–112. [CrossRef]
Jetiyanon K, Fowler W, Kloepper JW (2003) Broad-spectrum protection against several pathogens by PGPR mixtures under field conditions in Thailand. Plant Dis 87:1390–1394. [CrossRef]
Jones RA, Naidu RA (2019) Global dimensions of plant virus diseases: current status and future perspectives. Annu Rev Virol 6:387–409. https://doi.org/10.1146/annurev-virology-092818-015606
Ju YP, Si YY, Young CK, Jin-Cheol K, Quang LD, Jeong JK, In SK (2012) Antiviral peptide from Pseudomonas chlororaphis O6 against tobacco mosaic virus (TMV). J Korean Soc Appl Biol 55:89–94
Kandan A, Radjacommare R, Ramiah M, Ramanathan A, Samiyappan R (2003) PGPR induced systemic resistance in cowpea against tomato spotted wilt virus by activating defense against tomato spotted wilt virus by activating defense related enzymes and compound. In: Sarma YR (ed) Proceedings of the sixth international workshop on plant growth promoting rhizobacteria, Calicut, India, 5–10 October 2003. IISR Publishers, Calicut, pp 480–486
Kandan A, Ramiah M, Vasanthi VJ, Radjacommare R, Nandakumar R, Ramanathan A, Samiyappan R (2005) Use of *Pseudomonas fluorescens*-based formulations for management of tomato spotted wilt virus (TSWV) and enhanced yield in tomato. Biocontrol Sci Tech 15:553–569. [CrossRef]
Karthikeyan G, Doraisamy S, Rabindran R (2009) *Pseudomonas fluorescens* mediated systemic resistance against urdbean leaf crinkle virus in black gram (Vigna mungo). Arch Phytopathol Plant Protect 42:201–212. [CrossRef]
Kataoka M, Doke N, Hirai T (1969) Effects of antibiotics, inhibitors against protein synthesis on tobacco mosaic virus multiplication and the host metabolism. Ann Phytopathol Soc Japan 35:329–338. [CrossRef]
Kavino M, Harish S, Kumar N, Saravanakumar D, Samiyappan R (2008) Induction of systemic resistance in Banana (Musa spp.) against Banana bunchy top virus (BBTV) by combining chitin with root-colonizing *Pseudomonas fluorescens* strain CHA0. Eur J Plant Pathol 120:353–362. [CrossRef]
Kesiraju K, Sreevathsa R (2017) Apical meristem-targeted in planta transformation strategy: an overview on its utility in crop improvement. Agric Res Technol Open Access J 8(555734):19080. https://doi.org/10.19080/ARTOAJ.2017.08.555734
Kloepper JW, Ryu CM, Zhang S (2004) Induced systemic resistance and promotion of plant growth by Bacillus spp. Phytopathology 94:1259–1266. [CrossRef]
Kolase SV, Sawant DM (2007) Isolation and efficacy of antiviral principles from Trichoderma spp. against tobacco mosaic virus (TMV) on tomato. J Maharashtra Agric Univ 32:108–110
Kothari IL, Patel M (2004) Plant immunization [Review]. Indian J Exp Biol 42(3):244–252. Available at: http://nopr.niscpr.res.in/handle/123456789/23379
Kümmert J, Semal J (1971) Inhibition of the multiplication of bromegrass mosaic virus in barley by antibiotic blasticidin S. Phytopathology 61:10–14. [CrossRef]
Lam KS (2006) Discovery of novel metabolites from marine actinomycetes. Curr Opin Microbiol 9:245–255. [CrossRef]
Langenberg WG, Zhang L, Court DL, Giunchedi L, Mitra A (1997) Transgenic tobacco plants expressing the bacterial mc gene resist virus infection. Mol Breed 3:391–399. [CrossRef]
Latake SB, Borkar SG (2017) Characterization of marine actinomycete having antiviral activity against cucumber mosaic virus. Curr Sci 113:1442–1447. [CrossRef]
LeCocq K, Gurr SJ, Hirsch PR, Mauchline TH (2017) Exploitation of endophytes for sustainable agricultural intensification. Mol Plant Pathol 18:469–473. [CrossRef]
Lee GH, Ryu CM (2016) Spraying of leaf-colonizing Bacillus amyloliquefaciens protects pepper from cucumber mosaic virus. Plant Dis 100:2099–2105. [CrossRef] [PubMed]
Lee YM, Tscherne DM, Yun SI, Frolov I, Rice CM (2005) Dual mechanisms of pestiviral superinfection exclusion at entry and RNA replication. J Virol 79:3231–3242. [CrossRef]

Lehtonen P, Helander M, Wink M, Sporer F, Saikkonen K (2005) Transfer of endophyte-origin defensive alkaloids from a grass to a hemiparasitic plant. Ecol Lett 8:1256–1263. [CrossRef]
Li F, Wang A (2022) Transient expression–mediated gene silencing in plants and suppression of gene silencing with viral suppressors. Methods Mol Biol 2400:33–41. Springer. https://doi.org/10.1007/978-1-0716-1835-6_4
Li H, Ding X, Wang C, Ke H, Wu Z, Wang Y, Liu H, Guo J (2016) Control of tomato yellow leaf curl virus disease by Enterobacter asburiae BQ9 as a result of priming plant resistance in tomatoes. Turk J Biol 40:150–159. [CrossRef]
Liu H, Carvalhais LC, Crawford M, Singh E, Dennis PG, Pieterse CMJ, Schenk PM (2017) Inner plant values: diversity, colonization and benefits from endophytic bacteria. Front Microbiol 8:2552. [CrossRef]
Mace KD (1965) Effects of oxytetracycline and streptomycin on the metabolism of tobacco leaf tissue and tobacco mosaic virus multiplication. Diss Abstr 25:5220
Magarvey NA, Keller JM, Bernan V, Dworkin M, Sherman DH (2004) Isolation and characterization of novel marine-derived actinomycete taxa rich in bioactive metabolites. Appl Environ Microbiol 70:7520–7529. [CrossRef]
Makarova SS, Makarov VV, Taliansky ME, Kalinina NO (2017) Resistance to viruses of potato: current status and prospects. Vavilov, J Genet Breed 21:62–73. [CrossRef]
Maksimov IV, Veselova SV, Nuzhnaya TV, Sarvarova ER, Khairullin RM (2015) Plant growth-promoting bacteria in the regulation of plant resistance to stress factors. Russ J Plant Physiol 62:715–726. [CrossRef]
Maksimov I, Sorokan A, Burkhanova G, Veselova S, Alekseev V, Shein M et al (2019) Mechanisms of plant tolerance to RNA viruses induced by plant growth-promoting microorganisms. Plants (Basel) 8(12):575. https://doi.org/10.3390/plants8120575
Manjunatha L, Patil MS, Thimmegowda PR, Vijaya Mahantesha SR, Basamma. (2010) Effect of antiviral principle on groundnut bud necrosis virus. J Plant Dis Sci 5:12–15
Mauck K, Bosque-Pérez NA, Eigenbrode SD, DeMoraes CM, Mescher MC (2012) Transmission mechanisms shape pathogen effects on host–vector interactions: evidence from plant viruses. Funct Ecol 26:1162–1175. [CrossRef]
McGrann GR, Grimmer MK, Mutasa-Gottgens ES, Stevens M (2009) Progress toward the understanding and control of sugar beet rhizomania disease. Mol Plant Pathol 10:129–141. [CrossRef]
Meyer G, Audenaert K, Hofte M (1999) *Pseudomonas aeruginosa* 7NSK2-induced systemic resistance in tobacco depends on in planta salicylic acid accumulation but is not associated with PR1a expression. Eur J Plant Pathol 105:513–517. [CrossRef]
Milchenko KP (1969) Effects of culture liquids and extracts of some soil bacteria on tobacco mosaic virus. Microbiol J USSR 31:356–360
Mishra S, Jagadeesh KS, Krishnaraj PU, Prem S (2014) Biocontrol of tomato leaf curl virus (ToLCV) in tomato with chitosan supplemented formulations of Pseudomonas sp. under field conditions. Aust J Crop Sci 8:347–355
Mohamed SH, Galal AM (2005) Identification and antiviral activities of some halo tolerant Streptomycetes isolated from Qaroon lake. Int J Agric Biol 7:747–753
Montebianco CDB, Mattos BB, Silva TDF, Barreto-Bergter E, Vaslin MFS (2020) Cladosporium herbarum peptidogalactomannan triggers significant defense responses in whole tobacco plants. bioRxiv. [CrossRef]
Moses W, Rogers K, Mildred O-S (2017) Effect of thermotherapy duration, virus type and cultivar interactions on elimination of potato viruses X and s in infected seed stocks. African J Plant Sci 11(3):61–70. https://doi.org/10.5897/AJPS2016.1497
Mumford R, Macarthur R, Boonham N (2016) The role and challenges of new diagnostic technology in plant biosecurity. Food Secur 8:103–109. https://doi.org/10.1007/s12571-015-0533-y
Murphy JF, Zehnder GW, Schuster DJ, Sikora EJ, Polston JE, Kloepper JW (2000) Plant growth-promoting rhizobacterial mediated protection in tomato against Tomato mottle virus. Plant Dis 84:779–784. [CrossRef]

Murphy JF, Reddy MS, Ryu CM, Kloepper JW, Li R (2003) Rhizobacteria-mediated growth promotion of tomato leads to protection against cucumber mosaic virus. Phytopathology 93:1301–1307. [CrossRef] [PubMed]
Muvea AM, Subramanian S, Maniania NK, Poehling H-M, Ekesi S, Meyhöfer R (2018) Endophytic colonization of onions induces resistance against viruliferous thrips and virus replication. Front Plant Sci 9. https://doi.org/10.3389/fpls.2018.01785
Nagorskaia VP, Reunov AV, Lapshina LA, Ermak IM, Barabanova AO (2008) Influence of kappa/beta-carrageenan from red alga Tichocarpus crinitus on development of local infection induced by tobacco mosaic virus in Xanthi-nc tobacco leaves. Biol Bull 35:310–314. [CrossRef]
Niehl A, Wyrsch I, Boller T, Heinlein M (2016) Double-stranded RNAs induce a pattern-triggered immune signaling pathway in plants. New Phytol 211:1008–1019. [CrossRef] [PubMed]
Nowak J, Shulaev V (2003) Priming for transplant stress resistance in in vitro propagation. In Vitro Cell Dev Biol - Plant 39(2):107–124. https://doi.org/10.1079/ivp2002403
Ogawa T, Toguri T, Kudoh H, Okamura M, Momma T, Yoshioka M, Kato K, Hagiwara Y, Sano T (2005) Double-stranded RNA-specific ribonuclease confers tolerance against Chrysanthemum Stunt Viroid and Tomato Spotted Wilt Virus in transgenic Chrysanthemum plants. Breed Sci 55:49–55. [CrossRef]
Oves E, Gaytova N (2016) New elements in technology of improvement and obtaining of basic potato clones of promising cultivars and hybrids. Achievements Sci Technol Agro - Ind Complex 30(11):60–62
Pakniat-Jahromy A, Behjatnia SA, Dry IB, Izadpanah K, Rezaian MA (2010) A new strategy for generating geminivirus resistant plants using a DNA betasatellite/split barnase construct. J Virol Methods 170:57–66. [CrossRef]
Pan H, Chen G, Li F, Wi Q, Want S, Xie W, Liu B, Xu B, Zhang Y (2013) Tomato spotted wilt virus infection reduces the fitness of a nonvector herbivore on pepper. J Econ Entomol 106:924–928. [CrossRef]
Pandey RN, Gandhi KB, Manjunatha L (2021) In vitro evaluation of actinobacteria for biocontrol of dry root rot and Fusarium wilt diseases in chickpea (Cicer arietnum L.). J Food Legumes 34:245–247
Pardee KI, Ellis P, Bouthillier M, Towers GHN, French CJ (2004) Plant virus inhibitors from marine algae. Can J Bot 82:304–309. [CrossRef]
Pechinger K, Chooi KM, MacDiarmid RM, Harper SJ, Ziebell H (2019) A new era for mild strain cross-protection. Viruses 11:670. [CrossRef]
Pieterse CMJ, Zamioudis C, Berendsen RL, Weller DM, Van Wees SCM, Bakker PAHM (2014) Induced systemic resistance by beneficial microbes. Annu Rev Phytopathol 52:347–375. [CrossRef] [PubMed]
Pooggin MM (2018) Small RNA-omics for plant virus identification, virome reconstruction, and antiviral defense characterization. Front Microbiol 9:2779. [CrossRef] [PubMed]
Ramzan M, Bushra T, Idrees AN, Anwar KM, Tariq MFA, Naila S, Abdul QR, Muhammad UB, Nida T, Tayyab H (2016) Identification and application of biocontrol agents against cotton leaf curl virus disease in *Gossypium hirsutum* under greenhouse conditions. Biotechnol Equip 30:469–478. [CrossRef]
Ratcliff F, MacFarlane S, Baulcombe DC (1999) Gene silencing without DNA. RNA-mediated cross-protection between viruses. Plant Cell 11:1207–1216. [CrossRef]
Raychaudhuri SP (1970) Inhibition of plant viruses by growth regulators. Pl Dis Probl:489–500
Reddy A, Reddy L, Mallikarjuna N, Abdurahman MD, Reddy YV, Bramel PJ, Reddy DVR (2000) Identification of resistance to peanut bud necrosis virus (PBNV) in wild Arachis germplasm. Ann Appl Biol 137:135–139. [CrossRef]
Rochal KKL, Pierre E, Diane YY, Sahu KP, Vanessa ND, Herman KWT, Louise NW (2021) Biological elicitor potential of endospheric Trichoderma and derived consortia against pepper (Capsicum annuum L.) leaf curl virus. Arch Phytopathol Plant Prot 54:1926–1952. [CrossRef]

Rodrıguez M, Marín A, Torres M, Béjar V, Campos M, Sampedro I (2018) Aphicidal activity of surfactants produced by bacillus atrophaeus L193. Front Microbiol 9:3114. https://doi.org/10.3389/fmicb.2018.03114

Ryu CM, Murphy JF, Mysore KS, Kloepper JW (2004) Plant growth-promoting rhizobacteria systemically protect *Arabidopsis thaliana* against Cucumber mosaic virus by a salicylic acid and NPR1-independent and jasmonic acid-dependent signaling pathway. Plant J 39:381–392. [CrossRef]

Sahran BS, Nehra V (2011) Plant growth promoting rhizobacteria: a critical review. Life Sci Med Res LMSR 21:1–30

Salwan R, Sharma V (2018) Role of actinobacteria in production of industrial enzymes. In: Book volume on actinobacteria: diversity and biotechnological applications. Elsevier, Amsterdam. isbn:978-0-44463-995-0

Sanfaçon H (2015) Plant translation factors and virus resistance. Viruses 7:3392–3419. [CrossRef] [PubMed]

Schaeffer DJ, Krylov VS (2000) Anti-HIV activity of extracts and compounds from algae and cyanobacteria. Ecotoxicol Environ Saf 45:208–227. [CrossRef]

Schenk MF, Hamelink R, Van der Vlugt RAA, Vermunt AM, Kaarsenmaker RC, Stijger IC (2010) The use of attenuated isolates of Pepino mosaic virus for cross-protection. Eur J Plant Pathol 127:249–261. [CrossRef]

Selim SA, El-Alfy S, Al-Ruwaili M, Abdo A, Al-Jaouni S (2012) Susceptibility of imipenem-resistant *Pseudomonas aeruginosa* to flavonoid glycosides of date palm (Phoenix dactylifera L.) tamar growing in Al Madinah, Saudi Arabia. Afr J Biotechnol 11:416–422. [CrossRef]

Shekhar S, Sundaramanickam A, Balasubramanian T (2015) Biosurfactant producing microbes and their potential applications: a review. Crit Rev Environ Sci Technol 45:1522–1544. [CrossRef]

Shen S, Li W, Ouyang M, Wu Z, Lin Q, Xie L (2009) Identification of two marine fungi and evaluation of their antivirus and antitumor activities. Acta Microbiol Sin 49:1240–1246

Shen LL, Wang FL, Yang JG, Qian YM, Dong XW, Zhan HX (2014) Control of tobacco mosaic virus by *Pseudomonas fluorescens* CZ powder in greenhouses and the field. Crop Prot 56:87–90. [CrossRef]

Sorokan A, Cherepanova E, Burkhanova G, Veselova S, Rumyantsev S, Alekseev V, Mardanshin I, Sarvarova E, Khairullin R, Benkovskaya G et al (2020) Endophytic Bacillus spp. as a prospective biological tool for control of viral diseases and nonvector *Leptinotarsa decemlineata* Say. in *Solanum tuberosum* L. Front Microbiol 11:569457. [CrossRef]

Sorokina EV (2012) Toll-like receptors and primary pathogen recognition in infectious and noninfectious cutaneous pathology. Int J Immunopathol Allergol Infectol 2:6–15

Srinivasan K, Mathivanan N (2009) Biological control of sunflower necrosis virus disease with powder and liquid formulations of plant growth promoting microbial consortia under field conditions. BioControl 51:395–402. [CrossRef]

Stein T (2005) *Bacillus subtilis* antibiotics: structure, syntheses and specific functions. Mol Microbiol 56:845–857. [CrossRef]

Thapa SP, Lee HJ, Park DH, Kim SK, Cho JM, Cho S, Hur JH, Lim CK (2009) Antiviral effects of the culture filtrate from *Serratia marcescens* Gsm01, against cucumber mosaic virus (CMV). Plant Pathol J 25:369–375. [CrossRef]

Tollenaere C, Lacombe S, Wonni I, Barro M, Ndougonna C, Gnacko F, Sérémé D, Jacobs JM, Hebrard E, Cunnac S et al (2017) Virus-bacteria rice coinfection in Africa: field estimation, reciprocal effects, molecular mechanisms, and evolutionary implications. Front Plant Sci 8:645. [CrossRef]

Van Loon LC, Pierpoint WS, Boller T, Conejero V (1994) Recommendations for naming plant pathogenesis-related proteins. Plant Mol Biol Report 12:245–264. [CrossRef]

Vanthana M, Nakkeeran S, Malathi VG, Renukadevi P, Vinodkumar S (2019) Induction of in planta resistance by flagellin (Flg) and elongation factor-TU (EF-Tu) of Bacillus amyloliquefaciens (VB7) against groundnut bud necrosis virus in tomato. Microb Pathog 137:103757. [CrossRef] [PubMed]

Wang S, Zhong FD, Zhang YJ, Wu ZJ, Lin QY, Xie LH (2004) Molecular characterization of a new lectin from the marine algae Ulva pertusa. Acta Biochim Biophys Sin 36:111–117. [CrossRef]
Wang J, Wang HY, Xia XM, Li PP, Wang KY (2013) Inhibitory effect of sulfated lentinan and lentinan against tobacco mosaic virus (TMV) in tobacco seedlings. Int J Biol Macromol 61:264–269. [CrossRef]
Wang MR, Hamborg Z, Blystad DR, Wang QC (2021) Combining thermotherapy with meristem culture for improved eradication of onion yellow dwarf virus and shallot latent virus from infected in vitro-cultured shallot shoots. Ann Appl Biol 178(3):442–449. https://doi.org/10.1111/aab.12646
Wang P, Liu J, Lyu Y, Huang Z, Zhang X, Sun B et al (2022) A review of vector-borne rice viruses. Viruses 14(10):2258. https://doi.org/10.3390/v14102258
Xing Z, Li-Rong H, Da-Wei Z, Bing-Nian J, Jun-Tao F (2012) Anti-TMV activity of the fermentation product of actinomycetes strain ZX 01. J Northwest A F Univ 1:30–36
Yeh SD, Gonsalves D, Wang HL, Namba R, Chiu RJ (1988) Control of papaya ringspot virus by cross protection. Plant Dis 72:375–380. [CrossRef
Yi H-S, Yang JW, Ryu CM (2013) ISR meets SA Routside: additive action of the endophyte Bacillus pumilus INR7 and the chemical inducer, benzothiadiazole, on induced resistance against bacterial spot in field-grown pepper. Front Plant Sci 4:122. [CrossRef]
Zhang GQ, Han LR, Zhang GF, Zhang X, Feng JT (2015) Purification and characterization of a novel glycoprotein from Streptomyces sp. ZX01. Int J Biol Macromol 78:195–201. [CrossRef]
Zhang Z, Wang Q-C, Spetz C, Blystad D-R (2019) In vitro therapies for virus elimination of potato-valuable germplasm in Norway. Sci Hortic 249:7–14. https://doi.org/10.1016/j.scienta.2019.01.027
Zhao L, Hao XA, Wu YF (2015a) Inhibitory effect of polysaccharide peptide (PSP) against tobacco mosaic virus (TMV). Int J Biol Macromol 75:474–478. [CrossRef]
Zhao L, Feng CH, Hou CT, Hu LY, Wang QC, Wu YF (2015b) First discovery of acetone extract from cottonseed oil sludge as a novel antiviral agent against plant viruses. PLoS One 10:117496. [CrossRef]
Zhao L, Feng C, Wu K, Chen W, Chen Y, Hao X, Wu Y (2016) Advances and prospects in biogenic substances against plant virus: a review. Pestic Biochem Physiol 135:15–26. [CrossRef]
Zhou WW, Zhang XL, Zhang B, Wang F, Liang ZH, Niu TG (2008) Isolation and characterization of ZH14 with antiviral activity against tobacco mosaic virus. Can J Microbiol 54:441–449. [CrossRef]
Zvereva AS, Golyaev V, Turco S, Gubaeva EG, Rajeswaran R, Schepetilnikov MV, Srour O, Ryabova LA, Boller T, Pooggin MM (2016) Viral protein suppresses oxidative burst and salicylic acid dependent autophagy and facilitates bacterial growth on virus-infected plants. New Phytol 211:1020–1034. [CrossRef] [PubMed]

Chapter 16
Effects of Microbial Pesticides on Insect Control

Magda Mahmoud Sabbour

Abstract Microbial pesticides have emerged as effective and environmentally friendly alternatives to conventional chemical pesticides in the management of insect pests. These biopesticides, derived from naturally occurring microorganisms such as bacteria, fungi, viruses, and protozoa, target specific insect species while minimizing harm to nontarget organisms, including beneficial insects and humans. This paper explores the mechanisms through which microbial pesticides exert their effects, including toxin production, infection, and disruption of insect physiological processes. Recent studies have demonstrated that microbial agents such as *Bacillus thuringiensis*, *Beauveria bassiana*, and nucleopolyhedrovirus have shown significant success in agricultural and forestry applications. Additionally, the integration of microbial pesticides into integrated pest management strategies has improved pest control efficiency and reduced environmental impact. However, challenges such as inconsistent field performance, limited shelf-life, and regulatory hurdles remain. Overall, microbial pesticides represent promising tools for sustainable insect management, and further research into formulation and application methods can enhance their reliability and adoption.

Keywords Biopesticides · Beneficial insects · *Bacillus thuringiensis* (*"Bt"*) · Environmental impact

M. M. Sabbour (✉)
Pests and Plant Protection Department, Agriculture and Biological Institute, National Research Centre, Giza, Egypt

V. Gopikrishnan et al. (eds.), *Harnessing Microbial Resources for Sustainable Agriculture*, Microorganisms for Sustainability 61,
https://doi.org/10.1007/978-981-95-4817-0_16

16.1 Introduction

Microbial pesticides act as biopesticides, which might have lower toxicity than chemical pesticides do, are often more selective toward specific pests, breakdown rapidly, and have a reduced chance of leading to resistance. However, they may have drawbacks such as lower effectiveness and higher expenses. Biopesticides can include macroorganisms, natural substances, and microorganisms. Among these, microbial pesticides are a notable category. Microbial pesticides, or biopesticides, are a type of pesticide that relies on naturally occurring microorganisms to control pests (such as bacteria, fungi, viruses, or protozoa) or their byproducts to control pest populations, particularly insect pests. Unlike conventional chemical pesticides, which are often synthetic compounds, microbial pesticides offer a more environmentally sustainable alternative that presents lower risks to nontarget organisms, including beneficial insects, humans, and other wildlife (Klein 1992). The shift toward microbe-based pest control methods has gained significant momentum because of increasing concerns about the harmful effects of chemical pesticides on ecosystems, resistance development, and food safety (Klein and Kaya 1995).

Microbial pesticides are substances used to control pests, consisting of microorganisms that target and eliminate harmful organisms that naturally include microorganisms such as bacteria, fungi, viruses, protozoa, and nematodes, which are used to control and regulate pest populations.

These microorganisms can either infect or outcompete harmful pests, thereby reducing their numbers without the need for synthetic chemicals (Sabbour and Abd El-Aziz 2022).

Microbial pesticides are pest control agents derived from naturally occurring microorganisms, including bacteria, fungi, viruses, nematodes, and protozoa. These microorganisms can be harnessed to control insect pests by infecting them, producing toxic substances, or otherwise disrupting their life cycles. In contrast to chemical pesticides, microbial pesticides tend to be more selective in targeting specific pests, lowering the likelihood of affecting nontarget species (Sabbour and Abd-El-Aziz 2018).

16.2 Importance of Insect Control in Agriculture

*"Insects are among the most significant threats to agricultural production worldwide. Insect pests can damage crops, reduce yields, and spread diseases. Traditional insecticides, while effective, have many drawbacks, including environmental pollution, resistance development, and harm to beneficial insects. Microbial pesticides provide an alternative that aligns with sustainable agricultural practices".

Pests such as ants, rats, mice, and other food-infesting insects are common carriers of diseases that can harm public health. By controlling these pests, the risk is minimized, resulting in fewer wasted products, healthier consumers, and preserved reputations for farmers.

Agriculture has been the cornerstone of human civilization for thousands of years, and the protection of crops from pests is a central concern for farmers. Among the various threats to crop health, insects stand out as both beneficial and detrimental in agricultural systems.

Although certain insects, such as bees and ladybugs, are crucial for pollination and natural pest management, others, such as aphids, locusts, and weevils, cause extensive damage to crops and can lead to significant economic losses. This book explores the importance of insect control in agriculture, its economic implications, and the evolution of pest management practices.

Insect pests, including herbivores and pathogens, cause substantial damage to agricultural crops worldwide. In fact, it is estimated that insect pests destroy approximately 10–15% of global crop production annually (Oerke 2006). Therefore, effective insect control is essential to ensure food security and agricultural sustainability. Farmers must use a combination of strategies—ranging from traditional practices to advanced technologies—to mitigate the impacts of pests.

Historically, insect pests have been recognized as major threats to crops. In ancient civilizations, farmers relied on physical means, such as handpicking pests or using plant-based extracts, to control insects. The introduction of synthetic chemical pesticides in the twentieth century brought about a major shift in pest control methods, making pest management more efficient but also introducing new environmental and health challenges.

Insects have a complex relationship with agriculture. While many are pests that damage crops, others play essential roles in maintaining ecological balance and promoting agricultural productivity. Understanding this balance is crucial for effective pest management strategies. Pollinators such as honeybees, butterflies, and beetles are vital for the fertilization of a wide variety of crops. For example, approximately 75% of flowering plants and 35% of global food crops depend on insect pollination (Klein and Georgis 1994). In addition to pollination, some insects, such as earthworms and ants, aid in soil aeration and nutrient cycling, enhancing soil health and increasing crop productivity.

The economic costs of insect pests in agriculture are staggering. The direct damage caused by pests includes reduced yields, increased costs of pest management, and damage to quality. Indirect costs include the loss of biodiversity, soil degradation, and broader environmental impacts of pesticide use.

According to estimates, insect pests account for over $70 billion in annual crop losses globally (Ferro 2010). For example, in the United States alone, corn and soybean crops lose an estimated $1 billion annually because of insect infestations (Pimentel 2005). Moreover, pests can disrupt the food chain, affecting livestock and pollinators, which further results in economic losses in agriculture.

Effective insect control begins with a deep understanding of insect biology. Farmers must study the behavior, lifecycle, and ecological preferences of pests to develop effective management strategies.

Insects undergo different stages of life, such as egg, larva, pupa, and adult stages, and can present unique risks to crop types. For example, during the larval stage, the cotton bollworm feeds on cotton bolls, whereas adult weevils damage stored grains. Understanding these patterns enables farmers to time their pest control efforts for maximum efficacy. Before the widespread use of chemical pesticides, farmers employed a variety of traditional methods to control insect populations.

These methods involve manually removing pests from plants or using traps, barriers, and other mechanical devices to reduce pest infestations. While labor intensive, these methods are still used in organic farming systems. Biological control involves the use of natural enemies of pests, such as predators (e.g., ladybugs, spiders) or parasitoid insects, to keep pest populations in check. This approach can be highly effective and environmentally friendly, especially when natural predators are well suited to local ecosystems. In the twentieth century, the rise of synthetic chemical insecticides revolutionized pest management. However, these chemicals also cause concerns related to toxicity, resistance, and environmental damage.

Chemical insecticides, such as organophosphates and pyrethroids, have been widely used to combat insect pests. While they offer quick and effective solutions, they also pose potential dangers to human health, nontarget species, and the environment. Furthermore, excessive use of pesticides has contributed to the emergence of resistance in insect populations, rendering some chemicals ineffective over time.

Pesticides can pose risks to human health, nontarget organisms, and the environment. Additionally, overuse has led to the development of resistance in insect populations, which can be monitored to manage pest populations in an environmentally friendly manner. It emphasizes prevention, regular monitoring, and the application of diverse control strategies to minimize dependence on chemicals (Sabbour and Solieman2022). This recent study explored the role of fungal biocontrol agents in pest management, highlighting breakthroughs in their application and efficiency in integrated pest management (IPM) systems.

Sustainable agriculture aims to minimize the environmental effects of farming techniques, whereas sustainable pest control prioritizes natural and less damaging approaches.

Organic farming steers clear synthetic pesticides and fertilizers, opting instead for practices such as crop rotation, the use of natural predators, and the use of biological pesticides such as neem oil and diatomaceous earth for controlling insect pests. Organic methods aim to create a balanced ecosystem where pest populations are naturally regulated.

16.3 Role of Microbial Pesticides in Comprehensive Pest Management

Comprehensive pest management (CPM) provides a comprehensive examination of the increasing importance of microbial pesticides in modern agricultural practices. As the global agricultural industry faces increasing challenges related to pesticide resistance, environmental concerns, and the need for sustainable farming practices, microbial pesticides have emerged as critical CPM systems. This book explores the mechanisms, benefits, and challenges associated with microbial biocontrol agents and their integration into IPM frameworks (Sabbour and Solieman 2022). Microbial pesticides and their categories, such as bacteria, fungi, viruses, and nematodes, along with an exploration of their development and history, provide insights into their increasing application in crop protection. The role of microbial pesticides as eco-friendly alternatives to chemical pesticides has also been highlighted by Alvarez et al. (2020a) Klein and Georgis (1994).

Microbial pesticides constitute a cornerstone of CPM strategies, which are designed to manage pest populations via a combination of biological, physical, and chemical methods. They can be used in conjunction with other IPM techniques, such as cultural practices and chemical controls, to minimize the impact of pests while reducing reliance on synthetic chemicals. The application of microbial agents in pest control dates back to ancient practices, but modern development began in the twentieth century with the discovery of *Bacillus thuringiensis* (*Bt*) and other beneficial microbes. Over time, researchers have isolated and developed a wide range of microbial species with the potential to control various pests, marking a revolution in sustainable pest management (Klein and Kaya 1995).

A detailed exploration of how microbial pesticides work to control pests is provided. This section delves into the modes of action, such as pathogen-induced toxicity, parasitism, competition, and the production of secondary metabolites that inhibit pest development. Case studies of various microbial agents, such as the fungus *Beauveria bassiana* and the bacteria *Bacillus thuringiensis*, *Trichoderma* spp., and *Nosema* spp., illustrate the diversity of microbial pesticides and their modes of action.

The microbial pesticides fit within the broader IPM framework. It covers the principles of IPM, emphasizing the importance of using biological control methods alongside other management strategies, such as cultural, mechanical, and chemical control. It examines real-world examples of IPM success stories where microbial pesticides have played a key role in reducing pest populations without compromising environmental health.

Microbial pesticides are central to IPM strategies, offering more reference to practices, products, or actions that are designed to have a minimal negative impact on the environment. This includes reducing pollution, conserving natural resources, and promoting sustainability to protect the planet for future generations. Essentially, this means being mindful of nature and taking steps to reduce harm to ecosystems and biodiversity, substituting traditional pesticides that are made of chemicals.

These materials play a key role in promoting biodiversity, minimizing chemical contaminants in food and water, and decreasing reliance on chemical inputs in farming (Álvarez et al. 2020b).

This chapter explores the effects of microbial pesticides on insect control, delving into their mechanisms of action, effectiveness, advantages, and challenges.

Recent scientific advancements in microbial pesticide research are highlighted in this section, with a focus on cutting-edge technologies such as genetic modification, formulation improvements, and delivery systems. New research on the effectiveness of microbial agents against resistant pest populations has been conducted, alongside emerging trends in bioinformatics and genomics that are expanding the scope of microbial biocontrol (Sahab et al. 2014). This research examines the use of bioengineered microbial insecticides, providing insights into innovative solutions for pest control that are part of the next generation of sustainable agricultural practices.

While microbial pesticides offer many advantages, there are challenges to their widespread adoption. These include issues related to product stability, shelf time, environmental conditions for effective application, and regulatory hurdles. The book also addresses the need for a better understanding of microbial interactions in diverse ecosystems and the development of more cost-effective products for farmers (Thakur et al. 2020). The impact of microbial agents in managing soilborne pests contributes to broader IPM strategies and sustainable farming practices.

The promising future of microbial pesticides in agriculture involves the potential for combining microbial agents with other biotechnologies, such as RNA interference and clustered regularly interspaced short palindromic repeat. It also explores the role of microbial biocontrol agents in mitigating climate change impacts by promoting sustainable farming practices (Sabbour and Abd-El-Aziz 2021).

16.4 Understanding Microbial Pesticides

16.4.1 Microorganisms as Biological Control Agents

Microbial pesticides work by infecting pests, disrupting their growth, or inducing mortality through various biological processes. These microorganisms can produce toxins that target specific pests, multiply inside their hosts, or release enzymes that break down pest defenses.

The need for sustainable and eco-friendly solutions in agriculture has never been greater, particularly in pest management. Traditional chemical pesticides have contributed to environmental degradation, health risks, and the development of pest resistance. Microbial pesticides, which consist of naturally occurring microorganisms or their products, are alternatives that can address these challenges. These based pesticides have garnered attention for their specificity, safety to nontarget organisms, and minimal environmental impact.

This book will examine various types of microbial pesticides, how they work, their uses, and their potential for contributing to sustainable pest management in the future.

Microbial pesticides, or biopesticides, are pest control agents sourced from microorganisms such as bacteria, fungi, viruses, or algae. These agents work by affecting pest populations directly through pathogenic interactions or by producing substances that disrupt the pests' normal life cycles. Unlike synthetic chemical pesticides, microbial pesticides often have a narrower spectrum of activity, targeting specific pest species while leaving beneficial organisms largely unaffected.

Microbial pesticides are classified on the basis of the organism or the substance they produce. They are part of the broader category of biocontrol agents and have been used in agriculture for more than a century. The growing interest in microbial pesticides is a result of their potential to reduce environmental pollution and promote biodiversity in agroecosystems.

16.4.1.1 Types of Microbial Pesticides

Microbial pesticides are produced by a wide range of microorganisms, each with unique mechanisms of action. This chapter introduces the different categories of microbial pesticides—protozoal bacteria, viruses, nematodes, and fungi—and outlines their distinct characteristics and effectiveness against different insect pests (Klein and Kaya 1995).

How Microbial Pesticides Affect Insects

Microbial pesticides act on insects in several ways. Some produce toxins that disrupt the insect's digestive system or growth. Others, such as entomopathogenic fungi, infect the insect, eventually killing it through parasitism. Viruses can also infect insect cells, causing disease and death. Each group of microorganisms interacts with insects differently, providing a range of options for pest management.

Microbial pesticides are classified according to the type of microorganism they are derived from and used, including bacteria, fungi, viruses, and nematodes. Some of the most commonly used microbial agents in insect control are as follows:

16.4.2 Bacterial Pesticides

The bacteria *Bacillus thuringiensis* This bacterium generates proteins that are harmful to specific insect larvae, particularly those of moths, butterflies, and beetles. *Bt* is one of the most widely used microbial pesticides globally because of its specificity and safety for nontarget organisms.

Bacterial pesticides are derived from bacteria that have the ability to infect and kill specific pests. The most widely used bacterial biopesticides belong to the genus *Bacillus*. Notably, *Bt* has been demonstrated to be effective in controlling a variety of insect pests (Lacey and Shapiro-Ilan 2008).

Bacillus thuringiensis is perhaps the most famous bacterial pesticide. It produces a protein toxin known as delta-endotoxin, which targets specific receptors in the insect's digestive system, causing the gut to break down and ultimately leading to the death of the insect. The bacteria *B. thuringiensis* is especially effective against caterpillars, beetles, and flies. It is widely used in agriculture for pest control in crops such as cotton, corn, and potatoes.

In addition to *Bt*, other bacteria, such as *Bacillus sphaericus* and *Bacillus popilliae*, have also been used as microbial pesticides to target mosquitoes and soil-dwelling pests.

Bacillus thuringiensis is among the most well-known and commonly used microbial insecticides. It releases toxins that specifically target insects in the order Lepidoptera, including moths and butterflies. When ingested, these toxins damage the insect's gut, leading to death. This chapter explores the various strains of *Bt*, their modes of action, and their use in controlling pests such as the European corn borer and cotton bollworm. The spores produced by *Bt* harm the gut of insect larvae once ingested. For toxins to be activated, the insect gut needs a high pH ranging from 9.0 to 10.5, which contrasts with the lower, more acidic pH of the human gut. Once activated, the toxin disrupts the lining of the insect's gut. In addition to *B. thuringiensis*, other bacteria, such as *B. subtilis* and *Pseudomonas fluorescens,* are used in insect control (Klein 1994). These bacteria may work by antagonizing harmful pathogens, promoting plant health, or producing toxins that affect pest populations (Lacey and Shapiro-Ilan 2008).

Bacillus sphaericus This bacterium is used to control mosquito larvae and produces toxins that disrupt the larvae's digestive system, leading to their death.

Products containing *B. thuringiensis* consist of spores, toxin crystals, and additives. When the target species consumes these bioinsecticides, Cry toxins are activated enzymatically in the insect midgut. This activation causes the formation of pores in the gut lining, leading to osmotic lysis at an alkaline pH. Additionally, the spores germinate, and the bacteria, aided by Cry toxin-induced damage to the gut, invade the internal cavity of the insect, causing septicemia. As a result, the insect larva typically dies within 2–3 days (Fiuza et al. 2017; Adang et al. 2014). Importantly, the toxicity of Cry toxins primarily depends on the presence of specific receptors that can bind to them, with each Cry toxin family requiring its own distinct receptors (Rubio-Infante and Moreno-Fierros 2016). Numerous studies have shown that the gastrointestinal tract epithelial surfaces of nontarget insects and mammals, including humans, do not possess specific high-affinity receptors for Cry proteins.

Advantages

- Advantages: Unlike many insecticides that affect a wide variety of species, including both pests and beneficial insects, *Bt* is harmful to only a specific group of insects.
- Specific to target pests.
- It is safe for humans, animals, and beneficial insects.
- Effective in organic farming.

Disadvantages

- Its effectiveness may be reduced under harsh environmental conditions.
- Pests may develop resistance over time.

16.4.3 Fungal Pesticides

Fungal-based pesticides use fungi that act as natural pathogens to target pest organisms. Some of the most commonly used fungi in biopesticides include species of *Beauveria*, *Metarhizium*, and *Trichoderma*. Fungi constitute another important group of microbial pesticides. *Paecilomyces* sp., *B. bassiana*, *Trichoderma*, and *Metarhizium anisopliae* are among the most commonly used entomopathogenic fungi. These fungi infect insects by penetrating their exoskeletons and proliferating inside the host, ultimately killing it. This chapter delves into how fungi infect insects, the types of pests they target, and the environmental factors that influence their effectiveness (Jaronski 2010).

The fungus *B. bassiana* is a soil-dwelling fungus known to infect various insect pests, such as termites, aphids, and whiteflies. *B. bassiana* can infect a pest when conidia (fungal spores) land on its body. Once spores germinate, they penetrate the insect's exoskeleton and proliferate inside, killing the *Jaronski insect* Jaronski (2010).

Beauveria bassiana A soil-dwelling fungus affects a diverse variety of insects, "comprising" aphids, beetles, whiteflies, and many lepidopterous pests. *B. bassiana* infects the host by penetrating its cuticle and releasing toxins.

Metarhizium anisopliae Known for its efficacy against pests such as locusts, ants, and root weevils, this fungus infects insects via direct contact, releasing lethal toxins and enzymes that break down the insect's protective cuticle.

The fungus *Metarhizium anisopliae* is another widely used entomopathogenic fungus that targets soil-dwelling insect pests such as root-feeding grubs and weevils. It functions similarly to *Beauveria*, infecting pests via spore contact and killing them through internal fungal growth.

Advantages of Fungus

- A wide range of target pests.

- This material can be used for both soil and foliar applications.
- Environmental safety and sustainability.

Disadvantages of Fungus

- Proper moisture and temperature conditions are required for effectiveness.
- The shelf-life of spores in some formulations is limited.

16.4.4 Viral Pesticides

Baculoviruses, a type of virus that infects insects, are used as microbial pesticides for controlling insect pests, particularly caterpillars. These viruses can be species specific, making them excellent choices for targeting specific pests without affecting beneficial insects. This chapter explains how baculoviruses are used in pest control, the stages of infection, and the advantages they offer over other microbial pesticides. Viral biopesticides are derived from viruses that specifically infect insect pests. The most common viral biopesticides are nucleopolyhedroviruses (NPVs) and granuloviruses, both of which are effective against the caterpillars (Crook and Yamanaka 2012).

Nucleopolyhedrovirus These viruses affect a diverse range of insect pests, especially those within the Lepidoptera order. When ingested by the host, NPVs multiply inside the insect, resulting in the death of the pest. NPVs are baculoviruses that infect insects, particularly moths and butterflies. The virus infects insect larvae, causing them to die after the virus replicates inside their bodies. NPVs have been employed in both forestry and agriculture to manage pests, particularly spruce budworms and cotton bollworms (Crook and Yamanaka 2012).

Advantages

- This method is highly specific to target pests.
- It is effective even under challenging environmental conditions.

Disadvantages

- Slow-acting pesticides compared with chemical pesticides.
- The narrow host range limits their application.

16.4.5 Nematode Pesticides

Nematodes are microscopic worms that can be utilized to manage soil-inhabiting insect pests. Species such as *Steinernema* and *Heterorhabditis* parasitize insects such as root maggots, weevils, and other soil pests. This chapter covers how nematodes find and infect their hosts, their role in biological control, and the practical application methods used for effective pest management:

- *Steinernema and Heterorhabditis* species: These microscopic worms are parasitic to soil-dwelling insect larvae, such as rootworms. These bacteria enter the insect host, releasing bacteria that cause fatal infections.

16.4.6 Protozoan Pesticides

Protozoan pesticides are derived from protozoa that infect and kill pest organisms. One of the key examples is the protozoan *Nosema* spp., which infects insects, particularly bees and silkworms (Fries 2010).

Nosema spp.
Nosema is a genus of parasitic protozoa that infects insects, including honeybees. This disease is called Nosema disease and can lead to the death of infected insects. This organism is used as a biopesticide in beekeeping to control the spread of infections in honeybee colonies (Fries 2010).

Protozoa and Other Microorganisms Involved in Insect Control
While less commonly used, protozoa can act as biological control agents for insect pests. This chapter examines their role in pest management, focusing on their use against specific pests and how they fit into integrated pest management strategies.

Mechanisms of Action
Microbial pesticides act through various biological mechanisms, depending on the microorganism involved. The primary modes of action include the following:

Advantages

- They target specific pests without harming nontarget organisms.
- Can be integrated into the existing pest management systems.

Disadvantages

- It is slow-acting and difficult to apply effectively.

16.4.7 Algal Pesticides

Algal-based biopesticides are derived from algae that produce toxic compounds or release substances that inhibit the growth of pests. Some algae produce potent chemicals, such as toxins, that can be used for pest control (Singh and Chauhan 2012).

16.4.7.1 (Cyanobacteria) or Blue–Green Algae

Some algae or cyanobacteria, called blue–green algae, such as *Anabaena* or *Nostoc*, produce toxins that can target insect pests. These algal pesticides can be effective against specific insect pests, particularly in aquatic environments, where algae are introduced to manage populations of mosquito larvae (Singh and Chauhan 2012).

Advantages

- Can be incorporated into integrated pest management strategies.
- Safe for use inside aquatic environments.

Disadvantages

- There are limitations to certain pest types and ecosystems.

16.5 Mechanisms of Action

Microbial pesticides act on pests in several ways, including the following:

1. *Infection and parasitism:* Microorganisms infect the pest host, growing inside or on the host organism, ultimately causing death. This phenomenon is common in fungi, bacteria, and protozoa.
2. *Toxin production:* Some microorganisms, such as *Bacillus thuringiensis*, produce toxins that disrupt the pest's digestive system, leading to death.
3. *Immune system disruption:* Certain microorganisms trigger an immune response in a pest, causing it to succumb to infection.
4. *Reproductive interference:* Some microbes prevent pests from reproducing, thus reducing future pest populations.

Microbial pesticides act through various biological mechanisms, depending on the microorganism involved. The primary modes of action include the following:

1. *Toxic protein production*:
   - Many bacterial biopesticides, such as *Bacillus thuringiensis* (*Bt*), produce endotoxins, such as Cry proteins, which are toxic when ingested by insect larvae. These proteins target specific receptors in the insect gut, causing damage to the digestive system, paralysis, and ultimately death. The specificity of these toxins to certain pest species minimizes harm to nontarget organisms.
2. *Pathogenic infection*:
   - Fungal pesticides, such as *Beauveria bassiana*, infect insects by penetrating their cuticle and releasing enzymes that break down the pest's exoskeleton. The tested fungi then proliferate inside insect bodies, consuming them and leading to their death. This pathogenic infection often results in insect death via dehydration, as the fungus drains nutrients and moisture.

3. *Virus-induced lethality*:
   - Viruses, such as the NPV, cause infection when ingested by insects. The viral particles replicate inside the insect, eventually leading to tissue destruction, paralysis, and death. These viruses can spread rapidly in insect populations, particularly in crowded habitats.
4. *Parasitism by nematodes*:
   - Nematodes enter insects through natural openings, such as the mouth or anus, and once inside, they release symbiotic bacteria that rapidly destroy the host. Nematodes then feed on the insect's body fluids, growing and multiplying inside the carcass.

16.6 Mode of Action of Microbial Pesticides

Microbial pesticides affect pest populations in a variety of ways, including infection, paralysis, or growth inhibition. Viruses, such as baculoviruses, target insect larvae, inject their genetic material, and cause death. Bacteria such as *Bt* release toxins that are toxic to specific pests when ingested, making these solutions highly target specific.

This chapter explores the diverse applications of microbial pesticides across different crops, regions, and farming practices. This chapter also examines the application of microbial pesticides in areas such as horticulture, organic farming, and pest management for greenhouse crops. Microbial pesticides are vital components of IPM systems, where they are used in combination with other biological, mechanical, and cultural pest control methods. This section includes case studies of successful IPM programs and the growing acceptance of microbial pesticides in conventional farming.

While microbial pesticides have proven to be effective in many cases, challenges persist in areas such as resistance, environmental conditions, and regulatory hurdles. This chapter looks into these issues and provides solutions to improve their reliability and performance.

Regulation plays a crucial role in ensuring the safety and effectiveness of microbial pesticides. This chapter covers the regulatory processes governing their approval and highlights the importance of safety standards for human, animal, and environmental health. The future of microbial pesticides looks promising, with advancements in biotechnology and genetic engineering. New strains of microbes are being developed to target pests more efficiently, and microbial pesticides are expected to play an even larger role in climate-smart agriculture.

16.7 Effectiveness in Insect Control

The effectiveness of microbial pesticides in controlling insect pests is influenced by several factors, including the type of pest, environmental conditions, and mode of application. Some general insights into their effectiveness are as follows:

1. *Specificity*:
   - One of the greatest advantages of microbial pesticides is their specificity. For example, *Bacillus thuringiensis* is highly toxic to certain lepidopteran larvae but does not affect humans, animals, or beneficial insects such as bees. This specificity helps reduce the environmental impact and preserves natural pest control agents.
2. *Persistence and environmental factors*:
   - Compared with synthetic chemicals, microbial pesticides tend to have limited persistence. Temperature, humidity, and UV radiation are factors that can weaken microbes, reducing their effectiveness over time. However, some formulations, such as those protected by microencapsulation, can enhance persistence and control.
3. *Pest resistance*:
   - While resistance to microbial pesticides is less common than resistance to chemical pesticides, resistance can still occur, especially when the same pesticide is used repeatedly. Resistance mechanisms may include changes in the target sites (e.g., gut receptors for *Bt* toxins) or the development of an immune response in insects.

Impact on Nontarget Organisms
Microbial pesticides are typically regarded as safer than chemical pesticides for nontarget organisms. They tend to have limited or no effects on beneficial insects, wildlife, or humans. Importantly, however, microbial agents do not disrupt beneficial microorganisms in the ecosystem, such as soil bacteria or fungi.

16.8 Advantages of Microbial Pesticides

1. *Reduced environmental impact*:
   - Microbial pesticides are often biodegradable and do not persist in the environment, which reduces pollution risk. Unlike chemical pesticides, they do not leave toxic residues on crops, promoting safer food production.
2. *Targeted action*:
   - These biopesticides are highly specific to insect species, minimizing damage to nontarget species. This specificity ensures that beneficial insects such as pollinators, predators, and parasitoids are largely unaffected.

3. *Reduced resistance development*:

   - Because microbial pesticides operate through biological processes, their mode of action is often more complex than that of synthetic chemicals. This complexity reduces the likelihood of rapid resistance development in insect populations.

4. *Compatibility with integrated pest management*:

   - Microbial pesticides can be integrated into IPM programs, where they work in conjunction with other insect control methods, such as biological control agents and cultural practices. This integrated approach helps maintain long-term pest management effectiveness.

16.9 Challenges and Limitations

1. *Short shelf-life and storage*:

   - Microbial pesticides often require careful storage and handling conditions to maintain their viability. They may have shorter shelf-lives than chemical pesticides do, and improper storage (e.g., exposure to heat or moisture) can reduce their efficacy.

2. *Limited spectrum of control*:

   - While microbial insecticides or pesticides have the ability to successfully target and combat certain pests, their activity is often narrower than that of chemical pesticides. This limitation may require the use of multiple biopesticides for the effective control of different pest species.

3. *Environmental sensitivity*:

   - Microbial agents may be affected by environmental factors such as temperature, humidity, and sunlight, which can decrease their effectiveness, particularly if applied during adverse weather conditions.

4. *Cost and availability*:

   - Some microbial pesticides are more expensive to produce and apply than conventional chemicals are, and their availability may be limited in certain regions. Research and development to improve these products are ongoing to make them more cost-effective.

16.9.1 Advantages and Limitations

Advantages

- Selective toxicity to pests that have little effect on nontarget organisms.

- There is a reduced environmental impact and resistance issues in relation to chemical pesticides.
- Compatibility with organic farming practices.
- Reduced or lowered risk likelihood of pesticide residues in food and water.

Limitations

- Limited spectrum of activity.
- Slow action and potential for ineffective control under certain conditions.
- Shelf-life and stability issues in commercial formulations.
- Regulatory hurdles for approval and use.

16.9.2 Applications of Microbial Pesticides

Microbial pesticides are used in various sectors, including:

- *Agriculture:* Protecting crops from a wide range of pests.
- *Forestry:* Control of forest pests, such as bark beetles and caterpillars.
- *Public health:* Mosquito control in the fight against diseases such as malaria and dengue.
- *Aquaculture:* Managing pests in fish farming environments.

16.10 Regulatory Aspects

The approval of microbial pesticides requires rigorous assessment by regulatory bodies such as the Environmental Protection Agency in the United States or the European Food Safety Authority in Europe. These agencies assess safety, efficacy, and environmental impact before allowing the use of microbial pesticides.

16.10.1 Future Prospects

As global concerns about environmental sustainability continue to rise, the application of microbial pesticides is anticipated to grow. With advancements in biotechnology, the development of genetically modified microbial strains, and a better understanding of microbial ecology, microbial pesticides will likely play crucial roles in integrated pest management systems in the future.

Moreover, as technological advancements tailored to microbial products have progressed, these approaches should be integrated to detect and mitigate potential risks without relying on animal models. Additionally, standardizing testing and evaluation methods is essential for the successful global registration of microbial products.

16.11 Conclusion

Microbial pesticides represent environmentally friendly and sustainable replacements for traditional toxic insecticides made from harmful chemical materials. Their specificity, minimal environmental effects, and alignment with integrated pest management strategies make them an essential tool in modern pest control. However, challenges such as shelf time, a limited pest spectrum, and environmental sensitivity must be addressed to maximize their efficacy.

As scientific inquiry into microbial biopesticides as progress in this field continues, these products are expected to become more crucial in controlling insect pests, particularly in an era where chemical pesticide use is being scrutinized for its ecological and health impacts. By improving formulations, broadening their target range, and integrating them into comprehensive pest management systems, microbial pesticides can contribute significantly to more sustainable agricultural practices. They explore the environmental benefits and regulatory challenges of using microbial pesticides in agriculture, providing a balanced view of their practical implementation.

As agriculture continues to evolve, microbial pesticides are becoming an integral part of integrated pest management, offering a sustainable solution for pest control. By reducing reliance on synthetic chemicals, microbial pesticides provide farmers with effective, eco-friendly tools for protecting crops and enhancing biodiversity.

Insect control is a cornerstone of sustainable agricultural practices. By understanding the complex roles that insects play in crop ecosystems, recognizing the economic and environmental impacts of pest infestations, and adopting integrated, sustainable control strategies, we can ensure food security for future generations. Continued research and innovation in pest management are crucial for overcoming the challenges posed by insect pests and fostering a more sustainable, resilient agricultural system.

This chapter offers a detailed and up-to-date analysis of the contribution of microbial pesticides to IPM. This study provides essential insights for researchers, agricultural practitioners, and policymakers interested in sustainable pest management. With insights from top experts in the field, the chapter provides a valuable resource for those looking to understand the complex, evolving role of microbial biocontrol agents in global agriculture. Microbial pesticides present promising substitutes for traditional chemical pesticides, with the potential to support more sustainable farming practices and the potential to contribute to more environmentally friendly agricultural practices. By targeting particular pest populations and minimizing damage to the environment, microbial biopesticides are essential tools for pest control in the future.

This chapter provides a comprehensive exploration of microbial pesticides, offering both scientific insights and practical applications, with a focus on the potential of microorganisms to contribute to sustainable agriculture.

References

Adang MJ, Crickmore N, Jurat-Fuentes JL (2014) Chapter two diversity of bacillus thuringiensis crystal toxins and mechanism of action. In: Insect midgut and insecticidal proteins, pp 39–87

Álvarez M, Rodríguez A, Núñez F, Silva A, Andrade MJ (2020a) In vitro antifungal effects of spices on ochratoxin A production and related gene expression in *Penicillium nordicum* on a dry-cured fermented sausage medium. Food Control 114:107222

Álvarez M, Rodríguez A, Peromingo B, Núñez F, Rodríguez M (2020b) *Enterococcus faecium*: a promising protective culture to control growth of ochratoxigenic molds and mycotoxin production in dry-fermented sausages. Mycotoxin Res 36:137–145

Crook NE, Yamanaka J (2012) Baculovirus biology: a practical approach. Oxford University Press

Ferro D (2010) Economic impacts of insects on crop production. J Agric Sci 12(3):45–58

Fiuza LM, Polanczyk RA, Crickmore N (2017) *Bacillus thuringiensis* and *lysinibacillus sphaericus*. Springer International Publishing, Dordrecht

Fries I (2010) Nosema disease in honey bees: a review of current knowledge. Apidologie 41(3):252–270

Hajek AE, St. Leger RJ (2023) Microbial pesticides in integrated pest management: past, present, and future. Biocontrol Sci Tech 33(8):1151–1170

Jaronski ST (2010) Ecology of entomopathogenic fungi in agroecosystems. BioControl 55(1):3–34

Klein MG (1981) Advances in the use *of bacillus popilliae* for pest control. In: Burges HD (ed) Microbial control of pests and plant diseases 1970–1980. Academic, London, pp 183–192

Klein MG (1992) Use of *Bacillus popilliae* in Japanese beetle control. In: Jackson TA, Glare TR (eds) Use of pathogens in scarab pest management. Intercept Limited, Hampshire, UK, pp 179–189

Klein MG, Georgis R (1994) Application techniques for entomopathogenic nematodes. In: Proc. IV Int. Coll. on invertebrate pathology and microbial control, Montpelier, France, pp 483–484

Klein MG, Jackson TA (1992) Bacterial diseases of scarabs. In: Jackson TA, Glare TR (eds) Use of pathogens in scarab pest management. Intercept Limited, Hampshire, UK, pp 43–61

Klein MG, Kaya HK (1995) Bacillus and Serratia species for scarab control. Mem Inst Oswaldo Cruz 90:87–95

Lacey LA, Shapiro-Ilan DI (2008) Microbial control of insect and mite pests. Elsevier

Oerke EC (2006) Crop losses to pests. J Agric Sci 144(2):31–42

Pimentel D (2005) Environmental and economic costs of insect pests and pesticide use. Agric Ecosyst Environ 97(2):97–120

Rubio-Infante N, Moreno-Fierros L (2016) An overview of the safety and biological effects of *Bacillus thuringiensis* Cry toxins in mammals. J Appl Toxicol 36(5):630–648. https://doi.org/10.1002/jat.3252

Sabbour MM, Abd El-Aziz SE (2022) Bioinsecticidal and seed protectant effects of four essential oils against *Callosobruchus maculatus and Callosobruchus chinensis* during storage. Res Crop 23(3):676–681. https://doi.org/10.31830/2348-7542.2022.ROC-867

Sabbour MM, Abd-El-Aziz SE (2018) The combined effect of *Metarhizium anisopliae* and several natural oils against *Ephestia kuehniella* and *Ephestia cutella* (Lepidoptera--Pyralidae) under laboratory and storage conditions. Biosci Res 15:3480–3489

Sabbour MM, Abd-El-Aziz SE (2020) Plant essential oils for the management of two serious stored pests in Egypt. J Entomol Res 44:377–384

Sabbour MM, Abd-El-Aziz SE (2021) Bioefficacy of some essential oils and nanogel chitosan on two insect species of stored pea seeds. J Entomol Res 45:647–652

Sabbour MM, Solieman NY (2022) Effect of biological insecticides on three harmful sugarcane pests. J Posit Psychol Wellbeing 6(1):766–774. ISSN 2587-0130. http://journalppw.com

Sahab AF, Sabbour MM, Attallah AG, Abou-Serreh N (2014) Genetic analysis of the entomopathogenic fungus *Beauveria bassiana* to the corn borers tested by UV as physical mutagen. Int J ChemTech Res 6(5):2319–7064

Singh PP, Chauhan RS (2012) Biopesticides from algal sources. Springer

Thakur N, Kaur S, Tomar P, Thakur S, Yadav AN (2020) Microbial biopesticides: current status and advancement for sustainable agriculture and environment. In: New and future developments in microbial biotechnology and bioengineering. Elsevier, Amsterdam, pp 243–282. isbn:978-0-12-820526-6

Chapter 17
Value-Adding Agricultural Products by Microbial Intervention

Kamali Joshna Thayalan, Ponmurugan Karuppiah, Uma Eswaranpillai, Shyamkumar Rajaram, Manivannan Govindsamy, and Ganeshkumar Anbazhagan

Abstract The global farming sector not only produces foods, but it also generates substantial quantities of waste and byproducts, pretense environmental and economic challenges. This section describes the key role of microbial intervention as a sustainable approach for the valorization of these agricultural residual materials. Through harnessing the metabolic abilities of various microbes like bacteria, fungi, and yeasts, raw agricultural residues are to be competently converted into a varied range of value-added products. Significant microbial metabolic processes, like fermentation and bioconversion, are effective methods for making value-added products such as biofuels, organic acids, enzymes, biopolymers, vitamins, and probiotics. This approach improves the nutritional and functional quality of food and feed while also providing a sustainable way to recycle the waste valuable products. The chapter also highlights that microbial intervention is a keystone of a circular

K. J. Thayalan · U. Eswaranpillai (✉)
Department of Botany, PSGR Krishnammal College for Women, Peelamedu, Coimbatore, Tamil Nadu, India
e-mail: euma@psgrkcw.ac.in

P. Karuppiah
Department of Botany and Microbiology, College of Science, King Saud University, Riyadh, Saudi Arabia
e-mail: pkaruppiah@ksu.edu.sa

S. Rajaram
Department of Biotechnology, Kamaraj College of Engineering and Technology, Virudhunagar, Tamil Nadu, India

M. Govindsamy
Department of Psychiatry (Research), Saveetha Medical College and Hospital, Saveetha Institute of Medical and Technical Sciences (SIMATS), Saveetha University, Thandalam, Chennai, Tamil Nadu, India

G. Anbazhagan
Department of Microbiology, Saveetha Medical College and Hospital, Saveetha Institute of Medical and Technical Sciences (SIMATS), Saveetha University, Thandalam, Chennai, Tamil Nadu, India

V. Gopikrishnan et al. (eds.), *Harnessing Microbial Resources for Sustainable Agriculture*, Microorganisms for Sustainability 61,
https://doi.org/10.1007/978-981-95-4817-0_17

bioeconomy, presenting economically viable and biologically comprehensive metabolic pathways to recover agricultural sustainability, reduce waste, and create new profits.

Keywords Microbial intervention · Value-added products · Agricultural waste · Sustainability · Circular bioeconomy

17.1 Introduction

The food and agricultural sectors are among the most critical and fastest-growing sectors worldwide. Ongoing technical innovations and improvements in product quality maintain the high quality of our equipment, and we have also adopted a management system and established a quality control system. However, there is an urgent demand for agricultural practices to become more sustainable if they are applied not only for maximal yield production but also for minimal environmental pollution with chemicals, especially from chemical fertilizers and other agrochemicals. Similarly, food processing industries destroy the environment by discharging chemically treated waste. To maintain a sustainable environment and agricultural system, a circular economy is essential for crop production. This approach highlights the application of agro-industrial byproducts for the development of several value-added products. Value addition has a positive influence not only on satisfying high demand in the market but also on contributing greatly to the overall economic development and growth of the nation.

17.2 Definition and Significance of Value-Added Agricultural Products

Value-added products from agro-waste, a product made from agro-waste and a byproduct of food processing units, can also be classified as raw materials of agro-/food waste after they require processing or as products that are more ready to use and known to the public and commercial demand (Prinyawiwatkul 2024). The masses of agro-waste materials are vast and include crop residues such as wheat straw, corn straw, rice straw, barley straw, and aquaculture residues; livestock waste (Noman et al. 2020); and agro-industrial waste fruit peel, rice bran, husk, cotton stalk, sugarcane bagasse, etc. (Muhammad et al. 2022). Since these wastes cannot be directly used because they contain more complex structures and molecules, an oral process to convert complex structures into simpler forms and other reactions will be performed to make them useful. Pretreatments include mechanical treatment methods that involve shredding, grinding, and milling, which reduce the size of the material; chemical methods, which use acidic and basic solutions; and biological methods, which involve the degradation of complex particles by microbes (Bala et al. 2023).

1. Production of value-added products: The effort to produce value-added products will reduce the pressure of disposing of the arrearage and making use of those garbage for very conservative purposes, and the development of new products from agro- and food garbage is an objective of the United Nations Sustainable Development Goal, which is to conserve the environment and food security worldwide (Matumoto-Pintro and Saraiva 2023).
2. Utilization of value-added products from agro-waste can also lead to circular economies that minimize the occupational hazards from waste trapped in the ecosystem, as waste flows away from the ecosystem and is utilized in its usable state in a safe and reliable form (Oyedeji et al. 2025).
3. The efficient exploitation of the natural resources available in food processing factories into value-added products would be beneficial in terms of combating resource scarcity and environmental pollution.
4. Bioactive compounds separated from treated food effluents are some of the best products in the pharma and drug discovery sectors.

17.3 Microorganisms Intervention in Agriculture

Agriculture indirectly benefits from a favorable microbial group in which soil is more fertile, which automatically supports the growth of plants, and plants exhibit tolerance to pests, disease, and stress conditions. All these fungal and microbial processes will lead to modern microbially beneficial agriculture, which is associated with increased security and sustainable crop production, as well as further progress in the manipulation of microbes when these microbes are introduced into the soil, increasing their quality.

The production of the crop and capital of the crop enables the production of crops continuously in this rapidly growing human world. Microbial engineering is a growing industry that enhances the quality of microbial and plant characteristics via the modification of beneficial genes, clustered regularly interspaced short palindromic repeats (CRISPR), and next-generation sequencing techniques; in addition, microbial engineering aims to obtain a much-improved microbial consortium for crops (Nadarajah and Abdul Rahman 2023; Singh and Yadav 2022). The degree of root colonization will be strong, and the food and nutrient absorption of plants will also be relatively high, where they will slowly stop the use of chemical fertilizers, which will result in excessive root colonization and will be ideal where an eco-friendly process of cultivation occurs, which will ensure that all the naturally occurring bacteria that are present in the plant remain safe for consumption (Pascutti et al. 2024). Soil microbial inoculation aids agriculture through the production of biofilms and microbial-assisted biofortification, enriches the rhizosphere of the soil, and makes the ecosystem more robust. In addition to its advantages, there are several unwritten aspects to address when this type of process is used, such as the cost of production, the variability of different environmental soils, and other barriers (Upadhayay et al. 2024). We need to balance this innovation with the good old stuff, so we obtain a stable system of agricultural farming.

17.4 Scope and Relevance to Sustainable Agriculture

The awareness of food safety is increasing daily, and to shed light on this first step, people believe that instead of chemical products that have been used in agriculture, safe alternatives (farm inputs) should disseminate to the safest products, such as biopesticides (BP), biofertilizers (BF), fungal stomachs, etc., but the action of these materials should have low action compared with chemical products, and to have healthy ecosystems, progressing step by step, changing phases is very important (Jha et al. 2024). The current agro-practices continue to demonstrate negative effects on soil health; on the other hand, natural beneficial microbes are also declining, and the excessive use of inorganic fertilizer can result in the accumulation of some amount of toxic material in agricultural fields, which indirectly affects the health of humans and other living beings when they consume it, slowing the enrichment of the soil with the microbial community by shifting their regular farming practices to organic farming (Singh et al. 2024). Agroecosystems can be combined with new good agricultural practices (GAPs), such as the use of nanoparticles, which improve the biological availability of nutrients while they reach the soil. When they are mixed with biopesticides, they perform well against the target pest and do not affect other naturally beneficial insects or other soil particles. Research has revealed that gold nanoparticles are known to augment the plant growth-promoting bacteria (PGP) activities and biomass of crops, increasing yield, and in certain studies, they help increase the germination rate during shorter periods (Mahakham et al. 2016; Shukla et al. 2015).

17.5 Microbes as Resources for Value Addition

A wide range of value added products have been produced through microbes which are important in development and production of products like enzymes, organic acid, biofuel, lipids, biofuels and other pharmaceutical materials. In food industry the microbes were mostly used for the production of enzymes, sweetner, thickening agents, stabilizer, etc. Most prominently used microbes in food industry Candida, Saccharomyces, Penicillium, Rhizopus, Lactobacillus, Bacillus, Yarrowia, etc (Iram et al. 2023). For bioethanol production microbes likeS accharomyces, Zymomonas mobilis, Kluyveromyces marxianus, Thermoanaerobacterium aotearoense,etc

(Zhang et al. 2023; Wei et al. 2024; Sengupta et al. 2022; Dai et al. 2023). Escherichia coli, Saccharomyes were widely used in pharmaceutical industry and other than microbes like Streptomyces, Penicillium etc have been extensively used in the production of antibiotics and other pharmaceutical products. Some of the Bacillus sp like B. subtilis, B. cereus, B. lichenformis, B. thuringiensis, B. polymyxa. B. pumilus, B. megaterium, B. circulans, B. brevis were known to show antibiotic activity (Abbey et al. 2024).

17.6 Beneficial Microorganisms in Agriculture

The diversity of the soil microbial population promotes soil health; thus, the microbial community can be introduced directly and indirectly (presentation of a large bar on top vs the increasing number of white circles in the bottom three figures), naturally (as in "natural farming") or supplemented with a synthetic microbial consortium as a part of the overall yield and health of the plant and soil. The yield increase was proven to be quite substantial by screening both in vitro and in situ. Studies by Armanhi et al. (2018) also reported higher maize yields when bacteria such as *Lysinibacillus sphaericus*, *Paenibacillus alvei*, *Bacillus safensis*, *Bacillus pumilus*, *Brevundimonas vesicularis*, *Bacillus*, *Pseudomonas*, *Klebsiella*, *Acinetobacter*, *Streptomycete*, and *Azotobacter*, which are rhizospheric bacteria, were tested on major crops such as maize, sugarcane, and rice (Josephine and Thomas 2019; Ghorbani Nasrabadi et al. 2023). All these bacteria assist plants via different mechanisms, such as phosphate solubilization, nitrogen fixation, and potassium and zinc solubilization, which also increase the amount of indole-3-acetic acid (IAA) and siderophores and the production of enzymes; thus, these bacteria are resistant to pathogens. These bacteria, which are characterized as PGPs, colonize the roots of plants and synthesize phytohormones that are taken up by the plant for growth and development, which reduces the use of chemical fertilizers and suppresses the activity of the soil-inhabiting pathogenic organisms that are responsible for causing disease in the root region of the main crop plant. In addition to this role, PGP also contributes to lowering the bioremediation of toxins, which are toxic to plants, resulting in indirect improvements in food security and soil well-being (Verma et al. 2023).

17.7 Diversity of Microorganisms and Their Functions

The thriving microbial community of the soil interacts with the plant through endophytic activity, and the microbes that enter the plant increase phytohormonal production and accelerate growth. In the rhizosphere, microbes persist in the related root zone region and regulate the bioavailability of nitrogen, making it easy for the plant to absorb minerals and nutrients, which are essential components for growth, and in the phyllosphere, where microbes even colonize the surface of the fruit, flower, and epidermal stem (Yadav 2021). *Arthrobacter*, *Bacillus*, *Rhodococcus*, *Azotobacter*, *Azospirillum*, *Bacillus*, *Pantoea*, *Pseudomonas*, and *Rhizobium* have been identified (Pattnaik et al. 2021). These bacteria produce siderophores, which are vital phytohormones (Ferreira et al. 2019). All these microorganisms are associated in diverse ways with plant growth promotion by synthesizing enzymes, plant growth regulators, and major plant hormones, whereas some are indirectly associated with tolerance. Studies have indicated that microbial diversity is essential not only for carbon cycling but also for the decomposition of biomass in soil. In some cases, a reduction in the soil microbial community may lead to a decrease in decomposition (Maron et al. 2018).

The microbial diversity of soil was easily analyzed from a small soil aggregate that used to be sampled, and the microbe groups in the small aggregate included the following: *Fimbriimonadales*, *Acidimicrobiales*, *Actinomycetales*, *Alteromonadales*, *Burkholderiales*, *Gemmatimonadales*, *Rhodobacterales*, *Betigubrobacterales*, *Sphingobacteriales*, *Sphingomonadales*, *Spirobacillaes*, *Onygenales*, *Chaetosphaeriales*, *Trichosporanales*, *Pedosphaerales*, *Planctomycetales*, *Syntrophobacterales*, and *Glomeromycota* (arbuscular mycorrhizal fungi (AMF)) (Bach et al. 2018). Arbuscular mycorrhizal fungi are associated with approximately 80% of plant species; these fungi help in the control of soil erosion and participate in the anchorage of the plant root stem so that more nutrients gain access to the soil and that soil nutrients increase appreciably. Phosphate and nitrogen are the phosphate nutrients that fungi degrade and that become available to the plant, which are all carried out by all beneficial bacteria and fungi (Fall et al. 2022). All of these microorganisms increase as the health of the soil becomes healthier, and, then, the plants become healthier.

17.8 Isolation and Characterization Methods

The isolation of facultative bacteria is possible from soil, water, and a particular part of the plant, from the rhizosphere or from endophytic microorganisms. In accordance with need, microbes can also be taken from the sources mentioned above. Bacterial diversity in water bodies has been studied in one study via both culture-dependent and culture-independent methods; *Proteobacteria*, *Actinobacteria*, and *Bacteroidetes* were the bacterial groups most frequently found in both culture types (Vaz-Moreira et al. 2011). Culturally, identification, which is based on the culture, phenotypic, and functional characterization of microbes, suffers from low coverage of microbial diversity and slow growth. Culture-independent methods involve the denaturing gradient gel electrophoresis method, in which 16S rRNA is sequenced through which nonculturable and new strains of microbial community can be studied, and the disadvantage is that, sometimes, rare data can be skipped off, which is time-consuming, laborious, and expensive in comparison to other types of approaches (Park et al. 2023). Recently, culture-dependent methods have been combined with community-based culture systems, in which microbial populations from the rhizosphere environment and endophyte microbes are characterized and compared with the microbiomes of a plant of interest; in this way, we can identify the microflora restricted to a plant and their role in plant development (Armanhi et al. 2018). In all of these ways, these microbes can be isolated and identified, and we can look for the best microbe to make agricultural practices more sustainable.

17.9 Value Addition Through Fermentation

Traditional fermented foods are foods and beverages produced from raw agricultural products (fish, giblets, vegetables, fruits, cereals, roots, milk, and sugarcane) through fermentation, a spontaneous and natural biochemical process involving microorganisms from vegetable, animal, and human sources, including bacteria, yeasts, and molds. These microbes metabolize sugars and other substrates of the base, producing a spectrum of metabolites such as organic acids, alcohols, CO_2, and other compounds that affect the flavor, texture, shelf-life, and nutritional profile of the food (Sharma et al. 2020; Jimenez et al. 2022). Fermentation has several advantages: the preservation of perishable agricultural commodities by reducing the pH and the production of antimicrobial substances. An improvement in sensory characteristics, including flavor, aroma, and texture, is needed. Enhancement of nutrient availability and gut microbes can be easily gained by the consumption of fermented food because they contain enormous amount of probiotic microbes which directly benefits the consumers by improving their gut health (Siddiqui et al. 2023; Tachie et al. 2024).

17.10 Traditional Fermented Plant Foods

17.10.1 Fermented Dairy Products

A fermented milk drink is made elsewhere in the Caucasus. It is made by inoculating milk with kefir grains, which are a symbiotic culture of lactic acid bacteria (LAB) and yeasts in a matrix of proteins, lipids, and sugars. During fermentation, lactic acid, ethanol, CO_2, and flavor components are produced, giving the beverage a creamy consistency, a slightly sparkling impression, and the characteristic tart flavor. Another dairy-free type is water kefir, where water, sugar, and water kefir grains are used (Turkmen 2017). Other fermented dairy products are yogurt, some types of cheese, and fermented creams, which are consumed worldwide.

17.10.2 Fermented Vegetables

Cabbage, cucumbers, and peppers have long been used in fermentation to make sauerkraut, kimchi, and pickles. Generally, the process involves immersing chopped or whole fruits and vegetables in brine or water to promote the growth of lactic acid bacteria, which then ferment sugars in the food into lactic acid, increasing acidity and contributing to the preservation of the food. These fermentations are quite spontaneous with bound microorganisms. Fermentation not only enhances the overall flavor and texture of chilies but also increases their shelf stability and safety by preventing the effects of spoilage. Knowledge of the sciences contributes to the recognition and management of potential microbial hazards during the production process (Steinkraus, 1992; Behera et al. 2020).

17.10.3 Fermented Grain-Based Foods

Minor grain fermentation results in the generation of food products with high contents of nutrients, phytochemicals, and bioactive molecules, which can also provide probiotic properties. Examples include fermenting grain porridges, baking sourdough bread, and processing alcoholic beverages from grain. Nutrient bioavailability and the reduction in antinutritional factors in grains may be improved by this process (Qin et al. 2022; Tomar et al. 2025).

17.10.4 Fermented Fruits and Beverages

Traditional fermented alcohol products, such as wine and fruit beer, are made through the fermentation of fruit juice with yeast and sugar. The intense fermentation process is succeeded by ripening, and hygienic precautions are crucial to preserve the quality of the product. These drinks are shelf stable and are an important part of local food culture and livelihoods (Siddiqui et al. 2023; Avîrvarei et al. 2023).

17.10.5 Other Fermented Products

Traditionally people depend on the fermented food and get benefited by this type of food processing products. Fermented beans like temph, nattos and condiments like soy sauce and fish sauce were widely used by our ancestors. Few such fermented foods are listed below: (Table 17.1):

Table 17.1 Traditional fermented foods from agricultural products

Name of the food	Origin	Microbes involved	References
Ergo	Ethiopia	*Lactococcus lactis*, *Streptococcus thermophilus*, *Lactobacillus plantarum*, *Leuconostoc* spp.	Mulaw et al. (2019)
Rusip	Indonesia	*Lactobacillus* spp., *Leuconostoc* spp., *Streptococcus* spp.	Yuliana et al. (2018)
Kimchi	Korea	*Levilactobacillus brevis*, *Lactiplantibacillus plantarum*	Kim et al. (2021)
Khali	Nepal	*Lactobacillus plantarum*, *L. brevis*, *Leuconostoc fallax*, *Saccharomyces cerevisiae*	Tamang et al. (2009)
Sauerkraut	China	*Clostridium*, *Enterobacter*, *Lactobacillus*, *Leuconostoc*, *Weissella*	Yang et al. (2020)
Tempeh	Indonesia	*Enterococcus faecium*, Lactobacillus, *Escherichia coli*, Enterococcus	Dimidi et al. (2019)
Kefir	Europe	*Lactobacillus*, *Acetobacter*, *Leuconostoc*, *Gluconobacter*, *Kluyveromyces*, *Kazachstania*, *Dekkera*	Bourrie et al. (2021)

17.11 Microbial Ecology in Fermentation Mechanisms

Various indigenous fermented foods and beverages need different microbial communities: LAB are major indigenous microorganisms in vegetable and dairy fermentation and cause acidification. Yeasts are responsible for alcoholic fermentation and flavor compound formation. The other products, the microbial composition of which is quite diverse (e.g., kefir has a complex microbial community with symbiotic bacterial and yeast relationships) (Voidarou et al. 2020; Sionek et al. 2023).

17.12 Impact on Health and Nutrition

Traditional fermented foods provide functional benefits beyond their basic nutrients, such as promoting gut health via the presence of probiotics, facilitating nutrient bioavailability, and potentially immunomodulatory activities. They have been demonstrated to exert beneficial effects on the gut microbiota and well-being (Leeuwendaal et al. 2022; Pyo et al. 2024). Fermented traditional agricultural food products are sensitive and include milk, vegetables, grains, fruits, and legume-based products. They employ native or starter cultures of microorganisms to process raw ingredients in such a way as to produce goods with improved stability, taste, nutritional attributes, and health-promoting properties. Products such as kefir, sauerkraut, fermented minor grain foods, and traditional wines are integral parts of global food culture and have a strong potential for improving food security and livelihoods around the world (Dimidi et al. 2019; Sawant et al. 2025).

17.13 Biofertilizers and Biopesticides

There is now a surge in demand for biofertilizers and biopesticides due to the increasing realization of environmentally friendly, sustainable use, and deleterious effects of chemical agro-inputs. These biobased products provide an alternative to chemical fertilizers and pesticides by employing beneficial microbes for plant growth and protection against diseases. The manufacturing of these inputs at commercial levels undoubtedly needs meticulous planning, scientific accuracy, and an analytical approach as well as adherence to the standard operating norms of regulatory bodies for efficacy, safety, and market acceptance. Commercial biofertilizer production includes the cultivation, stabilization, and quality control of plant growth-promoting microorganisms. The process is based on vigorous fermentation upstream, highly controlled downstream techniques, different formulas and formulations techniques, and a high standard of quality control by using microorganisms for biofertilizers and biopesticides:

1. *Microbes for biofertilizers*

   Organisms fixing atmospheric nitrogen (*Rhizobium*, *Azospirillum*, and *Azotobacter*); phosphate mobilizers (*Bacteria–Bacillus*, *Pseudomonas*); and fungi–*Aspergillus* potassium solubilizers (*Frateuria aurantia*)
   Mycorrhizal fungi: AMF
   Zinc solubilizers: *Acinetobacter*, *Azotobacter*, *Azospirillum*, *Gluconacetobacter*, *Burkholderia*, *Thiobacillus*
   Sulfur oxidizers: *Beggiatoa*, *Thiobacillus*, *Acidithiobacillus*, and *Sulfurimonas*

2. *Microbes for biopesticides*

   Bacteria: *Bacillus thuringiensis*, *Pseudomonas fluorescens*
   Fungi: *Trichoderma* spp., *Beauveria bassiana*, *Metarhizium anisopliae*
   Viruses: Nuclear polyhedrosis virus (NPV)
   Antagonists: *Bacillus subtilis*, *Pseudomonas fluorescens*, *Stenotrophomonas maltophilia*, *Bdellovibrio* sp.
   Entomopathogenic microbes: *Bacillus* spp., *Paenibacillus* spp., *Clostridium* spp., *Pseudomonas*, *Serratia*, *Yersinia*, *Photorhabdus*, *Xenorhabdus*

17.14 Industrial Production

Manufacturing of the products generally scales up from laboratory shake flasks to pilot tanks, followed by large industrial bioreactors (García-Cabrera et al. 2021; Rojas et al. 2018), such as 30 L, 500 L, 6000 L, and 1500 L bioreactors. Sequential expansion is important for maintaining culture fidelity and a predictable growth profile. The temperature, pH, aeration rate, agitation, and dissolved oxygen (DO) were strictly monitored. Oxygen transfer (kLa) is a key scale transition criterion.

Economic substrates are focused on, e.g., industrial waste, including glycerol and starch wastewater, and yeast extract (Vassileva et al. 2021). Media formulation is typically optimized via statistical approaches such as central composite design, in which the optimum parameters are used for maximum viable cell or spore yield (Rouissi et al. 2010). Biomass/spore separation from fermentation broth is carried out via centrifugal harvesting (Rouissi et al. 2010; Xu et al. 2023). Parameters such as the g-force, temperature, and carbon source are optimized for the maximization of recovery. Product stabilization is accomplished by blending with carriers such as starch, skim milk, and protectants and drying by spray drying or freeze drying. The viability after drying and long-term ambient storage (in CFU/g, spore%) are important parameters for the choice of a commercial quality product. Some are provided as liquid-based concentrated inoculants, which should also be formulated with carrier oils or polymers whose shelf-life should be further improved (Parnell et al. 2016; Xu et al. 2023). The viable cell/spore concentration (CFU/mL), moisture content, product purity, and spore percentage are among the Critical Quality Attributes (CQAs). In-process control relies on plate counts, online DO/pH monitoring, and, occasionally, activity assays (e.g., insecticidal toxin units for *Bacillus thuringiensis*) (Rojas et al. 2018). Products must also frequently meet industry (ISO 9001, FAO/WHO guidelines) and national (biofertilizer regulations) standards. Consistency from batch to batch, contamination tracking, and documentation are also crucial for regulatory approval (Slininger et al. 2003; Parnell et al. 2016).

Factorial experimental designs and response surface methodologies are often employed to optimize fermentation conditions as well as downstream processing conditions. Industrial waste streams (such as starch industry effluent and lysed yeast cells) can be valorized to a great extent to minimize input costs with no concomitant reduction in product yield. Process monitoring tools such as kLa measurements for oxygen transfer, in-line measurements of dissolved oxygen and pH, and advanced viability/activity assays (e.g., spore count, toxin assay) are the core process tools (Rojas et al. 2018; García-Cabrera et al. 2021; Xu et al. 2023). These case studies illustrate the ability to scale several thousands of liters from 100-mL bench flasks, with complete technical parameter transparency (DO, agitation, feed rates, and bioreactor type in some instances). Certain studies have investigated their application in the field (e.g., application of continuous fermenter output via irrigation) (Steddom and Menge 2001), emphasizing the importance of robust formulations suitable for different agricultural scenarios. All these findings can help researchers and farmers optimize a cost-effective, environmentally friendly, regulatory-approved process of biofertilizer production in support of sustainable agriculture. The successful amalgamation of gas fermentation, tailor-made downstream processing, and stringent product validation is paramount for delivering gas-fermented biofertilizer products that are quality consistent, robust through distribution chains, and strong for on-farm application.

17.15 Microbial Enzymes and Metabolites

The development of enzymes from agro-industrial byproducts is important for sustainable biotechnological and industrial processes. Enzyme production by microbial fermentation of agricultural wastes is a promising source of low-cost, abundant, and renewable substrates. The use of these wastes not only decreases environmental contamination but also decreases production costs for industrial enzymes, significantly increasing the profitability of bioprocesses and the ecoproduction benefits they bring (Bharathiraja et al. 2017; Raina et al. 2022). Agro-waste materials include mainly lignocellulosic biomass from crop residues (rice bran, wheat bran and straw, sugarcane bagasse, corn cob, and other agro-industrial residues) (Gupta et al. 2022). This plant material contains large amounts of carbohydrates, cellulose, hemicellulose, and other organic components, which serve as excellent sources of carbon (C) for microbial growth and enzyme production.

Lignocellulosic Character Agricultural residues are mostly lignocellulosic and contain cellulose (polysaccharide), hemicellulose, and lignin. The carbohydrate richness of this biomass is appropriate for the production of enzymes, such as cellulases, xylanases, amylases, and ligninases (Blasi et al. 2023; Benatti and Polizeli 2023). Abundance and cheapness: Unwanted amounts of these wastes are produced globally in millions of tons, with a resulting disposal (landfills, burning, etc.) threat in the form of environmental problems, including greenhouse gas emissions. Their utilization as substrates results in waste upcycling and a circular bioeconomy (Mishra et al. 2023).

17.15.1 Industrial Enzymes Based on Agro-based Waste Products

Microbes such as fungi, yeasts, and bacteria (particularly actinobacteria) are used in fermentation for the synthesis of a wide range of industrial enzymes from agricultural waste:

Amylases are capable of degrading starch and are widely applied in industrial applications such as food, textiles, paper, and biofuels.

Cellulases digest cellulose to fermentable sugar, important in bioethanol production, the textile industry, and the paper industry.

Xylanases disintegrate hemicelluloses, applicable in baking, feed, and pulp bleaching.

Laccases and other oxidases are potential for dye color removal and bioremediation. The choice of enzyme is based on the microbial strain and the agricultural waste substrate.

17.15.2 *Methods for Enzyme Production Through Fermentation*

Two key modes of fermentation are known for the enzymes downstream of agro-wastes: solid-state fermentation (SSF), which uses solid substrates with a low content of free water, consequently resembling the natural ecosystems of a number of fungi and actinobacteria. It is especially suitable for lignocellulosic wastes and offers the advantages of a higher product concentration, a reduced risk of contamination, and lower water consumption. SSF can lead to increased enzyme titers and is inexpensive (Mattedi et al. 2023).

Submerged Fermentation Enzymes are synthesized in a liquid medium in which agro-waste materials or their extracts are dissolved. Submerged fermentation is more amenable to controlling the growth environment and scale-up than the SSF but might involve pretreatment of the raw materials (Savino et al. 2021).

17.15.2.1 Pretreatment and Optimization

Pretreatment of agricultural waste increases enzyme yields by degrading complex lignin and hemicellulose structures, which allows the substrate to be accessed by microbes. The methods used include physical (grind milling), chemical (acid, alkaline pretreatments), and biological (microbial decomposition) methods. Optimization parameters, including pH, temperature, moisture, inoculum size, and incubation period, are essential for increasing enzyme production (Chen et al. 2024).

17.15.2.2 Microbes Producing Enzymes

Actinobacteria are efficient producers of extracellular enzymes such as cellulases and amylases from agro-waste substrates. Fungi (e.g., species *of Aspergillus, Trichoderma*). This enzyme is often employed in SSF for cellulase and xylanase biosynthesis. Yeasts and bacteria are selected based on the target enzyme and the prevailing process conditions (Wen et al. 2005).

17.15.2.3 Applications and Significance of Industrial Enzymes in Agro-waste

Agro-waste-based enzymes are widely utilized in several industries:

Food industry: Food manufacturers use amylases and proteases to promote the pretreatment of foods and enhance flavor and preservation. Textiles: In textiles, cellulases and laccases are used in fabric finishing and dye fixing. Biofuel company: Biomass to be converted into fermentable sugar for ethanol production by cellulases and hemicellulases. Waste management: Waste management helps degrade waste pollutants and treat industrial effluents. Pharma and chemical industry: Enzymes such as streptokinase (agar plate) and chymosin (cheese) and regular insulin preparations are prepared in large amounts in reactors with immobilized cells.

17.16 Probiotics and Functional Foods

17.16.1 Contribution of Microbes to Food Fortification

Biofortification is the approach of increasing the content of essential nutrients in staple foods via biological means such as breeding, genetic engineering, or the use of mobilizing microbes. The approach is mostly targeted at hidden hunger—a lack of essential vitamins and minerals that affects more than 2 billion people globally, mostly in resource-poor settings (Bouis and Saltzman 2017). Unlike the conventional approach of providing nutrients to crops via fertilizers or adding nutrients to processed food, biofortification aims to produce more nutritious crop varieties that naturally accumulate more essential nutrients in their edible parts. Thus, biofortified crops are a more sustainable and long-lasting solution for overcoming nutrient deficiency.

The primary targets of biofortification are key micronutrients, including iron (Fe), zinc (Zn), vitamin A (a provitamin A carotenoid), and selenium (Se), which are indispensable for physiological functions, including immune function, cognitive development, and metabolism. The biofortification of crops, including rice, wheat, maize, cassava, and sweet potato, can offer enhanced nutrition with little effect on the habits of consumption or food systems (Saltzman et al. 2013). This is especially relevant in a remote setting where people's diet mainly consists of staple crops and where diverse diets and fortified foods are less accessible.

Biofortified crops have been supported by international health organizations and centers for agricultural research, such as HarvestPlus, the Center for International Maize and Wheat Improvement, and the International Food Policy Research Institute. Such institutions have made significant progress in producing and delivering types of iron-biofortified beans, zinc-rich wheat, and vitamin A-enhanced sweet potatoes in areas such as sub-Saharan Africa and South Asia.

17.16.1.1 Microbe-Mediated Biofortification Mechanisms

Microorganisms greatly contribute to the biofortification of food crops, which is accomplished through several biochemical interactions in the soil–plant continuum.

These activities lead to higher levels of nutrients that are available for uptake and consequently to higher nutritional values of agricultural products. The key mechanisms are as follows:

1. Siderophore production as an iron chelation strategy
2. Solubilization of nutrients by secretion of organic acids
3. Stimulated root growth by phytohormone production
4. Nutrient mobilization activity of enzymes
5. Increased nutrient absorption by mycorrhizal symbionts

1. *Iron Chelation by Siderophore Production*
   Siderophore-producing microorganisms have been reported to be potentially important in increasing the availability of iron in soil containing iron that is not available for plant growth. These microorganisms secrete low-molecular-weight compounds, i.e., siderophores, which form strong complexes with Fe^{3+}, making them more readily available to the plant root. This biological process is especially beneficial in calcareous or alkaline soils with naturally low iron solubility and the occurrence of iron chlorosis, which causes the loss of crop yields. The application of siderophore-producing microbial inoculants can thus be a long-term solution for alleviating crop iron deficiency. Zulfiqar et al. (2024) reported an impressive 67% foliar iron concentration response in wheat due to *Providencia* species inoculation compared with the control, indicating the potential of these microbes to increase the iron content via natural chelation processes. Similarly, Sharma and Johri (2003) reported that iron uptake in groundnuts was promoted by *Pseudomonas fluorescens* through siderophore production, which increased iron availability for roots and enhanced plant growth. These results highlight the feasibility of siderophore-producing bacteria as biofortification agents that enhance not only the nutritional status of the plant but also iron absorption in the human diet, which supplies fortified crop products.

2. *Solubilization of Nutrients by Secretion of Organic Acids*
   The release of essential nutrients through the microbial supplementation of organic acids is the main strategy of microbe-mediated biofortification. The soil microorganisms, such as *Pseudomonas*, *Bacillus*, *Aspergillus*, *Penicillium*, and *Rhizobium*, as well as the production of a series of organic acids (e.g., gluconic acid, citric acid, oxalic acid, lactic acid, etc.), can secrete some kind of organic acid, which has an important influence on the solubilization of insoluble nutrients, such as phosphorus, zinc, potassium, iron, etc. These nutrients are transformed into available forms for plants (Kalayu 2019). For example, gluconic acid is known to be highly effective at solubilizing zinc and phosphate minerals through the release of protons and the production of stable cation complexes. This metabolic activity is particularly crucial in alkaline and calcareous soils, where many nutrients are bound in insoluble forms. Singh and Yadav (2022) reported that *Pseudomonas protegens* CP17 increased the uptake and grain content of zinc in wheat by secreting organic acids, including gluconic acid. This isolate is also effective under saline stress and has two features: nutrient bioavailability and stress tolerance. Similarly, Vassilev

et al. (2001) reported that *Aspergillus niger* was able to solubilize phosphate in rock phosphate-treated soils and increased the absorption of phosphorus in maize. These results highlight the prospects of microbial inoculant alternatives to chemical fertilizers for eco-friendly practices. Rodriguez and Fraga (1999) reported the promiscuous role of plant growth-promoting rhizobacteria (PGPRs) in the release of organic acids for the mobilization of different nutrients at a time, which then facilitates plant growth and agricultural production with nutritional quality. The application of such microbes is more than just a soil fertility enhancer; it supports the sustainable biofortification of crops with micronutrients in an environmentally friendly way.

3. *Stimulated Root Growth by Phytohormone Production*

   PGPR, particularly *Bacillus subtilis*, are well known for their ability to promote plant growth via the production of phytohormones such as IAA. IAA is pivotal to control root architecture, as it stimulates cell elongation, root hair development, and the induction of lateral roots. These changes lead to a thicker and longer root system, which is essential for the plant because, as water and essential nutrients such as immobile Zn and Fe are moved less in plants, increasing their uptake is critical. This function is especially crucial in poor or marginal soils because the natural nutrient availability is low. In addition, the enhanced root system can provide plants with better tolerance to both drought and salinity stress by allowing them to access deeper water sources and by stabilizing the uptake of ions. In addition to phytohormone synthesis, *B. subtilis* influences plant growth through nitrogen fixation, phosphorus solubilization, and siderophore-mediated iron acquisition. It also secretes antifungal metabolites and enzymes to debilitate soil-borne pathogens, providing an additional advantage in terms of growth promotion and disease resistance. The combination of *B. subtilis*-based bioinoculants with cropping systems, in addition to encouraging plant health and productivity, may decrease reliance on chemical fertilizers and pesticides, thus achieving the goals of sustainable agriculture and environmental protection. Studies have indicated the potential of *B. subtilis* for increasing plant biomass, increasing nutrient use efficiency, and inducing resistance to abiotic stress, offering a holistic approach to improve crop productivity and food security (Upadhayay et al. 2022; Glick 2012; Vurukonda et al. 2016).

4. *Mobility of Nutrients by Enzymatic Action*

   Biofortification is an eco-friendly tool for improving the nutritional quality of crops by increasing the availability of essential nutrients via microbial enzymatic activities. The beneficial soil microorganisms, such as bacteria and fungi, produce enzymes, which increase the solubility and conversion of nutrient sources available to plants. For example, phosphate-solubilizing microorganisms synthesize phosphatase and phytase strains and promote the conversion of insoluble phosphorus compounds to plant-available orthophosphates that ultimately increase phosphorus uptake and crop yield (Ahmad et al. 2023). Nitrogen-fixing microorganisms, such as *Rhizobium* spp., possess nitrogenase enzymes that turn nitrogen in the air to ammonia, a form of nitrogen that can be used by plants (Mishra et al. 2023). Furthermore, potassium-solubilizing bacteria also secrete organic acids via enzymatic pathways, which help solubilize potassium-containing minerals (Mahadik et al. 2024). Microbes that produce siderophores promote Fe availability for the host (Li et al. 2025) through the nonribosomal peptide synthetase-dependent synthesis of siderophores to sequester Fe^{3+} and transfer it to the root of the host plant. Enzyme activities and plant growth (Zn and Mn) zinc-solubilizing and manganese-solubilizing bacteria use organic acid production and redox enzyme activity to mobilize Zn and Mn, two metals that are vital for enzymatic functions and plant development (Upadhayay et al. 2022; Khoshru et al. 2023). Furthermore, sulfur-oxidizing bacteria release enzymes such as sulfur oxidase and arylsulfatase, which convert sulfur into sulfate, a readily accessible form in plants. PGPRs also aid in nutrient uptake through the production of aminocyclopropane-1-carboxylate deaminase, which reduces ethylene production and promotes root growth, and IAA, which promotes root elongation (Mishra et al. 2023). These resident microbial processes play important roles in nutrient bioavailability and soil fertility and, ultimately, in the biofortification of staple crops.

5. *Enhanced Nutrient Uptake in Mycorrhizal Symbiosis*

AMF are a key part of microbe-based biofortification, as they form mutualistic symbiotic associations with the roots of the vast majority of terrestrial plants. These fungi expand the plant's effective absorptive surface by creating a profuse extraradical hyphal network that penetrates soil microsites that are not accessible to roots alone. This promotes the absorption of immobile and less available nutrients such as phosphorus (P), zinc (Zn), and selenium (Se) (Qin et al. 2022). AMF solubilize and mobilize phosphorus by secreting phosphatases and organic acids and increasing the bioavailability of micronutrients by affecting rhizosphere pH and chelation (Begum et al. 2019). Through this association, AMF symbiosis can improve plant tolerance to abiotic stressors such as drought and salinity, which indirectly aids in nutrient acquisition under suboptimum conditions. In terms of Se, AMF enhance Se uptake by mediating root transcriptome activity and the transformation of inorganic Se to organic Se forms that plants can absorb more easily (Qin et al. 2022). The synergistic advantages of AMF colonization not only promote plant growth and stress resistance but also increase the nutritional value of crops, thus representing an important factor in a sustainable agricultural biofortification strategy.

Agricultural and agro-industrial byproducts may be classified as intensive-derived, animal-derived, or food processing byproducts in general. Plant-derived byproducts refer to products such as the skin, stalks, seeds, shells, and husks, which are commonly discarded during processing, even as they contain nutrients and phytochemicals. Prime examples include tomato peels, coffee grinds, and grape skins. Animal waste products are derived mostly from the meat, dairy, and seafood industries. Leftovers from food processing also include sugarcane bagasse and cereal bran, which are rich in dietary fibers, enzymes, and antioxidants, among others, and may be highly valuable for the industry. These materials, once regarded as trash, are now bountifully used for the production of bioactive compounds for the food, pharmaceutical, and cosmetic agricultural industries (Ben-Othman et al. 2020; Debnath et al. 2022; Yadav et al. 2024).

Lignocellulose biowaste, including agricultural waste such as stalks, husks, and straw, is especially rich in cellulose and hemicellulose. Furthermore, the clothes of fruits and vegetables (skin, pomace, and stem) contain various health-friendly phytoconstituents, such as polyphenols, carotenoids, flavonoids, alkaloids, and essential oils. As reported by Yadav et al. (2024), these residues are produced at all stages in the supply chain, from cultivation to consumption, and constitute an appealing feedstock for sustainable valorization.

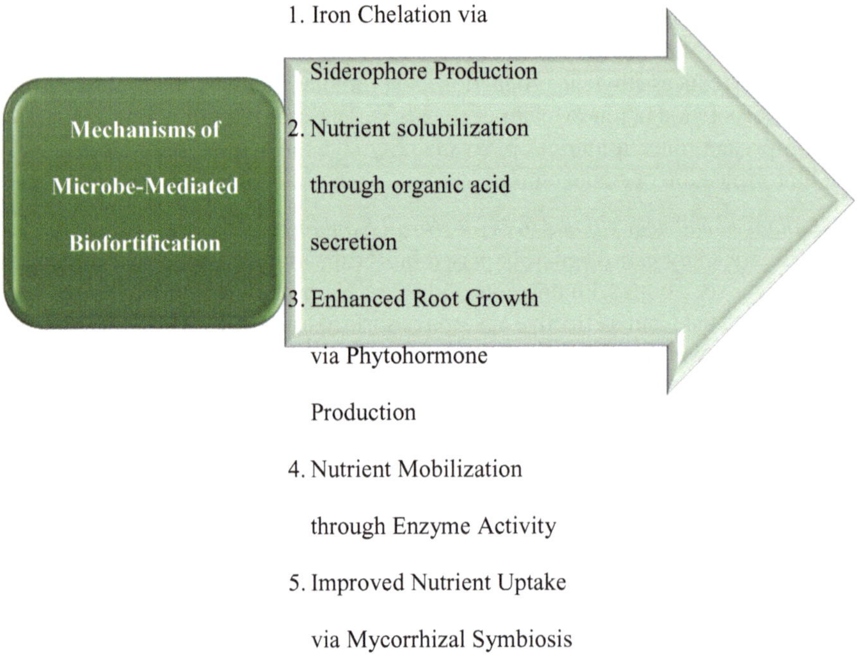

17.17 Functional Food Products from Agri-wastes

Agricultural wastes are now being turned into value-added functional foods and nutraceuticals. Owing to their high contents of antioxidants, dietary fibers, peptides, and pigments, these bioactive-rich extracts are added to food to obtain the desired nutritional or health profile. Grape seeds and coffee grounds contain polyphenols, which act as natural preservatives and antioxidants in several food systems. Similarly, peptides derived from soybean and *Jatropha* seed residues may be emerging to mitigate hypertension and inflammation (Beltrán-Ramírez et al. 2020; Debnath et al. 2022). These ingredients are also found in the product development of fortified foods, beverages, bakery products, and dairy products. For health and sensory reasons, functional drinks, for example, have advantages over citrus peel flavonoids or anthocyanins. In addition, the encapsulation (as nanoemulsions or polymer-based coatings) of these substances has increased their stability and bioavailability during processing and storage (Pérez-Marroquín et al. 2023; Yadav et al. 2024).

The use of the circular economy principle in the valorization of agro-waste is becoming more common, reflecting zero waste and sustainable development targets. The downward use of byproducts, as noted by Ben-Othman et al. (2020), permits the separation of several fractions (oils, dietary fibers, polyphenols, and organic acids) and induces the valorization of different components of this edible plant, leading to the enhancement of both the green economy and the social economy. This joint utilization is also in harmony with worldwide sustainability efforts in agriculture and food processing.

Both conventional and modern green technologies are utilized to recover bioactive compounds from agro-waste. The conventional extraction processes (such as maceration, Soxhlet, and hydrodistillation) are relatively inefficient in terms of solvent consumption and time required (i.e., they are less eco-efficient). The new methods, however, provide sustainable as well as higher-yield solutions. For example, in ultrasound-assisted extraction (UAE), cavitational forces are used to break plant cell walls, which allows solvent penetration and compound release. Microwave-assisted extraction uses electromagnetic energy to enhance molecular disruption, whereas enzyme-assisted extraction uses cell wall enzymes to release intracellular contents. Pressurized liquid extraction and supercritical fluid extraction (SFE) can increase the solubility and selectivity of target compounds such as terpenes at the end of optimized thermal conditions and pressures (Pérez-Marroquín et al. 2023; Yadav et al. 2024). These novel techniques are used to obtain a wide variety of phytochemicals, such as phenolic acids, flavonoids, carotenoids, and volatile oils. While UAE has been successfully applied to isolate flavonoids from tomato skins, SFE performs better in the extraction of nonpolar molecules, including quercetin and essential oils. Interestingly, such technologies are scalable and have a low environmental impact, and the produced products have enhanced stability (Sharma et al. 2020).

17.18 Agri-waste Functional Foods for Health

The use of bioactive compounds from agri-waste as ingredients in functional foods could provide health benefits:

1. Antioxidants: Free radicals, oxidative stress, and chronic diseases are neutralized by compounds such as flavonoids and phenolic acids.
2. Anti-inflammatory effects: Bioactive compounds, including quercetin and kaempferol, serve as anti-inflammatory agents by modifying the release of inflammation-causing enzymes, possibly reducing symptoms of arthritis and cardiovascular diseases.

3. Antimicrobial activity: Because of their potent antimicrobial activity, essential oils and phenolic compounds can be used for food safety purposes, increasing the shelf-life of any food that is protected.
4. Metabolic fitness: Agi-waste-derived dietary fibers lead to improved gut health, control of blood glucose, and increased body weight.
5. Cardiovascular benefits: Substances such as hydroxy tyrosol have been associated with heart benefits and lower cholesterol levels (Breitowich 2025).

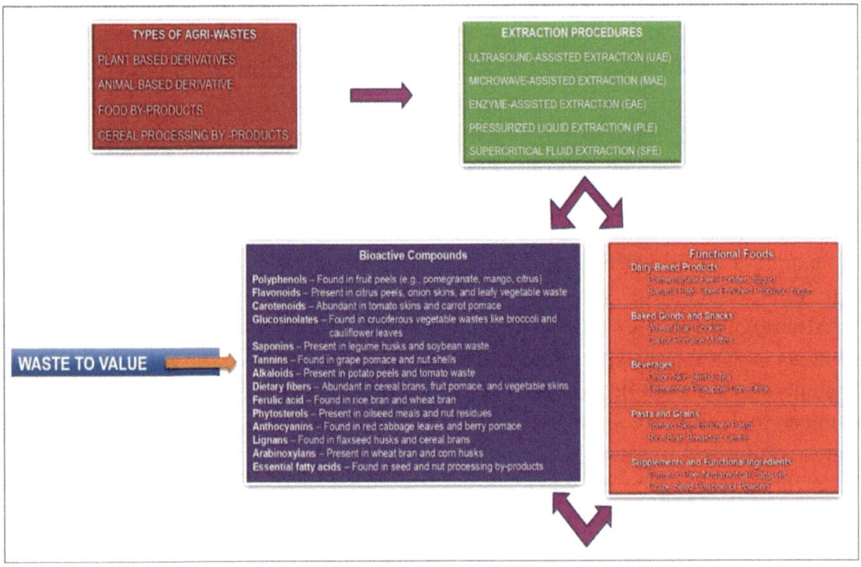

17.19 Bioplastics and Bioenergy Generated from Agricultural Waste

Increasing energy demand has led to energy scarcity and environmental problems, which have promoted the development of renewable and environmentally friendly energy resources. Among these methods, bioenergy production via agricultural waste represents an attractive route. Agricultural residues, such as the byproducts of farming, offer two advantages, as they can both generate renewable energy and solve waste disposal problems. Agricultural residues, such as straw, stalks, husks, and leaves, are parts of crops that are not consumed as food and remain in the field at harvest. These substrates contain abundant lignocellulosic materials, including cellulose, hemicellulose, and lignin. They can be used in a variety of bioenergy conversion mechanisms due to their constitution (Guddaraddi et al. 2023).

17.20 Bioenergy Conversion Systems

Multiple technologies have been researched to transform agricultural residues into bioenergy. These are as follows: (1) thermochemical processes ((a) pyrolysis, (b) gasification, (c) hydrothermal liquefaction (HTL), and (d) hydrothermal carbonization (HTC)), (2) biochemical conversion processes ((a) anaerobic digestion and (b) fermentation)), (3) nanotechnology in the conversion of biomass, and (4) development of biorefineries.

17.21 Thermochemical Processes

These include pyrolysis, gasification, and combustion. Pyrolysis is the thermal decomposition of biomass under anaerobic conditions to form bio-oil, syngas, and biochar. Gasification is the process that converts biomass into syngas by partial oxidation; it can be used to generate electricity or as a building block for chemicals. Combustion involves the direct burning of biomass and is used to generate heat and power (Guddaraddi et al. 2023):

(a) Pyrolysis is the thermolysis of biomass without supplementary oxygen, hydrocarbons such as biochar and bio-oil, and a mixture of carbon monoxide (CO), hydrogen (H_2), and CO_2 known as syngas is produced.
known as syngas is produced. It is carried out at a temperature between 300 °C and 700 °C. Biochar, a solid, carbon-rich product, is valuable, especially for soil amendment, with benefits to soil fertility and carbon sequestration (Lehmann and Joseph 2015).
(b) Gasification is a process in which the biomass is converted into syngas, i.e., carbon monoxide, hydrogen, methane, etc., by reacting it at high temperatures (800–1000 °C) in the presence of limited oxygen or steam. Syngas can be burned to produce electricity, used as fuel in IC engines or directly as a feedstock for chemicals (Basu 2010).
(c) HTL is a process that converts wet biomass at high pressures (5–25 MPa) and moderate temperatures (250–550 °C) to obtain biocrude oil. In contrast to pyrolysis, HTL can be used to treat feedstocks with high moisture levels without the need for drying, making it energy light for some biomass types (Toor et al. 2011).
(d) HTC is conducted at temperatures between 180 °C and 250 °C in a water-containing system under developed pressure on the biomass. The process produces hydrochar, a coal-like material that may be used as a solid fuel or soil conditioner. The HTC method is especially appropriate for wet biomass and has a lower energy requirement than the conventional drying process does (Funke and Ziegler 2010).

17.22 Biodegradation and Conversion Processes

Biological conversion degrades biomass into biofuels via the use of microorganisms and/or enzymes. The dominant biochemical methods are as follows:

(a) Anaerobic digestion is the process in which a material (usually organic) is decomposed by microorganisms in the absence of oxygen, which produces biogas—mainly methane (60–70%) and carbon dioxide (30–40%)—and detritus rich in nutrients known as digestate. The stages of this process include hydrolysis, acidogenesis, acetogenesis, and methanogenesis (Appels et al. 2008). This method has been extensively applied in the treatment of agricultural waste because of the dual benefits of waste reduction and renewable energy generation. The biogas produced can be used for a variety of energy uses, such as generating heat, electricity, or vehicle fuel, reducing reliance on fossil fuels. Moreover, the digestate is an eco-friendly fertilizer, resulting in a nutrient loop and sustainable farming practices (Taherdanak and Zilouei 2016).

(b) Fermentation: The fermentation process converts biomass-based sugars into ethanol via microbial activity and is an important operation in bioethanol production. Lignocellulosic biomass, such as agricultural residues, needs to be pretreated and subjected to enzymatic hydrolysis for sugar production, which is fermentable because of its complex structural constituents. The sugar yield and fermentation efficiency have been greatly improved by recently developed pretreatment technologies, such as the use of acid-functionalized magnetic nanoparticles (Wang et al. 2012). Therefore, agricultural byproducts can be efficiently transformed into bioethanol (a sustainable fuel to be utilized in place of fossil fuels, thus increasing energy security) (Pokhrel et al. 2010).

17.23 Nanotechnology in Biomass Conversion

Nanoscience has the potential to be transformational for biomass conversion. The use of nanoparticles in the pretreatment, hydrolysis, and fermentation stages can increase their efficiency. Magnetic nanoparticles, for example, have been used to enhance the digestion of lignocellulosic structures; hence, more sugar is released for biofuel production (Abdelsalam et al. 2017).

17.24 Development of Biorefineries

The construction of integrated biorefineries has expanded the production of multiple biobased products from agricultural biomass and has greatly improved the efficiency and sustainability of biomass processing. These biorefineries provide a variety of products, including biofuels, bioplastics, and other value-added products,

Table 17.2 Value-added products from agricultural waste

Category	Product	Descriptions
Bioenergy products	Biogas	Methane-rich gas from anaerobic digestion; used for heating, power, and fuel
	Bioethanol	Alcohol fuel is produced by the fermentation of sugars from crop residues
	Biodiesel	Produced indirectly via oil crops, residues support cultivation and processing
	Bio-oil	Liquid fuel from pyrolysis; a potential substitute for fossil fuels
	Syngas	Carbon monoxide and hydrogen mixture from gasification; used in power or synthesis
	Biochar	Solid carbon product from pyrolysis; enhances soil and stores carbon
	Hydrochar	Carbon-rich solid from HTC; used as fuel or soil conditioner
Industrial products	Bioplastics	Eco-friendly plastics synthesized from sugars or lignocellulosic biomass
	Platform chemicals	Includes furfural, levulinic acid; used in solvents, fuels, and polymers
	Activated carbon	Porous carbon material used in filtration and purification
	Composite materials	Natural fibers used in automotive, construction, and packaging sectors
Agricultural products	Organic fertilizers	Nutrient-rich digestate from anaerobic digestion used to fertilize crops
	Compost	Aerobically decomposed residue applied as soil amendment
	Mulch	Residue used to cover soil, improve moisture retention, and reduce weeds

which increase the economic feasibility of biomass conversion. They also help in waste reduction, as they make use of all the parts of the biomass feedstock, and wastes are produced at a relatively lower level of quantity and therefore a higher level of process sustainability (Begum et al. 2024) (Table 17.2).

17.25 Innovations and Case Studies

Microbial value addition is a potential strategy that aims to apply beneficial microorganisms, which are generated from agriculture and industry, to remediate organic waste products. Such microbial interventions increase productivity, minimize environmental damage, and provide an organic alternative to chemical-based strategies. Globally, successful examples support the economic and ecological resources of microbial value addition in agriculture, waste management, and bioindustry.

17.26 Successful Case Studies on Microbial Value Addition

1. *Biofertilizer from Food Waste Making—India Location:* Tamil Nadu, India

    Microorganisms: Azotobacter, Bacillus subtilis, Pseudomonas fluorescens
    Process: Aerobic composting of municipal food waste inoculated with PGPR.
    Results: The compost obtained was found to have high nitrogen, phosphorus, and microbial loads. Field application increased rice and maize yields by 18–25% relative to those in the untreated plots. Impact: Production of low-cost biofertilizer from waste reduces the reliance on chemical fertilizer and alleviates the landfill burden.

2. *Commercialization of Trichoderma as a Biopesticide—Kenya Location:* Nairobi, Kenya

    Microorganism: Trichoderma harzianum
    Application: Community startup businesses collaborated with agricultural cooperatives to develop and market biopesticides (*Trichoderma*) to control soil-borne fungi in vegetables.
    Outcome: Disease fell by as much as 40%, and farmers said that they made more money by selling higher-quality crops.
    Impact: Created rural jobs and reduced pesticide use, in line with organic farming standards.

3. *Phosphate-Solubilizing Microorganisms for Rock Phosphate Activation—Brazil Location:* São Paulo, Brazil

    Microorganisms: Aspergillus niger, Bacillus megaterium
    Background: Phosphate-solubilizing microbes have been applied to low-solubility rock phosphates.
    Results: Phosphorus solubilization increased significantly, and soybean plants absorbed more P from this P source.
    Impact: Less soluble fertilizers need to be imported, and more sustainable phosphorus use is encouraged.

4. *Anaerobic Digestion of Dairy Manure for Biogas and Digestate—Germany Location:* Bavaria, Germany

    Coevolutionary agents: Anaerobic methanogens.
    Application: A dairy farm-based integrated biogas plant utilizing cow manure and crop residues.
    Results: A steady flow of biogas was used for power generation and heating, and the digestate was used for biofertilizer.
    Impact: The farm is now energy self-sufficient and has closed its nutrient loops.

5. *Valorization of Agricultural Wastes Using Microbial Enzymes—China Location:* Shandong, China

Microbial products: Cellulase, pectinase, and *Bacillus* spp.
Substrate: Corn stalk and citrus peels.
Process: Crop residue fermentation for the production of enzymes for animal feed and the pulp industry.
Results: The conversion efficiency increased by more than 30%, and the residues were converted to value-added livestock feed.
Impact: Reducing waste and facilitating agri-industry convergence.

Introducing biovalues to waste values results in the conversion of waste to a resource (or "from waste to resource"). This helps reduce the carrying capacity of the environment and increases revenue. The scale-up of microbial technologies is likely to depend on local adaptation, stakeholder engagement, and cost-effective scaling. These success stories confirm that microbial innovations are "field-advanced" technologies, not just academic ideas.

17.27 Startups and Industrial Collaborations

The emergence of microbe-based biofertilizers, biopesticides, and bioenergy products has triggered opportunities for innovation-driven entrepreneurship and industry-academia associations. With the development of sustainable agriculture and a circular economy, startup companies are collaborating increasingly in microbial technology-related development, scaling up, and commercialization. These collaborations help facilitate the transition from lab development to application at the farm level, providing scalable and practical applications to real-world problems.

Startups help with at least the following: personalization of microbial consortia to crops and regions; development of innovative formulation and delivery systems (e.g., encapsulation, nanocarriers); this study provides biobased replacements for chemical fertilizers and pesticides; and applying digital applications and an IoT approach for tracking, applying, and monitoring microbial products.

For example:

AgBiome (USA): Applies genomics and machine learning to create microbial crop protection products.

BioPrime AgriSolutions (India): Provides microbial and nano-biostimulant products to improve stress tolerance.

Pivot Bio (USA): Selling genetically modified nitrogen-fixing microbes to replace synthetic N fertilizers.

Sustainable Green Technologies (Kenya): This method transforms organic waste into microbial biofertilizers via local bacteria.

17.27.1 Industry Collaboration Models

Discovery Partnerships Between Academia and Industry: Universities, research institutes, industry microbial strains and products jointly, pilot-scale infrastructure/fermentation facilities, and regulatory testing/field trials

Contract Manufacturing and Licensing: Many times, industries outlicense microbe technologies developed in research labs or startups for mass production and marketing.

Co-innovation with Agri-input Companies: Agri-tech companies and agrochemical giants are investing in or acquiring microbial startups (e.g., Bayer's acquisition of AgraQuest) while bundling microbial products with the existing seeds, fertilizers, and advisory platforms

17.27.2 Advantages of Startups and Industry Partnerships

With faster product development and early market access, field-level validation and user feedback loops, shared infrastructure reduces capital intensity, regulatory backing and certification ease, scalability, and brand confidence through industry networks. Startups and industrial partnerships are important facilitators for translating microbial research into commercial, farmer-friendly, and eco-sustainable applications. Their combined force will quicken the pace of innovation, enhance market reach, and turn microbial value addition into a true force for supporting global food security and planetary vitality.

17.28 Farmer-Led Innovations

Farmers are not only consumers of innovative agricultural technologies; they are themselves innovators. Around the world, especially under low-resource conditions, farmer-led innovations play a key role in the development and adaptation processes of microbe-based agricultural solutions. Such advances are frequently developed from low-cost, locally adapted ways to produce and apply BF, BP, compost teas, and microbe-based inoculants, utilizing indigenous knowledge and on-farm experimentation.

17.28.1 Features of Farmer-Led Innovations

1. *Site specific:* Focused on the local soil, crop, and climatic conditions
2. *Resource-light:* Typically, relies on locally available materials such as cow dung, plant extracts, or kitchen waste

3. *Experientially based:* Developed through application and observation over time
4. *Eco-friendly:* Encouraging the ecological balance and reducing chemical reliance
5. *Collaborative:* Often requires the sharing of knowledge among farmer groups or communities

17.28.2 Illustrative Cases of Farmer-Led Microbial Innovations

A. *On-farm microbial composting*

Application: Indian and Nepali farmers have developed aerobic composting methods in which cow dung is used as an inoculum with jaggery and gram flour as microbial activators.
Innovation: Microbial starters (e.g., *Trichoderma* and *Azotobacter*) are used in conventional composting piles.
Effect: Increased decomposition, increased nutrient content, and increased microbial activity in the soil.

B. *Jeevamrutha and Panchagavya (India)*

Details: Cow dung input-based biostimulant preparations from jaggery, pulse flour, and bund soil.
Microbes: These blends live with helpful bacteria and fungi that fix nitrogen, solubilize phosphorus, and stimulate plants.
Adoption: Zero Budget Natural Farming has been adopted because of its significant increase in yield and soil fertility.

C. *Biopesticide Brewing*

Areas of study: African and Southeast Asian smallholder farmers.
Applications: Fermented plant extracts (NEEMs, garlic, and chili) were diluted with cow urine and aged for antifungal and insecticidal activities.
Benefit: Affordable and compatible with organic production systems.

D. *Community Bioinput Units*

In Latin America and sub-Saharan countries, farmers set up small-scale bioreactors for stocks of nitrogen fixers (*Rhizobium*, *Azospirillum*) or *Trichoderma* for local distribution. This distributed approach increases the availability and diminishes the dependence on commercial solutions.

Effects of Farmer-Driven Innovations

1. Cost reduction with good profitability of the farm.
2. Improving the capacity of soils to resist or rebound from climate stress.

3. Microbial diversity and soil organic carbon also increased.
4. Greater ownership and confidence among farmers in eco-agriculture.
5. Formation of local knowledge networks and self-sufficiency.

Empowering Farmers to Innovate

1. Participatory research with universities and extensionists
2. Open-source libraries and starter cultures for those microbes
3. Development of certification guidelines for homemade bioinputs
4. Providing microgrants or technical assistance via rural innovation hubs

Grower-driven techniques are key to democratizing microbial technologies and ensuring that they are appropriate and used in practice. Such integrative approaches can ensure that these bottom-up innovations have the rigor to become integral parts of sustainable farming.

17.29 Challenges and Perspectives

The reclamation of agri-waste as a functional food is in keeping with the principles of the circular economy to enhance sustainability and resource utilization. The potential includes standardization of protocols for the incorporation of agri-waste-derived ingredients into mainstream food products and the development of efficient and scalable extraction-processing methods.

Future efforts should focus on technologies, as well as supportive policies and incentives, to promote the adoption of agri-waste in food applications. Additionally, systematic research must be conducted on the complete bioactive content of different agri-wastes and their probable therapeutic usage. Adopting these approaches can help convert agri-waste from being a threat to the environment into a resource for health and nutrition. Further challenges for the farmer include the lack of training in microbial manipulation and quality control, variability in efficacy depending on strains or substrates, limited access to the market or scale-up to be used more widely than at the farm level, limited scientific validation and acceptance, and recognition by the authority.

Future research should focus on formulating region-adapted microbial cocktails aimed at improving value-added traits. Linkage with precision agriculture for targeted application and monitoring. Success stories are scaled across geographic areas through the upscaling of public–private partnerships. Additionally, technological breakthroughs such as CRISPR and Synthetic Biology: Engineered bugs for hyperenzymes. AI for Enzyme Discovery: A Machine learning system for the discovery of new enzymes on agro waste microbiomes. Sustainability Integration: LCA to measure eco-efficiency potential.

17.30 Conclusion

The incorporation of microbial technologies in value-added agriculture represents a paradigm shift toward a sustainable and circular bioeconomy. This chapter exemplifies how various microbial applications, in agroecosystems (i.e., biofertilizers and biopesticides) as well as in food and nonfood fields (e.g., fermented food, industrial enzymes, and bioenergy), allow the value of crop and food wastes into economically and environmentally interesting materials. Useful microbes can be used for increasing soil fertility, increasing crop resistance, and increasing nutritional content and can also help reduce the environmental impact of agro-industrial byproducts. All these strategies will extends the beneficial use of microbes from on-farm composting and commercial-scale fermentation to cutting-edge innovations in strain development.

Selection, formulation, and process optimization are the keys to the successful addition of value through microbes. Furthermore, farmer-led, startup, and industrial collaborations have been essential in the scaling of those technologies and in developing them for local suitability and economic viability. Examples of commercial applications in the real world (India, Kenya, Brazil, Germany, and China) highlight the power and potential of microbial solutions to help increase productivity, reduce waste, and improve the lives of all.

However, many challenges need to be overcome to realize the promise of WASH, including unreliable performance of products, low demand due to low consumer awareness, a lack of supportive infrastructure links to regulate fair competition, and the ability to regulate the quality of products. Opportunities span microbiome consortia region importance, AI-powered enzyme discovery, CRISPR-mediated microbe design, and public–private partnerships that can deepen and replicate microbial intervention in multiple agroecological zones. At the end of the day, microbial value addition is just one of the many long-term investments needed to embed environmental health, food security, and economic resilience.

References

Abbey DT, Oganija AT, Ebube OL, Ifeanyi DE (2024) Harnessing Bacillus Species for Antibiotic Discovery: AReview. Nepal J Biotechnol 12(2):121-131

Abdelsalam E, Samer M, Attia YA, Abdel-Hadi MA, Hassan HE, Badr Y (2017) Comparison of nanoparticles effects on biogas and methane production from anaerobic digestion of cattle dung slurry. Renew Energy 87:592–598. https://doi.org/10.1016/j.renene.2015.10.064

Ahmad A, Moin SF, Liaqat I, Saleem S, Muhammad F, Mujahid T, Zafar U (2023) Isolation, solubilization ofinorganic phosphate, and production of organic acids by individual and co-inoculated microorganisms.Geomicrobiol J 40(1):111–121

Appels L, Baeyens J, Degrève J, Dewil R (2008) Principles and potential of the anaerobic digestion of waste-activated sludge. Prog Energy Combust Sci 34(6):755–781. https://doi.org/10.1016/j.pecs.2008.06.002

Armanhi JSL, de Souza RSC, Damasceno NB, de Araújo LM, Imperial J, Arruda P (2018) A community-based culture collection for targeting novel plant growth- promoting bacteria from the sugarcane microbiome. Front Plant Sci 8:2191. https://doi.org/10.3389/fpls.2017.02191

Avîrvarei AC, Salanţă LC, Pop CR, Mudura E, Pasqualone A, Anjos O, Barboza N, Usaga J, Dărab CP, Burja-Udrea C, Zhao H, Fărcaş AC, Coldea TE (2023) Fruit-based fermented beverages: contamination sources and emerging technologies applied to assure their safety. Foods 12(4):838. https://doi.org/10.3390/foods12040838

Bach EM, Williams RJ, Hargreaves SK, Yang F, Hofmockel KS (2018) Greatest soil microbial diversity found in microhabitats. Soil Biol Biochem 118:217–226

Bala S, Garg D, Sridhar K, Inbaraj BS, Singh R, Kamma S, Sharma M (2023) Transformation of agro-waste into value-added bio products and bioactive compounds: micro/nano formulations and application in the agri-food-pharma sector. Bioengineering 10(2):152

Basu P (2010) Biomass gasification and pyrolysis: practical design and theory. Academic Press, UK

Begum N, Qin C, Ahanger MA, Raza S, Khan M I, Ashraf M (2019). Role of arbuscular mycorrhizal fungi in plantgrowth regulation: implications in abiotic stress tolerance. Frontiers in Plant Science 10:1068. https://doi.org/10.3389/fpls.2019.01068

Begum YA, Kumari S, Jain SK, Garg MC (2024) A review on waste biomass-to-energy: integrated thermochemical and biochemical conversion for resource recovery. Environ Sci Adv 3:1197–1216. https://doi.org/10.1039/D4VA00109EScienceDirect+2

Behera SS, El Sheikha AF, Hammami R, Kumar A (2020) Traditionally fermented pickles: how the microbial diversity associated with their nutritional and health benefits? J Funct Foods 70:103971

Beltrán-Ramírez F, Orona-Tamayo D, Cornejo-Corona I, González-Cervantes JLN, Esparza-Claudio JJ, Quintana-Rodríguez E (2020) Agro-industrial waste revalorization: the growing biorefinery. In: Biomass for bioenergy - recent trends and future challenges. IntechOpen, London, UK. https://doi.org/10.5772/intechopen.83569

Benatti ALT, Polizeli MLTM (2023) Lignocellulolytic biocatalysts: Main players involved in multiple biotechnological processes for biomass valorization. Microorganisms 11(1):162. https://doi.org/10.3390/microorganisms11010162

Ben-Othman S, Jõudu I, Bhat R (2020) Bioactives from agri-food wastes: present insights and future challenges. Molecules 25(3):510. https://doi.org/10.3390/molecules25030510

Bharathiraja S, Suriya J, Krishnan M, Manivasagan P, Kim SK (2017) Production of enzymes from agricultural wastes and their potential industrial applications. In: Advances in food and nutrition research, vol 80. Academic, pp 125–148. Academic Press, UK

Blasi A, Verardi A, Lopresto CG, Siciliano S, Sangiorgio P (2023) Lignocellulosic agricultural waste valorization to obtain valuable products: an overview. Recycling 8(4):61. https://doi.org/10.3390/recycling8040061

Bouis HE, Saltzman A (2017) Improving nutrition through biofortification: A review of evidence from HarvestPlus,2003 through 2016. Glob Food Secur 12-49–58. https://doi.org/10.1016/j.gfs.2017.01.009

Bourrie BC, Ju T, Fouhse JM, Forgie AJ, Sergi C, Cotter PD, Willing BP (2021) Kefir microbial composition is a deciding factor in the physiological impact of kefir in a mouse model of obesity. Br J Nutr 125(2):129–138

Breitowich A (2025) Why scientists are calling this olive oil byproduct the next big superfood. Food & Wine. https://www.foodandwine.com/olive-mill-wastewater-superfood-11697763

Chen J, Ma X, Liang M, Guo Z, Cai Y, Zhu C, Wang Z, Wang S, Xu J, Ying H (2024) Physical–chemical–biological pretreatment for biomass degradation and industrial applications: a review. Waste 2(4):451–473. https://doi.org/10.3390/waste2040024

Dai K, Qu C, Feng J, Lan Y, Fu H, Wang J (2023) Metabolic engineering ofT hermoanaerobacterium aotearoensestrain SCUT27 for biofuels production from sucrose and molasses. Biotechnology for Biofuels and Bioproducts16(1):155

Debnath A, Sengupta A, Saha A, Das A (2022) Utilization of agro waste for beneficial product formulation. J Exp Biol Agric Sci 10(1):157–170. https://doi.org/10.18006/2022.10(1).157.170

Dimidi E, Cox SR, Rossi M, Whelan K (2019) Fermented foods: definitions and characteristics, impact on the gut microbiota and effects on gastrointestinal health and disease. Nutrients 11(8):1806. https://doi.org/10.3390/nu11081806

Fall AF, Nakabonge G, Ssekandi J, Founoune-Mboup H, Apori SO, Ndiaye A, Ngom K (2022) Roles of arbuscular mycorrhizal fungi on soil fertility: contribution in the improvement of physical, chemical, and biological properties of the soil. Front Fungal Biol 3:723892

Ferreira CM, Soares HM, Soares EV (2019) Promising bacterial genera for agricultural practices: an insight on plant growth-promoting properties and microbial safety aspects. Sci Total Environ 682:779–799

Funke A, Ziegler F (2010) Hydrothermal carbonization of biomass: a summary and discussion of chemical mechanisms for process engineering. Biofuels Bioprod Biorefin 4(2):160–177. https://doi.org/10.1002/bbb.198

García-Cabrera RI, Valdez-Cruz NA, Daniel-Vázquez A, Blancas-Cabrera A, Trujillo-Roldán MA (2021) Roles of culture media and oxygen transfer in the scale-up from shake flasks to pneumatic bioreactor of the plant growth-promoting bacterium rhizobium phaseoli. Rev Mex Ing Quimica 20(2):1091–1109

Ghorbani Nasrabadi R, Greiner R, Mayer-miebach E, Menezes-Blackburn D (2023) Phosphate solubilizing and phytate degrading Streptomyces isolates stimulate the growth and P accumulation of maize (*Zea mays*) fertilized with different phosphorus sources. Geomicrobiol J 40(4):325–336

Glick BK (2012) Plant Growth-Promoting Bacteria: Mechanisms and Applications. Scientifica 1–15. https://doi.org/10.6064/2012/963401

Guddaraddi A, Singh A, Amrutha G, Saikanth DRK, Kurmi R, Singh G, Chowdhury M, Singh BV (2023) Sustainable biofuel production from agricultural residues an eco-friendly approach: a review. Int J Environ Clim Chang 13(10):2905–2914. https://doi.org/10.9734/ijecc/2023/v13i102956

Gupta N, Mahur BK, Izrayeel AMD, Ahuja A, Rastogi VK (2022) Biomass conversion of agricultural waste residues for different applications: a comprehensive review. Environ Sci Pollut Res 29(49):73622–73647

Iram A, Ozcan A, Turhan I, Demirci A (2023) Production of value-added products as food ingredients via microbialfermentation. Processes 11(6):1715

Jha GK, Velayudhan PK, Bhatia A, Laishram C, Kumar D, Begho T, Eory V (2024) Transitioning toward sustainable agriculture: analyzing the factors and impact of adopting multiple sustainable inputs by paddy farmers in India. Front Sustain Food Syst 8:1447936. https://doi.org/10.3389/fsufs.2024.1447936

Jimenez ME, O'Donovan CM, de Ullivarri MF, Cotter PD (2022) Microorganisms present in artisanal fermented food from South America. Front Microbiol 13:941866. https://doi.org/10.3389/fmicb.2022.941866

Josephine R, Thomas J (2019) 16S microbial phylogeny of multifunctional plant-growth-promoting rhizobacteria from the rhizosphere of maize (*Zea mays* L.) for agricultural soil fortification. Biotechnologia (2)

Kalayu G (2019) Phosphate Solubilizing Microorganisms: PromisingApproach as Biofertilizers. InternationalJ Agronomy.1–7. https://doi.org/10.1155/2019/4917256

Khoshru B, Salek M, Asgari S (2023) Enhancing Manganese Availability for Plants through Microbial Potential:A Sustainable Approach for Improving Soil Health and Food Security. Diversity 15:537

Kim E, Yang SM, Kim HY (2021) Analysis of cultivable microbial community during kimchi fermentation using MALDI-TOF MS. Foods 10(5):1068

Leeuwendaal NK, Stanton C, O'Toole PW, Beresford TP (2022) Fermented foods, health and the gut microbiome. Nutrients 14(7):1527. https://doi.org/10.3390/nu14071527

Lehmann J, Joseph S (eds) (2015) Biochar for environmental management: science, technology and implementation, 2nd edn. Routledge, New York

Li J, Babcock-Adams L, McIlvin MR, Repeta DJ (2025) Distribution andcycling of siderophores in the easternNorth and Tropical Pacific Oceans. Geophys Res Lett, 52,e2024GL113874. https://doi.org/10.1029/2024GL113874

Mahadik SP, Patil SV, Kumudini BS (2024) Bioprospectingrhizosphere fungi endowed with multifarious plantgrowth-promoting potential to enhance finger millet growth under salinity stress. Plant Growth Regul 104(3):1483–1505

Mahakham W, Theerakulpisut P, Maensiri S, Phumying S, Sarmah AK (2016) Environmentally benign synthesis of phytochemicals-capped gold nanoparticles as nanopriming agent for promoting maize seed germination. Sci Total Environ 573:1089–1102

Maron PA, Sarr A, Kaisermann A, Lévêque J, Mathieu O, Guigue J et al (2018) High microbial diversity promotes soil ecosystem functioning. Appl Environ Microbiol 84(9):e02738–e02717

Mattedi A, Sabbi E, Farda B, Djebaili R, Mitra D, Ercole C, Cacchio P, Del Gallo M, Pellegrini M (2023) Solid-state fermentation: applications and future perspectives for biostimulant and biopesticides production. Microorganisms 11(6):1408. https://doi.org/10.3390/microorganisms11061408

Matumoto-Pintro PT, Saraiva BR (2023) High value-added products from food waste. Foods 12(21):4038

Mishra K, Siwal SS, Nayaka SC, Guan Z, Thakur VK (2023) Waste-to- chemicals: green solutions for bioeconomy markets. Sci Total Environ 887:164006

Muhammad S, Abdul Khalil HPS, Abd Hamid S, Albadn YM, Suriani AB, Kamaruzzaman S, Yahya EB (2022) Insights into agricultural-waste-based nanoactivated carbon fabrication and modifications for wastewater treatment application. Agriculture 12(10):1737

Mulaw G, Sisay Tessema T, Muleta D, Tesfaye A (2019) *In vitro* evaluation of probiotic properties of lactic acid bacteria isolated from some traditionally fermented Ethiopian food products. Int J Microbiol 2019(1):7179514

Nadarajah K, Abdul Rahman NSN (2023) The microbial connection to sustainable agriculture. Plants (Basel) 12(12):2307

Noman M, Shahid M, Ahmed T, Javed MR, Manzoor N, Mazoor I, Arshad M (2020) Lignocellulosic biomass andmicrobial genome engineering for sustainable ethanol production: an overview. Sustainable ethanol and climatechange: sustainability assessment for ethanol distilleries, 87-112

Steinkraus KH (1992) Lactic Acid Fermentations. In Applications of Biotechnology to Fermented Foods: Report of an Ad Hoc Panel of the Board on Science and Technology for International Development. National Research Council (US) Panel on the Applications of Biotechnology to Traditional Fermented Foods. Washington (DC): National Academies Press (US); 1992. https://doi.org/10.17226/1939

Oyedeji S, Patel N, Krishnamurthy R, Fatoba PO (2025) Agricultural wastes to value-added products: economic and environmental perspectives for waste conversion. Adv Biochem Eng Biotechnol 191:215–248

Parnell JJ, Berka R, Young HA, Sturino JM, Kang Y, Barnhart DM, DiLeo MV (2016) From the lab to the farm: an industrial perspective of plant beneficial microorganisms. Front Plant Sci 7:1110

Park J, Heo S, Na HE, Lee G, Kim T, Sung MH, Jeong DW (2023) Culture-dependent and-independentinvestigations of the effect of addition of jogi on the bacterialcommunity of kimchi. Food Biosci 54: 102832

Pascutti MCD, Silvestre RDSF, Ortiz TA (2024) The role of microorganisms in sustainable agriculture: a review. Revista delos 17(52):e1253–e1253

Pattnaik S, Mohapatra B, Gupta A (2021) Plant growth-promoting microbe mediated uptake of essential nutrients (Fe, P, K) for crop stress management: microbe–soil–plant continuum. Front Agron 3:689972

Pérez-Marroquín XA, Estrada-Fernández AG, García-Ceja A, Aguirre-Álvarez G, León-López A (2023) Agro-food waste as an ingredient in functional beverage processing: sources, functionality, market and regulation. Foods 12(8):1583. https://doi.org/10.3390/foods12081583

Pokhrel CP, Yadav RKP, Ohga S (2010) Agricultural waste residues as potential sources of bioethanol. Sci World 6(6):19–23. https://doi.org/10.3126/sw.v6i6.2628

Prinyawiwatkul W (2024) Value-added foods: composition, sensory, and consumer research. Foods 13(23):3952

Pyo Y, Kwon KH, Jung YJ (2024) Probiotic functions in fermented foods: anti-viral, immunomodulatory, and anti-cancer benefits. Foods 13(15):2386. https://doi.org/10.3390/foods13152386

Qin H, Wu H, Shen K, Liu Y, Li M, Wang H, Qiao Z, Mu Z (2022) Fermented minor grain foods: classification, functional components, and probiotic potential. Foods 11(20):3155. https://doi.org/10.3390/foods11203155

Raina D, Kumar V, Saran S (2022) A critical review on exploitation of agro-industrial biomass as substrates for the therapeutic microbial enzymes production and implemented protein purification techniques. Chemosphere 294:133712

Rodŕiguez H, Fraga R (1999) Phosphate solubilizing bacteria and their role in plant growth promotion.Biotechnol adv 17(4-5):319–339

Rojas NL, Lewkowicz ES, Nobile ML (2018) Alternative low-cost process for large-scale production of *Bacillus thuringiensis* in a simple and novel culture system. J Environ Sci Health B 53(11):719–728

Rouissi T, John RP, Brar SK, Tyagi RD, Prevost D (2010) Centrifugal recovery of rhizobial cells from fermented starch industry wastewater & development of stable formulation. Ind Biotechnol 6(1):41–49

Saltzman A, Birol E, Bouis HE, Boy E , De Moura FFa, Islam Y, Pfeiffer WH (2013) Biofortification: Progresstoward a more nourishing future. Glob Food Secur 2(1): 9–17. https://doi.org/10.1016/j.gfs.2012.12.003

Savino S, Bulgari D, Monti E, Gobbi E (2021) Agro-industrial wastes: a substrate for multi-enzymes production by *Cryphonectria parasitica*. Fermentation 7(4):279. https://doi.org/10.3390/fermentation7040279

Sawant SS, Park H-Y, Sim E-Y, Kim H-S, Choi H-S (2025) Microbial fermentation in food: impact on functional properties and nutritional enhancement—a review of recent developments. Fermentation 11(1):15. https://doi.org/10.3390/fermentation11010015

Sengupta P, Mohan R, Wheeldon I, Kisailus D, Wyman CE, Cai CM (2022) Prospects of thermotolerantKluyveromyces marxianus for high solids ethanol fermentation of lignocellulosic biomass. Biotechnology forBiofuels and Bioproducts 15(1):134

Sharma A, Johri BN (2003) Combat of iron deprivation through a plant growth promoting fluorescentPseudomonas strain GRP3A in mung bean (Vigna radiata L. Wilzeck). Microbiol Res I58 (l):77–81. https://doi.org/10.1078/0944-5013-00182

Sharma R, Garg P, Kumar P, Bhatia SK, Kulshrestha S (2020) Microbial fermentation and its role in quality improvement of fermented foods. Fermentation 6(4):106. https://doi.org/10.3390/fermentation6040106

Sharma S, Basu S, Shetti NP, Aminabhavi TM (2020) Waste-to-energy nexus for circular economy andenvironmental protection: Recent trends in hydrogen energy. Science of The Total Environment 713: 136633. https://doi.org/10.1016/j.scitotenv.2020.136633

Shukla SK, Kumar R, Mishra RK, Pandey A, Pathak A, Zaidi MGH et al (2015) Prediction and validation of gold nanoparticles (GNPs) on plant growth promoting rhizobacteria (PGPR): a step toward development of nano- biofertilizers. Nanotechnol Rev 4(5):439–448

Siddiqui SA, Erol Z, Rugji J, Taşçı F, Kahraman HA, Toppi V, Musa L, Di Giacinto G, Bahmid NA, Mehdizadeh M, Castro-Muñoz R (2023) An overview of fermentation in the food industry - looking back from a new perspective. Bioresour Bioprocess 10(1):85. https://doi.org/10.1186/s40643-023-00702-y

Singh B, Yadav JP (2022) Sustainable agriculture development through organic farming in India. ECS Trans 107(1):14919

Singh S, Ashoka P, Ahlawat U, Changdeo WB, Rehsawla R, Naruka A, Sharma D (2024) Mechanisms and applications of microbial biotechnology in soil health and agricultural productivity: a review. J Adv Biol Biotechnol 27:1420–1438

Sionek B, Szydłowska A, Küçükgöz K, Kołożyn-Krajewska D (2023) Traditional and new microorganisms in lactic acid fermentation of food. Fermentation 9(12):1019. https://doi.org/10.3390/fermentation9121019

Slininger PJ, Behle RW, Jackson MA, Schisler DA (2003) Discovery and development of biological agents to control crop pests. Neotrop Entomol 32:183–195

Steddom K, Menge JA (2001) Evaluation of continuous application technology for delivery of the biocontrol agent *Pseudomonas putida* 06909-rif/nal. Plant Dis 85(4):387–392

Tachie CY, Onuh JO, Aryee AN (2024) Nutritional and potential health benefits of fermented food proteins. J Sci Food Agric 104(3):1223–1233

Taherdanak M, Zilouei H (2016) Biogas production from agricultural residues: a review. Curr Biochem Eng 3(3). https://doi.org/10.2174/1570180813666160527120458

Tamang JP, Tamang B, Schillinger U, Guigas C, Holzapfel WH (2009) Functional properties of lactic acid bacteria isolated from ethnic fermented vegetables of the Himalayas. Int J Food Microbiol 135(1):28–33

Tomar T, Sachdeva A, Dutta J, Al Tawaha ARM, Karnwal A, Malik T, Selvaraj M (2025) Fermentation dynamics of millet beverages: microbial interactions, nutritional enhancements, and health implications. Food Chem X:102199

Toor SS, Rosendahl L, Rudolf A (2011) Hydrothermal liquefaction of biomass: a review of subcritical water technologies. Energy 36(5):2328–2342. https://doi.org/10.1016/j.energy.2011.03.013

Turkmen N (2017) Kefir as a functional dairy product. In: Dairy in human health and disease across the lifespan. Academic, pp 373–383. Academic Press, UK

Upadhayay VK, de los Santos Villalobos S, Aravindharajan STM, Kukreti B, Chitara MK, Jaggi V et al (2024) Microbial advancement in agriculture. In: Microbial inoculants: applications for sustainable agriculture. Singapore, Springer Nature Singapore, pp 95–125

Upadhayay VK, Singh AV, Khan A and Sharma A (2022) Contemplating the role of zinc-solubilizing bacteria incrop biofortification: An approach for sustainable bioeconomy. Front Agron 4:903321. https://doi.org/10.3389/fagro.2022.903321

Vassilev N, Vassileva M, Fenice M, Federici F (2001) Immobilized cell technology applied in solubilization ofinsoluble inorganic (rock) phosphates and P plant acquisition. Bioresource Technology 79(3):263–271

Vassileva M, Malusà E, Sas-Paszt L, Trzcinski P, Galvez A, Flor-Peregrin E, Shilev S, Canfora L, Mocali S, Vassilev N (2021) Fermentation strategies to improve soil bio-inoculant production and quality. Microorganisms 9(6):1254. https://doi.org/10.3390/microorganisms9061254

Vaz-Moreira I, Egas C, Nunes OC, Manaia CM (2011) Culture-dependent and culture-independent diversity surveys target different bacteria: a case study in a freshwater sample Antonie Van Leeuwenhoek 100(2):245–57. https://doi.org/10.1007/s10482-011-9583-0

Verma P, Meena RK, Supriya, Singh S (2023) Microbial-inoculants for sustainable agriculture. Vigyan Varta 4(7):91–94

Voidarou C, Antoniadou M, Rozos G, Tzora A, Skoufos I, Varzakas T, Lagiou A, Bezirtzoglou E (2020) Fermentative foods: microbiology, biochemistry, potential human health benefits and public health issues. Foods 10(1):69. https://doi.org/10.3390/foods10010069

Vurukonda SSKP, Vardharajula S, Shrivastava M, SkZ A (2016) Enhancement of drought stress tolerance incrops by plant growth promoting rhizobacteria. Microbiol Res 184:13–24. https://doi.org/10.1016/j.micres.2015.12.003

Wang H, Gurau G, Rogers RD (2012) Ionic liquid processing of cellulose. Chem Soc Rev 41(4):1519–1537. https://doi.org/10.1039/C2CS15311D

Wei Y, Li J, Wang C, Yang J, Shen W (2024) Development of a starch-fermentingZ ymomonas mobilis strain forbioethanol production. Microbial Cell Factories 23(1), 301

Wen Z, Liao W, Chen S (2005) Production of cellulase/β-glucosidase by the mixed fungi culture Trichoderma reesei and Aspergillus phoenicis on dairy manure. Process Biochem 40(9):3087–3094

Xu L, Zhan ZC, Du S, Wang S, Zhang Q, Wang C, Chen X (2023) Isolation, functional evaluation, and fermentation process optimization of probiotic *Bacillus coagulans*. PLoS One 18(11):e0286944

Yadav AN (2021) Beneficial plant–microbe interactions for agricultural sustainability. J Appl Biol Biotechnol 9(1):1–4

Yadav S, Malik K, Moore JM, Kamboj BR, Malik S, Malik VK, Arya S, Singh K, Mahanta S, Bishnoi DK (2024) Valorization of agri-food waste for bioactive compounds: recent trends and future sustainable challenges. Molecules 29(2055). https://doi.org/10.3390/molecules29092055

Yang X, Hu W, Xiu Z, Jiang A, Yang X, Saren G, Ji Y, Guan Y, Feng K (2020) Microbial community dynamics and metabolome changes during spontaneous fermentation of northeast sauerkraut from different households. Front Microbiol 11:1878

Yuliana N, Koesoemawardani D, Susilawati S, Kurniati Y (2018) Lactic acid bacteria during fish fermentation (Rusip). MOJ Food Process Technol 6(2):211–216

Zhang H, Zhang P, Wu T, Ruan H (2023) Bioethanol production based on Saccharomyces cerevisiae: opportunitiesand challenges. Fermentation 9(8):709

Zulfiqar U, Ayub A, Hussain S, Ahmad M, Rehman A, Ishfaq M, Ali MF, Shabaan M, Hong Yong JW (2024)Iron biofortification in cereal crops: Recent progress and prospects. Food Energy Secur 13(4):1–28. https://doi.org/10.1002/fes3.547

Chapter 18
Actinobacteria as Potential Biopesticides

Sangeetha Murthy and Balagurunathan Ramasamy

Abstract The backbone of many nations with agrarian economies worldwide is agriculture. There are many difficulties in feeding the world's growing population today. The attack of insect pests is a significant concern, among other factors, and contributes to the low production of agronomic crops. Currently, agronomic crop pests and diseases are controlled with a variety of chemical methods, including fungicides and insecticides. However, concerns about agronomic sustainability have led to the broad and extensive use of integrated techniques for disease and pest management. Biopesticides are essential for environmentally sound practices and sustainable agriculture. They provide a natural alternative to synthetic pesticides for pest management by targeting pests by living organisms such as fungi, bacteria, or plant extracts. Actinobacteria are among the most potent tools in sustainable agriculture, offering a natural and efficient substitute for synthetic/chemical pesticides while promoting plant resilience and soil health. This study revealed that the use of actinobacteria was the most effective way to control agronomic pests and increase the yield of agronomic crops.

Keywords Biopesticide · Agriculture · Actinobacteria · Microbial pesticide · *Streptomyces* · Pest

S. Murthy (✉)
Kriya Biosys Pvt. Ltd., Salem, Tamil Nadu, India

B. Ramasamy
Vivekanandha Educational Institutions for Women, Tiruchengode, Tamil Nadu, India

V. Gopikrishnan et al. (eds.), *Harnessing Microbial Resources for Sustainable Agriculture*, Microorganisms for Sustainability 61,
https://doi.org/10.1007/978-981-95-4817-0_18

18.1 Introduction

The deliberate effort to change a section of the Earth's crust by cultivating plants and other crops and breeding livestock for both economic and nutritional reasons is referred to as "agriculture," which is more frequently used to mean "farming." Agriculture not only has a significant economic influence but is also thought to be the cornerstone of economic systems in emerging countries. The development of vital food crops and agriculture has always been intertwined. Modern farming methods include forestry, poultry, apiculture, dairy, and conventional farming methods. It may also be referred to as marketing, distribution, processing, and development of livestock and agricultural products. Additionally, it offers a significant portion of the population job opportunities. To meet the global development goals, there must be inclusive, sustainable, and healthy food systems. Agricultural development is one of the most efficient ways to alleviate severe poverty, foster shared prosperity, and supply food for the 10 billion people who are predicted to inhabit this globe by 2050. Growth in the agriculture industry increases the incomes of the poor two to four times more efficiently than growth in other industries does. Furthermore, with agriculture accounting for 4% of the global GDP and up to 25% in some of the least developed countries, agriculture is crucial to economic growth. However, food security, poverty reduction, and agriculturally driven growth are all at risk: numerous shocks, such as pests, conflicts, severe weather, and human-caused disturbances, affect food systems. We are currently falling short of the 2030 deadline to eradicate world hunger. The production of food worldwide is severely hampered by pests. As a result, 20–40% of the global crop yield is lost each year. The Food and Agriculture Organization of the United Nations reported that invasive insects cost the global economy more than $70 billion a year, whereas plant diseases cost it more than $220 billion (Pardey et al. 2016; Parnell et al. 2017; Schneider et al. 2018; Sharma et al. 2019; Carney and Krause 2020; Tudi et al. 2021; FAO 2023).

Pests are thought to be growing and spreading quickly in agricultural and forestry ecosystems due in large part to climate change. Furthermore, finding an example does not require looking very far. Large swarms of migrating pests called desert locusts caused substantial crop damage and panic among farmers in India just a few years ago. This insect has caused many invasions in the western and northern regions of the country (Meena et al. 2021). There are two main categories of damage that insects can inflict. The first is direct harm to plants of the types described above, such as eating leaves or causing damage to roots. In the second type, an insect may transfer bacterial, fungal, or viral infection into a crop instead of causing direct harm. Owing to their rapid population growth, insects present many hazards to both production and ecology as a whole, which is one of the main problems associated with them. Insect pests that negatively impact agriculture will eventually adapt to shifting environmental conditions. For this reason, it is important to monitor them closely, check plants frequently for signs of infestation, and take targeted action to eliminate them. This is the point at which pesticides are mentioned. They enable farmers to produce food of excellent quality at competitive prices. Growing wholesome food all year round can also be facilitated by a farmer using advised pesticides in a safe and appropriate manner. Rather than more land usage, higher productivity will be the primary driver of the growth in global food production. Consequently, it is imperative to make investments in increased yields and better farm management (Lichtenberg 2013; Anand and Rajeshkumar 2022; Aida 2023). The ability of pesticides to prevent some dangerous pest species from growing and harming agricultural productivity is one of their most significant advantages. Similar to previous examples, pesticides can help plants flourish by preventing plant diseases and weeds, which increases the output of agricultural products when they are applied properly and effectively. Although the application of chemical pesticides and fertilizers is limited due to certain issues and hazards, it greatly enhances crop development (Pathak et al. 2022). Using biological inputs based on microorganisms is a solution that combines sustainability and efficiency with an increase in productivity.

18.2 Agronomic Crops

Agronomic crops are staples that are cultivated mainly for fuel, fiber, feed, and food. They are essential to agricultural economies and worldwide food security because of their extensive use and high nutritional content. Large-scale cultivation is the norm for these crops, and to maximize yield and quality, particular management techniques are frequently needed. Agronomic crops are frequently grown via intensive agricultural technologies such as fertilizers, pesticides, and irrigation to maximize production and revenue. The following examples involve agronomic crops. Cereal crops are rich in carbohydrates and form the basis of many food habits worldwide. Mostly consumed as staple foods such as rice, wheat, maize, barley, oats, rye, and sorghum, legume crops are essential in both human and animal diets. Through nitrogen fixation, they are essential for increasing soil fertility. Examples include chickpeas, beans, peanuts, lentils, soybeans, etc. Oilseed crops are used to produce edible oils, biodiesel, and other food products. Examples are canola, sunflower, safflower, and flaxseed; fiber crops, such as cotton, hemp, flax, and jute, are grown for their fibers, which are used in textiles, ropes, and industrial materials; sugar crops, such as sugarcane and sugar beet, are cultivated for sugar production, which is used in food; forage crops (alfalfa, clover, and grasses) are grown primarily as feed for livestock; and biofuel crops are processed into biofuels, which are renewable energy sources. Examples include corn, soybean, and switchgrass (FAO 2023). The need for agronomic crops for consumption/utilization, such as food, feed, fuel, and fiber, is increasing globally; estimates indicate that, in the face of natural resource shortages, pollution, ecological degradation, and climate change, the globe will require 50% more food by 2050 to sustain its growing population. To do this, we must increase productivity, promote healthy eating, decrease crop and food loss, conserve natural resources, and reduce the use of agricultural chemicals (Khan et al. 2019; Nazarov et al. 2020; Peng et al. 2021; Balestrini et al. 2024).

18.3 Agronomic Pest

Any living thing that adversely impacts crop productivity, livestock, or stored agricultural products is referred to as an agricultural/agronomic pest. Typically, these organisms include insects, weeds, microorganisms, and rodents. These pests have the potential to cause serious harm, which could result in lower yields, inferior production, or even total crop collapse. Additionally, they can spread disease to cattle and crops. Owing to their ability to reduce production, agronomic pests result in financial losses for agriculture. To manage insect pest populations and reduce economic damage and negative environmental side effects, an integrated pest management (IPM) program combines control strategies such as variety selection, biological control, cultural practices, and pesticides. IPM minimizes insect damage and ensures sustainable crop production by combining biological, cultural, chemical, and mechanical control techniques and is frequently used to manage agronomic pests. Identification and elimination of this risk are key components of pest control, which is necessary to reduce these losses (Agrios 2005; Ahmed et al. 2015; Bodah 2017; Doehlemann et al. 2017; Khan et al. 2019; Williamson-Benavides and Dhingra 2021; Bahadur 2022). Some common agronomic pests, such as bacteria, fungi, insects, nematodes, and viruses, are listed in Table 18.1.

Table 18.1 List of agronomic pests

Pest category	Name of the pathogen
Bacteria	*Erwinia chrysanthemi*, *Xanthomonas campestris pv. campestris*, *Pseudomonas syringae*, *P. corrugata*, *Pectobacterium carotovorum* subsp. *carotovorum*, *Ralstonia solanacearum*
Fungi	*Alternaria alternata*, *Botrytis cinerea*, *Colletotrichum gloeosporioides*, *Colletotrichum musae*, *Corynespora cassiicola*, *Fusarium solani*, *Fusarium verticillioides*, *Fusarium oxysporum*, *Fusarium moniliforme*, *Macrophomina phaseolina*, *Penicillium digitatum*, *Phytophthora cinnamomi*, *Pythium aphanidermatum*, *Rhizoctonia solani*, *Sclerotinia sclerotiorum*
Insect	*Blattodea*, *Coleoptera*, *Diptera*, *Hymenoptera*, *Isoptera*, *Lepidoptera*
Nematode	*Meloidogyne* spp., *Heterodera* spp., *Globodera* spp., *Bursaphelenchus xylophilus*, *M. incognita*, *M. javanica*
Virus	*Bean common mosaic virus* (*BCMV*), *Citrus tristeza virus* (*CTV*), *Cowpea aphid-borne mosaic virus* (*CABMV*), *Peanut bud necrosis virus* (*PBNV*), *Maize streak virus* (*MSV*), *Tobacco mosaic virus* (*TMV*)

18.4 Pesticide Era

A pesticide is a chemical/biological agent or combination of chemicals used to eliminate or control pests, prevent disease, and damage people, animals, and agricultural plants. Nonpersistent pesticides, which are thought to be safe for use in food, readily rupture when exposed to light. Pesticide safety concerns are of commercial importance and require constant assessment. With fewer safety concerns related to chemical pesticides, identifying natural sources of pesticides is now recommended. The chemical or physical characteristics of pesticides make them different from one another. Sulfur was the first pesticide used by Sumerians to control mites and insects at 2500 BCE; Chinese introduced ants in citrus trees to manage pests at 307 CE; the first applications of "para-pesticides," especially arsenic and mercury, appeared in the 1500s; and the Bordeaux mixture (copper sulfate and lime) was discovered by France in the 1880s. It is used in vineyards to protect grapes from downy mildew. As one of the earliest chemical fungicides in modern agriculture, the Bordeaux mixture was widely used as a result of this discovery. Paul Muller's 1939 discovery of dichlorodiphenyl-trichloroethane, the first modern pesticide, is a significant moment in the history of pesticides. As a result of this incident, the chemical industry started selling broad-spectrum pesticides, which were originally far more effective than previously available compounds and poisonous to the entire groups of living things. For decades, the use of pesticides has been steadily increasing worldwide. Pesticide intensities increased at varying rates between 1990 and 2022: usage per person increased by 35%, use per farmland area increased by 94%, and use per value of agricultural production increased by 5%. The health effects of pesticide use, particularly for children and elderly individuals, continue to concern the public today. On the other hand, the worldwide production of pesticides is increasing at the same rate as the population, which has grown exponentially in recent decades. Chemical pesticides are thought to be more harmful than biopesticides (Drum 1980; Buchel 1983; Zacharia 2011; Fishel and Ferrell 2013; Dara 2019; Beglaryan et al. 2022; Rehman et al. 2024). If the existing integrated pest management strategies are not effective, biopesticides may be a viable alternative.

18.4.1 Classification of Pesticides

Pesticides have several varieties on the basis of their physical and chemical properties. Pesticides are generally categorized in several ways, such as on the basis of their origin (organic sources, inorganic sources, and biological sources), the type of pest species they target (e.g., bactericide, herbicide, insecticide, larvicide, etc.), and their mode of action (e.g., feeding deterrent, repellent, fumigant, attractant, synergist, etc.) (Drum 1980; Zacharia 2011; Abubakar et al. 2020; Venkatasaichandrakanth and Iyapparaja 2023).

18.4.2 Biopesticide Era

There are several ways that biologicals are utilized to manage weeds, diseases, and pests. A disease or parasite that infects a target is one example of microbial biocontrol. On the other hand, they can compete with plants or cause resistance in their hosts. There are numerous ways that biochemical biocontrol can work. Some work by preventing a pest or disease from growing, feeding, developing, or reproducing. To prevent infection or feeding, additional biocontrols might be employed to create a barrier on the host (Swapan et al. 2024).

Plant extracts were likely the first agricultural biocontrols since nicotine (plant extracts) was used to reduce plum bugs as early as the seventeenth century. Since 1835, when Agostino Bassi demonstrated that the white-muscardine fungus (*Beauveria bassiana*) was potentially deployed to infect silkworms with an infectious disease, biological controls for agricultural insect pests have been in use. Other instances of experimentation with mineral oils as plant defenders date back to the nineteenth century. As agricultural research rapidly expanded institutionally in the early twentieth century, an increasing number of studies and proposals for biocontrol were produced (Bashan et al. 2016; Kumar et al. 2021a, b; Ayilara et al. 2023).

18.4.3 Microbial Pesticides

Microbial pesticides are a type of biopesticide that reduce/control pest populations by using microbes or their byproducts. This type of biocontrol agent is widely applied in agriculture/forestry to control pests and provides an eco-friendly substitute for chemical pesticides. Microbial pesticides are composed of microorganisms such as bacteria, fungi, viruses, or protozoa that specifically target pest species. *Bacillus thuringiensis* (Bt) spores were among the earliest and are now the most commonly employed biocontrols/biopesticides (Singh et al. 2021). The Japanese researcher Shigetane Ishiwata isolated Bt from a sick silkworm in 1901. Ten years later, Ernst Berliner in Thuringia, Germany, found it again in a flour moth larva that was unwell. The *Bacillus thuringiensis* pathogen, which was designated the type species in 1911, is still the most commonly employed biocontrol agent today (Singh et al. 2021). Bt was first used as a biological pesticide by France in the early 1920s. The first commercially marketed Bt product, sporeine, debuted in France in 1938. Numerous studies on the efficacy of Bt were published in the 1950s, which paved the way for the United States to widely adopt biocontrol. Submerged fermentation, which is an industrial technique, was developed by the commercial/private sector to manufacture Bt for large-scale production for commercial purposes in 1956. *Heliothis nuclear polyhedrosis virus* (*Heliothis* NPV) was exempted from tolerance in 1973. *Bacillus thuringiensis var. israelensis*, which is poisonous to flies, was identified in 1977; *Tenebrio*, which is toxic to beetles, was identified in 1983 (Chandler et al. 2011; Gouli et al. 2020; Varsha and Shahida 2020). The codling moth, a significant pest of tree fruits, including apples and pears, is naturally infected with *Cydia pomonella granulovirus*. It was used for commercial purposes in the 1980s and over the past decade, and its popularity has increased. Moreover, *Trichoderma* spp. were developed in the 1990s and have been widely used for commercial purposes in recent years. Furthermore, the success stories of commercial fields include *Agrobacterium radiobacter* (to protect woody crops against crown gall disease), *Pseudomonas fluorescens* (to protect orchards from fire blight disease), *Coniothyrium minitans* (to combat plant disease caused by *Sclerotinia sclerotiorum*), and a variety of microorganisms used against soil-borne diseases of agronomic crops. Research on the development of biocontrols has revived in both academia and industry. Additionally, adoption rates have risen quickly in tandem with the explosive growth of organic agriculture over the last 10 years. As a result, since the mid-1990s, the development of novel and practical biocontrols has continued to grow rapidly. On the basis of this evolution, the global market for biocontrol agents (BCIs) is expected to grow from its 2023 valuation of approximately USD 5.23 billion to USD 20.4 billion by 2032 (Bernard et al. 2017; Kumari et al. 2022; Kiran and Kunal 2024).

18.5 Actinobacteria: A Potential Source

Indeed, actinomycetes/actinobacteria are noteworthy for their bioactive qualities, and their potential for use in agriculture is becoming more widely acknowledged. These bacteria are a diverse group of microorganisms that include well-known genera such as *Streptomyces*, *Micromonospora*, and *Nocardia*. Actinobacteria are treasured for biological control and other agricultural uses since they are known to produce a variety of bioactive compounds, such as antibiotics, antifungal agents, and molecules that promote plant development. Actinobacteria are among the most numerous and varied types of bacteria that have been identified. These bacteria are mostly gram-positive and are found in freshwater, marine, and terrestrial habitats. They are members of the phylum *Actinobacteria* and the order *Actinomycetales*. The genome of actinobacteria is generally enriched in terms of the G+C (guanine and cytosine) content, which often exceeds 70% in many species. The high G+C content of actinobacteria underpins their versatility in producing bioactive compounds that support plant health, increase soil fertility, and reduce the reliance on chemical inputs in agriculture, aligning with the principles of sustainable and organic farming. Because of their strong and intricate secondary metabolism, *Streptomyces* species in particular have been regarded as reservoirs for a vast array of natural compounds. Seventy to eighty percentage of all naturally occurring bioactive compounds employed in pharmaceutical or agrochemical applications are the approximately 100,000 antibiotic compounds that *Streptomyces* produces (Ludwig et al. 2012; Palaniyandi et al. 2013; Viaene et al. 2016).

18.5.1 Actinobacteria: As Biopesticides

Pests are undesirable organisms that infest crops and reduce the balance between plant health and productivity, which has a continuous impact on agriculture. Typically, agricultural pest management involves the use of pesticides, compounds, or combinations of compounds whose element structures interact with undesirable organisms to reduce their impacts and cause damage. To increase agricultural output in a more sustainable and safe mode, it is crucial to search for alternatives that reduce the use of chemical pesticides. In this context, biologically controlled microorganisms, such as bacteria, nematodes, insects, and phytopathogenic fungi, have a great promise for controlling agricultural pests. Among the biopesticides, actinobacteria are potential sources because of their remarkable production of bioactive metabolites with unique properties, such as biocontrol potential, broad-spectrum activity, plant growth promotion, soil health improvement, and sustainability and eco-friendliness (Villaverde et al. 2016; Ebrahimi-Zarandi et al. 2022; Wahab et al. 2022). Some examples of potential actinobacterial strains that act as biopesticides are listed in a graphical illustration (Fig. 18.1).

Actinobacteria, particularly those in the genus *Streptomyces*, produce significant quantities of bactericidal secondary metabolites. Some examples of *Actinobacteria*-derived bactericides are briefly described here. *Streptomyces griseus* shows potential activity against bacterial pathogens such as *Xanthomonas* spp. (which causes bacterial leaf spot and blight) and *Erwinia amylovora* (which causes fire blight in apples and pears) (Le et al. 2022). These bacteria produce streptomycin, an antibacterial substance that binds to the 30S ribosomal subunit and prevents bacteria from synthesizing proteins. *Streptomyces kasugaensis* protects rice crops from *Xanthomonas oryzae pv. oryzae* (responsible for rice blight disease) and *Pseudomonas syringae* (causes bacterial specks in tomatoes). This actinobacterial strain produces the bioactive compound kasugamycin, which inhibits translation initiation in bacteria (Noura El-Ahmady El-Naggar 2021). Numerous species of *Streptomyces* spp. produce actinomycin, which was first identified from *Streptomyces*

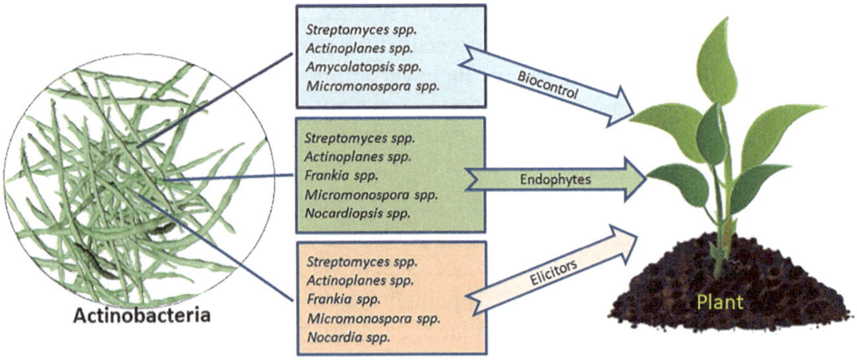

Fig. 18.1 Some potential actinobacterial strains as biopesticides

antibioticus. In the context of a wide range of bacterial and fungal infections, actinomycin has broad-spectrum effects. This type of molecule affects bacterial RNA synthesis by binding to DNA. Some types of actinomycin are used in crop protection as fungicides/bactericides. However, its use is limited by the potential toxicity of actinomycin D (Kulkarni et al. 2017). Furthermore, *Streptomyces* spp. produce polyoxins and validamycins. They act as dual-action pesticides against bacterial and fungal pathogens. These biomolecules inhibit chitin synthesis and other cellular processes. An analog of phosphoenolpyruvate, fosfomycin is a classical antibiotic agent that was discovered in 1969 and is produced by *Streptomyces* species, namely, *Streptomyces fradiae*. It has shown activity against the bacterial pathogens *Clavibacter michiganensis* (bacterial canker in tomatoes) and *Xanthomonas campestris* (black rot in crucifers) by inhibiting cell wall synthesis by blocking the enzyme MurA. It is used for soil treatment and foliar spray. *Streptomyces natalensis* is used in seed coatings and postharvest treatments. They primarily act as fungicides but are effective against certain pathogens. The biomolecule natamycin, which binds to ergosterol, disrupts membrane integrity in pathogens. *Streptomyces hygroscopicus* produces geldanamycin, which inhibits Hsp90, a heat shock protein critical for bacterial pathogen survival under stress. It is primarily used in agriculture and has a potential for disease resistance management against various bacterial pathogens. *Streptomyces griseus* disrupts bacterial proton pumps, inhibiting cellular energy production via the synthesis of bafilomycins. It has a potential against soil-borne pathogens (root disease) and is used as a soil amendment in agricultural fields. Virginiamycin inhibits protein synthesis by targeting the bacterial ribosome. This bioactive compound was isolated from *Streptomyces virginiae*. It is used in crop protection from *Erwinia* spp. (which causes bacterial wilt and soft rot). The unique mode of action of gramicidin S creates ion channels in pathogens, particularly bacterial membranes, leading to leakage and cell death. Therefore, this biomolecule producer, *Streptomyces gramicidinicus*, is applied as a biocontrol tool to protect against leaf- and soil-borne bacterial pathogens in integrated pest management systems. *Streptomyces platensis* protects plants from emerging threats caused by bacterial biofilms via the production of a potential biomolecule named pladienolide; it disrupts RNA splicing in bacterial cells. Owing to its potential application, the scientific community is working on the exploration of this actinobacterial strain for agricultural use. In addition, *Streptomyces roseosporus* has also been evaluated for its broad-spectrum antibacterial properties in agriculture. Its potential compound is named daptomycin, which is effective against gram-positive bacterial pathogens; it causes depolarization of bacterial membranes. *Streptomyces azureus* uses foliar sprays and root drenches to protect agricultural crops from *Xanthomonas* and *Pseudomonas* spp. This strain produces thiostrepton, which kills bacterial cells via the inhibition of protein synthesis. Rifamycin is used in soil treatments for systemic bacterial infections. The potential source of bioactive compounds is *Amycolatopsis rifamycinica*, which inhibits bacterial RNA polymerase of the bacterial strain *Ralstonia solanacearum* (which causes bacterial wilt) (Shimizu 2011; Villaverde et al. 2016; Shrivastava and Kumar 2018; Sharma and Salwan 2018; Köhl et al. 2019; Ebrahimi-Zarandi et al. 2022; Wahab et al. 2022).

When discussing actinobacteria-derived fungicide compounds, it is necessary to consider the compound polyoxin, which is a significant fungicide. It inhibits chitin synthase, leading to destruction of fungal cell wall synthesis. Polyoxin D is widely used to protect vegetables and rice crops from diseases such as powdery mildew (*Erysiphe* spp.) and rice sheath blight (*Rhizoctonia solani*). It is produced by *Streptomyces cacaoi* (Ebrahimi-Zarandi et al. 2022). Another agent used for rice crop protection is blasticidin-S, which is produced by *Streptomyces griseochromogenes*. It protects rice farms from the fungus rice blast (*Magnaporthe oryzae*). It inhibits cell protein synthesis by targeting ribosomal function (Li et al. 2013). Fungichromin has wide-spectrum antifungal activity against *Fusarium* spp. and *Phytophthora* spp. It interrupts fungal cell membranes. Hence, this compound's producer, *Streptomyces pandanus*, is effectively used in vegetable crops and horticultural plants. As biocontrol agents, nonpathogenic actinobacteria such as *Streptomyces lydicus* are marketed. Through the production of antifungal metabolites, they inhibit fungal infections in the rhizosphere and encourage plant growth. Lipopeptides, which are derived from organisms such as *Streptomyces scabies*, are substances that break down fungal biofilms and are being researched as environmentally friendly methods for controlling soil-borne infections. For integrated pest management, actinomycin—a bioactive peptide with potent antifungal activity—is being researched. The bioactive compounds of the commercial strain *Streptomyces lydicus* include fungicides and are used against *Fusarium*, *Phytophthora*, and other soil-borne pathogens. Another commercial strain, *Streptomyces griseoviridis*, is effective against damping-off diseases produced by *Pythium* and *Phytophthora* (Singh et al. 2018; Köhl et al. 2019; Subramaniam and Arumugam 2020; Ebrahimi-Zarandi et al. 2022).

Many actinobacteria-derived compounds target pests without harming beneficial insects. They are harmless to humans and animals and are biodegradable in the atmosphere. Hence, it is considered environmentally benign. Because of the various mechanisms of action, the chance of resistance is also decreased (Senthil-Nathan 2015). For example, spinosad is a well-known commercial product of organic farming. From *Saccharopolyspora spinosa*, it is extracted. It is applied as a foliar spray on ornamentals, fruits, and vegetable crops. Beetles, caterpillars, fruit flies, leaf miners, and thrips (*Ceratitis capitata*, *Cydia pomonella*, *Musca domestica*, *Plutella xylostella*, *Rhagoletis cerasi*, *Trichoplusia ni*, *Tuta absoluta*, etc.) are among the insects it effectively combats. This strain's biomolecule targets nicotinic acetylcholine receptors to interfere with neural function and γ-aminobutyric acid channels, leading to paralysis and death (Bridget et al. 2023). *Streptomyces avermitilis* targets *Tetranychus urticae*, *Helicoverpa armigera*, *Plutella xylostella*, *Trichoplusia ni*, *Myzus persicae*, *Bemisia tabaci*, *Panonychus citri*, *Scirtothrips citri*, *Helicoverpa zea*, *Liriomyza trifolii*, and *Liriomyza huidobrensis* (mites, aphids, and other sap-feeding insects) through the secretion of bioactive compounds called avermectins (El-Saber Batiha et al. 2020). It disrupts nerve and muscle function by increasing chloride ion influx. This commercial product is used in sprays or soil drenches. Milbemycins, which are produced by *Streptomyces hygroscopicus*, paralyze and ultimately kill target pests (*Aculus fockeui*, *Aculops lycopersici*, *Tetranychus urticae*, *Panonychus citri*, *Panonychus ulmi*, *Meloidogyne* spp., *Scirtothrips citri*, etc.) by binding to glutamate-gated chloride channels. They share a mode of action similar to avermectins but are chemically distinct and have longer residual activity. It is applied as a foliar treatment or seed coating. Piericidins are also produced by *Streptomyces* spp. These compounds exhibit specific activity against larvae and soil-dwelling pests such as fruit flies (*Drosophila melanogaster*), mosquito larvae (*Aedes aegypti*), beetle larvae (*Tenebrio molitor*), aphids (*Myzus persicae*), etc. Piericidins interfere with adenosine triphosphate synthesis by targeting the nicotinamide adenine dinucleotide + hydrogen oxidoreductase complex (Complex I) in the mitochondrial respiratory chain. Pests die as a result of this energy deprivation. Piericidins are being researched for wider insecticidal uses, including their ability to combat disease vectors and agricultural pests. Nevertheless, in contrast to other pesticides generated from actinobacteria, such as spinosad and avermectins, their practical application in agriculture is limited (Azad et al. 2022). Other biomolecules, such as actinomycins, lactones, and polyketides, are also under investigation for use against resistant pest populations. The remarkable production of compounds with insecticidal activities by actinobacteria makes them promising candidates. Avermectins, polynactins, spinosyns, tetramycins, and their analogs are among the primary classes of natural insecticides that actinobacteria produce: polyketide compounds (Shrivastava and Kumar 2018; Singh et al. 2018; Köhl et al. 2019; Ebrahimi-Zarandi et al. 2022).

Abamectin is the most widely used commercial nematicide. In vegetables, fruits, and ornamental crops, it controls burrowing nematodes (*Radopholus similis*), cyst nematodes (*Heterodera* spp., *Globodera* spp.), lesion nematodes (*Pratylenchus* spp.), and root-knot nematodes (*Meloidogyne* spp.). Many crops have root lesions caused by the same nematodes that produce root galls and undeveloped growth. To promote soil health, another commercial product that contains *Streptomyces* strains (such as *Streptomyces lydicus*) is offered for IPM systems (Pulavarty et al. 2021; Mendoza-de Gives 2022).

Actinobacteria-derived antiviral secondary metabolites are emerging as environmentally friendly ways to treat viral infections in plants. These substances support plant defense systems and inhibit virus multiplication. Among these examples are *Streptomyces* species, which create ribosome-manufactured peptides to prevent the formation of viral RNA. Polyketides and macrolides, which are secondary metabolites, have demonstrated antiviral properties against plant viruses such as cucumber mosaic virus and tomato yellow leaf curl virus. The chitinase enzymes of *Streptomyces* spp. reduce viral infection by causing systemic resistance in plants. Actinobacteria-derived antiviral compounds are still being studied for use in agriculture. Nonetheless, studies on biobased treatments such as the chitinase enzymes of *Streptomyces* spp. and antiviral peptides are progressing and may be used to treat viral illnesses. These remedies, which are frequently a component of IPM, may eventually be created for commercial usage. However, new products on the market, such as biological control agents and biopesticides, are designed to reduce the effects of plant viruses indirectly by enhancing plant health. Many of them are in the experimental stage or niche market and involve a combination of actinobacteria-based chemicals, such as chitinases and antiviral peptides (Veliz et al. 2017; Shrivastava and Kumar 2018; Köhl et al. 2019; Ebrahimi-Zarandi et al. 2022).

18.5.2 Actinobacteria: Application Methods

Actinobacteria are frequently employed as biopesticides because of their capacity to generate antimicrobial chemicals and strengthen plant defense responses, especially those belonging to the genera *Streptomyces*, *Micromonospora*, and *Nocardiopsis*. The product composition can influence the application techniques for biopesticides. Formulations are generally made to reduce the extra expenses associated with the use of chemical pesticides. Thus, using spraying methods, biopesticides are normally used on crops to protect the upper parts of the plants and shield the soil. The application methods, such as soil application, seed coating, foliar sprays, root drench/irrigation, aerosol/fog application, and integrated application, support sustainable, targeted disease management. *Streptomyces* spp. and *Micromonospora* spp. are used as soil drenches to colonize the rhizosphere of plants and provide protection against soil-borne pathogens such as *Fusarium*, *Pythium*, and *Rhizoctonia*, which are responsible for fungal root rot diseases. *Streptomyces* spp. and *Nocardiopsis* spp. are used in seed coating applications to prevent bacterial (e.g., *Xanthomonas*) and fungal (e.g., *Fusarium*) seed-borne diseases. Actinobacterial biopesticides such as *Streptomyces lydicus* are applied directly to aerial parts of plants to protect against fungal pathogens (e.g., *Alternaria* and *Botrytis*). For the root drench/irrigation method, biocontrol agents (*Streptomyces griseoviridis* and others) are applied with water to the root system of a plant to protect against root rot disease (e.g., *Phytophthora*). *Streptomyces* spp. and/or *Micromonospora* spp. are mixed with compost/biofertilizer that is applied to soil to enhance the soil microbiota and inhibit the growth of soil pathogens. *Streptomyces* spp. can be applied as aerosols or fog to prevent airborne disease (Shrivastava and Kumar 2018; Köhl et al. 2019; Silva et al. 2022).

18.5.3 Actinobacteria: Challenges as Biopesticides

Since the majority of biopesticides are made up of living products, their successful application depends upon both abiotic and biotic conditions as well as production technology. Abiotic factors: pH, salinity, light incidence, moisture, temperature, and organic and inorganic soil elements can drastically alter the bioactivity of actinobacteria, which act as biopesticides. This type of abiotic stress can disrupt the rhizosphere and its components, modifying the bioactivity of actinobacteria in relation to defense and plant growth-promoting activities. Certain species are able to withstand these abiotic stresses by generating piericidin-regulating biomolecules that inhibit cell membrane plasmolysis via controlled enzymatic activity and homeostasis (Azad et al. 2022). Under high osmolality conditions, several strains of *Streptomyces* and *Nocardiopsis* can synthesize major organic osmolytes, such as glycine betaine, proline, and trehalose, which cause osmotic adaptation and maintain biopotency even under stressful conditions (Ameur et al. 2011).

Biotic factors: A vast variety of biodiversity is found in soil. Interactions with other species in their surroundings can have an impact on the growth, survival, and functionality of biopesticides. When biopesticides are applied, biotic factors are crucial to consider since they dictate how successfully the biopesticide will communicate with the surrounding environment to control pests/pathogens of plants. For example, *Streptomyces* spp. and other actinobacteria frequently develop symbiotic interactions with plant roots. By generating antibiotics that inhibit harmful microorganisms in the soil, these actinobacteria can benefit plants by promoting plant development. For example, the colonization of *Streptomyces lydicus* protects plants from root diseases such as *Fusarium* spp. The relationship of actinobacteria with the host plant and its interactions with their environmental microbiota may be investigated for maximum product efficiency. The active component of the biopesticide may be required to colonize plant tissues to control leaf diseases. Furthermore, the rhizosphere colonization ability of biopesticides is essential. The ability of endophytic actinobacteria, which generate lytic enzymes and other compounds that increase their ability to colonize plants, can be achieved. The success of a biopesticide depends on the species and its activity; another aspect influencing microbial colonization is the application technique used. The crop must have access to live cells for the biopesticide to be effective; there must be sufficient cells to support plant protection and nutrition. Therefore, the microbial concentration has a direct effect on biopesticide effectiveness. Since actinobacterial cells reproduce more slowly than other bacteria do, one of the greatest difficulties in producing actinobacteria-based products is successful production on a large scale at the industrial level. The formulations include additional inert chemicals in addition to the active microbe to aid in field application and product shelf-life. To maintain the strain's bioactivity, these substances must interact. The nutritional needs of microbial cells and their vulnerability to abiotic influences may cause issues when developing formulations. Owing to these issues, the commercialization of actinobacterial biopesticides is less common than their bioactivity is (Sharma and Salwan 2018; Shrivastava and Kumar 2018; Köhl et al. 2019; Silva et al. 2022).

18.6 Future Perspectives and Conclusions

Actinobacteria-derived biopesticides have a great potential as environmentally acceptable substitutes for chemical pesticides, providing ways to manage a variety of pests and even certain viral infections. Compounds such as spinosad, abamectin, and milbemycin are gaining traction in agriculture; they are prized for their specificity, low environmental impact, and capacity to inhibit the development of pest resistance. The antimicrobial and nematicidal qualities of actinobacteria are still being studied, which indicates that these biopesticides will be developed further and support long-term sustainable pest control methods.

References

Abubakar Y, Tijjani H, Egbuna C et al (2020) Pesticides, history, and classification. In: Egbuna C, Sawicka B (eds) Natural remedies for pest, disease and weed control. Academic, pp 29–42

Agrios GN (2005) Plant pathology, 5th edn. Elsevier, Academic

Ahmed M, Sapp M, Prior T, Karssen G, Back M (2015) Nematode taxonomy: from morphology to metabarcoding. SOIL Discuss 2:1175–1220. https://doi.org/10.5194/soild-2-1175-2015

Aida T (2023) Integrating agricultural and industrial development. In: Estudillo JP, Kijima Y, Sonobe T (eds) Agricultural development in Asia and Africa, Emerging economy state and international policy studies. Springer, Singapore. https://doi.org/10.1007/978-981-19-5542-6_22

Ameur H, Ghoul M, Selvin J (2011) The osmoprotective effect of some organic solutes on Streptomyces sp. MADO2 and Nocardiopsis sp MADO3 growth. Braz J Microbiol 42(2):543–553. https://doi.org/10.1590/S1517-83822011000200019

Anand G, Rajeshkumar KC (2022) Challenges and threats posed by plant pathogenic fungi on agricultural productivity and economy. In: Rajpal VR, Singh I, Navi SS (eds) Fungal diversity, ecology and control management, Fungal biology. Springer, Singapore. https://doi.org/10.1007/978-981-16-8877-5_23

Ayilara MS, Adeleke BS, Akinola SA, Fayose CA, Adeyemi UT, Gbadegesin LA et al (2023) Biopesticides as a promising alternative to synthetic pesticides: a case for microbial pesticides, phytopesticides, and nanobiopesticides. Front Microbiol 14:1040901. https://doi.org/10.3389/fmicb.2023.1040901

Azad SM, Jin Y, Ser H-L, Goh B-H, Lee L-H, Thawai C, He Y-W (2022) Biological insights into the piericidin family of microbial metabolites. J Appl Microbiol 132(2):772–784. https://doi.org/10.1111/jam.15222

Bahadur A (2022) Nematodes diseases of fruits and vegetables crops in India. IntechOpen. https://doi.org/10.5772/intechopen.98850

Balestrini R, Bianciotto V, Ghignone S et al (2024) Chapter 11. In: Paul EA, Frey SD (eds) Plant–soil biota interactions, Soil microbiology, ecology and biochemistry, 5th edn. Elsevier, pp 303–328. https://doi.org/10.1016/B978-0-12-822941-5.00011-9

Bashan Y, de Bashan LE, Prabhu SR (2016) Agriculturally important microorganisms- commercialization and regulatory requirements in Asia. Springer, Singapore

Beglaryan M, Pipoyan D, Tepanosyan G, Sahakyan L (2022) Toxic element contents and associated multimedium health risk assessment in an area under continuous agricultural use. Environ Monit Assess 194(3):184. https://doi.org/10.1007/s10661-022-09853-5

Bernard GC, Egnin M, Bonsi C (2017) The impact of plant-parasitic nematodes on agriculture and methods of control. InTech. https://doi.org/10.5772/intechopen.68958

Bodah ET (2017) Root rot diseases in plants: a review of common causal agents and management strategies. Agric Res Technol Open Access J 5:555661

Bridget AF, Nguyen CT, Magar RT, Sohng JK (2023) Increasing production of spinosad in Saccharopolyspora spinosa by metabolic engineering. Biotechnol Appl Biochem 70(3):1035–1043. https://doi.org/10.1002/bab.2418

Buchel KH (1983) Chemistry of pesticides. Wiley, New York

Carney MA, Krause KC (2020) Immigration/migration and healthy publics: the threat of food insecurity. Palgrave Commun 6(1):93. https://doi.org/10.1057/s41599-020-0461-0

Chandler D, Bailey AS, Mark TG, Davidson G, Greaves J, Grant WP (2011) The development, regulation and use of biopesticides for integrated pest management. Philos Trans R Soc B 366:1987–1998. https://doi.org/10.1098/rstb.2010.0390

Dara SK (2019) The new integrated pest management paradigm for the modern age. J Integr Pest Manag 10(1):12. https://doi.org/10.1093/jipm/pmz010

Doehlemann G, Ökmen B, Zhu W, Sharon A (2017) Plant pathogenic fungi. Microbiol Spectr 5(1). https://doi.org/10.1128/microbiolspec.FUNK-0023-2016

Drum C (1980) Soil chemistry of pesticides. PPG Industries, USA

Ebrahimi-Zarandi M, Saberi Riseh R, Tarkka MT (2022) Actinobacteria as effective biocontrol agents against plant pathogens, an overview on their role in eliciting plant defense. Microorganisms 10(9):1739. https://doi.org/10.3390/microorganisms10091739

El-Saber Batiha G, Alqahtani A, Ilesanmi OB, Saati AA, El-Mleeh A, Hetta HF, Magdy Beshbishy A (2020) Avermectin derivatives, pharmacokinetics, therapeutic and toxic dosages, mechanism of action, and their biological effects. Pharmaceuticals (Basel) 13(8):196. https://doi.org/10.3390/ph13080196

FAO, IFAD, UNICEF, WFP and WHO (2023) The State of Food Security and Nutrition in the World 2023. Urbanization, agrifood systems transformation and healthy diets across the rural–urban continuum. FAO, Rome. https://doi.org/10.4060/cc3017en

Fishel FM, Ferrell JA (2013) Managing pesticide drift. Agronomy department. PI232 University of Florida, Gainesville

Gouli V, Marcelino J, Gouli S (2020) Microbial pesticides: biological resources, production and application. Academic

Khan SM, Ali S, Nawaz A, Bukhari SAH, Ejaz S, Ahmad S (2019) Integrated pest and disease management for better agronomic crop production. In: Hasanuzzaman M (ed) Agronomic crops. Springer, Singapore. https://doi.org/10.1007/978-981-32-9783-8_19

Kiran P, Kunal A (2024) Biocontrol agents market – by active substance (microbials, macroorganisms, biochemicals), by crops (fruits and vegetables, cereals & grains, pulses, others), & by application (seed treatment, On-Field, postharvest) & forecast, 2024–2032. Glob Mark Insights, GMI1551

Köhl J, Kolnaar R, Ravensberg WJ (2019) Mode of action of microbial biological control agents against plant diseases: relevance beyond efficacy. Front Plant Sci 10:845. https://doi.org/10.3389/fpls.2019.00845

Kulkarni M, Gorthi S, Banerjee G, Chattopadhyay P (2017) Production, characterization and optimization of Actinomycin D from *Streptomyces hydrogenans* IB310, an antagonistic bacterium against phytopathogens. Biocatal Agric Biotechnol 10:69–74. https://doi.org/10.1016/j.bcab.2017.02.009

Kumar J, Ramlal A, Mallick D, Mishra V (2021a) An overview of some biopesticides and their importance in plant protection for commercial acceptance. Plants (Basel) 10(6):1185. https://doi.org/10.3390/plants10061185

Kumar P, Kamle M, Borah R, Mahato DK, Sharma B (2021b) *Bacillus thuringiensis* as microbial biopesticide: uses and application for sustainable agriculture. Egypt J Biol Pest Control 31(1):95. https://doi.org/10.1186/s41938-021-00440-3

Kumari M, Srivastava A, Babu Sah S, Subhashini (2022) Biological control of agricultural insect pests. IntechOpen. https://doi.org/10.5772/intechopen.104464

Le KD, Yu NH, Park AR, Park DJ, Kim CJ, Kim JC (2022) *Streptomyces* sp. AN090126 as a biocontrol agent against bacterial and fungal plant diseases. Microorganisms 10(4):791. https://doi.org/10.3390/microorganisms10040791

Li L, Wu J, Deng Z, Zabriskie TM, He X (2013) *Streptomyces lividans* blasticidin S deaminase and its application in engineering a blasticidin S-producing strain for ease of genetic manipulation. Appl Environ Microbiol 79(7):2349–2357. https://doi.org/10.1128/AEM.03254-12

Lichtenberg E (2013) Economics of pesticide use and regulation. In: Shogren JF (ed) Encyclopedia of energy, natural resource, and environmental economics. Elsevier. https://doi.org/10.1016/B978-0-323-91013-2.00002-2

Ludwig W et al (2012) Road map of the phylum Actinobacteria. In: Goodfellow M, Kämpfer P, Busse H-J, Trujillo ME, Suzuki K-i, Ludwig W, Whitman WB (eds) Bergey's manual® of systematic bacteriology. Springer, New York. https://doi.org/10.1007/978-0-387-68233-4_1

Meena P, Binita S, Preeti K, Gresha S, Dipak K (2021) Desert locust invasion in Nepal and possible management strategies: a review. J Agric Food Res 5:100166. https://doi.org/10.1016/j.jafr.2021.100166

Mendoza-de Gives P (2022) Soil-borne nematodes: impact in agriculture and livestock and sustainable strategies of prevention and control with special reference to the use of nematode natural enemies. Pathogens 11(6):640. https://doi.org/10.3390/pathogens11060640

Nazarov PA, Baleev DN, Ivanova MI, Sokolova LM, Karakozova MV (2020) Infectious plant diseases: etiology, current status, problems and prospects in plant protection. Acta Nat 12(3):46–59. https://doi.org/10.32607/actanaturae.11026

Noura El-Ahmady El-Naggar (2021) Streptomyces-based cell factories for production of biomolecules and bioactive metabolites. In: Vijai S (ed) Microbial cell factories engineering for production of biomolecules. Academic. https://doi.org/10.1016/B978-0-12-821477-0.00011-8

Palaniyandi SA, Yang SH, Zhang L, Suh J (2013) Effects of actinobacteria on plant disease suppression and growth promotion. Appl Microbiol Biotechnol 97:9621–9636

Pardey PG, Chan-Kang C, Dehmer SP, Beddow JM (2016) Agricultural R&D is on the move. Nature 537(7620):301–303. https://doi.org/10.1038/537301a

Parnell S, van den Bosch F, Gottwald T, Gilligan CA (2017) Surveillance to inform control of emerging plant diseases: an epidemiological perspective. Annu Rev Phytopathol 55:591–610. https://doi.org/10.1146/annurev-phyto-080516-035334

Pathak VM, Verma VK, Rawat BS et al (2022) Current status of pesticide effects on environment, human health and its eco-friendly management as bioremediation: a comprehensive review. Front Microbiol 13:962619. https://doi.org/10.3389/fmicb.2022.962619

Peng Y, Li SJ, Yan J et al (2021) Research progress on phytopathogenic fungi and their role as biocontrol agents. Front Microbiol 12:670135. https://doi.org/10.3389/fmicb.2021.670135

Pulavarty A, Egan A, Karpinska A, Horgan K, Kakouli-Duarte T (2021) Plant parasitic nematodes: a review on their behavior, host interaction, management approaches and their occurrence in two sites in the Republic of Ireland. Plants (Basel) 10(11):2352. https://doi.org/10.3390/plants10112352

Rehman MU, Wali AF, Bajaj V (eds) (2024) Pesticide toxicology and risk assessment: linking exposure to biological effects. CRC Press. https://doi.org/10.1201/9781003222576

Schneider UA, Rasche L, McCarl BA (2018) Assessing the economic impacts of pesticide regulations. Agriculture 8(4):53. https://doi.org/10.3390/agriculture8040053

Senthil-Nathan S (2015) A review of biopesticides and their mode of action against insect pests. In: Thangavel P, Sridevi G (eds) Environmental sustainability. Springer, New Delhi. https://doi.org/10.1007/978-81-322-2056-5_3

Sharma V, Salwan R (2018) Biocontrol potential and applications of actinobacteria in agriculture. In: Singh BP, Gupta VK, Passari AK (eds) New and future developments in microbial biotechnology and bioengineering. Elsevier. https://doi.org/10.1016/B978-0-444-63994-3.00006-0

Sharma A, Kumar V, Shahzad B et al (2019) Worldwide pesticide usage and its impacts on ecosystem. SN Appl Sci 1(11):1–16. https://doi.org/10.1007/s42452-019-1485-1

Shimizu M (2011) Endophytic actinomycetes: biocontrol agents and growth promoters. In: Maheshwari DK (ed) Bacteria in agrobiology: plant growth responses. Springer, Berlin, Heidelberg. https://doi.org/10.1007/978-3-642-20332-9_10

Singh R, Dubey AK, Chaurasia S, (2018) Actinobacteria as plant growth-promoting and biocontrol agents. In: New and Future Developments in Microb Biotechnol Bioeng. Elsevier,Amsterdam, Netherlands 67–82

Shrivastava P, Kumar R (2018) Actinobacteria: eco-friendly candidates for control of plant diseases in a sustainable manner. In: Singh BP, Gupta VK, Passari AK (eds) New and future developments in microbial biotechnology and bioengineering. Elsevier. https://doi.org/10.1016/B978-0-444-63994-3.00005-9

Silva GC, Kitano IT, Ribeiro IA, Lacava PT (2022) The potential use of actinomycetes as microbial inoculants and biopesticides in agriculture. Front Soil Sci 2:833181. https://doi.org/10.3389/fsoil.2022.833181

Subramaniam G, Arumugam S (2020) Biocontrol of plant pathogens by actinomycetes: An eco-friendly approach for sustainable agriculture. Biotechnol Res Innov. 4(1):100032

Singh D, Samiksha, Thayil SM, Sohal SK, Kesavan AK (2021) Exploration of insecticidal potential of Cry protein purified from *Bacillus thuringiensis* VIID1. Int J Biol Macromol 174:362–369. https://doi.org/10.1016/j.ijbiomac.2021.01.143

Swapan C, Mainak B, Deewa B, Tanmoy M (2024) Natural pesticides for pest control in agricultural crops: an alternative and eco-friendly method. Plant Sci Today:433–450. https://doi.org/10.14719/pst.2547

Tudi M, Daniel Ruan H, Wang L et al (2021) Agriculture development, pesticide application and its impact on the environment. Int J Environ Res Public Health 18(3):1112. https://doi.org/10.3390/ijerph18031112

Varsha B, Shahida I (2020) Biopesticides: production, formulation and application systems. Int J Curr Microbiol App Sci 9(10):3931–3946. https://doi.org/10.20546/ijcmas.2020.910.453

Veliz EA, Martínez-Hidalgo P, Hirsch AM (2017) Chitinase-producing bacteria and their role in biocontrol. AIMS Microbiol 3(3):689–705. https://doi.org/10.3934/microbiol.2017.3.689

Venkatasaichandrakanth P, Iyapparaja M (2023) Review on pest detection and classification in agricultural environments using image-based deep learning models and its challenges. Opt Mem Neural Networks 32(4):295–309. https://doi.org/10.3103/S1060992X23040112

Viaene T, Langendries S, Beirinckx S, Maes M, Goormachtig S (2016) Streptomyces as a plant's best friend? FEMS Microbiol Ecol 92:fiw119

Villaverde JJ, Sandín-España P, Sevilla-Morán B, López-Goti C, Alonso-Prados JL (2016) Biopesticides from natural products: current development, legislative framework, and future trends. Bioresources 11(2):5618–5640. https://doi.org/10.15376/biores.11.2.Villaverde

Wahab S, Muzammil K, Nasir N et al (2022) Advancement and new trends in analysis of pesticide residues in food: a comprehensive review. Plants (Basel) 11(9):1106. https://doi.org/10.3390/plants11091106

Williamson-Benavides BA, Dhingra A (2021) Understanding root rot disease in agricultural crops. Horticulturae 7(2):33. https://doi.org/10.3390/horticulturae7020033

Zacharia TJ (2011) Identity, physical and chemical properties of pesticides. InTech. https://doi.org/10.5772/17513

Chapter 19
The Potential of Microbes as Biofertilizers and Biopesticides for Sustainable Agriculture

Yenny Sariasih, Zainal Muktamar, and Nanik Setyowati

Abstract This paper investigates the role of microbes as biofertilizers and biopesticides in advancing sustainable agriculture. Since 1974, research on sustainable agriculture has grown substantially, with a total of 75,788 publications to date. Microbial applications have gained attention as eco-friendly alternatives to chemical inputs, offering benefits such as enhanced soil health and reduced dependency on synthetic agrochemicals. The discussion is organized into two main sections: the first examines the role of microbes as biofertilizers, while the second focuses on their application as biopesticides. Additionally, a separate section addresses the advantages, limitations, and future prospects of microbial utilization in agriculture. Biofertilizers, which are preparations containing living microorganisms, increase plant growth by increasing nutrient availability. Examples include nitrogen-fixing bacteria, phosphate-solubilizing microorganisms (PSMs), and mycorrhizal fungi, all of which are critical for maintaining soil fertility and supporting plant development. Plant growth-promoting rhizobacteria (PGPR) offer dual benefits by both increasing plant growth and protecting against pathogens through mechanisms such as antibiosis, induced systemic resistance, and nutrient competition. PSMs are particularly important in converting insoluble phosphate in the soil into forms accessible to plants, thereby minimizing the need for chemical fertilizers. Biopesticides, including microbial agents such as spp. and *Beauveria bassiana*, have been exten-

Y. Sariasih (✉)
Department of Plant Protection, Faculty of Agriculture, University of Bengkulu, Bengkulu, Indonesia
e-mail: yennysariasih@unib.ac.id

Z. Muktamar
Department of Soil Science, Faculty of Agriculture, University of Bengkulu, Bengkulu, Indonesia
e-mail: muktamar@unib.ac.id

N. Setyowati
Department of Crop Production, Faculty of Agriculture, University of Bengkulu, Bengkulu, Indonesia
e-mail: nsetyowati@unib.ac.id

V. Gopikrishnan et al. (eds.), *Harnessing Microbial Resources for Sustainable Agriculture*, Microorganisms for Sustainability 61,
https://doi.org/10.1007/978-981-95-4817-0_19

sively studied for their biocontrol capabilities. These microbes act as antagonists of various plant pathogens, offering a natural and sustainable approach to pest and disease management. Overall, microbes present significant potential as biofertilizers and biopesticides, making them indispensable tools for sustainable agricultural practices.

Keywords Biofertilizer · Biopesticide · Microbes · Sustainable agriculture

Highlights

- This paper examines the role of microbes as biofertilizers and biopesticides in advancing sustainable agriculture.
- Microbial biofertilizers increase plant growth by increasing nutrient availability and soil health, reducing dependency on chemical inputs.
- This study discusses the roles of nitrogen-fixing and phosphate-solubilizing microorganisms and plant growth-promoting rhizobacteria in nutrient cycling and disease resistance.
- Additionally, it covers biopesticides such as *Trichoderma* and *Beauveria bassiana* for their biocontrol potential.
- Challenges, benefits, and future directions for microbial applications in sustainable agriculture are also addressed.

19.1 Introduction

Sustainable agriculture is a farming approach that aims to integrate a system of plant and animal production practices with site-specific applications that will, over the long term, satisfy human food and fiber needs. The environmental quality and the natural resource base upon which the agricultural economy depends should be enhanced. Make the most efficient use of nonrenewable resources and on-farm resources, and integrate, where appropriate, natural biological cycles and controls. The economic viability of farm operations should be maintained. The quality of life for farmers and society as a whole should be increased. The fundamental objectives of sustainable agriculture encompass environmental health, economic viability, and social equity. These three pillars collectively form the foundation of sustainability in agricultural systems (https://sustainableagriculture.net). Research on sustainable agriculture began in 1974 with the publication of an article titled "Studies Toward Sustainable Agriculture in Northern Ghana: An Internal Working Paper on Agro-Ecosystems" (Bennett and Schork 1979). Since then, the body of literature on sustainable agriculture has expanded significantly, with a total of 75,788 documents published to date. Notably, the 2000s saw a substantial increase in publications, surpassing 1000 new articles, culminating in a peak in 2024, with 10,127 publications with the most publication documents in the form of articles (62.2%), as indicated by an analysis conducted via Scopus (Fig. 19.1).

Research on sustainable agriculture has gained prominence because of the increasing recognition of the adverse effects associated with synthetic fertilizers and pesticides. Although these chemicals have been instrumental in increasing crop yields, their widespread use has led to significant environmental challenges, including soil degradation, water contamination, and a decline in biodiversity. In response, soil microbes are being explored as promising alternatives, serving as biofertilizers and biopesticides. These microorganisms not only improve crop productivity but also contribute to soil health and reduce dependence on chemical inputs, aligning with the principles of sustainable agricultural practices. This paper explores the potential of microbes as biofertilizers and biopesticides and discusses their benefits for sustainable agricultural practices. This paper is taken from a reference that has been widely published in Scopus as one of the indicators that the publication is produced by an internationally reputable journal. Publications on microbes as biofertilizers began in 1987 with the title "Microbial biotechnology—Integrated studies on the utilization of solid organic waste biopesticides" (Chang 1987) with a total of 523 publications until 2024, with the most publications in the form of book chapters (37.4%). Meanwhile, publications on microbes as biopesticides began to appear in 1995 with the title of the article "Semiautomated quantification of cytotoxic damage induced in cultured insect cells exposed to commercial Bacillus thuringiensis biopesticides" (Tayabali and Seligy 1995), with a total of 297 publications until 2024, with the most publication documents in the form of articles (50.5%). The published data concerning the use of microbes as biofertilizers and biopesticides are shown in Fig. 19.2.

The line graph (Fig. 19.2a) illustrates biofertilizer document trends over time, with the x-axis covering the years from 1987 to 2026 and the y-axis showing

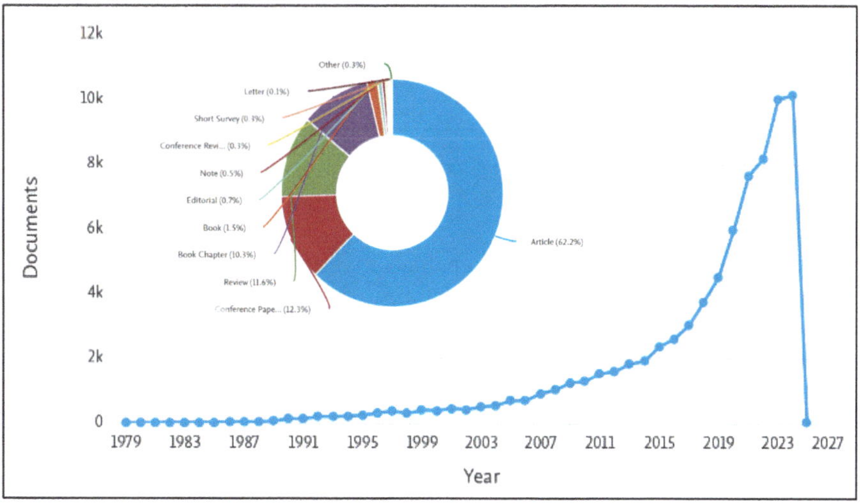

Fig. 19.1 Publications of sustainable agriculture have been published since 1979 (https://www.scopus.com)

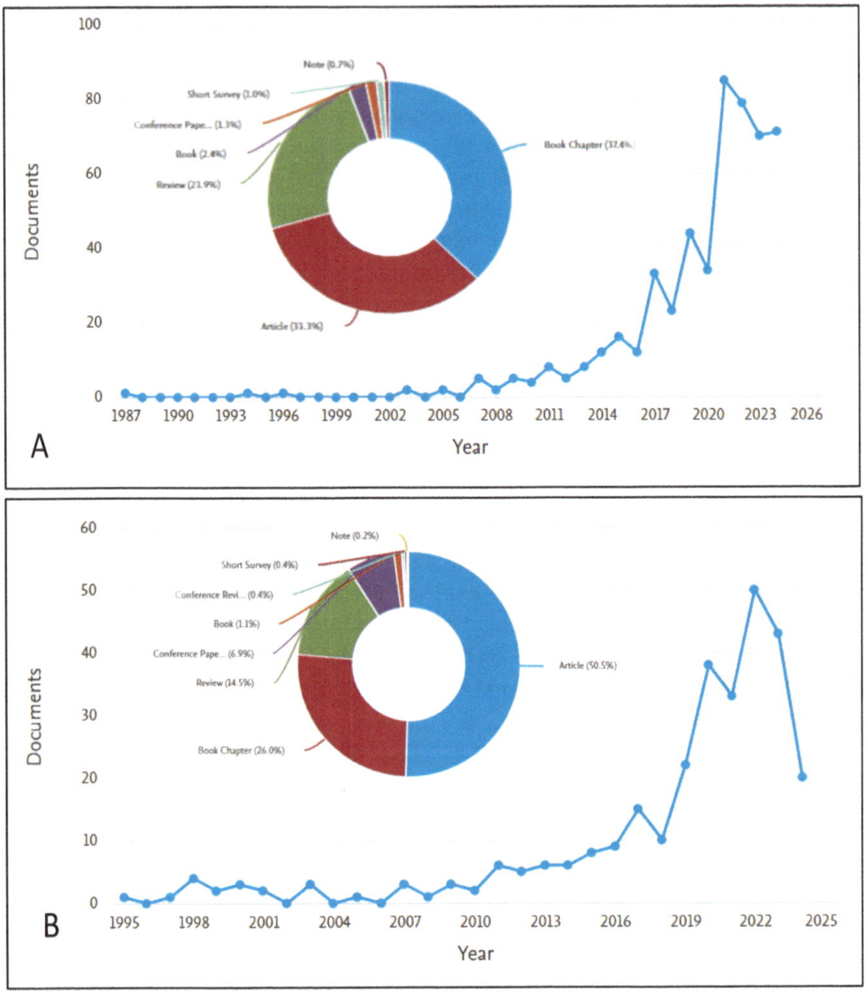

Fig. 19.2 Publications of microbes include (**a**) biofertilizer and (**b**) biopesticide (https://www.scopus.com)

document counts. A sharp increase in publications was observed after 2017, reaching a peak approximately 2023. The accompanying donut chart categorizes document types, with book chapters as the largest category (32.8%), followed by articles (31.3%) and reviews (21.9%). The smaller categories include conference papers (3.1%), books (2.3%), short surveys (0.8%), and notes (8.3%).

Figure 19.2b illustrates biopesticide document publication trends over time, with years on the x-axis ranging from 1995 to 2025 and a noticeable increase in publications after 2017, also peaking approximately 2023. However, the document-type distribution in Fig. 19.2b differs, with the article as the dominant category (50.5%), followed by book chapters (26.6%) and reviews (9.5%). Additional categories include conference papers (9.1%), books (1.3%), conference briefs (0.4%), and short surveys (0.4%).

Comparatively, both graphs indicate a sharp rise in publication activity after 2017, but the document-type distribution varies. Figure 19.2a is characterized by a predominance of book chapters, whereas Fig. 19.2b shows a greater proportion of articles. In the literature, the author sought to redefine the role of microbes as biofertilizers and biopesticides, highlighting their importance in sustainable agriculture. The initial section is divided into two subchapters: the first subchapter addresses the function of microbes as biofertilizers, whereas the second subchapter focuses on their role as biopesticides. In addition, the author added a subchapter on the benefits, challenges, and future directions of microbes as biofertilizers and biopesticides for sustainable agriculture.

19.2 Microbes as Biofertilizers

Biofertilizers are preparations comprising live microorganisms that increase plant growth by increasing nutrient availability. Microbial biofertilizers, including phosphate-solubilizing microorganisms (PSMs) and mycorrhizal fungi, are vital for maintaining soil fertility and supporting plant development.

19.2.1 Plant Growth-Promoting Rhizobacteria

Plant growth-promoting rhizobacteria (PGPR) employ various mechanisms, such as antibiosis, induced systemic resistance (ISR), and competition for nutrients and ecological niches, to protect plants from soilborne diseases. Additionally, PGPR can directly enhance plant growth through biofertilization, root stimulation, rhizoremediation, and stress mitigation (Lugtenberg and Kamilova 2009). The soil rhizosphere region is influenced by root exudates and provides a nutrient-rich environment that supports high microbial activity. Effective root colonization by PGPR is essential for their beneficial effects, with traits such as motility, adhesion, and efficient nutrient utilization playing critical roles in competitive colonization.

PGPR promote plant growth directly by fixing nitrogen, solubilizing phosphate, producing phytohormones such as auxins, and reducing ethylene levels in plants (Bashan et al. 2014; Bowman and Joubert 2020; Chaudhary et al. 2020; Kong and Liu 2022). Indirectly, they enhance plant growth by suppressing pathogens through antibiosis (antibiotic production), strengthening plant defense responses, and outcompeting pathogens for resources. The direct mechanisms by which PGPR facilitate plant growth are illustrated in Fig. 21.3.

Figure 19.3 illustrates the beneficial effects of PGPR on plants. PGPRs help plants alleviate stress, such as drought, salinity, and pathogen infection, by promoting resistance mechanisms. PGPR can improve a plant's overall health and productivity even under adverse conditions. PGPRs promote root growth and development (Lugtenberg and Kamilova 2009). By enhancing root stimulation, PGPR allows plants to absorb more water and nutrients from the soil (Kong and Liu 2022; Lugtenberg and Kamilova 2009; Nadeem et al. 2014), leading to better growth and yield. Some PGPRs can fix atmospheric nitrogen (N_2), converting it into a form that plants can use, such as ammonia (NH_3). This reduces the need for synthetic nitrogen fertilizers. PGPRs that fix nitrogen produce ammonium (NH_4^+), a form of nitrogen that plants can readily absorb and use to build proteins and other essential molecules. PGPR can release organic acids and enzymes that solubilize phosphate in the

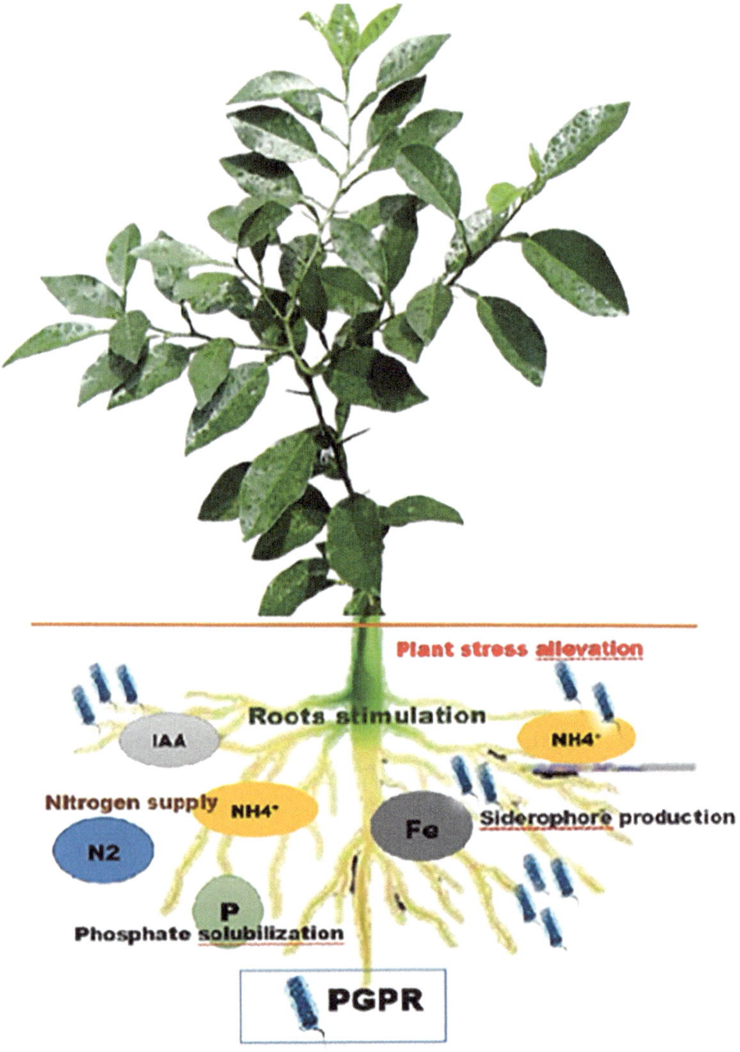

Fig. 19.3 Mechanisms by which PGPRs improve plant growth

soil, making it available to plants (Barea and Azc 2002; Bashir et al. 2024; Varma and Prasad 2020). Other benefits of PGPR include the ability to produce indole-3-acetic acid and help plants acquire iron through various mechanisms, particularly through siderophore production (Rawat et al. 2021; Vassilev et al. 2006). This promotes growth and development. PGPRs play a crucial role in sustainable agriculture by reducing the need for chemical fertilizers and pesticides. Figure 19.3 highlights how PGPRs contribute to a healthier and more efficient growth environment for plants, supporting both agricultural productivity and environmental sustainability.

After the extensive benefits of PGPR in sustainable agriculture are outlined, the author will discuss the specific microbes that fall into the group of PGPR. The first microbe is *Rhizobium* spp., which form symbiotic relationships with leguminous plants (Bashir et al. 2024; du Jardin 2015), fixing atmospheric nitrogen into a form that plants can use. This enhances soil fertility and supports the growth of host plants. The next microbe was *Pseudomonas fluorescens*. This microbe is known for its ability to suppress various plant pathogens; this bacterium produces antifungal compounds and promotes plant health (Bashir et al. 2024; Bona et al. 2018; Yang and Ancona 2021). It can also increase nutrient availability and root development. The third microbe of PGPR is *Bacillus subtilis*. This bacterium helps plants produce hormones that stimulate growth and enhance resistance to diseases. It also improves soil health by promoting beneficial microbial activity. The fourth microbe is *Azospirillum* spp., which are associated with a wide range of plants, particularly grasses (Mar Vázquez et al. 2000). They promote root growth and increase nutrient uptake, particularly nitrogen, thus increasing overall plant vigor. The last microbe is *Enterobacter* spp. Some species in this genus can increase plant growth by producing growth-promoting hormones and improving nutrient solubilization, making them valuable for various crops (Ngalimat et al. 2021; Sun et al. 2024). These examples illustrate the diverse roles of PGPR in supporting plant growth and health, contributing to sustainable agricultural practices.

19.2.2 Mycorrhiza

The term mycorrhiza comes from two Greek words, namely, *myco*, which means fungus, and *rhiza,* which means root, and the true meaning of mycorrhizae is a symbiosis between fungi and roots (Alizadeh 2011; Tahat et al. 2010). Mycorrhizae are described as sharing relationships in life, where fungi are the main partners of crops that provide a source of food and growth hormones and protect crop roots from pathogen attack; in return, the host crop provides a source of energy to the fungus. Most higher crops are associated with one or more mycorrhizae that coexist (Alizadeh 2011). Why do arbuscular mycorrhizal fungi (AMF) exist? Because microbes of AMF are abundant in nature, they are easy to obtain and easy to propagate, and the success rate of their association with host crops is quite high. Mycorrhizae are microorganisms from the kingdom of fungi, the phylum *Glomeromycota*, and the class *Glomeromycetes*, with four orders, namely,

Archaeosporales, *Glomerales*, *Diversisporales*, and *Paraglomerales*. The most recent development of mycorrhizal taxonomy consists of 12 families and 41 genera (Schüßler and Walker 2010).

The grouping of mycorrhizae is based on the form of chlamydospores or resting spores formed by mycorrhizae, hyphal characters, auxiliary cells, subtending hyphae, germination, and spore wall biochemical and molecular structures (Tahat et al. 2010). These mycorrhizal spores subsequently germinate or germinate and form an appressorium when attached to the host crop cell. Some mycorrhiza spores from several genera are shown in Fig. 21.4 (Sariasih et al. 2023).

The three major taxa present in the soil rhizosphere are represented by the three spores in Fig. 19.4. In addition to being rather easy to grow for the production of commercial biofertilizers, all three are highly plentiful in nature. Mycorrhizal spores adhere to the roots of plants and sift through the cells. Mycorrhizae colonize the host crop rapidly and penetrate the living cells of the host crop without damaging them and form distinctive organs, such as arbuscules, hyphae, and vesicles, in the roots (Ortas 2012). The appressorium develops into vesicle and arbuscular structures, as well as into intracellular hyphae. Hyphae that resemble roots and are used to spread from one cell to another are known as arbuscular hyphae. Even so, in a matter of seconds, these hyphal formations transform into intercellular hyphae that penetrate the cells. On the other hand, a vesicle acts as a location in the host crop cell where food reserves or mycorrhiza energy are stored. The existence of hyphae, arbuscules, and/or vesicles in the root cells indicates mycorrhizal colonization. Figure 19.5 shows the colonization of roots by mycorrhizae.

Mycorrhizal symbiosis and the roots of the host crop can cause changes in genetic expression locally and systemically, including the induction of crop resistance responses to pathogens. This finding is similar to the results of research that revealed that AMF symbiosis with crop roots could activate crop resistance locally and systemically (Demir et al. 2015; Khallal 2007). The role of arbuscular mycorrhizae in crops is to encourage the absorption of minerals, especially P, Zn, Cu, and NH_4, increase water absorption, stimulate crop growth, improve the quality of fruit

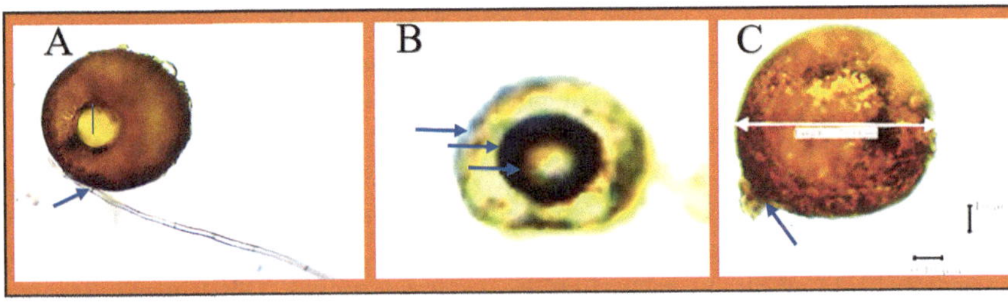

Fig. 19.4 The spores of mycorrhizae from the rhizosphere. (**a**) Yellow spore with attached hyphae (arrow) of *Funneliformis* or *Glomus*. (**b**) Yellow spore of *Acaulospora* with three layers of wall cells (arrow). (**c**) Golden yellow spore of *Gigaspora* with bulbous suspensors (arrow) and scale of measurement. The magnification power of the microscope was 400× (Sariasih et al. 2023)

Fig. 19.5 Mycorrhizal colonization of citrus roots. (**a**) Vesicle (arrow). (**b**) Intracellular hyphae (arrow). (**c**) External hyphae (arrow). 400× magnification power of microscope (Sariasih et al. 2023)

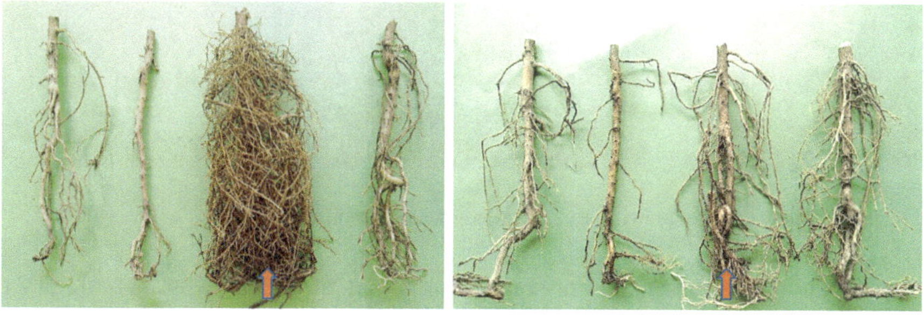

Fig. 19.6 Root morphology of citrus plants treated with and without mycorrhizae. ⇧: with mycorrhizae (Sariasih 2024)

produced by crops, increase crop resistance to environmental stress, and increase crop resistance to pathogens (Srivastava 2012).

Mycorrhizal fungi can stimulate the emergence of local and systemic resistance to pathogens in roots and activate the systemic resistance of crops to pathogens in leaves. In addition, AMF are capable of negating the formation of systemic compounds of salicylic acid (SA) and jasmonic acid (JA), increasing crop resistance to salinity and drought and helping crops obtain nutrients from the soil (Jacott et al. 2017). Mycorrhizal colonization of roots also affects the root morphology of plants. Plants treated with mycorrhizae have more hair roots with greater root volume than do plants without mycorrhizae, according to the results of the Sariasih (2024) study shown in Fig. 19.6.

Figure 19.6 shows that among the four roots of the rough lemon variety orange plant, the third root from the left has been treated with mycorrhizae and has a greater density of root hairs. Compared with nonmycorrhizal plants, mycorrhizal-colonized root systems also presented significantly greater root wet weights. Increased root development and increased surface area enable these plants to access a broader range of nutrients, thereby positively influencing their growth (Hao et al. 2012; Khallal 2007; Wu 2017). Plants that exhibit optimal growth conditions are likely to display improved resistance to both biotic and abiotic stresses, including pathogen

Fig. 19.7 Plant response to mycorrhizal and nonmycorrhizal treatments

attacks and environmental stressors (Dhiman et al. 2022; Mitra et al. 2021; Schouteden et al. 2015). Figure 19.7 provides a visual representation of the differences between the application of mycorrhizal inoculum and the absence of mycorrhizae.

As shown in Fig. 19.7, compared with nonmycorrhizal plants, mycorrhizal-colonized plants clearly presented increased growth and greater biomass. Mycorrhizae also promote the synthesis of resistant compounds such as SA and JA, thereby increasing the overall resistance of mycorrhizal plants. During root development, mycorrhizae form hyphae and filaments that expand the root surface area, facilitating the optimal absorption of essential water and minerals.

19.2.3 Phosphate-Solubilizing Microorganisms

The amount of phosphorus (P) that is available to plants is just a small percentage of the total amount of P in the soil. Phosphorus is an essential component of biological molecules and is one of the primary factors that restricts the creation of biomass (Bashir et al. 2024; Hii et al. 2020). The ever-increasing need for food on a worldwide scale, along with the modern agricultural industry's consumption of phosphate fertilizers, may result in excessive inputs of inorganic phosphorus in croplands that are intensively managed, leading to an increase in phosphate losses and the continued eutrophication of surface waters.

PSMs are a group of beneficial microbes that play crucial roles in making phosphorus available to plants (Bashir et al. 2024; Hii et al. 2020; Pang et al. 2024; Wang et al. 2023). Phosphorus is a critical nutrient for plant development and plays a key role in energy transfer, photosynthesis, and nucleic acid synthesis. However, a significant portion of soil phosphorus exists in forms that are not readily accessible to plants. PSMs help convert these forms into soluble phosphates that plants can absorb. The types of PSMs are bacteria, fungi, and actinomycetes (Bashir et al. 2024; Hii et al. 2020; Pang et al. 2024; Wang et al. 2023). The bacterial genera of PSMs are *Bacillus* (Bashir et al. 2024; Ngalimat et al. 2021), *Pseudomonas* (Bashir et al. 2024; Bona et al. 2018), *Enterobacter* (Ngalimat et al. 2021; Sun et al. 2024), and *Rhizobacterium* (Bashir et al. 2024).

These bacteria release organic acids and enzymes that convert insoluble phosphate compounds, such as calcium phosphate and rock phosphate, into forms available to plants. Among phosphate-solubilizing fungi, AMF, particularly species such as *Glomus* or *Funneliformis*, play a significant role (Pang et al. 2024; Wang et al. 2023). These fungi establish symbiotic associations with plant roots, extending their hyphae into the soil to access and solubilize phosphorus, which is then transferred to the plant in exchange for carbohydrates. The role of mycorrhizae was described in the preceding subchapter. PSMs use several mechanisms to solubilize phosphorus and produce organic acids. They secrete organic acids (e.g., citric, acetic, and oxalic acids) that lower the soil pH, converting insoluble phosphates into soluble forms. For the production of PSMs, certain enzymes, such as phosphatases, break down organic phosphorus compounds, releasing available phosphate. In the case of mycorrhizal fungi, their association with plant roots enhances phosphorus uptake and overall nutrient absorption. The importance of PSMs in agriculture lies in nutrient availability, sustainable practices, and soil health. An illustration of PSMs in agricultural cultivation is shown in Fig. 19.8.

Figure 19.8 illustrates the process of phosphate solubilization in the soil by various PSMs, including fungi, bacteria, and actinomycetes. Inorganic phosphate (Pi) solubilization, which is often present in insoluble forms in soil, makes it difficult for plants to absorb. PSMs help convert these insoluble forms into soluble phosphate, which plants can readily take up and use for growth and development. They secrete various compounds that facilitate phosphate solubilization (Hii et al. 2020; Pang et al. 2024). The image highlights two main types of secretions, acids and siderophores, both of which play key roles in solubilizing phosphate.

The roles of each group of PSM microbes are as follows: Fungi produce compounds such as hydrogen sulfide (H_2S) and carbonic acid (H_2CO_3), which contribute to phosphate solubilization. These compounds acidify the soil environment, helping release phosphate from bound forms. In other ways, bacteria release ammonium (NH_4^+) and organic acids that lower the soil pH, promoting the release of phosphate. The acids dissolve phosphate compounds, increasing the availability of phosphorus to plants. The last microbe is the filamentous bacteria actinomycetes, which produce siderophores and chelators (Bashir et al. 2024; Zhang et al. 2016). Siderophores are specialized molecules that bind to metal ions (such as iron) and increase their

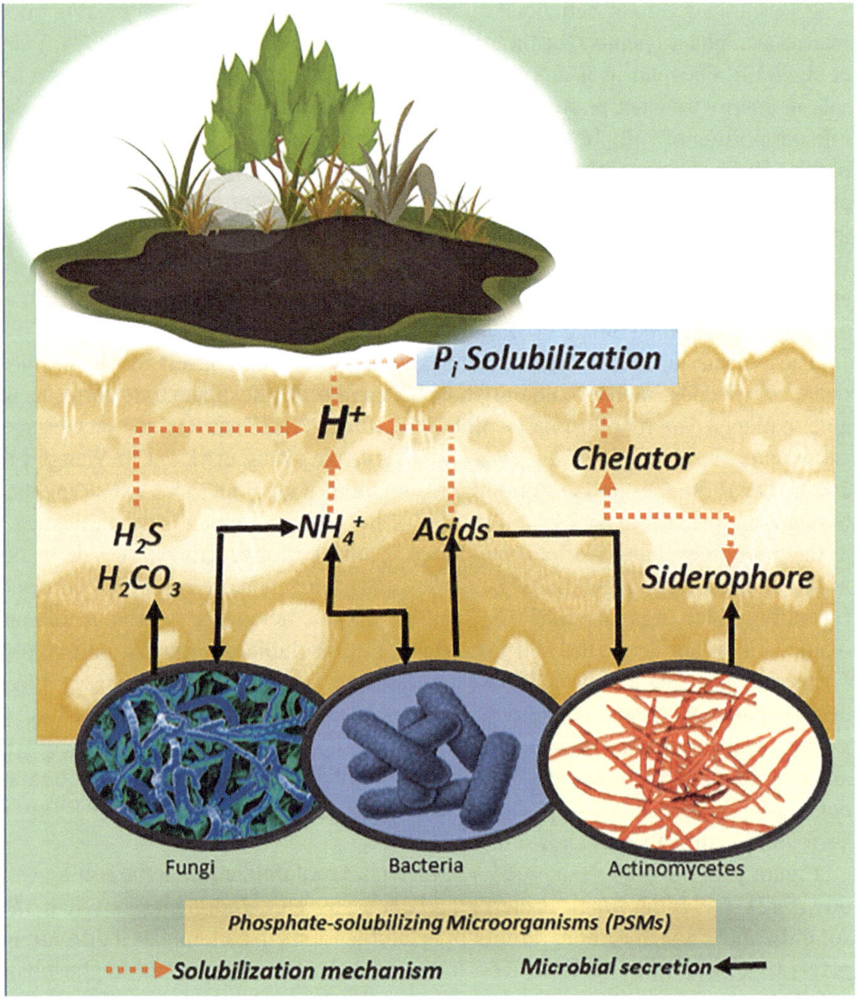

Fig. 19.8 Illustration of PSMs in agricultural cultivation

availability to plants. Chelators bind tightly to metals, helping free phosphate from metal complexes in the soil.

The solubilization mechanism (red-dotted lines) indicates the solubilization pathway, where various compounds (such as H^+ ions from acids) facilitate the breakdown of phosphate compounds in the soil. This process makes phosphate ions accessible to plants. Microbial secretion (black arrows) indicates the substances secreted by each group of microorganisms. These secretions directly contribute to the release of phosphate from insoluble forms. In summary, these images illustrate how fungi, bacteria, and actinomycetes work together to increase phosphate availability in the soil. Through the secretion of acids, chelators, and siderophores, these microorganisms solubilize inorganic phosphate, supporting plant growth and reducing the need for chemical phosphorus fertilizers.

By using PSMs as biofertilizers, which significantly increase the availability of phosphorus in the soil, promote healthier plant growth and higher crop yields. Farmers can reduce their reliance on chemical fertilizers, which can have negative environmental impacts and contribute to soil microbial diversity and health, which are critical for maintaining ecosystem balance and fertility. Phosphate-solubilizing microorganisms are essential for increasing phosphorus availability in soil, supporting plant growth, and promoting sustainable agricultural practices. Their role in nutrient cycling makes them valuable for improving soil fertility and crop productivity.

19.3 Microbes as Biopesticides

Biopesticides are microbial agents used to manage agricultural pests and diseases. Unlike synthetic pesticides, which often harm nontarget organisms and disrupt ecosystems, biopesticides target specific pests and pathogens, making them safer and more sustainable options. The author will explain some microbes that have the potential to become biopesticides and have gone through various tests and research in the laboratory.

19.3.1 Trichoderma

The biocontrol properties of *Trichoderma* have been extensively studied (De and Roy 2024; Mahfut et al. 2023; Meddad-Hamza et al. 2023; Mkhwanazi et al. 2024; Vinale et al. 2008; Ziedan et al. 2011). *Trichoderma* species serve as antagonists of pathogenic fungi by either parasitizing them or competing for available nutrients. This genus of soil fungi is commonly utilized as a biocontrol agent because of its ability to protect plants and inhibit pathogens through mechanisms such as mycoparasitism (Meddad-Hamza et al. 2023; Vinale et al. 2008), antibiosis (Prismantoro et al. 2024), and competition (Sinha et al. 2024). Fungi produce cell wall-degrading enzymes and secondary metabolites that hinder the growth of pathogens. *Trichoderma* can be grown on potato dextrose agar (PDA) media, and the results of this cultivation are shown in Fig. 21.9.

As shown in Fig. 19.9, *Trichoderma* was isolated from natural sources via standard PDA media and subsequently purified on fresh PDA media. The *Trichoderma* colonies formed are circular in shape, radiating outward from the center of the Petri dish and displaying green pigmentation. To assess the ability of *Trichoderma* to inhibit plant growth, a pure *Trichoderma* isolate was subjected to dual-culture testing with the pathogen *Sclerotium rolfsii*. The outcomes of this dual-culture experiment are presented in Fig. 19.10.

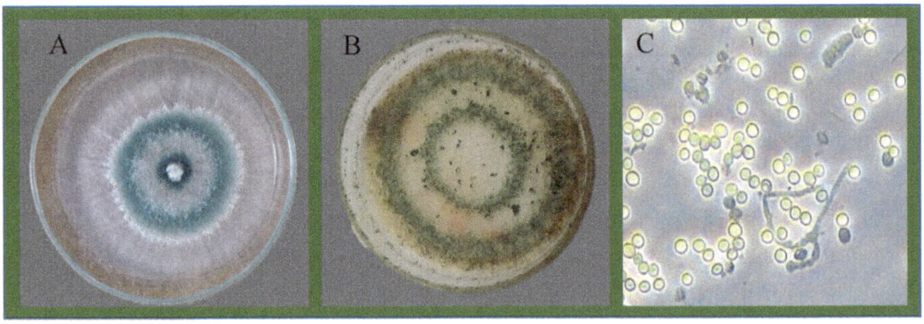

Fig. 19.9 *Trichoderma* on PDA. (**a**) Green colony with white edges. (**b**) Dark green colony. (**c**) Conidia and hyphae (Hendri Bustamam's isolate, Bengkulu University, Indonesia)

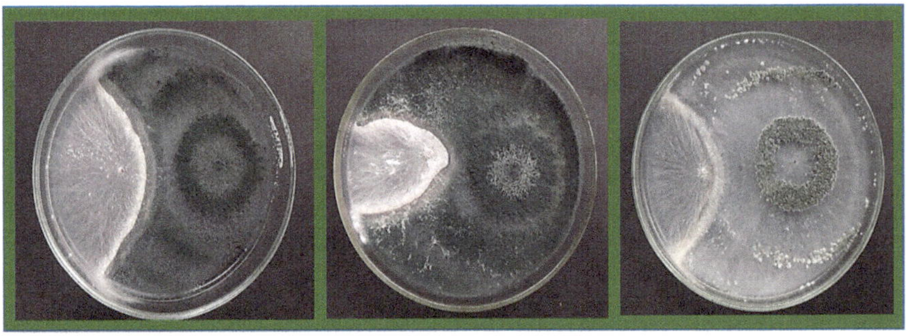

Fig. 19.10 Three Petri dishes of dual-culture *Trichoderma* (green) with *Sclerotium rolfsii* (white) soilborne pathogens on three PDA media (Hendri Bustamam's isolate, Bengkulu University, Indonesia) were used

Figure 19.10 demonstrated the inhibitory effects of *Trichoderma* on the growth of pathogenic soilborne fungi in a dual-culture assay. These results align with those of prior studies that highlighted the potential of *Trichoderma* as a biopesticide in agricultural applications. The mechanisms underlying the action of *Trichoderma* include mycoparasitism, antibiotic production, and competitive exclusion. In the mycoparasitic mechanism, *Trichoderma* exerts its effects by targeting and penetrating the cell walls of pathogens through the action of carbohydrate-active enzymes (CWDEs) Cell Wall–Degrading Enzymes, which subsequently release oligosaccharides that stimulate further enzyme synthesis. In terms of antibiotic activity, *Trichoderma* synthesizes a range of antibiotics and secondary metabolites capable of inhibiting pathogenic growth; these substances include volatile compounds, water-soluble metabolites, and peptaibols. Furthermore, via competitive strategies, *Trichoderma* effectively competes for essential nutrients and physical space, further diminishing the viability of pathogenic organisms.

The indirect effects of *Trichoderma* on plant health include its ability to colonize plant roots (Meddad-Hamza et al. 2023), which enhances growth and induces systemic resistance (ISR) to pathogens. This interaction involves molecular signaling and the production of elicitors that trigger plant defense mechanisms. Additionally, the effectiveness of *Trichoderma* is affected by soil conditions and the presence of organic matter, both of which influence its growth and biocontrol activity.

19.3.2 Beauveria bassiana

Beauveria bassiana is a filamentous fungus that belongs to the family *Cordycipitaceae*. It is a well-known entomopathogenic fungus, meaning that it infects and causes disease in insects (El-Maraghy et al. 2023; Sain et al. 2024; Smith and Koprivnikar 2024), making it a significant agent for biological control in pest management. *B. bassiana* is present in a variety of environments, including soil and plant surfaces. It can be cultured on standard media such as PDA, Sabouraud dextrose agar (SDA), and SDA with yeast extract (SDAY). For biopesticide applications, *B. bassiana* can be propagated on rice that has been either steamed or boiled. The fungi *B. bassiana* in SDAY media and rice media are shown in Fig. 19.11.

B. bassiana can infect a wide range of insect hosts, including agricultural pests such as aphids, whiteflies, and beetles. It enters the host through the cuticle, germinates, and proliferates inside the host, leading to its death. The mode of action of *B. bassiana* is that the pathogen produces extracellular enzymes that break down the host cuticle. After infection, it can release toxic metabolites, which contribute to the mortality of the host insect. Owing to its ability to target specific insect pests, *B. bassiana* is used as a biopesticide for the biocontrol of *B. bassiana* (Fig. 19.12).

Pathogenicity tests conducted on these two pest species demonstrated that *B. bassiana* has a significant potential as a biopesticide, offering considerable advantages for farmers. Its application can reduce reliance on chemical pesticides, promoting more sustainable agricultural practices. Studies have shown that this fungus effectively controls pest populations in both field and controlled environments. However, factors such as humidity, temperature, and the developmental stage of the

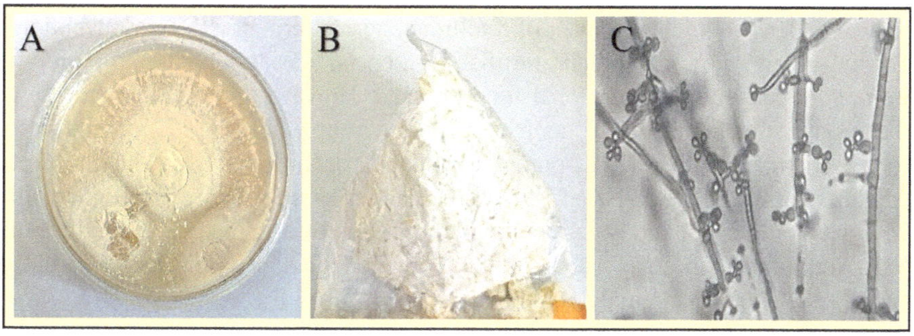

Fig. 19.11 *Beauveria bassiana*. (**a**) Pure isolate in SDAY media. (**b**) Propagation of the isolate with rice media. (**c**) Conidia and hyphae of *B. bassiana* (Nadrawati isolate, Bengkulu University, Indonesia)

Fig. 19.12 Pathogenicity test of *Beauveria bassiana* as a pest biopesticide (entomopathogen). (**a**) *B. bassiana* effectively kills *Physomerus grossipes*. (**b**) Test of the ability of *B. bassiana* to infect *Tenebrio molitor*. (**c**) *Spodoptera litura* infected with *B. bassiana* (Nadrawati's isolate, Bengkulu University, Indonesia)

host organism can influence its effectiveness. Furthermore, *B. bassiana* is generally considered safe for nontarget organisms, including humans and beneficial insects, making it a viable candidate for integrated pest management approaches. Overall, *B. bassiana* is a promising biocontrol agent that supports sustainable agriculture by managing pest populations while reducing the need for chemical pesticides.

19.4 Benefits of Microbial Biofertilizers and Biopesticides

This subchapter discusses the advantages of microbial biofertilizers and biopesticides in promoting sustainable agriculture. The widespread use of these microbial products by farmers worldwide can contribute to the establishment of long-term sustainable agricultural practices, benefiting future generations. The specific benefits are outlined in the following sections:

1. Environmental benefits

Microbial biofertilizers and biopesticides provide a sustainable solution to reduce the environmental impact of traditional agricultural practices. Conventional farming often relies on synthetic fertilizers and pesticides to increase crop production and meet the growing global food demand. However, the widespread use of these synthetic chemicals has contributed to environmental degradation and poses risks to human health, such as potential poisoning from long-term pesticide exposure. In contrast, microbe-based biofertilizers and biopesticides are environmentally benign, avoiding soil and water contamination and reducing the risk of biodiversity loss. Additionally, these microbial products are biodegradable, limiting the potential for persistent environmental contamination.

2. Improved soil health

Microbes are crucial for nutrient cycling, improving soil structure, and increasing soil fertility. The application of biofertilizers supports a balanced soil microbiome, which in turn promotes long-term soil health. This is in contrast to synthetic fertilizers, which can lead to soil acidification and degradation over time. Healthy soil will be very beneficial to humans because the soil will continue to be able to be used for agricultural cultivation.

3. Reduced dependence on chemical inputs

The integration of microbial biofertilizers and biopesticides reduces the reliance on chemical fertilizers and pesticides. Farmers typically incur significant costs for purchasing synthetic fertilizers and pesticides. This, in turn, burdens overall production costs, reducing the potential profitability of agricultural practices. By using microorganisms as biofertilizers and biopesticides, which farmers can independently cultivate, production costs can be lowered, thus increasing the profitability of farming operations. This approach not only reduces expenses for farmers but also promotes healthier ecosystems and safer food production systems.

4. Target-specific action

Unlike chemical pesticides, which often affect nontarget organisms, microbial biopesticides are highly specific, reducing harm to beneficial insects and soil organisms. This specificity is especially important in preserving pollinators such as bees, beetles, or butterflies and other beneficial species such as natural enemies of pests (ladybugs, parasitic wasps) that contribute to agricultural productivity. This approach is highly beneficial for the environment and supports the long-term sustainability of agriculture, aligning closely with the goals of achieving sustainable farming practices.

19.5 Challenges and Future Directions of Microbial Biofertilizers and Biopesticides

Despite these benefits, there are several challenges to the widespread adoption of microbial biofertilizers and biopesticides. In detail, the challenges will be explained below:

1. Consistency and efficacy

The effectiveness of microbial products can be affected by environmental variables such as soil pH, temperature, and moisture. A common challenge is the inconsistent performance observed in the field compared with controlled laboratory settings. Since microorganisms are living organisms sensitive to numerous factors that impact their efficacy, researchers face the challenge of developing stable strains that maintain their activity when used as biofertilizers or biopesticides. This task is complex and demands continuous research to ensure optimal results before these findings are made available to farmers for practical use in their fields.

2. Shelf-life and storage

Microbial products often have a limited shelf-life and require specific storage conditions to maintain their viability. Innovations in formulation technology are needed to improve the shelf-life and ease of use of these products. Microorganisms used as biofertilizers and biopesticides generally have relatively short shelf-lives, as their ability to function optimally is time dependent. Bacteria-based microorganisms, in particular, often need to be rejuvenated to maintain their efficacy, as do fungal microorganisms, which have varying active periods that tend to be shorter than those of synthetic fertilizers and pesticides.

3. Regulatory hurdles

The regulatory framework for biofertilizers and biopesticides is often stringent, slowing the approval process for new products. Streamlining regulations while ensuring safety and efficacy will be critical for promoting microbial solutions in agriculture. A collaborative effort among farmers, biofertilizer and biopesticide producers, and government agencies, such as the Ministry of Agriculture, is essential to streamline legal processes and provide guidance for the release of new biofertilizer and biopesticide products that have undergone rigorous research and field trials. Farmers will have easier access to biofertilizer and biopesticide products with lower prices and more environmental advantages if guidelines and regulations are clear and accessible. This strategy immediately contributes to meeting the aims of sustainable agriculture.

19.6 Conclusion

This paper examines the application of microbes as biofertilizers and biopesticides in promoting sustainable agriculture, with the aim of enhancing environmental health, economic viability, and social equity. Soil microbes are being explored as environmentally friendly alternatives that improve crop productivity and soil health and reduce the dependence on chemical inputs. Biofertilizers, such as nitrogen-fixing bacteria, PSMs, and mycorrhizal fungi, increase plant growth by increasing nutrient availability, whereas biopesticides such as *Trichoderma* serve as biocontrol agents against pathogenic fungi. Microbial solutions show a significant promise for fostering sustainable agricultural practices, offering a natural and eco-friendly alternative to chemical inputs that enhances soil fertility, pest control, and overall plant health. However, to fully capitalize on their potential, further research is necessary to optimize product consistency, efficacy, and farmer adoption. With continued advancements, microbial technologies can play a pivotal role in shifting global agriculture toward more sustainable and resilient systems.

References

Alizadeh O (2011) Mycorrhizal symbiosis. Mycorrhizal symbiosis 3:273–281. https://doi.org/10.1016/B978-0-12-370526-6.X5001-6

Barea M, Azc R (2002) Mycorrhizosphere interactions to improve plant fitness and soil quality. Antonie Van Leeuwenhoek 81(1–4):343–351

Bashan Y, de-Bashan LE, Prabhu SR, Hernandez JP (2014) Advances in plant growth-promoting bacterial inoculant technology: formulations and practical perspectives (1998-2013). Plant Soil 378:1–33. https://doi.org/10.1007/s11104-013-1956-x

Bashir Z, Hamid B, Yatoo AM, Nisa M, Sultan Z, Popescu SM (2024) Phosphorus solubilizing microorganisms: an eco-friendly approach for sustainable plant health and bioremediation. J Soil Sci Plant Nutr. https://doi.org/10.1007/s42729-024-02007-1

Bennett, A., Schork, W., 1979. Studies toward a sustainable agriculture in northern Ghana: an internal working paper on agro- ecosystems

Bona E, Todeschini V, Cantamessa S, Cesaro P, Copetta A, Lingua G, Gamalero E, Berta G, Massa N (2018) Combined bacterial and mycorrhizal inocula improve tomato quality at reduced fertilization. Sci Hortic (Amsterdam). 234:160–165. https://doi.org/10.1016/j.scienta.2018.02.026

Bowman KD, Joubert J (2020) Citrus rootstocks. Genus Citrus:105–127. https://doi.org/10.1016/B978-0-12-812163-4.00006-1

Chang ST (1987) Microbial biotechnology - integrated studies on utilization of solid organic wastes. Resour Conserv 13:75–82. https://doi.org/10.1016/0166-3097(87)90051-4

Chaudhary T, Dixit M, Gera R, Shukla AK, Prakash A, Gupta G, Shukla P (2020) Techniques for improving formulations of bioinoculants. 3 Biotech 10:1–9. https://doi.org/10.1007/s13205-020-02182-9

De T, Roy A (2024) Antagonistic potential of mutants of Trichoderma species antagonistic potential of mutants of Trichoderma species. https://doi.org/10.12692/ijb/25.1.104-114

Demir S, Şensoy S, Ocak E, Tüfenkçi Ş, Demirer Durak E, Erdinç Ç, Ünsal H (2015) Effects of arbuscular mycorrhizal fungus, humic acid, and whey on wilt disease caused by verticillium dahliae kleb. In three solanaceous crops. Turk J Agric For 39:300–309. https://doi.org/10.3906/tar-1403-39

Dhiman M, Sharma L, Kaushik P, Singh A, Sharma MM (2022) Mycorrhiza: an ecofriendly bio-tool for better survival of plants in nature. Sustainability 14. https://doi.org/10.3390/su141610220

du Jardin P (2015) Plant biostimulants: definition, concept, main categories and regulation. Sci Hortic (Amsterdam) 196:3–14. https://doi.org/10.1016/j.scienta.2015.09.021

El-Maraghy SSM, Abdel-Rahman MAA, Hassan SHM, Hussein KA (2023) Pathogenicity and other characteristics of the endophytic Beauveria bassiana strain (Bals.) (Hypocreales: Cordycipitaceae). Egypt J Biol Pest Control 33. https://doi.org/10.1186/s41938-023-00690-3

Hao ZH, Fayolle L, van Tuinen D, Chatagnier O, Li X, Gianinazzi S, Pearson VG, Cosenza I, Gianinazzi-Pearson V (2012) Local and systemic mycorrhiza-induced protection against in Posidonia oceanica cadmium induces changes in DNA the ectoparasitic nematode Xiphinema index involves methylation and chromatin patterning priming of defense gene responses in grapevin. J Exp Bot 63:695–709. https://doi.org/10.1093/jxb/err313

Hii YS, Yen San C, Lau SW, Danquah MK (2020) Isolation and characterization of phosphate solubilizing microorganisms from peat. Biocatal Agric Biotechnol 26:101643. https://doi.org/10.1016/j.bcab.2020.101643

Jacott CN, Murray JD, Ridout CJ (2017) Trade-offs in arbuscular mycorrhizal symbiosis: disease resistance, growth responses and perspectives for crop breeding. Agronomy 7:1–18. https://doi.org/10.3390/agronomy7040075

Khallal SME (2007) Induction and modulation of resistance in tomato plants against fusarium wilt disease by bioagent fungi (arbuscular mycorrhiza) and/or hormonal elicitors (Jasmonic Acid& Salicylic Acid): 1- changes in growth, some metabolic activities and endogenous Hormo. Aust J Basic Appl Sci 1:691–705

Kong Z, Liu H (2022) Modification of rhizosphere microbial communities: a possible mechanism of plant growth promoting rhizobacteria enhancing plant growth and fitness. Front Plant Sci 13. https://doi.org/10.3389/fpls.2022.920813

Lugtenberg B, Kamilova F (2009) Plant-growth-promoting rhizobacteria. Ann Rev Microbiol 63:541–556. https://doi.org/10.1146/annurev.micro.62.081307.162918

Mahfut, Setiawan A, Sari M, Sijabat VE, Siregar VAPP, Ahmad Z (2023) Molecular characterization of mycorrhiza and its potential as biocontrol. Sabrao J Breed Genet 55:2092–2104. https://doi.org/10.54910/sabrao2023.55.6.21

Mar Vázquez M, César S, Azcón R, Barea JM (2000) Interactions between arbuscular mycorrhizal fungi and other microbial inoculants (Azospirillum, pseudomonas, Trichoderma) and their effects on microbial population and enzyme activities in the rhizosphere of maize plants. Appl Soil Ecol 15:261–272. https://doi.org/10.1016/S0929-1393(00)00075-5

Meddad-Hamza A, Benzina F, Meddad C, Hamza N, Reghmit A, Ziane H, Ksentini H (2023) Biological control of arbuscular mycorrhizal fungi and Trichoderma harzianum against *Fusarium oxysporum* and Verticillium dahliae induced wilt in tomato plants. Egypt J Biol Pest Control 33. https://doi.org/10.1186/s41938-023-00737-5

Mitra D, Djebaili R, Pellegrini M, Mahakur B, Sarker A, Chaudhary P, Khoshru B, Del Gallo M, Kitouni M, Barik DP, Panneerselvam P, Mohapatra PKD (2021) Arbuscular mycorrhizal symbiosis: plant growth improvement and induction of resistance under stressful conditions. J Plant Nutr 44:1993–2028. https://doi.org/10.1080/01904167.2021.1881552

Mkhwanazi GJM, Halleen F, Mostert L (2024) Fungicide sensitivity of Trichoderma atroviride and the application of this biocontrol fungus to protect grapevine sucker. Wounds 45:66–74

Nadeem SM, Ahmad M, Zahir ZA, Javaid A, Ashraf M (2014) The role of mycorrhizae and plant growth promoting rhizobacteria (PGPR) in improving crop productivity under stressful environments. Biotechnol Adv 32:429–448. https://doi.org/10.1016/j.biotechadv.2013.12.005

Ngalimat MS, Hata EM, Zulperi D, Ismail SI, Ismail MR, Zainudin NAIM, Saidi NB, Yusof MT (2021) Plant growth-promoting bacteria as an emerging tool to manage bacterial rice pathogens. Microorganisms 9:1–23. https://doi.org/10.3390/microorganisms9040682

Ortas I (2012) Mycorrhiza in citrus. In: Advances in citrus nutrition. https://doi.org/10.1007/978-94-007-4171-3

Pang F, Li Q, Solanki MK, Wang Z, Xing YX, Dong DF (2024) Soil phosphorus transformation and plant uptake driven by phosphate-solubilizing microorganisms. Front Microbiol 15:1–13. https://doi.org/10.3389/fmicb.2024.1383813

Prismantoro D, Akbari SI, Permadi N, Dey U, Anhar A, Miranti M, Mispan MS, Doni F (2024) The multifaceted roles of Trichoderma in managing rice diseases for enhanced productivity and sustainability. J Agric Food Res 18:101324. https://doi.org/10.1016/j.jafr.2024.101324

Rawat P, Das S, Shankhdhar D, Shankhdhar SC (2021) Phosphate-solubilizing microorganisms: mechanism and their role in phosphate solubilization and uptake. J Soil Sci Plant Nutr 21:49–68. https://doi.org/10.1007/s42729-020-00342-7

Sain SK, Kranthi S, Kranthi KR, Monga D, Paul D, Prasad YG (2024) Diversity study of Beauveria bassiana species for finding the most virulent strain to manage Bemisia tabaci in cotton. Appl Microbiol Biotechnol 108. https://doi.org/10.1007/s00253-024-13188-1

Sariasih Y (2024) Growth and health response of citrus against arbuscular mycorrhizal fungi (AMF) and Candidatus Liberibacter asiaticus inoculation in rootstock and Scion combinations. Diss. Univ. Gadjah Mada Yogyakarta

Sariasih Y, Subandiyah S, Widyaningsih S, Khurshid T, Mo J (2023) Identification and molecular detection of arbuscular mycorrhizal fungi colonized naturally in citrus roots in nurseries. Biodiversitas 24:6271–6278. https://doi.org/10.13057/biodiv/d241150

Schouteden N, De Waele D, Panis B, Vos CM (2015) Arbuscular mycorrhizal fungi for the biocontrol of plant-parasitic nematodes: a review of the mechanisms involved. Front Microbiol 6:1–12. https://doi.org/10.3389/fmicb.2015.01280

Schüßler A, Walker C (2010) A species list with new families and new genera. Glomeromycota 57

Sinha B, Haldhar SM, Sarangi PK, Chakrapani K, Mishra LK, Sandhya RP, Tampakleima W, Konjengbam R, Th B, Ralte Z (2024) Trichoderma role as an essential plant growth promotion-a review. J Agric Ecol 19:1–25. https://doi.org/10.58628/jae-2419-201

Smith TR, Koprivnikar J (2024) Influences of compound age and identity in the effectiveness of insect quinone secretions against the fungus Beauveria bassiana. Parasitol Res 123. https://doi.org/10.1007/s00436-024-08145-w

Srivastava AK (2012) Advances in citrus nutrition. In: Advances in citrus nutrition. https://doi.org/10.1007/978-94-007-4171-3

Sun W, Shahrajabian MH, Soleymani A (2024) The roles of plant-growth-promoting rhizobacteria (PGPR)-based biostimulants for agricultural production systems. Plants (Basel) 13:1–37. https://doi.org/10.3390/plants13050613

Tahat MM, Sijam K, Othman R (2010) Mycorrhizal Fungi As Biocontrol.pdf. Plant Pathol

Tayabali AF, Seligy VL (1995) Semiautomated quantification of cytotoxic damage induced in cultured insect cells exposed to commercial bacillus thuringiensis biopesticides. J Appl Toxicol 15:365–373. https://doi.org/10.1002/jat.2550150505

Varma A, Prasad R (2020) Plant microbe symbiosis. https://doi.org/10.1007/978-3-030-36248-5

Vassilev N, Vassileva M, Nikolaeva I (2006) Simultaneous P-solubilizing and biocontrol activity of microorganisms: potentials and future trends. Appl Microbiol Biotechnol 71:137–144. https://doi.org/10.1007/s00253-006-0380-z

Vinale F, Sivasithamparam K, Ghisalberti EL, Marra R, Woo SL, Lorito M (2008) Trichoderma-plant–pathogen interactions. Soil Biol Biochem 40:1–10. https://doi.org/10.1016/j.soilbio.2007.07.002

Wang C, Pan G, Lu X, Qi W (2023) Phosphorus solubilizing microorganisms: potential promoters of agricultural and environmental engineering. Front Bioeng Biotechnol 11:1–5. https://doi.org/10.3389/fbioe.2023.1181078

Wu QS (2017) Arbuscular mycorrhizas and stress tolerance of plants. In: Arbuscular mycorrhizas and stress tolerance of plants. https://doi.org/10.1007/978-981-10-4115-0

Yang C, Ancona V (2021) Metagenomic analysis reveals reduced beneficial microorganism associations in roots of foot-rot-affected citrus trees. Phytobiomes J 5. https://doi.org/10.1094/PBIOMES-07-20-0049-R

Zhang MQ, Guo Y, Powell CA, Doud MS, Yang CY, Zhou H, Duan YP (2016) Zinc treatment increases the titer of "Candidatus Liberibacter asiaticus" in huanglongbing-affected citrus plants while affecting the bacterial microbiomes. J Appl Microbiol 120:1616–1628. https://doi.org/10.1111/jam.13102

Ziedan ES, Elewa I, Mostafa M, Sahab A (2011) Application of mycorrhizae for controlling root diseases of sesame. J Plant Prot Res 51:355–361. https://doi.org/10.2478/v10045-011-0058-0

Chapter 20
Portraying Plant Drought Responses and Its Mechanisms of Tolerance by Rhizobacteria

Sakthi Uma Devi Eswaran and Lalitha Sundaram

Abstract Drought, an ecological stress, poses significant challenges for agricultural practices. Various strategies are being explored to improve the resilience of plants during drought conditions, with an emphasis on plant–microbe interactions to effectively increase drought tolerance in crops and maintain yield under such challenging circumstances. Plant growth promoting rhizobacteria (PGPRs) can facilitate plant growth via phytostimulation and the production of osmolytes, antioxidative enzymes, and exopolysaccharides under drought conditions. Nonetheless, our knowledge regarding how PGPRs influence plant drought resistance and their interactions with plants during drought conditions remains insufficient. Therefore, this chapter summarizes the significant impact of drought on plant growth within agricultural ecosystems and the role of drought-tolerant microbes, reveals the initiation of tolerance mediated by rhizobacterial inoculation through plant transduction pathways, and elucidates the interaction mechanisms of plants with drought-tolerant microbes.

Keywords Drought · Plant growth promoting rhizobacteria · Plant–microbe interaction · Mitigation · Sustainable · Agriculture

S. U. D. Eswaran
Research Scholar, Soil Biology and PGPR Lab, Department of Botany, Periyar University, Salem, Tamil Nadu, India

L. Sundaram (✉)
Assistant Professor, Soil Biology and PGPR Lab, Department of Botany, Periyar University, Salem, Tamil Nadu, India
e-mail: lalithabot@periyaruniversity.ac.in

V. Gopikrishnan et al. (eds.), *Harnessing Microbial Resources for Sustainable Agriculture*, Microorganisms for Sustainability 61,
https://doi.org/10.1007/978-981-95-4817-0_20

20.1 Introduction

Drought is one of the most deleterious climatic factors and varies in severity and frequency in all-weather zones (Chen et al. 2024). Owing to the changing climate, the demand for irrigation has increased with increased evaporation and transpiration as a result of global warming in almost all regions of the world. These elements, combined with the decline in precipitation, cause a serious demand for future food safety and contribute to ecological stress (Corwin 2021). Drought is expected to have a major impact on plant development and, eventually, productivity worldwide by 2050 (Seleiman et al. 2021).

In agriculture, drought is characterized as a period of below-normal precipitation during which the levels of accessible water in the plant rhizosphere fall below the thresholds necessary for efficient development and biomass production. This process has a significant effect on all essential physiological functions, including photosynthesis, respiration, and mineral nutrient intake (Zia et al. 2021). Drought has an impact on soil fertility, crop yield, and, ultimately, economic outcomes. The main objective of sustainable agriculture is to increase the resilience of crops to stress and increase their yield to satisfy the dietary requirements of the expanding population (Kumari et al. 2022). As a result, increasing agricultural output through the use of scientific and improved techniques is critical to meet the rising demand. Drought tolerance and other scientific and socioeconomic strategies are important for preserving agriculture in the future.

Over the past few decades, the use of microbe-based plant biotechnology in pioneer research has improved with respect to agricultural drought resistance because of its field operability and environmental friendliness (Zhang et al. 2022). Plants feed carbon (C) to the soil food web through their below-ground carbon inputs. In turn, the activities of the rhizosphere microbiome release nutrients for plant growth and determine the balance between C respiration and soil stabilization (Ayangbenro et al. 2022). Furthermore, microbes interact with plants directly in the rhizosphere by developing symbiotic relationships, increasing plant growth through phytohormone synthesis, or reducing plant stress signals (Khan et al. 2021). Therefore, the relative importance of plants, the microbiome, and drought is undeniable.

20.2 Drought-tolerant and Plant Growth Promoting Rhizobacteria

Drought is recognized worldwide as one of the major constraints on agricultural productivity. Plants use three major strategies to address water scarcity: (i) drought escape, (ii) drought avoidance, and (iii) drought tolerance. In general, all three approaches have an impact on the development of drought resistance, which refers to the capacity to sustain a water balance and turgor pressure in arid conditions.

Consequently, drought resistance involves a complicated process that necessitates a worldwide perspective to understand its fundamental mechanisms. As a result, a sophisticated multilevel regulatory network is needed to manage plant adaptive responses to drought stress (Bashir et al. 2021). Although many networks have been resolved, there is still a significant gap. Thus, predicting solutions for the effects of drought on food security is valuable and important. Crop yields are being improved through breeding techniques, but there are ethical and time constraints.

One such drought resistance strategy could be the application of stress-tolerant PGP rhizobacteria, which play critical roles in maintaining plant growth performance under adverse conditions (Mohamed et al. 2024). Plants adapt to water scarcity by changing their morphology and physiology but require the assistance of associated microbiomes to alleviate stress. Furthermore, inoculation with PGPR is one of the most effective techniques for enhancing plant drought resistance (Al-Turki et al. 2023). Inoculating PGPR into plant roots is an effective and simple technology. Plant growth-promoting rhizobacteria (PGPR) have attracted the attention of numerous agrochemical firms, which have selected the most effective PGPR for creating and bringing bioproducts to markets. To alleviate stress, these microbes increase nodulation, enzyme production, and a variety of physiological processes in the rhizosphere (Bittencourt et al. 2023). The interaction of PGPR with plant roots, as well as their ideal effects on crop development and yield under stressed and unstressed conditions, has been extensively studied and understood. Despite advances in our understanding of interactions in plant–microbe pathways, our knowledge of these mechanisms under drought conditions remains limited. In this study, an attempt was made to elucidate the plant response to drought and all the mechanisms adopted by PGPR to mitigate it.

20.3 Identification of markers of drought stress in plants

20.3.1 Plant morphological responses as markers of drought

Water is essential for life on Earth, whereas drought or osmotic stress has enormous effects on numerous aspects of plant morphology and physiology. Drought stress often causes major declines in the development and yield of several food crops (Chada et al. 2023). Plants subjected to severe drought typically exhibit a slight shift in growth pattern, as well as a small increase in the root mass fraction (RMF). Under moderate drought, plants sustain aboveground growth and thus compete for resources as much as possible.

Compared with the control plants, the plants exposed to extreme drought presented a greater than 50% reduction in biomass and responded with a significant increase in RMF. To avoid dehydration, the architecture and resource allocation of roots are adjusted during moderate drought stress (Zia et al. 2021; Kou et al. 2022).

Under extreme drought, roots shrink, and photosystem II (PSII) in leaves becomes dysfunctional (Ma et al. 2020). Osmotic stress further restricts the diffusion and transport of water-soluble nutrients (Sati et al. 2023). Plants have evolved numerous stress-responsive signaling pathways and defense mechanisms to adapt to drought.

20.3.2 Physiological and biochemical markers of drought stress

Early responses to drought induce water stress, leading to major alterations in the growth, photosynthetic activity, and primary metabolism of plants. A category of physiological and biochemical markers ideally accompanies drought. The dynamic characterization of plant adaptation responses is greatly aided by these markers (Wahab et al. 2022). It is thought that roots recognize dryness and chemically communicate this information to the shoot. A key component of this signaling pathway is abscisic acid (ABA). This effector is transported to leaf epidermal cells, resulting in stomatal closure due to the response of hydraulic signals in vascular tissues. As a result, xylem transport, turgor pressure, and root growth are suppressed (Dos Santos et al. 2022). These responses are characterized as physiological and biochemical markers, which are discussed in the following sections (Fig. 20.1).

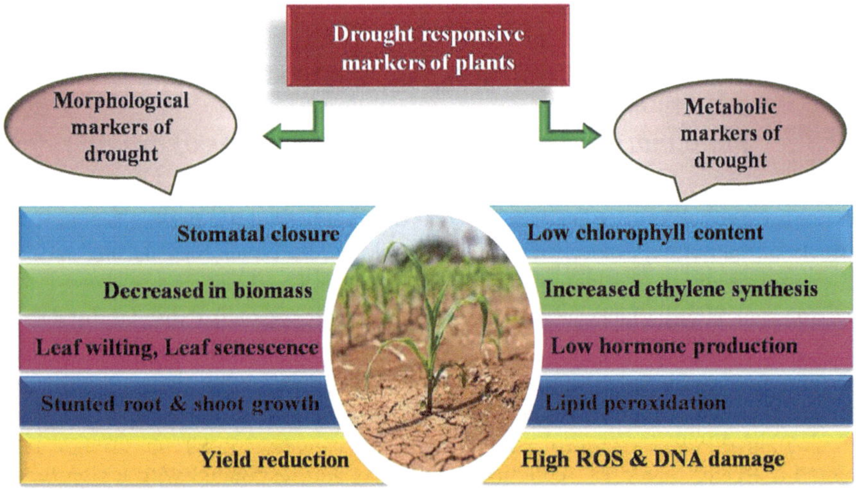

Fig. 20.1 Responsive markers of drought stress in plants

20.3.3 Water status and photosynthetic parameters as markers of drought stress

The desiccation of plant parts, characterized by a reduction in water potential and leaf turgor, is one of the first apparent indicators of drought (Munné-Bosch and Villadangos 2023). Stomatal closure is a fundamental plant response to dehydration that is meant to minimize transpiration as a drought escape approach (Lucas et al. 2023). Drought reduces photosynthetic efficiency by reducing carbon assimilation and light reactions because stomatal closure disturbs the delivery of carbon dioxide to parenchyma cells (Kapoor et al. 2020). Dehydration, in addition to degrading photosynthetic pigments, has a deleterious effect on the entire photosynthetic apparatus.

20.3.4 Differential phytohormone patterns as markers of drought stress

Plant adaptation to environmental variations is a complex process that is precisely adjusted by multiple regulatory systems. More precisely, desiccation triggers the activation of signal transduction pathways, which include phytohormone-mediated long-distance transport stages (Govind et al. 2020). ABA-dependent signaling coordinates the closing of stomata during drought (Ahmad et al. 2023). In addition to ABA, the regulation of stomatal conductance during water scarcity is influenced by a few additional hormones and the networks of interactions between them. Thus, ethylene, cytokinin, and auxin are committed to blocking the ABA-mediated mechanism of stomatal closure, whereas brassinosteroids, isoleucinyl jasmonate, jasmonic acid, and salicylic acid increase the actions of ABA (Salvi et al. 2021; Zhao et al. 2021). Therefore, the aforementioned phytohormones exhibit various patterns of action irrespective of their influence on the drought response.

20.3.5 Metabolites as markers of drought stress

The elevated level of osmotically active and metabolically neutral solutes, including various sugars, amino acids (glycine and proline), betaine, polyamines, and organic acids, during dry circumstances is a well-documented process of metabolic adjustment (Kaur et al. 2024). Metabolic adjustment, the second stage of plant adaptation to drought following stomatal closure, is essential for preserving the water availability and biological activity of plant cells. Metabolites serve as both species-specific and universal markers of drought stress (Takahashi et al. 2020).

20.3.6 Proteins as protective markers of drought

Long-term drought adaptation in plants results in a significant increase in the expression of genes relevant to drought (Moursi et al. 2021). The upregulation of a wide range of drought-protective proteins, including chaperones, late embryogenesis abundant (LEA) proteins, and antioxidant defense enzymes, has also been noted (Osmolovskaya et al. 2018). The group of proteins known as chaperones is responsible for the synthesis and maintenance of heat-shock proteins involved in the shock response and in an array of stress adaptation responses (Camaille et al. 2021). A special emphasis is now focused on the function of heat-shock proteins in drought resistance. Another kind of polypeptide associated with water deficit adaptation is LEA proteins. Drought resistance can be identified by determining the intrinsic rate of LEA proteins as a marker (Chaffai et al. 2024).

20.3.7 Oxidative stress associated with drought

A shortage of water generates oxidative stress and increases the production of reactive oxygen species (ROS) by altering the equilibrium between ROS formation and detoxification (Zia et al. 2021). The saturated state of electron transport chains in mitochondria and chloroplasts with reduced forms of nucleotides is the primary cause of the excessive production of cellular ROS in response to plant dehydration (Nadarajah 2020). The degree of drought stress is positively correlated with ROS generation. This makes it possible to employ oxidative stress-related molecules as biochemical indicators of drought. Lipid hydroperoxides are generated when ROS attack the double bonds of polyunsaturated fatty acids (PUFAs). As a result, these compounds break down to produce shorter and more reactive carbonyl compounds, including malondialdehyde, which is a reliable indicator of lipid oxidative damage (Hasanuzzaman et al. 2020; Kumar et al. 2021). Different plant species and cultivars may be screened for drought resistance on the basis of the expression of certain antioxidant enzymes.

20.4 PGPR–Plant interactions and mechanisms for alleviating drought stress

Microbe–plant interactions are considered essential factors in determining ecological processes. In the root system, plants closely coexist with microbial populations (Aqeel et al. 2023). Plant-associated microbes, particularly plant growth-promoting microorganisms (PGPMs), can influence the growth, nutritional status, development, and fitness of host plants, whereas roots generally define the niche or environment where microbial communities establish and survive (Chauhan et al. 2023).

Host plants may benefit from drought-resistant PGPM and become more capable of withstanding a variety of abiotic challenges. Agriculture losses under unfavorable climate change circumstances may be reduced by the isolation, characterization, and identification of microbes with the capacity to increase plant tolerance to drought.

The growth, nutrient absorption, and systemic stress resistance of their host plants are improved by PGPR, which live in the rhizosphere or as epiphytes in roots (Eswaran et al. 2023). The majority of these rhizobacteria are gram-negative bacteria that may attach themselves to plants by fixing nitrogen, solubilizing phosphate, producing phytosiderophores, and adjusting phytohormones.

Plant growth promoting rhizobacteria can increase agronomic sustainability and reduce drought stress. These microorganisms are the most prevalent microorganisms that reduce drought stress. According to Chieb and Gachomo (2023), these compounds are essential for plant stress responses and produce chemicals that have a significant potential as biostimulants. By causing biochemical alterations, PGPRs reduce drought stress through rhizobacterium-induced drought endurance and resilience (RIDER) (Eswaran et al. 2024).

Different RIDER mechanisms involve enhancing the antioxidant defense system, the production of bacterial exopolysaccharides (EPSs), the synthesis of phytohormones, and adherence to cyclic metabolic pathways. These processes result in the accumulation of various carbon-based compounds, including sugars, amino acids, and polyamines, along with the formation of heat-shock proteins (Fadiji et al. 2022).

Molecular interactions occur between host plants and PGPR to modify rhizosphere communities in response to drought stress (Patel et al. 2023). Numerous previous studies have demonstrated that PGPRs are associated with the drought tolerance of host plants (Table 20.1). There are three primary ways in which PGPRs contribute to enhancing plant drought resistance: (1) enhancing stress alleviators, (2) reducing stress aggravators, and (3) gene alterations and genomic responses under drought conditions are shown in Fig. 20.2.

20.4.1 Enhancing the Stress Alleviators

20.4.1.1 Phytohormonal modification by PGPR under drought

Drought stress conditions can be mitigated by plant interactions with phytohormone-producing rhizobacteria. Plant growth-promoting bacteria play crucial roles in encouraging the production of phytohormones, which are essential for increasing plant growth, development, and tolerance to drought stress. Phytohormones released by bacteria influence the metabolism and respiration rate of host plants. The major signaling and developmental cascades involved in stress tolerance are reprogrammed by phytohormones, which serve as essential links in this process (Gupta et al. 2022; Kumar et al. 2024a, b).

Table 20.1 PGPR mechanisms involved in diverse crops under drought mitigation

Crop inoculated	PGP microbes	Mechanism of action	References
Cucumis sativus (cucumber)	*Acinetobacter*, *Pseudomonas*, *Burkholderia*	Gibberellins synthesis	Kang et al. (2015)
Arachis hypogaea L. (groundnut)	*Acinetobacter calcoaceticus* AC01	Production of hormones, increased plant physiological and biochemical activity	Eswaran and Sundaram (2024)
Setaria italica (foxtail millet)	*Acinetobacter* EU- LRNA-72, *Penicillium* sp. EU-FTF-6	Enhanced accumulation of glycine betaine, proline, chlorophyll content, and reduction of lipid peroxidation	Kour et al. (2020)
Triticum aestivum (wheat)	*Azospirillum*	Drought tolerant by phytohormones and exopolysaccharide	Priyanka et al. (2019)
Trema micrantha (nettle tree)	*Azospirillum brasilense*	Protein and starch content, photosynthesis, reduced lipid peroxidation	Tiepo et al. (2020)
Zea mays (maize)	*Azospirillum brasilense*	Higher biomass and lower levels of abscisic acid and ethylene	Cura et al. (2017)
Glycine max (soybean)	*Bacillus* strains UFGS1, UFGS2, UFGRB2, and UFGRB3	Increased stomatal conductance, transpiration, and decreased lipid peroxidation	Martins et al. (2018)
Triticum aestivum (wheat)	*Bacillus*, *Enterobacter*, *Moraxella*, and *Pseudomonas*	Auxin accumulation	Raheem et al. (2018)
Zea mays (maize)	*Bacillus* sp. (12D6) and *Enterobacter* sp. (16i)	IAA and salicylic acid production	Jochum et al. (2019)
Triticum aestivum (wheat)	*Bacillus thuringiensis*	Production of volatile organic compounds	Khan et al. (2018)
Piper nigrum (pepper)	*Bacillus licheniformis*	Stress-related protein and genes	Lim and Kim (2013), Ali and Xie (2020)
Oryza sativa (rice)	*Bacillus altitudinis*	Increased relative water content, chlorophyll stability, and membrane stability index	Kumar et al. (2017)
Zea mays (maize)	*Klebsiella variicola* F2, *Pseudomonas fluorescens* YX2, *Raoultella planticola* YL2	Accumulation of choline	Gou et al. (2015)
Triticum aestivum (wheat)	*Pseudomonas azotoformans* FAP5	ACC deaminase, EPS, and biofilm activity	Ansari et al. (2021)
Mentha piperita (peppermint)	*Pseudomonas fluorescens* and *Bacillus amyloliquefaciens* GB03	Increase in phenolic compounds and antioxidant defense	Chiappero et al. (2019)

(continued)

Table 20.1 (continued)

Crop inoculated	PGP microbes	Mechanism of action	References
Ziziphus jujube (jujube)	*Pseudomonas lini*, *Serratia plymuthica*	Induced choline accumulation	Zhang et al. (2020)
Arabidopsis thaliana (thale cress)	*Pseudomonas putida*	Growth parameters, chlorophyll, and proline content	Ghosh et al. (2017)
Glycine max (soybean)	*Pseudomonas simiae*	Phytohormone biosynthesis and modulation of gene expression profile	Vaishnav and Choudhary (2019)
Sorghum bicolor (great millet)	*Pseudomonas* sp.	Modulation of protein profiles	Dhawi (2020)
Vigna radiata (mung bean)	*Staphylococcus* sp.	Production of acetic acid and ACC deaminase	Jayakumar et al. (2020)
Triticum aestivum (wheat)	*Streptomyces pactum* Act12	Over expression of stress-related genes *EXPA2*, *EXPA6*, and *P5SC*	Li et al. (2020)

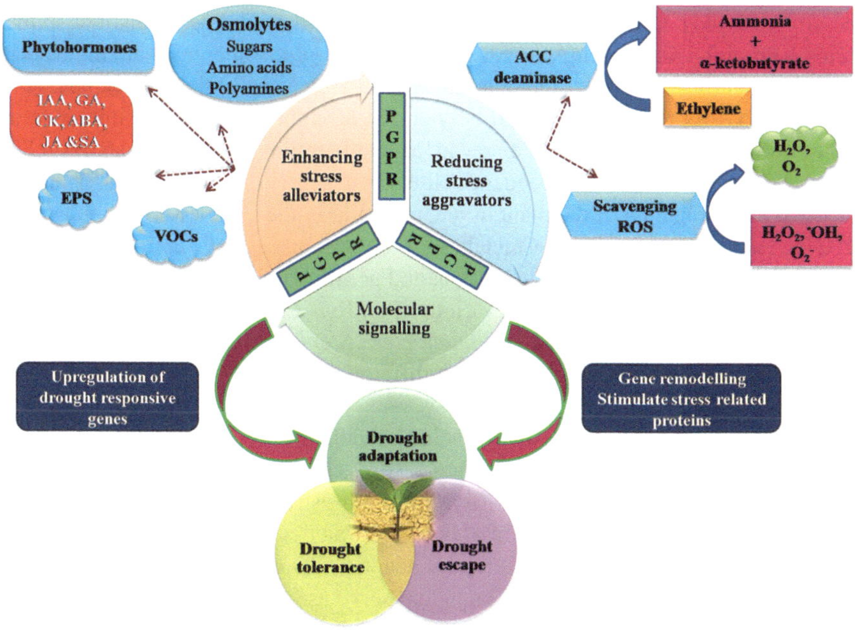

Fig. 20.2 Mechanism of PGPR-mediated drought mitigation. PGPR plant growth-promoting rhizobacteria, IAA indole acetic acid, GA gibberellins, CK cytokinins, ABA abscisic acid, JA jasmonic acid, SA salicylic acid, EPS exopolysaccharide, VOCs volatile compounds, ROS reactive oxygen species

The most effective form of auxin in plants is indole acetic acid (IAA), a significant hormone that controls development and expansion. It helps in germination, root growth, and plant cell differentiation and division (Gomes and Scortecci 2021). According to estimates, 80% of soil bacteria can secrete IAA as a secondary metabolite. The majority of known PGPR are now capable of synthesizing IAA. In the presence of tryptophan, PGPRs produce indole-3-acetic acid, an auxin phytohormone (Koza et al. 2022).

Bacteria belonging to the genera *Azospirillum*, *Pseudomonas*, *Xanthomonas*, *Rhizobium*, *Enterobacter*, *Serratia*, *Mycobacterium*, *Burkholderia*, *Azotobacter*, *Bacillus*, and *Bradyrhizobium* reportedly synthesize auxins. Diverse metabolic pathways, including the indole-3-pyruvic acid, indole-3-acetamide, tryptamine, and indole-3-acetonitrile pathways, are involved in the production of bacterial IAA (Meena et al. 2020).

Auxin-producing bacteria inoculated with different species significantly facilitate the propagation of lateral root hairs, increase nutrient and water intake, and help plants survive drought stress (Uzma et al. 2022). The genes involved in auxin production, signaling, or transport machinery are impacted by PGPR. Depending on the strain, PGPRs affect auxin homeostasis through a range of strategies. To successfully employ each PGPR strain in agriculture, understanding its distinctive features is crucial. Under drought stress conditions, *B. subtilis* increases auxin levels in wheat by altering the expression of auxin-responsive genes (Mal and Panchal 2024).

Since PGPRs regulate IAA and ethylene in drought tolerance, its influence on other hormone levels cannot be disregarded. Gibberellins (GAs) are a special class of phytohormones that are frequently linked to alterations in plant morphology through the thickening of the plant tissue, particularly the stem tissue (Bhattacharya 2021). They elicit a variety of metabolic processes necessary for flowering, fruit production, senescence, stem elongation, and germination of seeds. The production of GA has been reported in various PGPRs, including *Enterobacter* sp., *Pseudomonas*, *Sphingomonas*, *Acinetobacter*, *Staphylococcus*, *B. pumilus*, and *B. licheniformis*, which can produce four different forms of GA (Ranjan et al. 2024; Lotfi et al. 2022).

Abscisic acid (ABA) and jasmonic acid (JA) dehydration trigger the biosynthesis of ABA, which is popularly referred to as a "stress hormone" because of its massive build-up under drought conditions. By controlling stress signal transduction pathways and stomatal closure, ABA maintains the water content (Kumar et al. 2024a). The endogenous levels of ABA and JA adapt to changes in response to abiotic stresses. ABA and jasmonates are known to serve as the main signaling molecules that control plant defense. JA contributes to the antioxidant activities caused by dehydration (Ali and Baek 2020). ABA plays a role in balancing the water status of plants by controlling guard cells and inducing genes encoding proteins and enzymes linked to dehydration (Muhammad Aslam et al. 2022).

20.4.1.2 Role of Volatile producing PGPR in promoting drought tolerance

When plants undergo drought stress, PGPRs release an array of secondary metabolites in combination with volatile organic compounds (VOCs), increasing plant development and reducing drought stress (Sunita et al. 2020). Metabolites with high vapor pressure and lipophilic characteristics are referred to as volatile compounds. They act as signaling molecules for cell-to-cell communication as well as interactions between inter and intrakingdoms. These volatile compounds can be distinguished by their small size (300 Da). A wide variety of VOCs, including esters, alcohols, aliphatic aldehydes, ethers, ketones, and sulfur compounds, are produced by PGPR (Chandrasekaran et al. 2022; Harkhani and Sharma 2024).

The volatile substances generated by PGPR cause plant resistance inadvertently. The released compounds serve as impulses for priming and stimulating plants (Ojuederie et al. 2019). Volatile molecules function as signals for plants to prime and respond systematically to stress. To assess agricultural drought stress and its relief quickly and noninvasively during stress development, VOCs are known to be prospective candidates. Additionally, a few of these substances have been shown to activate the production of genes that encode ROS-scavenging enzymes such as catalase, glutathione reductase, ascorbate reductase, and superoxide dismutase (Midzi et al. 2022; Laller et al. 2023).

20.4.1.3 Osmolytes: Biomolecules of PGPR that endure drought stress in plants

The accumulation of osmolyte molecules, also known as osmotic adjustment, is frequently considered a way to counteract the detrimental effects of water scarceness on agricultural production. One of the key physiological mechanisms for increasing resistance to conditions of water scarcity in plant cells is osmotic adaptation (Zulfiqar et al. 2020). Plants collect a variety of suitable osmolytes/solutes, such as amino acids, polyamines, betaines, sugars, polyhydric alcohols, quaternary ammonium and certain significant proteins, and dehydrins, to ease the effects of water stress (Ali and Khan 2021).

Microorganisms may acquire these substances directly from the environment or indirectly through de novo production. Osmolyte production in plants relies on physiology, growth conditions, and environmental factors. It is synthesized without disrupting metabolic processes, maintaining the osmotic balance between the cell's surroundings and the cytosol (Ghosh et al. 2021). In PGPR, the production of osmolytes is clearly characterized as a survival mechanism under adverse environmental conditions. The application of these secondary metabolites together has been more effective than PGPR alone in reducing the effects of drought stress (Gamalero and Glick 2022).

PGPRs produce a variety of secondary metabolites that are beneficial for plant productivity. With increasing drought stress, high quantities of osmolytes build and keep cells hydrated. In response to osmotic stress, PGPR strains and plants synthesize compatible solutes that work in concert to promote plant growth and survival, maintain cell turgor, and reduce water potential (Eswaran et al. 2024). Additionally, osmolytes are crucial for preserving the structural integrity of the membrane. Proline, trehalose, glycine betaine, mannitol, sorbitol and fructans are some of the main osmolytes (Yasmin et al. 2021; Ahmad et al. 2022).

Trehalose is a nonreducing sugar composed of two glucose units linked by glycosidic bonds (Hassan et al. 2023). It functions as a signaling molecule, and its compatibility with cellular metabolism, even at high concentrations, makes it a versatile and valuable compound in various biological processes. It functions as an osmoprotectant in microbes, protecting membranes and dehydrating enzymes, but it is uncommon in vascular plants (Sharma et al. 2020). It is known to exist in a wide variety of anhydrobionts, including several plant, invertebrate, fungal, and bacterial species. Owing to the involvement of both reducing subunits in the formation of glycosidic bonds, trehalose has a number of distinctive properties that make it an effective molecular, membrane-bound, and protein preservative (Raza et al. 2023).

Despite acidic pH circumstances, it has the capacity to maintain durability in a soluble form at extreme temperatures. Owing to its inability to form internal hydrogen bonds, it is highly hydrophilic. It has the ability to form a gel phase as cells dehydrate, causing it to be very stable and less prone to cause cellular harm under abiotic stress conditions such as salinity and drought. Trehalose replaces water molecules during dehydration by forming hydrogen bonds with the surrounding macromolecules and membranes (Kour and Yadav 2022).

Proline is an exceptionally rigid conformational proteinogenic amino acid that is crucial for basic metabolism. To counteract stress, proline is one of the key osmolytes that is produced and accumulates through protein hydrolysis (Farooqi et al. 2020). It is known to accumulate under conditions of drought, salt, exposure to heavy metals, and low temperature. Proline acts as a critical signaling molecule to regulate various mitochondrial processes, influence cell division and programmed cell death, and stimulate the recovery of expression from stressful conditions. The exceptional structural rigidity of proline endows it with great stability under adverse circumstances (Ahluwalia et al. 2021).

Excessive proline synthesis helps stabilize the subcellular structure (proteins and membranes), eliminate reactive oxygen species, buffer the cellular redox potential, serve as osmolytes for osmotic adjustments, and ultimately defend plants under stress (Abdela et al. 2020). It has been proposed that the proline accumulation of PGPMs can increase their survival and drought resistance. In response to osmotic stress, plants treated with rhizobacteria presented greater proline levels, which, in addition to the previously reported proline levels, help plants combat drought stress. Cucumber, potato, and great millet plants all presented increased proline contents after treatment with PGP microbes (Asghari et al. 2020; Li et al. 2020).

Glycine betaine (GB) is a quaternary ammonium osmoprotectant that is present extensively in various animals, microbes, and higher plants. It is derived from glycine and fully substituted with a methyl group. GB is a vital compound categorized under compatible solutes in plant physiology. These small organic metabolites play crucial roles in plant growth and development because of their unique solubility in water and nontoxic nature at high concentrations. These solutes play crucial roles in cellular osmotic regulation. Choline is important for the synthesis and build-up of glycine betaine for plant resilience (Mäkelä et al. 2019; Sharma et al. 2024).

GB exhibits a broad spectrum of protective functions, with the level of its accumulation being linked to the degree of stress tolerance. Dehydration stress is widely recognized as the primary cause of GB. Glycine betaine has distinctive characteristics, including excellent solubility, low molecular weight, and low viscosity, which contribute to its effectiveness as an efficient osmoregulatory agent (Backer et al. 2018; Singh et al. 2019). Notably, GB plays a crucial role in enhancing plant stress tolerance by stabilizing membranes and improving RuBisCO activity. GB safeguards the plant by regulating osmotic balance and maintaining the equilibrium of external proteins in PSII (Annunziata et al. 2019).

Two distinct biosynthetic routes have been identified that give rise to the production of GB. Betaine is commonly synthesized from choline through a two-step oxidation process in the majority of animals, microbes, and plants. In the first step, in animals and bacteria, this process involves the oxidation of choline to betaine aldehyde, which is catalyzed by choline dehydrogenase, whereas in plants, it is catalyzed by choline monooxygenase. The subsequent stage involves the conversion of betaine aldehyde to GB through oxidation facilitated by an enzyme called NAD^+-dependent dehydrogenase. GB is also involved in controlling the functions of antioxidant enzymes (Sharma et al. 2023).

20.4.1.4 PGPR alleviation of drought stress by exopolysaccharides

Drought stress impacts the ability of PGPR to release EPSs, which are composed of multiple organic compounds, saccharides, proteins, and uronic acid. Rhizobacterial EPS in the form of slime or capsules is discharged into the soil and contains homo- or heteropolysaccharides. EPSs occur on the microbial cell surface, where they safeguard and stabilize the membrane (Morel and Castro-Sowinski 2013). They are also actively present in soil organic matter, where they protect cells from water stress by assisting in the attraction and colonization of bacteria, the development of biofilms, and, ultimately, the interaction of plants and microbes.

The formation of EPS results in the creation of a smart, hydrated microenvironment guarding the plant roots from the drying effects of drought (Latif et al. 2022). During the stationary phase of bacterial growth, EPS is produced intracellularly. Numerous factors, including the stress of a drought, affect its production rate. By enhancing aggregate stability and the ratio of root-adhering soil to root tissue, EPS has been shown to promote microaggregates that enhance the development of drought stressed plants (Ansari et al. 2021; Ilyas et al. 2020).

The ability of rhizobacterial strains to conserve water depends on the polysaccharide composition of EPS. When EPSs are released into the soil-by-soil bacteria as slime and capsules, the clay surface can absorb it through the actions of van der Waal forces, hydrogen bonds, and anion absorption mechanisms, creating a protective capsule (Saha et al. 2020). Since the EPSs of capsules are resistant to dehydration, they can shield bacterial growth and plant roots from drought stress. EPSs produced by microbes shield them from hostile environments and aid in their survival, indirectly affecting the soil structure (Niu et al. 2017). In general, plants treated with EPS-producing microorganisms can tolerate water stress by altering the soil structure or encouraging soil aggregation.

20.4.1.5 ACC Deaminase producing rhizobacteria in plant stress tolerance

The plant produces ethylene in response to a variety of environmental factors, such as organic contaminants, salt, high temperatures, drought, and hazardous metals. Ethylene controls plant homeostasis (Salehi-Lisar and Bakhshayeshan-Agdam 2016; Vaseva et al. 2018); its accumulation limits water and food intake by reducing root and shoot growth under stress. S-adenosylmethionine is transformed into 1-aminocyclopropane-1-carboxylate (ACC) by 1-aminocyclopropane-1-carboxylate synthase during the production of ethylene (Abeles et al. 2012).

Plants release ACC into the rhizosphere, absorb it back at the root, and produce it into ethylene. To minimize ethylene accumulation and improve the ability of the root system to withstand environmental challenges, ACC is converted into ketobutyrate and ammonia by PGPR in the rhizosphere, which uses it as a source of energy and nitrogen (Chieb and Gachomo 2023). The plants produced less ethylene due to the low level of ACC.

PGPR strains that have ACC deaminase activity help reduce the negative impacts of drought stress and mineral deficiencies on plant growth and development. *Agrobacterium*, *Azospirillum*, *Achromobacter*, *Burkholderia*, *Enterobacter*, *Pseudomonas*, *Rhizobium*, and *Ralstonia* are PGPR strains that contain the ACC deaminase gene (Jabborova et al. 2021). IAA derived from plants or associated bacteria may proceed to function in any of these two ways. It might stimulate cell elongation or increase ACC synthase transcription to increase the quantity of ethylene in plants (Devarajan et al. 2021).

20.4.2 Reducing the Stress Aggravators

20.4.2.1 Regulating the levels of Ethylene

Ethylene, a plant growth regulator, has been a crucial factor in farming practices. It is involved in the normal growth of plants and their reactions to stress. Ethylene affects the development of vegetative parts such as stems, roots, and petioles, along with all stages of plant development. It is referred to as the "wounding hormone" because its production is easily triggered by physical or chemical disruption of plant tissues (Vaseva et al. 2018; Barnawal et al. 2019).

Ethylene triggers the stress response by causing hypertrophy, accelerating aging, promoting senescence and abscission, and inhibiting auxin transport, nodulation, and root elongation. Plant growth promoting (PGP) microbes contain a crucial enzyme called ACC deaminase with a molecular mass of approximately 35–42 kDa. This enzyme converts ACC, a precursor of ethylene biosynthesis, into α-ketobutyrate and ammonia, thus modulating the levels of ethylene in plants.

ACC deaminase reduces ethylene production in plant roots, resulting in root extension (Gamalero and Glick 2015; Abeles et al. 2012).

ACC deaminase-producing bacteria play a vital role in protecting plants from ethylene-induced damage during drought stress. Inoculating crops with PGP bacteria that carry ACC deaminase characteristics may help maintain plant growth and development under stress conditions (Glick 2012). *Bacillus* sp., *Methylobacterium* sp., and *Streptomyces* sp. all produce the ACC deaminase enzyme (Gupta and Pandey 2019; Devarajan et al. 2021).

20.4.2.2 Mechanisms of ROS scavenging by PGPR

In the normal course of aerobic metabolism in plant cells, the formation of ROS is typically moderate across various plant organelles. However, when plants experience drought stress, excessive production of ROS can disturb the redox equilibrium of plant cells (Kaur and Asthir 2017). An increased ROS content causes oxidative stress, which impairs metabolic function and eventually leads to programmed cell death (Gill and Tuteja 2010).

Drought stress causes plants to produce ROS, such as hydrogen peroxide (H_2O_2), hydrogen radicals (OH), singlet oxygen (O_{12}), superoxide anion radicals (O_2), and alkoxy radicals (RO), which are all partially reduced forms of oxygen that are highly reactive. As a result, it can be described as the generation of ROS at a certain threshold pool, causing inequity in the cell's redox homeostasis (Kaur and Asthir 2017).

Lower levels of ROS are required for plant signaling events to activate several defense pathways. Under stress, the formation of ROS, a byproduct of cellular metabolism, is exacerbated. ROS tend to rapidly reduce water, leading to the oxidation of nucleic acids, proteins, and carbohydrates, and may also affect nearby molecules, disrupting normal cellular functions in plants and resulting in oxidative harm. Additionally, alterations to bases, deletions of DNA bases, the formation of pyrimidine dimers, and breaks in strands can also occur (Barnawal et al. 2019).

Lipids are hotspots for ROS destruction during drought stress. ROS attack PUFAs, which are significant components of the plasma membrane, causing lipid peroxidation and aggravating oxidative stress. This results in the breakdown of fatty acids, resulting in the production of a variety of chemicals, including aldehydes (Mukami et al. 2019). It is critical to manage ROS through coordinated control of ROS formation and quenching mechanisms to control oxidative stress damage and signaling functions.

To counteract these effects, plants develop and initiate an antioxidant system that includes both enzymatic and nonenzymatic machineries to ease oxidative damage and prevent ROS accumulation (Bashir et al. 2021). SOD is the prime line of defense against superoxide stress in plant cells (Rajput et al. 2021).

The incorporation of PGP microbes into plants can increase their antioxidant enzyme activities and antioxidant build-up, providing plants with defense mechanisms to withstand water stress. This strategy offers a sustainable solution to mitigate the adverse effects of drought on plant health. Numerous studies have shown that plants inoculated with rhizobacteria have increased levels of antioxidant enzymes. These enzymes contribute to plant development and survival during oxidative stress by regulating antioxidant activity (Heidari and Golpayegani 2011).

20.4.3 Modifying the molecular pathways

The response of plants to drought stress is mediated by a diverse group of genes. Plant gene expression is regulated at several stages, including transcription, posttranscription, translation, and posttranslation. Gene expression analysis via diverse molecular approaches is a powerful tool for understanding and revealing the mysterious feedback responses of organisms to environmental situations (Mukarram et al. 2021). Many gene expression patterns are known to change when plants are subjected to drought stress. Plant responses to drought stress are controlled at the transcriptional and translational levels. Drought stress gene upregulation has recently been characterized by molecular techniques, as have the physiological functions of these genes in tolerance caused by PGP bacteria (Wu et al. 2022). Rhizobacterial treatment of plants increases the expression of genes that improve stress resistance.

A macroarray analysis of 7200 expressed sequence tags from common bean nodules exposed to *Rhizobium etli* indicated that trehalose signaling increased the expression of genes related to carbon and nitrogen metabolism (Kaushal 2019a). In the context of drought stress, a comprehensive RNA analysis technique, known as RNA differential display, was utilized to identify the upregulation of key genes involved in the response of plants to dehydration. This innovative approach led to the discovery of two pivotal drought-responsive genes, namely, ERD15 (early response to dehydration 15) and RAB18 (LEA), which both play crucial roles in the adaptation and survival of plants under water-deficient conditions. Rhizobacterial inoculation significantly increased the expression of the stress-specific genes Cadhn, VA, sHSP, and CaPR-10 in the treated plants. This upregulation indicates that plants are better equipped to respond to various stressors, such as dehydration, heat shock, and pathogen attack (Thampi et al. 2022).

20.5 Conclusion and Future prospects

Over the last few decades, significant progress has been made in understanding how abiotic factors impact plant growth and yield, as well as how plants respond/adapt to these challenges. Plants have developed in an ingenious manner with microbes to combat drought stress and its negative consequences within cells. Drought tolerance has evolved tremendously as a result of bacterial activity. Microbes may offer a long-term and cost-effective solution to plant productivity losses caused by changing climatic variables and help in the optimization of human inputs in agroecosystems. The innate stress tolerance mechanisms of plants are strengthened by microbes.

Given the fundamental knowledge of the processes of plant and rhizobacterial interactions, the use of plant growth-inducing microbes in farming is expected to accelerate plant survival in the face of environmental change. The lack of an integrative universal regulatory framework impairs product promotion and may prevent the progress of innovative products. Future studies should aim to develop more targeted solutions, particularly by exploring the interactions between microbial bio stimulants and the microbiomes associated with native plants.

Competing Interests The authors disclose no conflicts of interest that pertain to the content of this chapter.

References

Abdela AA, Barka GD, Degefu T (2020) Coinoculation effect of *Mesorhizobium ciceri* and *Pseudomonas fluorescens* on physiological and biochemical responses of Kabuli chickpea (*Cicer arietinum* L.) during drought stress. Plant Physiol Rep 25:359–369

Abeles FB, Morgan PW, Saltveit ME (2012) Ethylene in plant biology. Academic

Ahluwalia O, Singh PC, Bhatia R (2021) A review on drought stress in plants: implications, mitigation and the role of plant growth promoting rhizobacteria. Resour Environ Sustain 5:100032

Ahmad HM, Fiaz S, Hafeez S, Zahra S, Shah AN, Gul B, Aziz O, Ur-Rahman M, Fakhar A, Rafique M (2022) Plant growth-promoting rhizobacteria eliminate the effect of drought stress in plants: a review. Front Plant Sci 1965:875774

Ahmad S, Belwal V, Punia SS, Ram M, Dalip, Rajput SS, Kunwar R, Meena MK, Gupta D, Kumawat GL, Hussain (2023) Role of plant secondary metabolites and phytohormones in drought tolerance: a review. Gesunde Pflanzen 75(4):729–746

Ali MS, Baek KH (2020) Jasmonic acid signaling pathway in response to abiotic stresses in plants. Int J Mol Sci 21(2):621

Ali S, Khan N (2021) Delineation of mechanistic approaches employed by plant growth promoting microorganisms for improving drought stress tolerance in plants. Microbiol Res 249:126771

Ali S, Xie L (2020) Plant growth promoting and stress mitigating abilities of soil born microorganisms. Recent Pat Food Nutr Agric 11:96–104. https://doi.org/10.2174/2212798410666190515115548

Al-Turki A, Murali M, Omar AF, Rehan M, Sayyed RZ (2023) Recent advances in PGPR-mediated resilience toward interactive effects of drought and salt stress in plants. Front Microbiol 14:1214845

Annunziata MG, Ciarmiello LF, Woodrow P, Dell'Aversana E, Carillo P (2019) Spatial and temporal profile of glycine betaine accumulation in plants under abiotic stresses. Front Plant Sci 10:230

Ansari FA, Jabeen M, Ahmad I (2021) *Pseudomonas azotoformans* FAP5, a novel biofilm-forming PGPR strain, alleviates drought stress in wheat plant. Int J Environ Sci Technol 3:1–6

Aqeel M, Ran J, Hu W, Irshad MK, Dong L, Akram MA, Eldesoky GE, Aljuwayid AM, Chuah LF, Deng J (2023) Plant–soil-microbe interactions in maintaining ecosystem stability and coordinated turnover under changing environmental conditions. Chemosphere 318:137924

Asghari B, Khademian R, Sedaghati B (2020) Plant growth promoting rhizobacteria (PGPR) confer drought resistance and stimulate biosynthesis of secondary metabolites in pennyroyal (*Mentha pulegium* L.) under water shortage condition. Sci Hortic 263:109132

Ayangbenro AS, Chukwuneme CF, Ayilara MS, Kutu FR, Khantsi M, Adeleke BS, Glick BR, Babalola OO (2022) Harnessing the rhizosphere soil microbiome of organically amended soil for plant productivity. Agronomy 12(12):3179

Backer R, Rokem JS, Ilangumaran G, Lamont J, Praslickova D, Ricci E et al (2018) Plant growth-promoting rhizobacteria: context, mechanisms of action, and roadmap to commercialization of biostimulants for sustainable agriculture. Front Plant Sci 9:1473

Barnawal D, Singh R, Singh RP (2019) Role of plant growth promoting rhizobacteria in drought tolerance: regulating growth hormones and osmolytes. In: Singh AK, Kumar A, Singh PK (eds) PGPR amelioration in sustainable agriculture. Woodhead Publishing, Sawston, pp 107–128

Bashir SS, Hussain A, Hussain SJ, Wani OA, Zahid Nabi S, Dar NA, Baloch FS, Mansoor S (2021) Plant drought stress tolerance: understanding its physiological, biochemical and molecular mechanisms. Biotechnol Equip 35(1):1912–1925

Bhattacharya A (2021) Role of plant growth hormones during soil water deficit: a review. In: Bhattacharya A (ed) Soil water deficit and physiological issues in plants. Springer, Singapore, pp 489–583

Bittencourt PP, Alves AF, Ferreira MB, da Silva Irineu LE, Pinto VB, Olivares FL (2023) Mechanisms and applications of bacterial inoculants in plant drought stress tolerance. Microorganisms 11(2):502

Camaille M, Fabre N, Clément C, Ait Barka E (2021) Advances in wheat physiology in response to drought and the role of plant growth promoting rhizobacteria to trigger drought tolerance. Microorganisms 9(4):687

Chada S, Asiedu SK, Ofoe R (2023) An overview of plant morpho-physiology, biochemicals, and metabolic pathways under water stress. Hortic Int J 7(4):115–125

Chaffai R, Ganesan M, Cherif A (2024) Gene expression regulation in plant abiotic stress response. In: Chaffai R, Ganesan M, Cherif A (eds) Plant adaptation to abiotic stress: from Signaling pathways and microbiomes to molecular mechanisms. Springer Nature Singapore, Singapore, pp 275–302

Chandrasekaran M, Paramasivan M, Sahayarayan JJ (2022) Microbial volatile organic compounds: an alternative for chemical fertilizers in sustainable agriculture development. Microorganisms 11(1):42

Chauhan P, Sharma N, Tapwal A, Kumar A, Verma GS, Meena M, Seth CS, Swapnil P (2023) Soil microbiome: diversity, benefits and interactions with plants. Sustainability 15(19):14643

Chen D, Guo Y, Zhao Y, Zhang J, Liu X, Tong Z, Zhao C (2024) Dynamic evolution characteristics and hazard assessment of compound drought/waterlogging and low temperature events for maize. Sci Total Environ 946:174427

Chiappero J, Del Rosario Cappellari L, Alderete LGS, Palermo TB, Banchio E (2019) Plant growth promoting rhizobacteria improve the antioxidant status in *Mentha piperita* grown under drought stress leading to an enhancement of plant growth and total phenolic content. Industr Crops Prod 139:111553. https://doi.org/10.1016/j.indcrop.2019.111553

Chieb M, Gachomo EW (2023) The role of plant growth promoting rhizobacteria in plant drought stress responses. BMC Plant Biol 23(1):407

Corwin DL (2021) Climate change impacts on soil salinity in agricultural areas. Eur J Soil Sci 72(2):842–862

Cura JA, Franz DR, Filosofia JE, Balestrasse KB, Burgueno LE (2017) Inoculation with *Azospirillum* sp and *Herbaspirillum* sp bacteria increases the tolerance of maize to drought stress. Microorganisms 5(3):41. https://doi.org/10.3390/microorganisms5030041

Devarajan AK, Muthukrishanan G, Truu J, Truu M, Ostonen I, Kizhaeral SS et al (2021) The foliar application of Rice Phyllosphere bacteria induces drought-stress tolerance in *Oryza sativa* (L). Plants (Basel) 10(2):387

Dhawi F (2020) Plant growth promoting rhizobacteria (PGPR) regulated phyto and microbial beneficial protein interactions. Open Life Sci 28(1):68–78

Dos Santos TB, Ribas AF, de Souza SG, Budzinski IG, Domingues DS (2022) Physiological responses to drought, salinity, and heat stress in plants: a review. Stress 2(1):113–135

Eswaran SUD, Sundaram L (2024) Bioencapsulation of microbial biostimulant for drought resilience in groundnut (*Arachis hypogaea* L.). Commun Soil Sci Plant Anal 55:1–21

Eswaran SUD, Sundaram L, Siang TC, Alharbi SA, Alahmadi TA, Kadam SK (2023) Multifarious microbial biostimulants promote growth in *Arachis hypogaea* L. Front Sustain Food Syst 7:1170374

Eswaran SUD, Sundaram L, Perveen K, Bukhari NA, Sayyed RZ (2024) Osmolyte-producing microbial biostimulants regulate the growth of *Arachis hypogaea* L. under drought stress. BMC Microbiol 24(1):165

Fadiji AE, Babalola OO, Santoyo G, Perazzolli M (2022) The potential role of microbial biostimulants in the amelioration of climate change-associated abiotic stresses on crops. Front Microbiol 12:829099. https://doi.org/10.3389/fmicb.2021.829099

Farooqi ZUR, Ayub MA, Rehman MZ, Sohail MI, Usman M, Khalid H, Naz K (2020) Regulation of drought stress in plants. In: Plant life under changing environment. Academic, Cambridge, MA, pp 77–104

Gamalero E, Glick BR (2015) Bacterial modulation of plant ethylene levels. Plant Physiol 169(1):13–22

Gamalero E, Glick BR (2022) Recent advances in bacterial amelioration of plant drought and salt stress. Biology 1(3):437

Ghosh D, Sen S, Mohapatra S (2017) Modulation of proline metabolic gene expression in *Arabidopsis thaliana* under water stressed conditions by a drought-mitigating *Pseudomonas* putidastrain. Ann Microbiol 67(10):655–668. https://doi.org/10.1007/s13213-017-1294-y

Ghosh UK, Islam MN, Siddiqui MN, Khan MA (2021) Understanding the roles of osmolytes for acclimatizing plants to changing environment: a review of potential mechanism. PSB 16(8):1913306

Gill SS, Tuteja N (2010) Reactive oxygen species and antioxidant machinery in abiotic stress tolerance in crop plants. Plant Physiol Biochem 48:909–930

Glick BR (2012) Plant growth-promoting bacteria: mechanisms and applications. Scientifica (Cairo) 2012:963401

Gomes GL, Scortecci KC (2021) Auxin and its role in plant development: structure, signaling, regulation and response mechanisms. Plant Biol 6:894–904

Gou W, Tian L, Ruan Z, Zheng P et al (2015) Accumulation of choline and glycine betaine and drought stress tolerance induced in maize (*Zea mays*) by three plant growth promoting rhizobacteria (PGPR) strains. Pak J Bot 47:581–586

Govind G, Harshavardhan VT, Hong CY (2020) Phytohormone signaling in response to drought. In: Govind G, Harshavardhan VT, Hong C-Y (eds) Salt and drought stress tolerance in plants, Signaling networks and adaptive mechanisms. Springer, Cham, pp 315–335

Gupta S, Pandey S (2019) ACC deaminase producing bacteria with multifarious plant growth promoting traits alleviates salinity stress in french bean (*Phaseolus vulgaris*) plants. Front Microbiol 10:1506

Gupta A, Mishra R, Rai S, Bano A, Pathak N, Fujita M, Kumar M, Hasanuzzaman M (2022) Mechanistic insights of plant growth promoting bacteria mediated drought and salt stress tolerance in plants for sustainable agriculture. Int J Mol Sci 23(7):3741

Harkhani K, Sharma AK (2024) Alleviation of drought stress by plant growth-promoting rhizobacteria (PGPR) in crop plants: a review. Commun Soil Sci Plant Anal 55(5):735–758

Hasanuzzaman M, Bhuyan MB, Parvin K, Bhuiyan TF, Anee TI, Nahar K, Hossen MS, Zulfiqar F, Alam MM, Fujita M (2020) Regulation of ROS metabolism in plants under environmental stress: a review of recent experimental evidence. Int J Mol Sci 21(22):8695

Hassan MU, Nawaz M, Shah AN, Raza A, Barbanti L, Skalicky M, Hashem M, Brestic M, Pandey S, Alamri S, Mostafa YS (2023) Trehalose: a key player in plant growth regulation and tolerance to abiotic stresses. J Plant Growth Regul 42(8):4935–4957

Heidari M, Golpayegani A (2011) Effects of water stress and inoculation with plant growth promoting rhizobacteria (PGPR) on antioxidant status and photosynthetic pigments in basil (*Ocimum basilicum* L). J Saudi Soc Agric Sci 11:57–61

Ilyas N, Mumtaz K, Akhtar N, Yasmin H, Sayyed RZ, Khan W, Enshasy HAE, Dailin DJ, Elsayed EA, Ali Z (2020) Exopolysaccharides producing bacteria for the amelioration of drought stress in wheat. Sustainability 12:8876

Jabborova D, Kannepalli A, Davranov K, Narimanov A, Enakiev Y, Syed A, Elgorban AM, Bahkali AH, Wirth S, Sayyed RZ, Gafur A (2021) Coinoculation of rhizobacteria promotes growth, yield, and nutrient contents in soybean and improves soil enzymes and nutrients under drought conditions. Sci Rep 11(1):22081

Jayakumar A, Padmakumar P, Nair IC, Radhakrishnan E (2020) Drought tolerant bacterial endophytes with potential plant probiotic effects from *Anana scomosus*. Biologia 75:1769–1778. https://doi.org/10.2478/s11756-020-00483-1

Jochum MD, Mcwilliams KL, Borrego EJ, Kolomiets MV, Niu G, Pierson EA et al (2019) Bioprospecting plant growth-promoting rhizobacteria that mitigate drought stress in grasses. Front Microbiol 10:2106. https://doi.org/10.3389/fmicb.2019.02106

Kang SM, Radhakrishnan R, You YH, Khan AL, Park JM, Lee SM et al (2015) Cucumber performance is improved by inoculation with plant growth-promoting microorganisms. Acta Agric Scand Sect B Soil Plant 65:36–44. https://doi.org/10.1080/09064710.2014.960889

Kapoor D, Bhardwaj S, Landi M, Sharma A, Ramakrishnan M, Sharma A (2020) The impact of drought in plant metabolism: how to exploit tolerance mechanisms to increase crop production. Appl Sci 10(16):5692

Kaur G, Asthir B (2017) Molecular responses to drought stress in plants. Biol Plant 61(2):201–209

Kaur G, Sanwal SK, Kumar A, Pundir RK, Yadav M, Sehrawat N (2024) Role of osmolytes dynamics in plant metabolism to cope with salinity induced osmotic stress. Discov Agric 2(1):59

Kaushal M (2019a) Microbes in cahoots with plants: MIST to hit the jackpot of agricultural productivity during drought. Int J Mol Sci 20(7):1769

Kaushal M (2019b) Portraying rhizobacterial mechanisms in drought tolerance: a way forward toward sustainable agriculture. In: Singh AK, Kumar A, Singh PK (eds) PGPR amelioration in sustainable agriculture. Woodhead Publishing, pp 195–216

Khan N, Bano A, Zandi P (2018) Effects of exogenously applied plantgrowth regulators in combination with PGPR on the physiology and rootgrowth of chickpea (Cicer arietinum) and their role in drought tolerance. J Plant Interact 13:239–247. https://doi.org/10.1080/17429145.2018.1471527

Khan N, Ali S, Shahid MA, Mustafa A, Sayyed RZ, Curá JA (2021) Insights into the interactions among roots, rhizosphere, and rhizobacteria for improving plant growth and tolerance to abiotic stresses: a review. Cells 10(6):1551

Kou X, Han W, Kang J (2022) Responses of root system architecture to water stress at multiple levels: a meta-analysis of trials under controlled conditions. Front Plant Sci 13:1085409

Kour D, Yadav AN (2022) Bacterial mitigation of drought stress in plants: current perspectives and future challenges. Curr Microbiol 79(9):248

Kour D, Rana KL, Yadav AN, Sheikh I, Kumar V, Dhaliwal HS et al (2020) Amelioration of drought stress in Foxtail millet (*Setaria italica* L.) by P-solubilizing drought-tolerant microbes with multifarious plant growth promoting attributes. Environ Sustain 3:23–34. https://doi.org/10.1007/s42398-020-00094-1

Koza NA, Adedayo AA, Babalola OO, Kappo AP (2022) Microorganisms in plant growth and development: roles in abiotic stress tolerance and secondary metabolites secretion. Microorganisms 10(8):1528

Kumar AS, Sridar R, Uthandi S (2017) Mitigation of droughtin rice by a phyllosphere bacterium *Bacillus altitudinis* FD48. Afr J Microbiol Res 11(45):1614–1625. https://doi.org/10.5897/AJMR2017.8610

Kumar M, Kumar Patel M, Kumar N, Bajpai AB, Siddique KH (2021) Metabolomics and molecular approaches reveal drought stress tolerance in plants. Int J Mol Sci 22(17):9108

Kumar A, Naroju SP, Kumari N, Arsey S, Kumar D, Gubre DF, Roychowdhury A, Tyagi S, Saini P (2024a) The role of drought response genes and plant growth promoting bacteria on plant growth promotion under sustainable agriculture: a review. Microbiol Res 127827

Kumar P, Ashique S, Kumar N, Jain A, Sharma H, Pandey SN, Singh A (2024b) Regulation of plant hormones under abiotic stress conditions in plants. In: Plant secondary metabolites and abiotic stress, pp 243–276

Kumari VV, Banerjee P, Verma VC, Sukumaran S, Chandran MA, Gopinath KA, Venkatesh G, Yadav SK, Singh VK, Awasthi NK (2022) Plant nutrition: an effective way to alleviate abiotic stress in agricultural crops. Int J Mol Sci 23(15):8519

Laller R, Khosla PK, Negi N, Avinash H, Thakur N, Kashyap S, Shukla SK, Hussain I (2023) Bacterial volatiles as PGPRs: inducing plant defense mechanisms during stress periods. S Afr J Bot 159:131–139

Latif M, Bukhari SAH, Alrajhi AA, Alotaibi FS, Ahmad M, Shahzad AN, Dewidar AZ, Mattar MA (2022) Inducing drought tolerance in wheat through exopolysaccharide-producing rhizobacteria. Agronomy 12:1140

Li H, Guo Q, Jing Y, Liu Z, Zheng Z, Sun Y, Xue Q, Lai H (2020) Application of *Streptomyces pactum* Act12 enhances drought resistance in wheat. J Plant Growth Regul 39:122–132

Lim JH, Kim SD (2013) Induction of drought stress resistance by multifunctional PGPR *Bacillus licheniformis* K11 in pepper. Plant Pathol J 29:201–208. https://doi.org/10.5423/PPJ.SI.02.2013.0021

Lotfi N, Soleimani A, Çakmakçı R, Vahdati K, Mohammadi P (2022) Characterization of plant growth-promoting rhizobacteria (PGPR) in Persian walnut associated with drought stress tolerance. Sci Rep 12(1):12725

Lucas JA, Garcia-Villaraco A, Montero-Palmero MB, Montalban B, Ramos Solano B, Gutierrez-Mañero FJ (2023) Physiological and genetic modifications induced by plant-growth-promoting rhizobacteria (PGPR) in tomato plants under moderate water stress. Biology 12(7):901

Ma Y, Dias MC, Freitas H (2020) Drought and salinity stress responses and microbe-induced tolerance in plants. Front Plant Sci 11:591911

Mäkelä PSA, Jokinen K, Himanen K (2019) Roles of endogenous Glycinebetaine in plant abiotic stress responses. In: Hossain M, Kumar V, Burritt D, Fujita M, Mäkelä P (eds) Osmoprotectant-mediated abiotic stress tolerance in plants. Springer, Cham

Mal S, Panchal S (2024) Drought and salt stress mitigation in crop plants using stress-tolerant auxin-producing endophytic bacteria: a futuristic approach toward sustainable agriculture. Front Plant Sci 15:1422504

Martins SJ, Rocha GA, De Melo HC, De Castro Georg R, Ulhôa CJ, De Campos Dianese É et al (2018) Plant-associated bacteria mitigate drought stress in soybean. Environ Sci Pollut Res 25:13676–13686. https://doi.org/10.1007/s11356-018-1610-5

Meena M, Swapnil P, Divyanshu K, Kumar S, Harish, Tripathi YN, Zehra A, Marwal A, Upadhyay RS (2020) PGPR-mediated induction of systemic resistance and physiochemical alterations in plants against the pathogens: current perspectives. J Basic Microbiol 60(10):828–861

Midzi J, Jeffery DW, Baumann U, Rogiers S, Tyerman SD, Pagay V (2022) Stress-induced volatile emissions and signaling in interplant communication. Plants 11(19):2566

Mohamed T, Saad AM, Mohammed DM, Fahmy MA, Elesawi IE, Ahmed AE, Algopishi UB, Elrys AS, Desoky ES, Mosa WF, Abd El-Mageed TA (2024) Drought-tolerant plant growth-promoting rhizobacteria alleviate drought stress and enhance soil health for sustainable agriculture: a comprehensive review. Plant Stress 14(10063):2

Morel MA, Castro-Sowinski S (2013) The complex molecular signaling network in microbe–plant interaction. In: Plant microbe symbiosis: fundamentals and advances. Springer India, New Delhi

Moursi YS, Dawood MF, Sallam A, Thabet SG, Alqudah AM (2021) Antioxidant enzymes and their genetic mechanism in alleviating drought stress in plants. In: Moursi YS, Dawood MFA, Sallam A, Thabet SG, Alqudah AM (eds) Organic solutes, oxidative stress, and antioxidant enzymes under abiotic stressors. CRC Press, pp 233–262

Muhammad Aslam M, Waseem M, Jakada BH, Okal EJ, Lei Z, Saqib HS, Yuan W, Xu W, Zhang Q (2022) Mechanisms of abscisic acid-mediated drought stress responses in plants. Int J Mol Sci 23(3):1084

Mukami A, Ngetich A, Mweu C, Oduor RO, Muthangya M, Mbinda WM (2019) Differential characterization of physiological and biochemical responses during drought stress in finger millet varieties. Physiol Mol Biol Plants 25:837–846

Mukarram M, Choudhary S, Kurjak D, Petek A, Khan MMA (2021) Drought: sensing, signaling, effects and tolerance in higher plants. Physiol Plant 172:1291–1300

Munné-Bosch S, Villadangos S (2023) Cheap, cost-effective, and quick stress biomarkers for drought stress detection and monitoring in plants. Trends Plant Sci 28(5):527–536

Nadarajah KK (2020) ROS homeostasis in abiotic stress tolerance in plants. Int J Mol Sci 21(15):5208

Niu X, Song L, Xiao Y, Ge W (2017) Drought-tolerant plant growth-promoting rhizobacteria associated with Foxtail millet in a semiarid agroecosystem and their potential in alleviating drought stress. Front Microbiol 8:2580

Ojuederie OB, Olanrewaju OS, Babalola OO (2019) Plant growth promoting rhizobacterial mitigation of drought stress in crop plants: implications for sustainable agriculture. Agronomy 9(11):712

Osmolovskaya N, Shumilina J, Kim A, Didio A, Grishina T, Bilova T, Keltsieva OA, Zhukov V, Tikhonovich I, Tarakhovskaya E, Frolov A (2018) Methodology of drought stress research: experimental setup and physiological characterization. Int J Mol Sci 19(12):4089

Patel P, Sayyed RZ, Patel H (2023) PGPR: a sustainable agricultural mitigator for stressed agro-environments. In: Patel P, Sayyed RZ, Patel H (eds) Plant growth promoting microorganisms of arid region. Springer Nature Singapore, Singapore, pp 303–318

Priyanka JP, Goral RT, Rupal KS, Saraf M (2019) Rhizospheric microflora: a natural alleviator of drought stress in agricultural crops. In: Sayyed RZ (ed) Plant growth promoting rhizobacteria for sustainable stress management. Springer, Singapore, pp 103–115

Raheem A, Shaposhnikov A, Belimov AA, Dodd IC, Ali B (2018) Auxin production by rhizobacteria was associated with improved yield of wheat (*Triticum aestivum* L.) under drought stress. Arch Agron Soil Sci 64:574–587. https://doi.org/10.1080/03650340.2017.1362105

Rajput VD, Harish, Singh RK, Verma KK, Sharma L, Quiroz-Figueroa FR, Meena M, Gour VS, Minkina T, Sushkova S et al (2021) Recent developments in enzymatic antioxidant defense mechanism in plants with special reference to abiotic stress. Biology 10:267

Ranjan A, Rajput VD, Prazdnova EV, Gurnani M, Sharma S, Bhardwaj P, Shende SS, Mandzhieva SS, Sushkova S, Minkina T, Chauhan A (2024) Augmenting abiotic stress tolerance and root architecture: the function of phytohormone-producing PGPR and their interaction with nanoparticles. S Afr J Bot 167:612–629

Raza A, Bhardwaj S, Rahman MA, García-Caparrós P, Habib M, Saeed F, Charagh S, Foyer CH, Siddique KH, Varshney RK (2023) Trehalose: a sugar molecule involved in temperature stress management in plants. Crop J 12(1):1–16

Saha I, Datta S, Biswas D (2020) Exploring the role of bacterial extracellular polymeric substances for sustainable development in agriculture. Curr Microbiol 77(11):3224–3239

Salehi-Lisar SY, Bakhshayeshan-Agdam H (2016) Drought stress in plants: causes, consequences, and tolerance. In: Hossain M, Wani S, Bhattacharjee S, Burritt D, Tran LS (eds) Drought stress tolerance in plants. Springer, Cham

Salvi P, Manna M, Kaur H, Thakur T, Gandass N, Bhatt D, Muthamilarasan M (2021) Phytohormone signaling and crosstalk in regulating drought stress response in plants. Plant Cell Rep 40:1305–1329

Sati D, Pande V, Pandey SC, Samant M (2023) Recent advances in PGPR and molecular mechanisms involved in drought stress resistance. J Soil Sci Plant Nutr 23(1):106–124

Seleiman MF, Al-Suhaibani N, Ali N, Akmal M, Alotaibi M, Refay Y, Dindaroglu T, Abdul-Wajid HH, Battaglia ML (2021) Drought stress impacts on plants and different approaches to alleviate its adverse effects. Plants (Basel) 10(2):259

Sharma MP, Grover M, Chourasiya D, Bharti A, Agnihotri R, Maheshwari HS, Pareek A, Buyer JS, Sharma SK, Schütz L, Mathimaran N (2020) Deciphering the role of trehalose in tripartite symbiosis among rhizobia, arbuscular mycorrhizal fungi, and legumes for enhancing abiotic stress tolerance in crop plants. Front Microbiol 11:509919

Sharma A, Pathania A, Sharma P, Bhardwaj R, Sharma I (2023) Role of glycine betaine in regulating physiological and molecular aspects of plants under abiotic stress. In: The role of growth regulators and phytohormones in overcoming environmental stress, pp 327–353

Sharma J, Kumar S, Singh P, Kumar V, Verma S, Khyalia P, Sharma A (2024) Emerging role of osmoprotectant glycine betaine to mitigate heavy metals toxicity in plants: a systematic review. Biol Futur 6:1–8

Singh A, Kumar A, Yadav S, Singh IK (2019) Reactive oxygen species-mediated signaling during abiotic stress. Plant Gene 18:100173

Sunita K, Mishra I, Mishra J, Prakash J, Arora NK (2020) Secondary metabolites from halotolerant plant growth promoting rhizobacteria for ameliorating salinity stress in plants. Front Microbiol 11:567768

Takahashi F, Kuromori T, Urano K, Yamaguchi-Shinozaki K, Shinozaki K (2020) Drought stress responses and resistance in plants: from cellular responses to long-distance intercellular communication. Front Plant Sci 11:556972

Thampi M, Varghese EM, Jisha MS (2022) *Bacillus* spp. facilitated abiotic stress mitigation in Rice. In: *Bacilli* in agrobiotechnology: plant stress tolerance, bioremediation, and bioprospecting. Springer International Publishing, Cham, pp 285–318

Tiepo AN, Constantino LV, Madeira TB, Goncalves LSA, Pimenta JA, Bianchini E, de Oliveira ALM, Oliveira HC, Stolf-Moreira R (2020) Plant growth-promoting bacteria improve leaf antioxidant metabolism of drought-stressed neotropical trees. Planta 251(4):1–11. https://doi.org/10.1007/s00425-020-03373-7

Uzma M, Iqbal A, Hasnain S (2022) Drought tolerance induction and growth promotion by indole acetic acid producing *Pseudomonas aeruginosa* in *Vigna radiata*. PLoS One 17(2):e0262932

Vaishnav A, Choudhary DK (2019) Regulation of drought responsive gene expression in *Glycine max* L Merrill is mediated through *Pseudomonas simiae* strain AU. J Plant Growth Regul 38(1):333–342. https://doi.org/10.1007/s00344-018-9846-3

Vaseva II, Qudeimat E, Potuschak T, Du Y, Genschik P, Vandenbussche F et al (2018) The plant hormone ethylene restricts *Arabidopsis* growth via the epidermis. Proc Natl Acad Sci 115(17):E4130–E4139

Wahab A, Abdi G, Saleem MH, Ali B, Ullah S, Shah W, Mumtaz S, Yasin G, Muresan CC, Marc RA (2022) Plants' physio-biochemical and phyto-hormonal responses to alleviate the adverse effects of drought stress: a comprehensive review. Plants (Basel) 11(13):1620

Wu J, Wang J, Hui W, Zhao F, Wang P, Su C, Gong W (2022) Physiology of plant responses to water stress and related genes: a review. Forests 13:324

Yasmin H, Rashid U, Hassan MN, Nosheen A, Naz R, Ilyas N, Sajjad M, Azmat A, Alyemeni M (2021) Volatile organic compounds produced by *Pseudomonas pseudoalcaligenes* alleviated drought stress by modulating defense system in maize (*Zea mays* L.). Physiol Plant 172:896–911

Zhang X, Su C, Liu X et al (2020) Effect of plant-growth-promoting rhizobacteria on phytoremediation efficiency of Scirpustriqueter in pyrene-Ni cocontaminated soils. Chemosphere 241:125027

Zhang H, Sun X, Dai M (2022) Improving crop drought resistance with plant growth regulators and rhizobacteria: mechanisms, applications, and perspectives. Plant Commun 3(1)

Zhao B, Liu Q, Wang B, Yuan F (2021) Roles of phytohormones and their signaling pathways in leaf development and stress responses. J Agric Food Chem 69(12):3566–3584

Zia R, Nawaz MS, Siddique MJ, Hakim S, Imran A (2021) Plant survival under drought stress: implications, adaptive responses, and integrated rhizosphere management strategy for stress mitigation. Microbiol Res 242:126626

Zulfiqar F, Akram NA, Ashraf M (2020) Osmoprotection in plants under abiotic stresses: new insights into a classical phenomenon. Planta 251(1):3

GPSR Compliance

The European Union's (EU) General Product Safety Regulation (GPSR) is a set of rules that requires consumer products to be safe and our obligations to ensure this.

If you have any concerns about our products, you can contact us on ProductSafety@springernature.com

In case Publisher is established outside the EU, the EU authorized representative is:

Springer Nature Customer Service Center GmbH
Europaplatz 3
69115 Heidelberg, Germany

Batch number: 09467310

Printed by Printforce, the Netherlands